# Subcellular Biochemistry

## Volume 15

## Virally Infected Cells

Edited by

**J. R. Harris**

*North East Thames Regional Transfusion Centre*
*Brentwood, Essex, England*

PLENUM PRESS • NEW YORK AND LONDON

The Library of Congress cataloged the first volume of this title as follows:

Sub-cellular biochemistry.

London, New York, Plenum Press.
v. illus. 23 cm. quarterly.
Began with Sept. 1971 issue. Cf. New serial titles.
1. Cytochemistry – Periodicals. 2. Cell organelles – Periodicals.
QH611.S84 574.8'76 73-643479

ISBN 0-306-43138-6

This series is a continuation of the journal *Sub-Cellular Biochemistry*, Volumes 1 to 4 of which were published quarterly from 1972 to 1975


A Division of Plenum Publishing Corporation
233 Spring Street, New York, N.Y. 10013



Printed in the United States of America

# Subcellular Biochemistry

## Volume 15
## Virally Infected Cells

# SUBCELLULAR BIOCHEMISTRY



---

*Recent Volumes in This Series:*

| | |
|---|---|
| Volumes 5–11 | Edited by Donald B. Roodyn |
| Volume 12 | **Immunological Aspects**<br>Edited by J. R. Harris |
| Volume 13 | **Fluorescence Studies on Biological Membranes**<br>Edited by H. J. Hilderson |
| Volume 14 | **Artificial and Reconstituted Membrane Systems**<br>Edited by J. R. Harris and A.-H. Etémadi |
| Volume 15 | **Virally Infected Cells**<br>Edited by J. R. Harris |

# Contributors


**S. AbuBakar** Department of Microbiology, University of Texas Medical Branch, Galveston, Texas 77550

**Thomas Albrecht** Department of Microbiology, The University of Texas Medical Branch, Galveston, Texas 77550

**W. W. Au** Department of Preventive Medicine and Community Health, University of Texas Medical Branch, Galveston, Texas 77550

**L. Babiuk** Department of Veterinary Microbiology and the Veterinary Infectious Diseases Organization, University of Saskatchewan, Saskatoon, Saskatchewan S7N 0W0, Canada

**I. Boldogh** Department of Microbiology, University of Texas Medical Branch, Galveston, Texas 77550

**D. S. Bowden** MacFarlane Burnet Centre for Medical Research, Fairfield Hospital, Fairfield, Victoria 3078, Australia

**Kristian Dalsgaard** State Veterinary Institute of Virus Research, Lindholm, Kalvehave, Sweden

**Bryan T. Eaton** Australian Animal Health Laboratory, CSIRO, Geelong, Victoria 3220, Australia

**Kenneth H. Eckels** Department of Biologics Research, Walter Reed Army Institute of Research, Washington, D.C. 20307

**M. Fons** Department of Microbiology, University of Texas Medical Branch, Galveston, Texas 77550

**Tatsuo Hase** Department of Ultrastructural Studies, Walter Reed Army Institute of Research, Washington, D.C. 20307

**Stefan Höglund** Institute of Biochemistry, Biomedical Center, Uppsala, Sweden

**Alex D. Hyatt** Australian Animal Health Laboratory, CSIRO, Geelong, Victoria 3220, Australia

**Shin-ichi Ito** Shin-Etsu Chemical Co., Ltd., Ohtemachi 2-6-1, Chiyoda-ku, Tokyo 100, Japan

**Abraham Karpas** Department of Haematology, University Clinical School, Cambridge, England

**C. Kurstak** Faculty of Medicine, University of Montreal and Hospital Hôtel Dieu de Montreal, Montreal, Quebec H3C 3J7, Canada

**E. Kurstak** Faculty of Medicine, University of Montreal and Hospital Hôtel Dieu de Montreal, Montreal, Quebec H3C 3J7, Canada

**C. H. Lee** Department of Microbiology, University of Texas Medical Branch, Galveston, Texas 77550

**Karin Lövgren** The National Veterinary Institute, Department of Virology, Biomedical Center, Uppsala, Sweden

**Marja Makarow** Institute of Biotechnology, University of Helsinki, Valimotie 7, 00380 Helsinki, Finland

**R. Marusyk** Viral Pathogenesis Research Unit, Department of Medical Microbiology and Infectious Diseases, Faculty of Medicine, University of Alberta, Edmonton, Alberta T6G 247, Canada

**Thomas Mayer** Institute for Medical Virology, Justus-Liebig University, D-6300 Giessen, Federal Republic of Germany

**Bror Morein** The National Veterinary Institute, Department of Virology, Biomedical Center, Uppsala, Sweden

**Heiner Niemann** Institute for Medical Virology, Justus-Liebig University, D-6300 Giessen, Federal Republic of Germany

**Kazukiyo Onodera** Department of Agricultural Chemistry, The University of Tokyo, Yayoi 1-1, Bunkyo-ku, Tokyo 113, Japan

**Ab Osterhaus** National Institute of Public Health and Environmental Hygiene, Bilthoven, The Netherlands

**Joseph R. Putnak** Department of Viral Diseases, Walter Reed Army Institute of Research, Washington, D.C. 20307

**J. M. Russell** Department of Physiology and Biophysics, University of Texas Medical Branch, Galveston, Texas 77550

**A. Salmi** Viral Pathogenesis Research Unit, Department of Medical Microbiology and Infectious diseases, Faculty of Medicine, University of Alberta, Edmonton, Alberta T6G 247, Canada

**Hannele Sareneva** Institute of Biotechnology, University of Helsinki, Valimotie 7, 00380 Helsinki, Finland

**Kazuya Sasaki** Tokyo Institute of Kaken Pharmaceutical Co., Ltd., Honkomagome, Bunkyo-ku, Tokyo 113, Japan

**Otto Schmidt** Institute for Biology III, University of Freiburg, D-7800 Freiburg, Federal Republic of Germany

**Imke Schuchmann-Feddersen** Institute for Biology III, University of Freiburg, D-7800 Freiburg, Federal Republic of Germany

**Hisatoshi Shida** Institute for Virus Research, Kyoto University, Kyoto 606, Japan

**Peter L. Summers** Department of Biologics Research, Walter Reed Army Institute of Research, Washington, D.C. 20307

**Bo Sundquist** The National Veterinary Institute, Division of Vaccine Research, Uppsala, Sweden

**Teruko Tamura** Institute for Medical Virology, Justus-Liebig University, D-6300 Giessen, Federal Republic of Germany

**M. Van Regenmortel** Institute of Molecular and Cellular Biology, 6700 Strasbourg, France

**E. G. Westaway** Department of Microbiology, Monash University, Clayton, Victoria 3168, Australia

# Preface

Cellular virology has made tremendous advances in the past decade due to the availability and application of new immunological techniques together with the vast range of biochemical techniques and the continued impact of transmission electron microscopy. The chapters contained in this volume provide significant coverage of the subject of cellular virology as a whole. Considerable overall emphasis is placed upon the membrane biochemistry of viral proteins and glycoproteins within the infected cell.

In the opening chapter Edouard Kurstak and his colleagues provide a useful survey on the detection of viral antigens and antibodies by immunoassays. This chapter, with its emphasis on the important role of immunology in present-day virology, sets the scene for the volume. Following this is an exciting presentation from Stefan Höglund and his colleagues on ISCOMs and immunostimulation with viral antigens. This unique approach is already proving to be of value, particularly in animal virology. A somewhat pharmacological diversion appears in Chapter 3, by Kazukiyo Onodera and his colleagues, in which the biological activity of the damavaricin C derivatives is discussed. This chapter provides a link between the biochemical and the chemotherapeutic approach in cellular virology. Yet another specialist area is covered in Chapter 4 by Otto Schmidt and Imke Schuchmann-Feddersen, who discuss the role of virus-like particles in parasite–host interactions of insects. Contributing a strong biomedical emphasis to the volume is the provocative chapter by Abraham Karpas on human leukemia and retroviruses. Along more fundamental biochemical lines is the chapter by Thomas Albrecht and his colleagues, who present a detailed account of cell activation responses to cytomegalovirus infection; indeed, this chapter contains an impressive combination of ultrastructural and biochemical data. Scott Bowden and E. G. Westaway have written an assessment of rubella virus products and their distribution in infected cells that contains a good combination of biochemistry, immunofluorescence microscopy, and transmission electron microscopy. In a similar vein, Bryan Eaton and Alex Hyatt present a survey of the association of bluetongue virus with the cytoskeleton. In this instance, much greater emphasis is placed upon the ultrastructural aspects that have been so very important for the expansion of knowledge of cytoskeletal structure and of the nuclear matrix and nuclear structure. The ultrastructural approach also receives considerable attention in the chapter by Tatsuo Hase and his colleagues, who deal with the morphogenesis of flaviviruses.

The three remaining chapters, one by Heiner Niemann and colleagues on signals for membrane-associated transport in eukaryotic cells, one by Hannele Sareneva and Marja Makarow on membrane biology in yeast as probed with enveloped viruses, and one by Hisatoshi Shida on vaccinia virus hemagglutinin, form a group of chapters that complement one another rather well. A strong biochemical and molecular biological emphasis runs throughout these three chapters, which nicely redresses the imbalance, the ultrastructural approach having been somewhat more dominant in a number of the earlier chapters.

Overall, the combination of high-quality biochemical, immunological, and ultrastructural data that is presented in this volume provides a unique insight into current research on virally infected cells. By bringing together these exciting specialized, yet related, contributions on cellular virology, this volume continues the new theme of the *Subcellular Biochemistry* series that was established earlier in Volume 12. It is hoped that this book will prove to be of value and interest to both biochemists and virologists and to those who are working in other areas and wish to update their knowledge of current progress in cellular virology.

Robin Harris

*Brentwood, Essex, United Kingdom*

# Contents

*Chapter 3*

**Biological Activity of Damavaricin Derivatives**

Kazukiyo Onodera, Shin-ichi Ito, and Kazuya Sasaki

*Chapter 4*

**Role of Virus-like Particles in Parasitoid–Host Interaction of Insects**

Otto Schmidt and Imke Schuchmann-Feddersen



*Chapter 5*

**Human Leukemia and Retroviruses**

Abraham Karpas

*Chapter 7*

**Rubella Virus Products and Their Distribution in Infected Cells**

D. S. Bowden and E. G. Westaway



*Chapter 8*

**Association of Bluetongue Virus with the Cytoskeleton**

Bryan T. Eaton and Alex D. Hyatt

*Chapter 9*

**Morphogenesis of Flaviviruses**

Tatsuo Hase, Peter L. Summers, Kenneth H. Eckels, and Joseph R. Putnak

*Chapter 10*

**Signals for Membrane-Associated Transport in Eukaryotic Cells**

Heiner Niemann, Thomas Mayer, and Teruko Tamura

*Chapter 11*

**Membrane Biology in Yeast as Probed with Enveloped Viruses**

Hannele Sareneva and Marja Makarow



*Chapter 12*

**Vaccinia Virus Hemagglutinin**

Hisatoshi Shida

*Chapter 1*

# Detection of Viral Antigens and Antibodies

## Enzyme Immunoassays

E. Kurstak, R. Marusyk, A. Salmi, L. Babiuk,
C. Kurstak and M. Van Regenmortel

## 1. APPLICATION OF ENZYME IMMUNOASSAYS IN HUMAN AND ANIMAL VIROLOGY

### 1.1. Introduction

Serological techniques have been the most widely used means of diagnosing viral diseases for years, but several of their shortcomings have shifted the emphasis toward the direct detection of antigens. Nevertheless, serological testing is still useful if it is impractical or impossible to demonstrate the presence of a viral antigen, e.g., measles virus, togavirus, rubella virus, and Epstein–Barr virus (EBV). Moreover, serological approaches are important in distinguishing between serotypes or isolates. Developments in the design of enzyme immu-

**E. Kurstak and C. Kurstak** Faculty of Medicine, University of Montreal and Hospital Hôtel Dieu de Montréal, Montreal, Quebec H3C 3J7, Canada. **R. Marusyk and A. Salmi** Viral Pathogenesis Research Unit, Department of Medical Microbiology and Infectious Diseases, Faculty of Medicine, University of Alberta, Edmonton, Alberta T6G 247, Canada. **L. Babiuk** Department of Veterinary Microbiology and the Veterinary Infectious Disease Organization, University of Saskatchewan, Saskatoon, Saskatchewan S7N 0W0, Canada. **M. Van Regenmortel** Institute of Molecular and Cellular Biology, 6700 Strasbourg, France.

noassays (EIA) and the production of monoclonal antibodies have permitted the development of techniques that have a specificity approaching or surpassing that required for clinical applications but that do not suffer from the serious drawbacks of classic methods.

The ideal EIA should permit the direct detection of viral antigens and thus offer the possibility of rapid disease management. Unfortunately, this is often not possible because of low concentrations or different serotypes of the viral agents or viremia at certain stages of the disease. Serological testing can be helpful in some situations, such as (1) the case of virus detection but with equivocal interpretation; (2) cases in which virus is suspected but the direct antigen detection is difficult or very time consuming, e.g., enteroviruses, rubella virus, mumps, viral encephalitides, lymphocytic choriomeningitis virus, EBV; (3) heterophile-negative mononucleosis syndromes, i.e., cytomegalovirus (CMV), EBV; and (4) in situations in which the immune status should be established from single sera, i.e., rubella, hepatitis B, varicella zoster.

The EIA has become a useful tool in the research laboratory as well, and virtually all current studies in viral structure and function and viral immunopathogenesis rely on variations of the technique (Mäkelä *et al.*, 1987). The availability and characterization of libraries of monoclonal antibodies directed against specific viral epitopes has come about directly as a result of the application of EIA. Another recent development is the use of chemically defined antigens in immunoassays. Specific viral antigens produced by recombinant DNA technology *in vitro* will have many advantages in different EIA applications in the future. Similarly, synthetic peptides with viral antigenic sites will also be used as reagents in both the basic research and clinical applications of EIA.

Recent developments in recombinant DNA technology permit the use of nucleic acid probes in diagnostic virology. Most probes used in this technique are radiolabeled with [$^{32}$P]nucleotides. An almost equivalent sensitivity can be achieved by incorporating biotinylated nucleotides into the probes, e.g., by nick translation, and by detecting these probes using succinylated avidin or streptavidin and biotinylated enzymes in a manner similar to the enzyme immunoassays (Langer, 1982). The use of nucleic acid probes obtained by molecular cloning techniques permits the preparation of large amounts of detector molecules for viruses that cannot be cultivated *in vitro,* e.g., hepatitis virus B (HBV) and human papilloma virus. An added advantage of the technique is that, in addition to free nucleic acid, integrated DNA can be detected (Scotto *et al.*, 1983).

Nucleic acid hybridization is particularly useful in the study of herpesviruses can be readily distinguished at low concentrations (Stålhandske and Pettersson, 1982). The technique has also permitted the direct detection of CMV in the urine of immunosuppressed patients (Chou and Merigan, 1983). Similarly,

EBV has been detected at a level of 0.1–0.5 genome equivalents per cell (Brandsma and Miller, 1980; Diaz-Mitoma *et al.*, 1987).

## 1.2. Humoral Immune Response in Viral Diseases

The humoral immune response to viral infections generally involves the production of immunoglobin G (IgG) antibodies, but with individual variation of the effect of these antibodies on the course and convalescence of the infection (Cradock-Watson *et al.*, 1979). However, in the initial response to a primary viral infection transient IgM antibodies are produced (Schluederberg, 1965). The detection of antiviral IgM antibodies is therefore widely used in diagnosis (Meurman, 1983).

Many exceptions exist to the supposition that a primary viral infection leads to generation of IgM and IgG antibodies and a secondary infection to an IgG antibody population with increased affinity. For example, an IgM response to a primary infection is often absent or low in young children, in immunocompromised persons, in local infections (e.g., respiratory), and in reinfections or reactivation of latent infections. Meurman (1983) noted that IgM tests for confirmed respiratory infections of young children by parainfluenza virus types 1–3, respiratory syncytial virus, and adenovirus were positive in only 25–63% of patients. Moreover, Welliver *et al.* (1980) observed that the IgM response to respiratory syncytial virus infections of 1- to 3-month-old children was much weaker than that of children aged 3–12 months. These viral infections are often too localized to generate systemic immune response (Cranage and Gardner, 1980; Roggendorf *et al.* 1982; Ukkonen *et al.*, 1980). Furthermore, IgM responses are absent in about 25% of influenza virus A or B infections (Goldwater *et al.*, 1982), and similar phenomena have been observed with echovirus, coxsackievirus, and rotavirus infections (McLean *et al.*, 1980; Reiner and Wecker, 1981). This observation may reflect the fact that these patients earlier had a primary infection with a related virus, and the measured response is therefore secondary in nature.

Although the absence of an IgM response after rubella infections has been regarded as indicative of reinfection (Boué *et al.*, 1971), an IgM response has been recorded in about 20% of vaccinated subjects reinfected with live rubella virus (Harcourt *et al.*, 1980). The IgM antibody response to reactivation of latent infections of herpesvirus types 1 and 2 (HSV-1 and HSV-2) is rare, but frequent (70–80%) in cases of herpesvirus zoster (HSV) (Meurman, 1983). Reactivation of CMV sometimes results in high IgM responses (Cappel *et al.*, 1978), although low values have also been reported (Kangro, 1980; van Loon *et al.*, 1981).

Temporal variations in the IgM antibody response may also lead to difficulties in interpreting the results of IgM tests. Complicated infections often result in a prolonged IgM response (Burke and Nisalak, 1982); for example, a rubella virus IgM response of up to 4 years has been observed (Stallman *et al.*, 1974). Subacute sclerosing panencephalitis patients with chronic measles infection very rarely have IgM antibodies (Ziola *et al.*, 1979). After renal transplantation in immunosuppressed patients, CMV, and BK papovavirus are often reactivated followed by a prolonged IgM response (Cappel *et al.*, 1978; Flower, 1977). Patients with chronic hepatitis virus B (HBV) or healthy carriers also frequently have a prolonged IgM antibody response (Roggendorf *et al.*, 1981).

It should be noted that the detection of these IgM responses may simply arise because of the high sensitivity of the assays compared with less sensitive tests. The value of IgM tests may also be decreased for infections with related serotypes of viruses. Viruses with uniform antigenicity that produce long-lasting immunity are characterized by a constant and transient IgM response (Meurman, 1983), However, viruses for which several strains or serotypes exist may lead to an unpredictable IgM response, e.g., absence of IgM response, or false-positive reactions. The increased specificity and sensitivity of enzyme immunoassays can have a profound impact on the determination of specific IgM antibodies for heterogeneous groups of viruses such as enteroviruses, togaviruses, parainfluenza virus, and adenoviruses.

In recent years, EIA have permitted the study of the response of subclasses of IgG antibodies to viral infections. These subclasses differ in their biological properties, and subclass profiles of antibodies may indicate the state of infection. IgG1 is the major subclass; antibodies to viruses are usually found in this subclass, although subclass IgG3 antibodies are also frequently found (Linde, 1983; Morell *et al.*, 1983; Sundqvist *et al.*, 1984). As a whole, IgG2, the second subclass, appears mainly to contain antibodies to polysaccharides, e.g., lipopolysaccharides of bacteria. Subclass IgG3, which does not react with protein A, has a rapid turnover and is the most active subclass in activating the C1 component of complement; it is frequently associated with recurrent illnesses (Gilljam *et al.*, 1985). Subclass IgG4 antibodies have been detected after herpesvirus infections. Interestingly, IgG4 responses to HSV are common, but for CMV and varicella zoster virus (VZV), a familial relationship has been observed (Gilljam *et al.*, 1985; Vejtorp *et al.*, 1980), indicating that IgG4 responses are related to allergic disorders that may have a genetic component. The principles of class-capture assays to detect antibodies belonging to specific classes or subclasses are shown in Fig. 1.

### 1.3. Technical Aspects of IgM Antibody Assays

Early methods still widely used to distinguish IgM from IgG antibodies were based on the physical separation of IgM from IgG by sucrose-density gradient fractionation, gel filtration, affinity or ion-exchange chromatography.

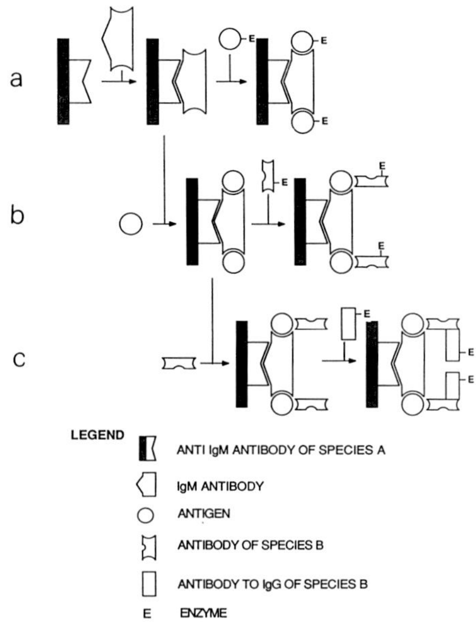


**FIGURE 1.** Principles of class-capture assays. The anti-immunoglobulin class antibody (e.g., IgG or IgM) is adsorbed on the solid phase and will trap molecules of that class. If antibodies are contained within this Ig class, antigen added subsequently will also be immobilized (a). This antigen can then be detected either directly (b) or indirectly (c).

Alternatively, IgM molecules can be selectively dissociated by thiols to produce subunits with lower activity. However, to produce significantly different titers for untreated and treated serum, at least 75% of total antibodies needs to be of the IgM class. However, only a few viruses cause such marked IgM immune responses, e.g., tick-borne encephalitis (Kunz and Hofman, 1971). Nevertheless, this method can be used in conjunction with physical separation methods.

The basic EIA for IgM antibodies employs viral antigens captured on a solid phase. Specific viral antibodies to all subclasses bind to this antigen in the assay. Bound antibodies of a specific subclass are detected by class-specific enzyme-labeled antibodies. If the amount of specific IgM antibody is low compared with IgG antibody, the positive signal in the IgM test is low. Another problem in this type of direct assay is interference by rheumatoid factors. These problems can be reduced using reversed assays in which all the IgM antibodies are first captured on a solid phase and the specificity then determined by use of specific antigens (Figure 2). Duermeyer and van der Veen (1978) first reported such a class-capture EIA (Fig. 1) that avoided the problems mentioned. A drawback is that the sensitivity of IgM detection is influenced by the fraction of specific IgM antibodies in the total IgM pool (Heinz *et al.*, 1981).

Nonspecific adsorption of IgM to a solid phase seems to be higher than that of IgG (Meurman, 1983) because of the law of mass action and the physical properties of IgM. Rheumatoid factors (RF) have a steep dose–response curve (Meurman and Ziola, 1978), and dilution is one way of decreasing their inter-

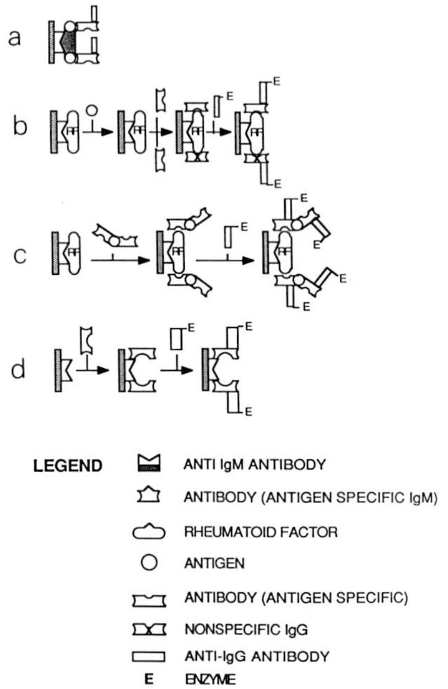


**FIGURE 2.** Class-capture assay (a) and possible causes of nonspecific interactions (b,c,d). Anti-immunoglobin M (IgM) assays may cause the nonspecific adsorption of rheumatoid factors (RF) that adsorb IgG without the presence of antigen (b) or immune complexes (c). Moreover, nonspecific protein–protein interactions (d) may cause background staining.

ference. RF are produced in patients with autoimmune and infectious disease. Pregnancy is also often accompanied by the activation of RFs (Meurman, 1978). Current methods to decrease RF interference in immunoassays involves the adsorption of specimens with heat-aggregated IgG (Kurstak *et al.*, 1977) or glutaraldehyde (Krishna *et al.*, 1980), with latex particles covered with IgG (Vejtorp *et al.*, 1980), or excess normal IgG (Milan and Stigbrand, 1981). None of these methods is entirely satisfactory. The use of $F(ab')_2$ instead of intact IgG is, however, often effective in reducing nonspecific reactions. The principles employed in some frequently used EIA are shown in Fig. 3.

## 1.4. Viral Antigen Detection in Clinical Specimens

A large number of virus particles are known to be secreted during the acute phase of viral diseases. It was shown that viral antigens can be detected in nasopharyngeal cells by the fluorescent antibody technique (Gardner and McQuillin, 1968). Electron microscopic observation that patients with rotavirus infection excrete a large number of virus particles (Flewett *et al.*, 1974) led to the first successful attempts to detect viral antigens in clinical specimens by immunoassay (Middleton *et al.*, 1977; Yolken *et al.*, 1977). It was shown later that a number of acute viral diseases could be efficiently diagnosed by the demonstration of virus antigens in clinical specimens (Yolken, 1982a).

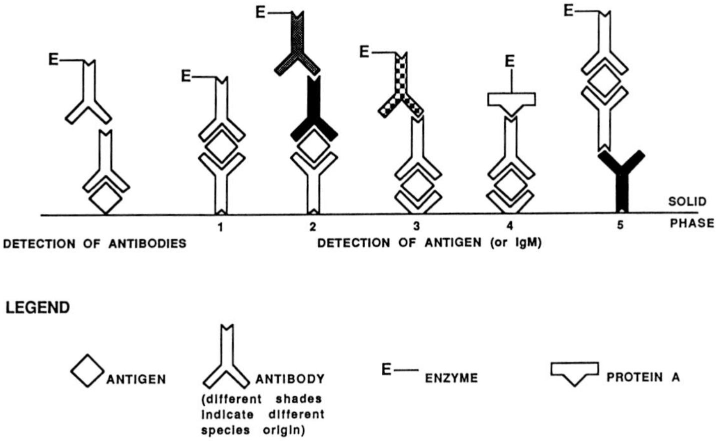


**FIGURE 3.** Frequently used enzyme immunoassays for the detection of antibodies, antigen (1–4), or antibodies belonging to specific immunoglobulin classes (5). The molecules to be detected are always found in the second layer.

Different technical variations of the antigen detection method have been described (Fig. 3). Provided that specific, high-titered antisera are available, the establishment of a reliable antigen assay for detection of virus antigen is technically feasible. The threshold sensitivity of the assay system depends mainly on the quality of the antiserum and the enzyme used. Antigen concentration of ~1 ng/ml has been reached in most of the antigen assays but it is theoretically possible to measure attomoles ($10^{-18}$ M) of viral antigens by EIA (Richman *et al.*, 1984). Such ultrasensitive tests will ultimately be developed, but even the present EIA are sensitive enough for the diagnosis of most acute viral infections.

The major problems in antigen-detection systems may be in collection of the specimens. The best method for obtaining specimens from respiratory infections is the collection of nasopharyngeal secretions. Collection of stool specimens from gastroenteritis patients does not require any special technique. As the virus secretion is short-lived during an acute viral infection, timing of the specimen collection is critical. Information about the period of antigen excretion in different viral diseases is fragmentary. It can be safely assumed that during the early phases of the disease the likelihood for antigen-positive specimens is high. Therefore, the specimens should be collected as early as possible during the infection. Another problem in specimens collected late during viral infection is the formation of immune complexes between the antigen and antibodies. Detection of virus antigens in such complexes cannot be performed by ordinary antigen assays but requires laborious separation techniques (Lund and Salmi, 1982) or capture EIA designed to detect antigen–antibody complexes (Crouch *et al.*, 1985).

## 1.5. Detection by EIA of Human and Animal Virus Infections

### 1.5.1. Hepatitis

The presence of hepatitis B surface antigen (HBsAg) in blood is frequently used as an indicator of HBV infection and concentrations may range from 0.001 to 50 μg/ml. HBsAg is, however, a poor marker for infections, as serum may transmit HBV in the absence of detectable HBsAg. In model situations, DNA probes have enhanced sensitivities and may be the only method of detecting HBV infection (Feinman, 1985; Harrison *et al.*, 1983; Thomas *et al.*, 1985).

Detection of a weak anti-hepatitis B core (HBc) IgM response in HBsAg-positive patients cannot always be considered a definitive marker of recent infection. However, quantitation of HBc IgM in HBsAg carriers using class-capture EIA increases the clinical value of the test and permits differentiation between acute HBV infections and jaundice due to other causes (Kryger, 1985). Anti-HBc IgM class capture EIA tests are important, since HBsAg carriers generally exhibit very high total antibody titers for anti HBc.

Papaevangelou *et al.* (1984) showed that the diagnostic value of anti-HBc in areas in which the prevalence of HBV is high is important because complex situations arising from other forms of acute liver injury may complicate the diagnosis of a patient with chronic HBV liver disease.

Hepatitis A virus (HAV) can be detected in cell culture by electron microscopy with immunoperoxidase staining (Asher *et al.*, 1987), a technique that is particularly useful, considering the difficulty in obtaining suitable culture systems for this virus.

HDV (delta agent, δ) can be detected by the immunoperoxidase method (Recchia *et al.*, 1981) and by EIA for determination of total anti-HDV antibodies (Shattock and Morgan, 1984) or IgM using a class-capture assay (Smedile *et al.*, 1982). The potential importance of HDV has been stressed by Smedile *et al.* (1982), who found the incidence of HDV markers higher in patients with fulminant hepatitis than in those with benign hepatitis. The use of antigen derived from detergent-treated serum increases the sensitivity of the EIA for HDV (Shattock and Morgan, 1984).

### 1.5.2. Viral Gastroenteritis

Viral gastroenteritis is common worldwide, its prevalence being second only to that of respiratory illnesses (Smedile *et al.*, 1982). Its clinical features vary and, after a rapid onset, include nausea, vomiting, abdominal cramps, headache, anorexia, myalgia, and malaise. The maladsorption triggered by gastroenteritis-associated viruses may increase morbidity and mortality in poorly nourished populations.

Rotaviruses infect the gastrointestinal (GI) tracts of many species, including man, and produce clinical and serological effects that are age dependent. Subgroup specificity of the particular rotaviruses is determined by the major inner capsid protein and the serotype specificity by an outer capsid glycoprotein (Kalica *et al.*, 1981). Cultivation of human rotaviruses in the laboratory is not satisfactory for diagnostic purposes (low isolation rates and the need for several passages). Solid-phase EIA for rotavirus antigen have proved useful in this respect (Yolken, 1982b); however, nonspecific reactions frequently occurred in early studies (Krause *et al.*, 1983). Nevertheless, various modifications, such as the use of chelating agents, protease inhibitors, pH neutralization, and the inclusion of normal serum, greatly improved the reliability of these diagnostic tests (Beards *et al.*, 1984; Hammond *et al.*, 1982; Hogg and Davidson, 1982; Hovi *et al.*, 1982). The availability of a monoclonal antibody to the common rotavirus antigen has increased the sensitivity and specificity of commercially available tests.

Another important gastroenteritis virus is the Norwalk virus. The low shedding of the Norwalk virus has so far hampered its characterization and the preparation of a hyperimmune serum. Infantile gastroenteritis is often caused by adenovirus serotypes 40 and 41 (Brandt, 1983; De Jong *et al.*, 1983), which are readily identifiable by EIA.

A radioimmunoassay (RIA) has been developed to Norwalk agent and has been proved more efficient than immunoelectron microscopy (Blacklow *et al.*, 1979) for antigen detection. Similarly, a RIA-blocking test has been used for serological investigations into Norwalk virus infections. Similar studies should be possible with EIA.

Using a capture EIA assay, it was possible to study the persistent shedding of bovine enteric coronavirus antigen–antibody complexes in clinically normal cows over extended periods (Crouch *et al.*, 1985). Similarly, EIA can be used to detect other GI infections, such as astroviruses, caliciviruses, toroviruses, and other enteric infections, that may be important in causing GI infections in man and animals (Ashley *et al.*, 1978; Spratt *et al.*, 1978).

### 1.5.3. Herpesviruses

Numerous serological tests and other procedures have been developed for the specific detection of antibodies to HSV-1 and HSV-2 in human sera, e.g., neutralization kinetics, microneutralization, indirect hemagglutination, immune lysis of infected cells, immunofluorescence, immunoperoxidase assay, solid-phase RIA, immunoelectrophoresis, and polyacrylamide gel electrophoresis (PAGE) after immunoprecipitation (Hampar *et al.*, 1985; Kurstak *et al.*, 1977; Kurstak, 1986; Kurstak *et al.*, 1984a; Kurstak *et al.*, 1984b). Microneutralization is most frequently used in seroepidemiological studies (Pauls and Dowdle,

1967; Rawls *et al.*, 1979), but EIA methods are faster and easier to perform (Coleman *et al.*, 1983; Katz *et al.*, 1986; Vestergaard, 1982). Hampar *et al.* (1985) developed a modified EIA test that uses as target antigens highly immunogenic glycoproteins purified with monoclonal antibodies; the method uses heterologously infected cell extracts for the absorption (30 min) of cross-reacting antibodies from test sera. The index of agreement with microneutralization is close to 100%. Monoclonal antibodies have been used in immune affinity chromatography to purify each of the three major glycoproteins of varicella zoster virus (VZV) and bovine herpesvirus 2 (BHV-2) for subsequent use as solid-phase reagents in EIA for detection of VZV specific antibodies (Keller *et al.*, 1986).

Using purified bovine herpesvirus-1 (BHV-1) glycoproteins, a solid-phase EIA has been established to test the kinetics of the development of the immune response as well as the specificity of the immune response to the individual BHV-1 glycoproteins (Babiuk *et al.*, 1987; van Drunen Littel-van den Hurk and Babiuk, 1986). These assays in addition to testing the efficacy of specific glycoproteins and immunogens also have the potential of being used to differentiate animals which are immunized with individual glycoproteins from animals immunized with conventional viruses or those exposed to field strains of virus. This differentiation is becoming important as producers and countries try to eliminate herpesvirus infections from their herds. A similar approach is being used in the control and eradication of pseudorabies virus (PRV) in pigs. A subunit vaccine is being marketed in conjunction with a card EIA. Using this approach, producers test animals for the presence of antibody to the glycoprotein being used in the vaccine, as well as to a second major glycoprotein not present in the vaccine. All animals that possess antibodies to the glycoprotein that is not in the vaccine are considered to be potential carriers of latent pseudorabies virus. These animals are then eliminated from the herd and replaced with animals that are seronegative to that specific test glycoprotein. Within a short time, it is hoped that such an approach would eliminate all latent carriers of PRV. Although a similar test has been conducted with BHV-1 under laboratory conditions, to the authors' knowledge, this is not being developed commercially.

It has been suggested that IgE contributes to the pathogenesis of herpesvirus infection (Ida *et al.*, 1983); however, there have been few reports of the production and significance of antiviral IgE antibodies. This may be because of difficulties arising from competitive inhibition of antiviral IgE binding to antigenic determinants caused by the presence of excess IgM, IgA, or IgG antibodies. To circumvent this problem, van Loon *et al.* (1981) devised a sensitive direct EIA test based on the antibody-capture principle (sequence: affinity-purified, anti-human IgE-coated microtitration plates; sample with IgE antibodies; peroxidase-labeled viral antigens). This method may become an important tool for investigating the role played by the IgE response in immunological defense and viral pathogenesis.

Cytomegalovirus infections are serious if recurrent, particularly in immunocompromised patients (Whelchel *et al.*, 1979; van Loon *et al.*, 1987). Several factors prevent EIA detection of CMV in fresh urine by binding to the virus, masking its antigenic determinants (McKeating *et al.*, 1986; El-Mekki *et al.*, 1987). A competitive EIA method was recently developed for detecting CMV antibody in the sera of cardiac transplant patients. The method is five times more sensitive than the complement-fixation test and twice as sensitive as indirect EIA (Wreghitt *et al.*, 1986). Ida *et al.* (1983) obtained evidence that the IgE response to herpesvirus infection is important in patients with decreased numbers of T-suppressor cells because of immunosuppressive therapy.

### 1.5.4. Pestiviruses

Maternal rubella is transmitted to the fetus by way of viremia, particularly during early pregnancy (Miller *et al.*, 1982), resulting in an infection rate of 80–90% for the placenta and 60–70% for the fetus. In about 10% of cases, spontaneous abortion results and in 25–35% there is malformation of the fetus.

Several problems are encountered in the serological diagnosis of rubella: accidental rubella vaccination during pregnancy, reinfection of previously vaccinated women during pregnancy, acute rubella infection or contact with infected individuals, and rubella immunity during pregnancy.

Antibodies to rubella are present in the newborn up to the age of about 6 months. Cell-mediated immunity is also important in the control of rubella infections. The level of IgM antibodies declines rapidly after the onset of symptoms. The detection of rubella specific IgM antibodies indicates recent infection by rubella. For standard hemagglutination inhibition tests, serum fractionation is necessary but is impractical for large-scale screening. Considerable research has gone into developing class capture assays of the EIA type for rubella (Bellamy *et al.*, 1986; Isaac and Payne, 1982; Vejtorp *et al.*, 1980). Physical separation of IgG from IgM antibodies need not be carried out and only 5 μl of sample suffices. In this respect, the $F(ab')_2$ conjugates have proved particularly reliable and sensitive probes.

One virus that is a problem for the cattle industry due to its ability to infect animals and cause disease, as well as in anyone working with tissue culture is the bovine viral diarrhea (BVD) virus. This virus can often infect immunologically naive animals; if the animal is pregnant, the virus can be transferred across the placenta, infecting the fetus. Depending on the stage of gestation, a number of scenarios may occur. First, the animal may be infected with a virus and become immunotolerant to the virus. This generally occurs if animals are infected before the 130th day of gestation. Second, the animal may develop antibodies and clear the virus before birth. This occurs if the animal is infected after 130 days of gestation. The animals infected prior to 130 days of gestation are born and are persistent shedders of BVD virus and are immunotolerant to the virus. To prevent continued infection and spread of the virus, it is important to identify these carriers by detecting the presence of virus in the blood. This can be done either by EIA or fluorescent antibody tests (Bielefeldt Ohmann *et al.*, 1982; Bolin *et al.*, 1985; McClurkin *et al.*, 1984). If animals are killed during pregnancy and fetal bovine serum (FBS) is collected for use in tissue culture, the serum may contain either BVD or virus–antibody complexes. FBS containing either BVD or BVD antibody, or both, is unsatisfactory as a component of cell-culture media for growing viruses, for research, or for vaccine production. Noncytopathic BVD present in the serum may infect cell cultures, resulting in viral interference with replication of other viruses or alter the kinetics of viral protein synthesis. There are EIA and competitive blocking EIA for detecting either BVD antibodies or antigen in the serum (Katz and Hanson, 1987).

### 1.5.5. Measles

In many countries, the prevalence of measles has decreased dramatically since the introduction of measles immunization (Hinman *et al.*, 1983; Larke, 1983; Peradze and Smorodintsev, 1983; Sejda, 1983). Nevertheless, a diagnostic screening test to identify nonimmune persons is important because of the severe complications that can occur after infection with measles virus (Morgan and Rapp, 1977), and the occurrence of measles virus outbreaks in supposedly protected populations (Marusyk, 1984). Hemagglutination inhibition tests have been widely used for this purpose but are inconvenient, as sera should be pretreated to remove nonspecific inhibitors or agglutinins that would otherwise influence the reliability or specificity of the test.

Measles may be associated with complications such as prolonged diarrhea, encephalitis, pneumonia, otitis media, and sometimes subacute sclerosing panencephalitis (SSPE). The global case fatality rate in developing countries has been estimated to approach 2%, i.e., 1.5 million deaths per year (Marusyk, 1984).

Enzyme immunoassays are as sensitive as other serological techniques in detecting seroconversion to measles virus (Rapicetta *et al.*, 1983; Tuokko and Salmi, 1983). A convenient and flexible EIA method for detecting antibody to measles antigen was introduced by Rice *et al.* (1983). In this approach, the wells of microtitration plates are coated with infected cells. Prepared in this way, the plates can be stored desiccated for months at room temperature; the major viral epitopes are preserved and can be detected by conventional EIA procedures. Measles-capture assays for IgM using monoclonal antibodies are useful and produce little or no background values (Forghani *et al.*, 1983; Kramer and Cremer, 1948; Tuokko, 1984).

Forghani *et al.* (1982) produced monoclonal antibodies to human IgM that were evaluated in capture-class assays for measles and rubella IgM antibodies. This approach avoided cross-reactivity with IgG and eliminated the need to include antibody-negative human sera in the diluents for test reagents (Isaac and Payne, 1982). False-positive reactivity caused by the presence of rheumatoid factors was virtually eliminated using glutaraldehyde-aggregated IgG as an absorbent. Latex agglutination tests are poor indicators of rheumatoid factors and give high levels of false-positive reactivity.

### 1.5.6. Infections of the Respiratory Tract

Most infections of the upper respiratory tract are caused by viruses, but bacteria may account for up to one third of such infections. Conventional hemagglutin inhibition and neuraminidase inhibition tests for influenza antibodies are relatively insensitive (Turner *et al.*, 1982). However, EIA methods are much more sensitive and permit the assay of antibodies in nasal washings. Murphy *et al.* (1981) developed an EIA test for hemagglutinin instead of for the whole virus, whereas Khan *et al.* (1982) reported an EIA test for neuraminidase-specific antibodies. A pitfall in the determination of titers for influenza viruses was pointed out by Madore *et al.* (1983), who observed that dose–response curves could be displaced upward, reflecting an increase in antibody affinity, without a change in the end point (steeper curves). Differentiation of influenza virus strains requires the use of the constituent proteins rather than of whole virus (Al-Kaissi and Mostratos, 1983).

Parainfluenza viruses are associated with many common respiratory illnesses of children. The main problems in standard methods for the determination of parainfluenza viruses is the rapid loss of infectivity and the high cost of cell-culture assay systems. Immune responses may take several weeks before a detectable antibody titer is present. In contrast to aspiration with a mucous extractor (through the nostrils), the use of nasopharyngeal washings gives low yields of parainfluenza antigen. Cross-reactivity between antibodies to parainfluenza virus types 1 and 3 as well as with antibodies to mumps virus can be significant (Julkunen, 1984). Live virus is not required for EIA.

Respiratory syncytial virus (RSV) is also commonly diagnosed by cell-culture techniques. EIA have been introduced and are an important alternative (Sarkkinen *et al.*, 1981). Compared with cell-culture methods, the EIA has a sensitivity of more than 90% and a specificity of more than 96% (Flanders *et al.*, 1986). EIA methods have also been used to detect coronaviruses successfully (Kraaijeveld *et al.*, 190; Macnaughton *et al.*, 1983) at a level of $10^5$ particles/ml. Adenovirus antigen can also be detected in nasopharyngeal specimens from patients with respiratory tract infections (Sarkkinen *et al.*, 1981).

An EIA that detects rhinovirus-specific antibody in human sera and unconcentrated nasal secretions is 100- to 1000-fold more sensitive than standard neutralization assays (Barclay and Al-Nakib, 1987).

### 1.5.7. Rabies

The classic method for detecting neutralizing antibodies for rabies is seroneutralization using an *in vivo* mouse system. However, the results obtained are variable, and the method is time consuming (up to 21 days). Wiktor *et al.* (1973) improved the reproducibility and speed of the method by using tissue-culture and inhibition immunofluorescence, while Atanasiu and Perrin (1979) developed an immunoenzymatic micromethod for the titration of antirabies antibodies using as antigens either the rabies virus or a constituent glycoprotein. Significant results were obtained within a few hours (Sureau *et al.*, 1982).

Both rabies vaccine and laboratory-adapted rabies virus replicate in cell culture to give high titers. However, wild strains of the virus grow poorly in cell cultures, and it is difficult to produce the amount of virus required for large scale screening. Smith *et al.* (1984) developed a procedure in which cells infected with $\gamma$-irradiated rabies virus rather than the purified virus are used as target antigens. A single freeze-thawing cycle permits reaction of the antibody with both internal and external (nucleocapsid and membrane) viral antigens.

### 1.5.8. Arboviruses and Viral Fevers

Attempts to isolate arboviruses from patients or animals are rarely successful, and the diagnosis usually depends on serological testing. Although complement fixation, hemagglutination inhibition, and neutralization are sometimes useful in this respect, each has important disadvantages (Calisher and Poland, 1980). Specific antibody-capture immunoassays have proved useful for detecting an immune response to arboviruses, e.g., Japanese encephalitis (Burke *et al.*, 1982; Burke and Nisalak, 1982; Chang *et al.*, 1984; Konishi and Yamaoka, 1982; Zhang *et al.*, 1979; Roggendorf *et al.*, 1981), St. Louis encephalitis (Monath *et al.*, 1984; Wolff *et al.*, 1981), and yellow fever (Duebel *et al.*, 1983).

Kurstak *et al.* (1980) detected arbovirus antigens in infected cells using the immunoperoxidase method. The use of this method permitted McLean *et al.* (1979) to demonstrate the intracytoplasmic replication of California encephalitis virus in both domestic and wild mosquitoes.

An illustrative example is the detection of Lassa virus antigens and Lassa virus-specific antibodies. Lassa fever is a severe, often fatal, human disease of considerable importance in some regions of West Africa. Almost all patients with Lassa fever are viremic upon admission, but tests designed to determine Lassa virus infectivity require 2–7 days, a high biological containment laboratory, and tissue-culture facilities. Niklasson *et al.* (1984) developed an EIA using $\beta$-propiolactone-inactivated sera that provides a definitive diagnosis within a few hours.

### 1.5.10. Acquired Immune Deficiency Syndrome

The need for a rapid and accurate diagnosis of acquired immune deficiency syndrome (AIDS) by a reliable and sensitive test has resulted in a burst of commercially available enzyme-linked immunosorbent assays (ELISA) for antibodies to HIV-1 virus (Cooper *et al.*, 1978; Evans *et al.*, 1987; Oldham *et al.*, 1987). The tests are based on HIV-1 antigen attached to a solid-phase support. The antibodies in human specimens bind to the antigen-coated solid phase and are visualized by enzyme-tagged antibodies to human Ig. Most tests for HIV antibodies require reading by a photometer, but simpler bedside EIA requiring only visual observation have also been described (Ivo dos Santos *et al.*, 1987).

Detection of antibodies in serum or cerebrospinal fluid (CSF) specimens (Goudsmit *et al.*, 1986) is the only practical large-scale routine method for definite viral diagnosis. HIV antigens can also be detected by immunoassays (Higgins *et al.*, 1986), but the amount of antigen is too low for direct measurement of these antigens in clinical specimens from AIDS patients. Although the reliability of the EIA for HIV antibodies is good, a small percentage of the test results should be confirmed by Western blotting on electrophoresed virus polypeptides. The strain variation of HIV-1 virus is not an obstacle to antibody determination, as some of the strong antigenic determinants are conserved. A more serious problem in HIV-antibody determinations is the existence of one or more antigenically distinct HIV viruses, such as HIV-2 (Clavel *et al.*, 1986). Low-level cross-reactivity between these viruses may be seen in EIA with the HIV-1 antigen, but sensitive antibody tests for these viruses require a homologous antigen on the solid phase (Brun-Vezinet *et al.*, 1987).

The antigens used in the first tests for HIV antibodies were produced in tissue culture. Most tests still employ purified HIV virions on solid phase, but some newer tests are based on antigens produced by recombinant DNA technology or, recently, on synthetic peptides with predicted sequences of HIV proteins (Burke *et al.*, 1978; Shoeman *et al.*, 1987). Antibody response against both the internal (p24 and p55) and envelope (gp41, gp120) proteins occurs in infected persons, but the development of all these antibodies is slow, beginning 1–2 months after the infection (Cooper *et al.*, 1987; Gaines *et al.*, 1987). The best response occurs against the internal components and the transmembrane envelope protein gp41 (Schüpach *et al.*, 1985). The peak immune response may be relatively weak in some patients or may decline when the disease progresses. The optimal type of virus antigen to be employed in the diagnostic EIA for HIV antibodies has not been determined, but it is likely that a combination or more than one polypeptide or synthetic peptide will give the most reliable results.

As the HIV antigens produced in T-cell lines contain cellular material, particularly molecules with function in the immune system (Hozie *et al.*, 1987), autoreactivity against HLA antigens and other molecules on lymphocytes causes false-positive reactions. Patients with a history of multiple blood transfusions or multiple pregnancies and patients with organ transplants or past dialysis treatments are especially susceptible to development of antilymphocyte antibodies (DeSanto *et al.*, 1987). Variations in the virus antigens used in the tests also have an effect on the false-positive reactions (Wartick *et al.*, 1987). The problems caused by lymphocyte antigens in the viral preparations can probably be avoided when recombinant proteins or synthetic peptides are employed in the EIA for HIV antibodies.

The false-positive results in the studies of a larger number of healthy populations is less than 1%. As the EIA for HIV antibodies are made to be very sensitive, false-negative results are rare. However, EIA antibody tests may give negative results in HIV-infected persons in two principal circumstances. First, antibodies against HIV develop slowly (Gaines *et al.*, 1987), and specimens taken only a few weeks after infection may remain negative. Second, HIV antibodies may also disappear at later stages of the disease (Weber *et al.*, 1987).

In addition to using the EIA in detecting oncogenic viruses in man, EIA are used extensively in detecting RNA tumor viruses in animals. These are used both to detect free virus, virus antigen–antibody complexes, and viral antigen in the serum. There is some difference of opinion on the relative accuracy of EIA versus immunofluorescent (IFA) procedures (Lutz *et al.*, 1983). It must be emphasized that when both tests are performed properly, the correlation between them is more than 95% (Lutz *et al.*, 1980). The 5% discrepancy arises in that EIA detects cats that are persistently antigenemic but IFA negative. These are not false-positive EIA results, as cats in this category often have high antibody titers to FELV and OCMA. In some of these animals, infectious FELV was found in several organs, including salivary gland and urinary bladder. These cats belong to a category of silent shedders that are not detected by the IFA procedure. In addition to being more sensitive and identifying a category of cats that are negative by IFA, the advantage of EIA is its greater sensitivity and the ease with which it can be conducted. EIA also lends itself to use in a clinician's office. A disadvantage of the procedure is that false-positive results may result when the test is improperly done. Most false-positive results are associated with improper or too rapid washing of the reaction wells at the end of the incubation steps, and with the use of whole blood or badly hemolyzed serum. In the case of hemolyzed serum, it is very difficult to wash away, resulting in false-positive results. Thus, if clinicians are to use this procedure, they must be aware of the problems associated with the assay.

Results with EIA using tears or saliva give similar results to blood tests (Hawkins *et al.*, 1986). If this is the case, it may reduce some of the false-negative results seen with improper washing procedures. Although it is often not a problem to detect gp27 in feline leukemia virus (FLV) cats by EIA, a micro-EIA has also been developed to investigate immune complexes present in the blood of leukemic cats (Tuomari *et al.*, 1984). This assay uses a solid-phase C1q protein A peroxidase EIA to quantitate circulating immune complexes during the course of infection. Increases in circulating immune complexes occur early in the infection and are often cleared following recovery. A persistent elevation of circulating immune complexes is an indication of a cat with persistent viremia.

Bovine leukemia virus (BLV), similar to HIV, has a long incubation period with clinical signs occurring late in the disease. One of the prominent early features of infection with BLV is the presence of antibody against unglycosylated components, mainly the major internal protein p24 and against virus envelope glycoproteins gp51 and gp30. These antibodies are detectable within 2 or 3 weeks after experimental infection and constitute the earliest possible indication of BLV infection (Portelle *et al.*, 1983, 1984). To reduce the horizontal spread of virus between animals within a herd, and for elimination of the disease from herds, it is important to have a method for early detection of infection (Manz *et al.*, 1981). Both direct binding of BLV virus to microtiter plates or, more recently, the development of monoclonal antibodies to individual glycoproteins can be used as a capture EIA (Bruck *et al.*, 1981). To date, the BLV gp51 EIA appears to be the most sensitive method for early detection of BLV antibodies.

## 2. APPLICATION OF ENZYME IMMUNOASSAYS IN PLANT VIROLOGY

### 2.1. Introduction

The EIA technique was used to detect plant viruses for the first time by Voller *et al.* (1976). The following year, Clark and Adams (1977) described in detail the double antibody sandwich (DAS) form of EIA, a format that has become the one most commonly used by plant virologists. In this method, the wells of a microtiter plate are first coated with antivirus globulins, and the virus in the test sample is then trapped by the solid-phase antibody. The presence of virus is demonstrated by an enzyme-labeled antivirus conjugate prepared from the same antiserum used for coating the plate (procedure 4, Table I). In spite of its great popularity with plant virologists, this test suffers from two limitations: (1) it requires the preparation of a different antivirus enzyme conjugate for each virus to be tested; and (2) it is extremely serotype specific, which means that enzyme conjugates prepared with antibodies against one virus strain often fail to react with other strains of the same virus species (Koenig, 1978; Lister and Rochow, 1979; Van Regenmortel and Burckard, 1980; Rybicki and Von Wechmar, 1981; Devergne *et al.*, 1981; Koenig and Paul, 1982; Wetter *et al.*, 1984).

Applications of DAS–EIA in plant virology have been described in several

**Table I**
**Types of EIA Used in Plant Virology**[a]

| Procedure | Types of assays |
|---|---|
| 1 | AG, $Ab^R$-E |
| 2 | AG, $Ab^R$, anti$R^G$-E |
| 3 | Ag, $Ab^R$, protein A–E |
| 4 | $Ab^R$, Ag, $Ab^R$-E |
| 5 | $Ab^C$, Ag, $Ab^R$, anti$R^G$-E |
| 6 | $F(ab')^R$, Ag, $Ab^R$, antiFc$R^G$-E |
| 7 | $F(ab')^R$, Ag, $Ab^R$, protein A–E |
| 8 | $Ab^R$, Ag, $MAb^M$, anti$M^R$-E |
| 9 | Ag, $Ab^M$, anti$M^R$, anti$R^G$-E |
| 10 | Anti$M^R$, $MAb^M$, Ag, $Ab^C$, anti$C^R$-E |
| 11 | $Ab^C$, Ag, $Ab^M$, anti$M^R$, anti$R^G$-E |
| 12 | $MAb^M$, Ag, $MAb^M$-Bio, Avi-E |
| 13 | Protein A, $Ab^R$, Ag, $Ab^R$, protein A–E |

[a]Ag, antigen; Ab, antibody; R, rabbit; G, goat; C, chicken; E, enzyme; MAb, monoclonal antibody; M, mouse; Bio, biotin; Avi, avidin; F(ab'), fragment of immunoglobulin.

reviews (Bar-Joseph and Garnsey, 1981; Clark, 1981; Torrance and Jones, 1981; Clark and Bar-Joseph, 1984); this topic is not reviewed again here. Instead, this section concentrates on the other EIA formats used with plant viruses. Attention is also given to some recent methodological advances.

## 2.2. Types of EIA Used in Plant Virology

Enzyme immunoassay methods can be divided into direct and indirect procedures. In the first type of procedure, the antivirus antibody is itself labeled with an enzyme, while in the second type, the enzyme conjugate is an anti-Ig reagent. This points to two advantages of indirect EIA methods: (1) a single enzyme conjugate, such as a goat antirabbit globulin conjugate can be used to detect any number of different viruses, eliminating the need for separate conjugates for each virus system; and (2) since antiglobulin enzyme conjugates are commercially available at very competitive prices, the need for preparing conjugates is eliminated.

Some of the EIA methods used with plant viruses are listed in Table I. The simplest forms are procedures 1–3, in which the viral antigen is adsorbed directly to the solid phase. In many cases, there is enough viral antigen present in crude leaf extracts to make it possible to identify infected plants by incubating crude plant juice directly in the wells (Lommel *et al.*, 1982; Mowat and Dawson, 1987). Leaf extracts can be prepared with pH 9.6 carbonate buffer, with pH 7.4

phosphate-buffered saline containing 0.05% Tween-20 and 2% polyvinylpyrollidone or simply with water (Lommel *et al.*, 1982; Ehlers and Paul, 1984). The adsorbed antigen can be detected with an enzyme-labeled antivirus conjugate (procedure 1), indirectly with an antiglobulin conjugate (procedure 2), or a protein A enzyme conjugate (procedure 3) (Mowat and Dawson, 1987). This simple approach is unsuitable for detecting viruses such as luteoviruses and geminiviruses that are present at low concentration in infected tissue. The method can be used when purified preparations of viral antigen are available, although it should be remembered that different viruses may have widely different capacities for sticking to polystyrene or polyvinyl plates. In addition, some viruses may be degraded in the presence of pH 9.6 carbonate buffer commonly used during the coating procedure; the viral subunits produced as a result of such degradation may become preferentially adsorbed to the solid phase, which could prevent any effective adsorption of virus particles. If a viral antiserum reacting both with the intact virions and dissociated subunits is used, positive detection will still occur, although a quantitative titration of the different viral antigens may become difficult. If monoclonal antibodies specific for either the polymerized or dissociated form of viral protein are used in the assay, results may become erratic (Van Regenmortel, 1986). Furthermore, it is now generally recognized that proteins become at least partly denatured or distorted when they are adsorbed to a layer of plastic during a solid-phase assay (Soderquist and Walton, 1980; Kennel, 1982; Friguet *et al.*, 1984; Altschuh *et al.*, 1985; McCullough *et al.*, 1985). The increasing use of monoclonal antibodies which are often specific for a conformation of the protein antigen has shown that this phenomenon is very common (Al Moudallal *et al.*, 1985; Vaidya *et al.*, 1985).

The most commonly used indirect EIA methods are the procedures 5 and 6 in Table I. Viral antiserum must be obtained from two different animal species. The most convenient animals for this purpose are rabbits, chickens, mice, rats, and goats. Since chicken globulins do not cross-react with mammalian Ig (Leslie and Clem, 1969), they are particularly useful, as they will not be recognized, for instance, by a goat antirabbit globulin conjugate (procedure 5, Table I). Another advantage of using chicken antibody is that the antivirus Ig can be obtained easily from the eggs of hens immunized with the virus (Polson *et al.*, 190; Van Regenmortel, 1982). Laying hens are convenient animals to house, and as much as 100 mg Ig can be extracted from one egg yolk (Gardner and Kaye, 1982). Although some investigators consider the need for a second virus antiserum obtained from another animals species a major inconvenience of the indirect immunoassay approach, this requirement can in fact be easily met. In the case of most plant viruses, there is no difficulty in obtaining quantities of the order of 1–10 mg purified virus, sufficient to immunize several rabbits, as well as a few chickens or even a goat.

The need for employing antibodies from two animal species can be circumvented by using $F(ab')_2$ fragments of antivirus Ig for coating microtitre plates (Barbara and Clark, 1982). The virus is trapped by the $F(ab')_2$ fragments and can be detected with the corresponding whole immunoglobulin followed by a protein A conjugate or by an anti-Fc fragment conjugate (procedures 6 and 7, Table I). It is a matter of opinion whether the need to prepare $F(ab')_2$ fragments is a simpler requirement than the preparation of antibodies in two animal species. However, the method has been used successfully to detect relationships among carlaviruses that were overlooked when direct EIA was used (Adams and Barbara, 1982). The use of the C1q component of complement instead of $F(ab')_2$ fragments for circumventing the need for two different virus antisera (Torrance, 1980) has not become popular, probably because C1q is not readily available commercially.

Recently, another EIA method using protein A was also been shown to overcome the excessive specificity of direct EIA procedures (Edwards and Cooper, 1985). Using this method, the wells of the microtiter plate are first coated with 1 μg/ml protein A, before incubation with the virus antiserum (procedure 13, Table I). The layer of protein A traps the antibody molecules by the Fab end. After incubation with antigen, the trapped virus is detected by another layer of antibodies (from the same antivirus serum), in turn detected by labeled protein A. This method has been shown to work satisfactorily with crude virus antiserum (diluted 1 : 1000) and eliminates the need to prepare Ig.

### 2.3. Use of Monoclonal Antibodies

Monoclonal antibodies (MAbs) have been produced against about 40 different plant viruses (Van Regenmortel, 1986). These reagents are especially interesting in the case of viruses, such as the luteoviruses or geminiviruses, which are present in only very low concentration in infected plants and which can be purified only with difficulty. Compared with conventional polyclonal antisera, the advantages of MAbs are (1) production of homogeneous and biochemically defined reagents in practically unlimited quantity, (2) increased specificity, and (3) easier immunization requirements, as mice can be immunized with virus preparations that are only partially purified.

The immunoassay most commonly used with MAb is some form of EIA. During the production of hybridomas, the clones secreting the desired antibody are mostly selected by an EIA that is adjusted to ensure maximum sensitivity of antibody detection. In a competitive study of several EIA procedures used for screening hybridoma cell lines, it was found that multilayered sandwich procedures (e.g., procedure 11, Table I) were usually the most sensitive (Al Moudallal *et al.*, 1984). However, it should be emphasized that the quality of the reagents used in such comparisons is of crucial importance, since the alleged superiority of a certain procedure may simply reflect the use of a reagent of exceptionally high titre.

The type of EIA used in the selection of the hybridoma can determine the properties of the MAb obtained. This is illustrated in Tables II and III in the case of MAbs raised against tobacco mosaic virus (TMV) and TMV coat protein, respectively (Altschuh *et al.*, 1985; Al Moudallal *et al.*, 1985). When clones were selected by an EIA procedure with antibody-coated plates (procedures 8 and 11, Table I), most MAb were obtained that recognized the native structure of the antigen. When the selection was done by means of antigen-coated plates (procedure 9), a large number of MAb were found to be specific for dissociated viral subunits or for a partly denatured form of the viral protein. Different EIA procedures will thus preferentially select MAb that recognize epitopes specific for the quaternary structure of the viral protein, i.e., the so-called neotopes (Van Regenmortel, 1966, 1982) or alternatively MAb that recognize epitopes specific only for the dissociated viral protein, i.e., the so-called cryptotopes (Dekker *et al.*, 1987). In addition, MAb can also be obtained that recognize epitopes common to both virions and subunits, i.e., metatopes (Table III).

## 2.4. Biotin Labeling

When the aim is to develop a double antibody sandwich assay using only MAb, the format of procedure 12 should be used (Diaco *et al.*, 1985). Biotin labeling of antibodies is made particularly simple using N-hydroxysuccinimido-biotin (Sigma, St. Louis). This activated biotin (0.2 mg) in dimethylformamide is simply incubated with the globulin (1 mg) in phosphate-buffered saline for 4 hr at 25°C (Zrein *et al.*, 1986). Labeling of the antibodies with biotin instead of with enzymes is not only advantageous in the case of MAb (Diaco *et al.*, 1985); biotin is a small molecule (244 $M_r$) that can be covalently linked to antibody

**Table II**
**Reactivity of 18 Anti-TMV Monoclonal Antibodies with Virion and Subunit Preparations in Two EIA Procedures**[a,b]

| Monoclonal antibodies | EIA procedure 1[c] | | EIA procedure 2[d] | |
|---|---|---|---|---|
| | Virions | Subunits | Virions | Subunits |
| Group 1 (8 clones) | + | − | − | − |
| Group 2 (5 clones) | + | + | − | − |
| Group 3 (5 clones) | + | + | + | + |

[a]Data from Altschuh *et al.* (1985).
[b]For explanation of symbols, see Table I.
[c]EIA procedure 1 uses antibody-coated plates and consists of the following steps: $Ab^R$, Ag, MAb, anti$M^R$-E.
[d]EIA procedure 2 uses antigen-coated plates and consists of the following steps: Ag, MAb, anti$M^G$-E.

**Table III**
**Reactivity of 30 Anti-TMVP Monoclonal Antibodies with Virion and Subunit Preparations in Two EIA Procedures[a]**

| Monoclonal antibodies | EIA procedure 1[b] | | EIA procedure 2[c] | |
|---|---|---|---|---|
| | Virions | Subunits | Virions | Subunits |
| Group 1 (3 antineotope MAb) | + | − | − | − |
| Group 2 (5 anticryptotope MAb) | − | + | + | + |
| Group 3 (22 antimetatope MAb) | + | + | − | − |

[a]Data from Al Moudallal *et al.* (1985).
[b]EIA procedure 1: $Ab^G$, Ag, MAb, anti$M^G$-E.
[c]EIA procedure 2: Ag, MAb, anti$M^G$-E.

without affecting its antigen-binding capacity (Guesdon *et al.*, 1979; Kendall *et al.*, 1983). The attachment of several biotin molecules to viral antibody is less detrimental to its capacity to recognize a broad range of serologically related viruses than is conjugation to an enzyme. In a recent study using biotinylated anti-TMV antibodies, Zrein *et al.* (1986) showed that tobamoviruses differing serologically from TMV by SDI values of $\leq$4.0 could be readily detected. By contrast, when enzyme-labeled viral antibodies were used, tobamoviruses differing from TMV by SDI values of only 2.0 could not be detected (Van Regenmortel and Burckard, 1980). One of the major drawbacks of DAS–EIA, i.e., excessive strain specificity, is thus overcome when the viral antibodies are labelled with biotin instead of with enzyme. Another advantage of using the biotin–avidin system is that the test is rendered more sensitive (Adler-Storthz *et al.*, 1983; Kendall *et al.*, 1983; Yolken *et al.*, 1983). A disadvantage of the biotin system is that it requires an additional incubation step with avidin. However, this can be avoided by incubating the biotinylated antibody together with the enzyme-labeled avidin; the resulting slight decrease in sensitivity can be compensated for by using the avidin conjugate at a higher concentration.

It has been claimed (Hill *et al.*, 1984; Diaco *et al.*, 1986) that when viruses are tested by DAS–EIA, it is necessary to use coating antibodies and labeled antibodies specific for different epitopes of the virus. However, since virions always possess a large number of identical epitopes (one per subunit), enough accessible epitopes remain free after trapping of the virus on the solid phase to permit subsequent binding of biotin-labeled antibodies of the same specificity (Dekker *et al.*, 1987). It should be remembered that different MAb may lose some of their binding activity after adsorption to plastic or after labelling; as a result, a test using the same antibody in both positions may appear inferior if the stability of the antibody is affected in one or the other position (Torrance and Pead, 1986).

## 2.5. Specific Applications of EIA in Plant Virology

### 2.5.1. Diagnosis

Enzyme immunoassay has replaced most of the older serological techniques used for the diagnosis of plant viruses (Van Regenmortel, 1982), mainly because it is much more convenient for large-scale testing. The major limitation of analyzing large numbers of samples by EIA resides in the preparation of extracts of infected plant material. Tissue grinding can be done in mortars or with electrically driven devices (Banttari, 1980; McLaughlin *et al.*, 1984). Alternatively, leaf homogenates can be replaced by discs of leaf tissue cut out with a paper punch (Marco and Cohen, 1979; Romaine *et al.*, 1981).

The use of crude plant extracts in EIA may lead to a high level of nonspecific reactions. These can be diminished by diluting the extract 10- or 100-fold or by the addition of substances such as 1–2% polyvinylpyrrolidone (Clark and Adams, 1977), mercaptoethanol, or dithiothreitol (Tamada and Harrison, 1980), diethyldithiocarbamate (McLaughlin *et al.*, 1984), 1 M urea (Gugerli, 1979), and nicotine (Walter *et al.*, 1984). There are no rules about which substances will be found advantageous or deleterious in a particular instance, and the best extraction medium must be selected empirically (Huss *et al.*, 1986). Instead of performing EIA in the usual microtiter plates, other solid-phase materials such as polystyrene and frosted glass beads have also been used for detecting plant viruses (Felgner, 1978; Chen *et al.*, 1982; Polson *et al.*, 1985).

Enzyme immunoassay procedures are widely used for large-scale investigations of the incidence of viral infections in various crops such as potatos (Gugerli, 1979; Tamada and Harrison, 1980), deciduous fruit (Barbara *et al.*, 1978; McMorran and Cameron, 1983), citrus (Lommel *et al.*, 1983; Stein *et al.*, 1979), vegetables (Jafarpour *et al.*, 1979; Marco and Cohen, 1979), grain (Lister *et al.*, 1981; Rochkow, 1979), and grapevines (Gonsalves, 1979; Walter *et al.*, 1984).

### 2.5.2. Virus Detection in Vectors

Because of their sensitivity, EIA procedures are particularly suitable for detecting plant viruses in their insect vectors (Gera *et al.*, 1978; Clarke *et al.*, 1980; Tamada and Harrison, 1981; Carlebach *et al.*, 1982; Ghabrial and Schultz, 1983; Caciagli *et al.*, 1985). In this connection, increased sensitivity of antigen detection by the use of a fluorogenic substrate (Torrance and Jones, 1982) and the amplified EIA technique (Torrance, 1987) has been beneficial. The ability of detecting virus in single viruliferous insects has opened the way to new epidemiological investigations.

### 2.5.3. Quantitative Analysis of Antigenic Relationships

Until recently, there have been few attempts to use EIA for analyzing the antigenic relationships between plant viruses (Rybicki and Von Wechmar, 1981; Powell and Derr, 1983). Because of their sensitivity, EIA procedures will tend to bring out any reactions with contaminating antigens; this problem can often be circumvented using relatively high dilutions of the reagents. In all types of solid-phase immunoassays, it is important to avoid the use of too highly concentrated reagents, as this tends to emphasize nonspecific interactions (Altschuh and Van Regenmortel, 1983; Muller *et al.*, 1985). Electrostatic interactions between Ig and viral proteins can be prevented by the addition of heparin to incubation buffers (Dietzgen and Francki, 1987).

The degree of antigenic relatedness between two plant viruses is commonly expressed by the serological differentiation index (SDI), which corresponds to the average number of twofold dilution steps separating homologous from heterologous precipitation titers (Van Regenmortel and Von Wechmar, 1970). Unfortunately, SDI values calculated from different antisera to the same antigen can vary considerably, and a reliable estimate of the degree of cross-reactivity between two viruses can only be obtained by studying many bleedings from different immunized animals (Van Regenmortel, 1982). Jaegle and Van Regenmortel (1985) showed that reliable SDI values, corresponding to those measured in precipitation tests, could be obtained by indirect EIA measurements. The economy of reagents obtainable in EIA makes it easier to examine a large number of different antisera; furthermore, since EIA titers are usually much higher than precipitation titers (of the order of $10^{-6}$ instead of $10^{-3}$), EIA can detect much more distant serological relationships than can the precipitin test (Hoffman and Van Regenmortel, 1984). For instance, a distant relationship between turnip crinkle virus and several tombusviruses (SDI = 10–13) that remained undetected by precipitin tests was shown by indirect EIA (Jaeglé and Van Regenmortel, 1985).

An indirect EIA method suitable for calculating relative titers of plant virus antisera and for deriving SDI values was recently described (Clark and Barbara, 1987). Data from the proportional response region of each experimental curve were converted to logarithmic values and subjected to regression analysis to obtain relative titers. The method permitted closely related viruses to be discriminated in a reproducible manner and demonstrated the potential value of EIA for quantitative antigenic analysis.

### 2.5.4. Analysis of Antigenic Structure

The advent of EIA has made it possible to analyze the antigenic structure of plant viruses by fairly simple procedures. By means of EIA inhibition tests using peptide fragments (Altschuh and Van Regenmortel, 1982), the location of continuous epitopes in viral proteins can be easily determined (Quesniaux *et al.*, 1983). Synthetic peptides coupled to protein carriers have also been used in EIA to identify the epitopes of plant virus proteins (Al Moudallal *et al.*, 1985). When used in conjunction with MAb, various formats of EIA represent the most powerful methodology available for elucidating the antigen structure of plant viruses (Al Moudallal *et al.*, 1982, 1985; Briand *et al.*, 1982; Altschuh *et al.*, 1985; Doughtery *et al.*, 1985; Tremaine *et al.*, 1958; Koenig and Torrance, 1986).

## 3. CONCLUSIONS

The introduction of the EIA technique has been exceedingly important for the diagnosis of viral diseases, and its use in detecting molecules immobilized on a solid phase has revolutionized both clinical and research laboratories. Early methods were adopted because the tests were simple, inexpensive, fast, and sensitive. However, with the advent of highly specific reagents, such as monoclonal antibodies, the specificity of the methods has increased considerably. Moreover, techniques have been developed for the detailed study of the immunopathogenesis of viral infections.

The introduction of class-capture EIA has permitted a much better follow-up of management of the IgM and IgG immune responses than had been possible with classic methods and shown many erroneous assumptions about the relation between primary infection and the IgM response, particularly for viruses that lack antigenic uniformity (Meurman, 1983), such as enteroviruses, adenoviruses, parainfluenzaviruses, and togaviruses. Class-capture assays are also increasingly used for studying the response of the IgG subclasses of antibodies. Assays of this type indicate that antibodies often occur in one subclass after the primary infection but in another after a recurrent infection.

The trend toward developing highly specific reagents and innovative tests can be expected to continue and should lead to a better comprehension of the process of infection and of the body's defense systems. In addition, an increasing number of commercial EIA kits are available for the detection of infectious agents or their antibodies. The assays have become more rapid and simpler to perform.

Advances in recombinant DNA technology have permitted the use of nucleic acid probes in the diagnosis of human viral diseases. However, wide application of these probes will occur only if enzymes or other nonradioactive detectors are used to determine whether the probes have hybridized to the specimen nucleic acid. This methodology is suitable both for quantitation and localization purposes and is a valuable adjunct to the enzyme immunoassay.

The examples described illustrate the application of EIA methods, or DNA probes, to the study of viral diseases that were previously difficult to diagnose. A multitude of papers have been published in recent years with EIA in the titles, attesting to the popularity of the technique and to the difficulties of reviewing this topic comprehensively.

## 4. REFERENCES


Adams, A. N. and Barbara, D. J., 1982, The use of $F(ab')_2$-based ELISA to detect serological relationships among carlaviruses, *Ann. Appl. Biol.* **101**:495–500.

Adler-Storthz, K., Kendall, C., Kennedy, R. C., Henkel, R. D. and Dreesman, G. R., 1983, Biotin–avidin amplified enzyme immunoassays for detection of herpes simplex virus antigen in clinical specimens, *J. Clin. Microbiol.* **18**:1329–1334.

Al-Kaissi, E., and Mostratos, A., 1983, Assessment of substrates for horseradish peroxidase in enzyme immunoassay, *J. Immunol. Methods* **58**:127–132.

Al Moudallal, Z., Briand, J. P. and Van Regenmortel, M. H. V., 1982, Monoclonal antibodies as probe of the antigenic structure of tobacco mosaic virus, *EMBO J.* **1**:1005–1010.

Al Moudallal, Z., Altschuh, D., Briand, J. P. and Van Regenmortel, M. H. V., 1984, Comparative sensitivity of different ELISA procedures for detecting monoclonal antibodies, *J. Immunol. Meth.* **68**:35–43.

Al Moudallal, Z., Briand, J. P. and Van Regenmortel, M H. V., 1985, The major part of the polypeptide chain of tobacco mosaic virus protein is antigenic, *EMBO J.* **4**:1231–1235.

Altschuh, D. and Van Regenmortel, M. H. V., 1982, Localization of antigenic determinants of a viral protein by inhibition of enzyme immunosorbent essay (ELISA) with tryptic peptides, *J. Immunol. Meth.* **50**:99–108.

Altschuh, D. and Van Regenmortel, M. H. V., 1983, Refutation of the alleged antigenic relationship between an isometric and a rod-shaped plant virus, *Int. J. Microbiol.* **1**:13–17.

Altschuh, D., Al Moudallal, Z., Briand, J. P. and Van Regenmortel, M. H. V., 1985, Immunochemical studies of tobacco mosaic virus, VI. Attempts to localize viral epitopes with monoclonal antibodies, *Molec. Immunol.* **22**:329–337.

Asher, L., Binn, L., and Marchwicki, R., 1987, Demonstration of hepatitis A virus in cell culture by electron microscopy with immunoperoxidase staining, *J. Virol. Methods* **15**:323–328.

Ashley, C., Caul, E., and Paver, W., 1978, Astrovirus-associated gastroenteritis in children, *J. Clin. Pathol.* **31**:939–943.

Atanasiu, P., Favre, S., Sisman, J., and Perrin, P., 1979, Glycosylation et propriétés isoélectriques des virus rabiques complets et défectifs, *Ann. Microbiologie* **130A**:85–101.

Babiuk, L., L'Italien, J., van Drunen Littel-van den Hurk, S., Zamb, T., Lawman, J., Hughes, G., and Gifford, G., 1987, Glycosylation et propriétés isoélectriques des virus rabiques complets et défectifs, *Virology* **159**:57–66.

Banttari, E. E., 1980, Sap extraction device for sampling potato leaves and tubers, *Am. Potato J.* **57**:345–348.

Barbara, D. J. and Clark, M. F., 1982, A simple indirect ELISA using $F(ab')_2$ fragments of immunoglobulin, *J. Gen. Virol.* **58**:315–322.

Barbara, D. J., Clark, M. F., Thresh, J. M. and Casper, R., 1978, Rapid detection and serotyping of prunus necrotic ringspot virus in perennial crops by enzyme-linked immunosorbent assay, *Ann. Appl. Biol.* **90**:395–399.

Barclay, W. S., and Al-Nakib, W., 1987, An ELISA for detection of rhinovirus specific antibody in serum and nasal secretion, . *Virol Methods* **15**:53–64.

Bar-Joseph, M. and Garnsey, S. M., 1981, Enzyme-linked immunosorbent assay (ELISA): Principles and applications for diagnosis of plant viruses, in: Plant Diseases and Vectors: Ecology and Epidemiology (K. Maramorosch and K. F. Harris, eds), pp. 35–39, Academic Press, New York.

Bar-Joseph, M., Garnsey, S. M., Gonsalves, D., Moscovitz, M., Purcifull, D. E., Clark, M. F. and Loebenstein, G., 1979, The use of enzyme-linked immunosorbent assay for the detection of citrus tristeza virus, *Phytopathology* **69**:190–194.

Beards, G., Campbell, A., Cottrell, N., Peiris, J., Rees, N., Sanders, R., Shirley, A., Wood, H., and Flewett, T., 1984, Enzyme-linked immunosorbent assays based on polyclonal and monoclonal antibodies for rotavirus detection, *J. Clin. Microbiol.* **19**:248–254.

Bellamy, K., Rousseau, S., and Gardner, P., 1986, The development of an M antibody capture ELISA for rubella IgM. *J. Virol. Methods* **14**:243–251.

Blacklow, N., Cukor, G., Bedigian, M., Echeverria, P., Greenberg, H., Schreiber, D., and Trier, J., 1979, Immune response and prevalence of antibody to Norwalk enteritis virus as determined by radioimmunoassay, *J. Clin. Microbiol.* **10**:903–909.

Bolin, S., McClurkin, A., and Coria, M., 1985, Effects of bovine viral diarrhea virus on the percentages and absolute numbers of circulating B and T lymphocytes in cattle, *Am. J. Vet. Res.* **46**:2385–2387.

Boué, A., Nicolas, A., and Montagnon, B., 1971, Reinfection with rubella in pregnant women. *Lancet* **1**:1251–1253.

Brandsma, J., and Miller, G., 1980, Nucleic acid spot hybridization: rapid quantitative screening of lymphoid cell lines for Epstein-Barr viral DNA, *Proc. Natl. Acad. Sci.* (USA) **77**:6851–6855.

Brandt, C., Kim, H., Rodriguez, W., Arrobio, J., Jeffries, B., Stallings, E., Lewis, C., Miles, A., Chanock, R., Kapikian, A., and Parrot, R., 1983, Pediatric viral gastroenteritis during eight years of study, *J. Clin. Microbiol.* **18**:71–78.

Briand, J. P., Al Moudallal, Z. and Van Regenmortel, M. H. V., 1982, Serological differentiation of tobamoviruses by means of monoclonal antibodies, *J. Virol. Meth.* **5**:293–300.

Bruck, C., Portetelle, D., Glineur, C., and Bollen, A., 1982, One-step purification of mouse monoclonal antibodies from ascitic fluid by DEAE Affi-Gel blue chromatography, *J. Immunol. Methods* **53**:313–319.

Brun-Vezinet, F., Katlama, C., Roulot, D., Lenoble, L., Alizon, M., Madjar, J., Rey, M., Girard, P., Yeni, P., Clavel, F., Gadelle, S., and Harzic, M., 1987, Lymphadenopathy-associated virus type 2 in AIDS and AIDS-related complex, *Lancet* **1**:128–132.

Burke, D., Nisalak, A., and Ussery, M., 1982, Antibody capture immunoassay detection of japanese encephalitis virus immunoglobulin M and G antibodies in cerebrospinal fluid, *J. Clin. Microbiol.* **16**:1034–1042.

Burke, D., Brandt, B., Redfield, R., Tun-Hou, L., Thorn, R., Beltz, G., and Chung-Ho, H., 1987, Diagnosis of human immunodeficiency virus infection by immunoassay using a molecularly cloned and expressed virus envelope polypeptide. Comparison to Western blot on 2707 consecutive serum samples, *Ann. Int. Med.* **106**:671–676.

Burke, D., and Nisalak, A., 1982, Detection of Japanese encephalitis virus immunoglobulin G antibodies in serum by antibody capture radioimmunoassay, *J. Clin. Microbiol.* **15**:353–361.

Caciagli, P., Roggero, P. and Luisoni, E., 1985, Detection of maize rough dwarf virus by enzyme-linked immunosorbent assay in plant hosts and in the planthopper vector, *Ann. Appl. Biol.* **107**:463–471.

Calisher, C., and Poland, J., 1980, In: Monath, T. (ed.). St. Louis Encephalitis. p. 371. Am. Public Health Assoc, Washington.

Cappel, R., Hestermans, O., Touissant, C., Vereerstraeten, P., van Beers, D., de Braekeleer, J., and Schoutens, E., 1978, Cytomegalovirus infection and graft survival in renal graft recipients, *Arch. Virology* **56:**149–156.

Carlebach, R., Raccah, B. and Loebenstein, G., 1982, Detection of potato virus Y in the aphid Myzus persicae by enzyme-linked immunosorbent assay (ELISA), *Ann. Appl. Biol.* **101:**511–516.

Chang, H-C., Takashima, I., Arikawa, J., and Hashimoto, N., 1984, Biotin-labeled protein A enzyme-linked immunosorbent assay for the detection of Japanese encephalitis antibody in sera from humans, swine, and several animal species, *J. Virol. Methods* **9:**143–151.

Chen, L.-C., Durand, D. P. and Hill, J. H., 1982, Detection of pathogenic strains of sobyean mosaic virus by enzyme-lnked immunosorbent assay with polystyrene plates and beads as the solid phase, *Phytopathology* **72:**1177–1181.

Chou, S., and Merigan, T., 1983, Rapid detection and quantitation of human cytomegalovirus in urine through DNA hybridization, *New Engl. J. Med.* **308:**921–925.

Clark, M. F., 1981, Immunosorbent assays in plant pathology, *Annu. Rev. Phytopathology* **19:**83–106.

Clark, M. F. and Adams, A. N., 1977, Characteristics of the microplate method of enzyme-linked immunosorbent assay for the detection of plant viruses, *J. Gen. Virol.* **34:**475–483.

Clark, M. F. and Barbara, D. J., 1987, A method for the quantitative analysis of ELISA data, *J. Virol. Meth.* **15:**213–222.

Clark, M. F. and Bar-Joseph, M., 1984, Enzyme immunosorbent assays in plant virology, *Meth. Virol.* **7:**51–85.

Clarke, R. G., Converse, R. H. and Kojima, M., 1980, Enzyme-linked immunosorbent assay to detect potato leaf roll virus in potato tubers and viruliferous aphids, *Plant Dis.* **64:**43–45.

Clavel, F., Guétard, D., Brun-Vézinet, F., Chamaret, S., Rey, M., Santos-Ferreira, M., Laurent, A., Dauguet, G., Katlama, C., and Rouzioux, C., 1986, Isolation of a new human retrovirus from West African patients with AIDS, *Science* **233:**343–346.

Coleman, R., Pereira, L., Bailey, P., Dondero, D., Wickliffe, C., and Nahmias, A., 1983, Determination of herpes simplex virus type-specific antibodies by enzyme-linked immunosorbent assay, *J. Clin. Microbiol.* **18:**287–291.

Cooper, D., Imrie, A., and Penny, R., 1987, Antibody response to human immunodeficiency virus after primary infection, *J. Infect. Dis.* **155:**1113–1118.

Cradock-Watson, J., Ridehalgh, M., and Chantler, S., 1979, Specific immuno-globulins in infants with the congenital rubella syndrome, *J. Hyg.* **76:**109–123.

Cranage, M., and Gardner, P., 1980, Systemic cell-mediated and antibody responses in infants with respiratory syncytial virus infections, *J. Med. Virol.* **5:**161–170.

Crouch, C., Ohmann, H., Watts, T., and Babiuk, L., 1985, Chronic shedding of bovine enteric coronavirus antigen-antibody complexes by clinically normal cows, *J. Gen. Virology* **66:**1489–1500.

De Jong, J., Wigand, R., Kidd, A., Wadell, G., Kapsenberg, J., Muzerie, C., Wermenbol, A., and Firtzlaff, R.-G., 1983, Candidate adenoviruses 40 and 41: Fastidious adenoviruses from human infant stool, *J. Med. Virol.* **11:**215.

Dekker, E. L., Dore, I., Porta, C. and Van Regenmortel, M. H. V., 1987, Conformational specificity of monoclonal antibodies used in the diagnosis of tomato mosaic virus, *Arch. Virol.* **94:**191–203.

DeSanto, N., Tarro, G., Giordano, C., Capodicasa, G., Esposito, C., Flaminio, G., Sodano, G., Caropreso, M., Pluvio, M., and Giordano, D., 1987, Screening hemodialysis patients for infection with human immune deficiency virus (HIV), *J. Infect.* **14:**229–232.

Devergne, J. C., Cardin, L., Burckard, J. and Van Regenmortel, M. H. V., 1981, Comparison of direct and indirect ELISA for detecting antigenically related cucumoviruses, *J. Virol. Meth.* **3:**193–200.

Diaco, R., Hill, J. H., Hill, E. K., Tachibana, H. and Durand, D. P., 1985, Monoclonal antibody-based biotin-avidin ELISA for the detection of soybean mosaic virus in sobybean seeds, *J. Gen. Virol.* **66:**2089–2094.

Diaco, R., Lister, R. M., Hill, J. H. and Durand, D. P., 1986, Demonstration of serological relationships among isolates of barley yellow dwarf virus by using polyclonal and monoclonal antibodies, *J. Gen. Virol.* **67:**353–362.

Diaz-Mitoma, F., Preiksaitis, J., Leung, W., and Tyrell, D., 1987, DNA-DNA dot hybridization to detect Epstein-Barr virus in throat washings, *J. Inf. Dis.* **155:**297–303.

Dietzgen, R. G. and Francki, R. I. B., 1987, Nonspecific binding of immunoglobulins to coat proteins of certain plant viruses in immunoblots and indirect ELISA, *J. Virol. Meth.* **15:**159–164.

Dougherty, W. G., Willis, L. and Johnston, R. E., 1985, Topographic analysis of tobacco etch virus capsid protein epitopes, *Virology* **144:**66–72.

Duebel, V., et al., 1983, Comparison of the enzyme-linked immunosorbent assay (ELISA) with standard tests used to detect yelow fever virus antibodies, *Am. J. Trop. Med. Hyg.* **32:**565.

Duermeyer, W., and van der Veen, J., 1978, Specific detection of IgM-antibodies by ELISA, applied in hepatitis-A, *Lancet* **2:**684–685.

Edwards, M. L. and Cooper, J. I., 1985, Plant virus detection using a new form of indirect ELISA, *J. Virol. Meth.* **11:**309–319.

Ehlers, U. and Paul, H. L., 1984, Binding of viruses from crude plant extracts to glutaraldehyde-treated plates for indirect ELISA. *J. Virol. Meth.* **8:**217–224.

El-Mekki, A., W. Al Nakib, and R. Bibi, 1987, Factors affecting the detection of cytomegalovirus in urine by sandwich enzyme immunoassay, *J. Virol. Methods* **15:**75–83.

Evans, R., Shanson, D., and Mortimer, P., 1987, Clinical evaluation of Abbott and Wellcome enzyme linked immunosorbent assays for the detection of serum antibodies to human immunodeficiency virus (HIV), *J. Clin. Pathol.* **40:**552–555.

Feinman, S. V., 1984, Berris, B., Guha, A., Sooknanan, R., Bradley, D. W., Bond, W. W., and Maynard, J. E., 1984, DNA: DNA hybridization method for diagnosis of hepatitis B infection. *J. Virol. Methods* **8:**199–206.

Felgner, P., 1978, A new technique of heterogeneous enzyme-linked immunosorbent assay, stick-ELISA. II. Application of stick-ELISA to antigens of various infective agents. *Zbl. Bakt. Hyg. I. Abt. Orig. A.* **240:**118–122

Flanders, R., Lindsay, P., Chairez, R., Brawner, T., Kumar, M., Swenson, P., and Bromberg, K., 1986, Evaluation of clinical specimens for the presence of respiratory syncytial virus antigens using an enzyme immunoassay, *J. Med. Virology* **19:**1–9.

Flewett, T., Davies, H., Bryden, A., and Robertson, M., 1974, Acute gastroenteritis associated with reovirus-like particles, *J. Clin. Pathol.* **27:**608–614.

Flower, A., Banatvala, J., and Chrystie, I., 1977, BK antibody and virus-specific IgM responses in renal transplant recipients, patients with malignant disease, and healthy people, *Brit. Med. J.* **2:** 220–223.

Forghani, B., Myoraku, C., and Schmidt, N., 1982, Production of monoclonal antibodies to human IgM for assay of viral IgM antibodies, *J. Virol. Method* **5:**317–327.

Forghani, B., Myoraku, C., and Schmidt, N., 1983, Use of monoclonal antibodies to human immunoglobulin M in "capture" assays for measles and rubella immunoglobulin M, *J. Clin. Microbiol.* **18:**652–657.

Friguet, B., Djavadi-Ohaniance, L. and Goldberg, M. E., 1984, Some monoclonal antibodies raised with a native protein bind preferentially to the denatured antigen, *Molec. Immunol.* **21:**673–677.

Gaines, H., Sönnerborg, A., Czajkowski, J., Chiadi, F., Fenyö, E., Forsgren, M., Von Sydow, M.,

Pehrson, P., Albert, J., Moberg, L., and Asjo, B., 1987, Antibody response in primary human immunodeficiency virus infection, *Lancet* **1:**1249–1253.

Gardner, P., and McQuillin, J., 1968, Application of immunofluorescent antibody technique in rapid diagnosis of respiratory syncytial virus infection, *Brit. Med. J.* **3:**340–343.

Gardner, P. S., and Kaye, S., 1982, Egg globulins in rapid virus diagnosis, *J. Virol. Meth.* **4:**257–262.

Gera, A., Loebenstein, G. and Raccah, B., 1978, Detection of cucumber mosaic virus in viruliferous aphids by enzyme-linked immunosorbent assay, *Virology* **86:**542–545.

Ghabrial, S. A. and Schultz, F. J., 1983, Serological detection of bean pod mottle virus in bean leaf beetles, *Phytopathology* **73:**480–483.

Gilljam, G., Sundqvist, V., Linde, A., Pihlstedt, P., Eklund, A., and Wahren, B., 1985, Sensitive analytic ELISA for subclass herpes virus IgG, *J. Virol. Methods* **10:**203–214.

Goldwater, P., Webster, M., and Banatvala, J., 1982, Use of a simple, new test for virus-specific IgM to investigate an outbreak of influenza B in a hospitalized aged community, *J. Virol. Methods* **4:**9–18.

Gonsalves, D., 1979, Detection of tomato ringspot virus in grapevines: A comparison of *Chenopodium quinoa* and enzyme-linked immunosorbent assay (ELISA), *Plant Dis. Rep.* **63:**962–965.

Goudsmit, J., Woltors, E., Bakker, M., Smit, L., Van der Noordaa, J., Hische, E., Tutuarima, J., and van der Helm, H., 1986, Intrathecal synthesis of antibodies to HTLV-III in brains of patients without AIDS or AIDS-related complex, *Brit. Med. J.* **292:**1231–1234.

Guesdon, J.-L., Ternyck, T. and Avrameas, S., 1979, The use of avidin–biotin interaction in immunoenzymatic techniques. *J. Histochem. Cytochem.* **27:**1131–1139.

Gugerli, P., 1979, Potato virus A and potato leafroll virus: Purification, antiserum production and serological detection in potato and test plants by enzyme-linked immunosorbent assay (ELISA), *Phytopath. Z.* **96:**97–107.

Hammond, G., Ahluwalia, G., Barker, F., Horsman, G., and Hazelton, P., 1982, Comparison of direct and indirect enzyme immunoassays with direct ultracentrification before electron microscopy for detection of rotaviruses, *J. Clin. Microbiol.* **16:**53–59.

Hampar, B., Zwieg, M., Showalter, S., Bladen, S., and Riggs, C., 1985, Enzyme-linked immunosorbent assay for determination of antibodies against herpes simplex virus types 1 and 2 in human sera, *J. Clin. Microbiol.* **21:**496–500.

Harcourt, G., Best, J., and Banatvala, J., 1980, Rubella-specific serum and nasopharyngeal antibodies in volunteers with naturally acquired and vaccine-induced immunity after intranasal challenge, *J. Infect. Dis.* **142:**145–155.

Harrison, T., Tsiquaye, K., and Zuckerman, A., 1983, Assay of HBV DNA in the plasma of HBV-carrier chimpanzees superinfected with non-A, non-B hepatitis, *J. Virol. Methods* **6:**295–302.

Hawkins, E., Johnson, L., Pedersen, N., and Winston, S., 196, Use of tears for diagnosis of feline leukemia virus infection, *J. Am. Vet. Med. Assoc.* **188:**1031–1034.

Heinz, F., Roggendorf, M., Hofmann, H., Kunz, C., and Deinhardt, F., 1981, Comparison of two different enzyme immunoassays for detection of immunoglobulin M antibodies against tick-borne encephalitis virus in serum and cerbrospinal fluid, *J. Clin. Microbiol.* **14:**141–146.

Higgins, J., Pedersen, N., and Carlson, J., 1968, Detection and differentiation by sandwich enzyme-linked immunosorbent assay of human T-cell lymphotropic virus type III/lymphadenopathy associated virus- and acquired immunodeficiency syndrome-associated retrovirus-like clinical isolates, *J. Clin. Microbiol.* **24:**424–430.

Hill, E. K., Hill, J. H. and Durand, D. P., 1984, Production of monoclonal antibodies to viruses in the Potyvirus group: Use in radioimmunoassay, *J. Gen. Virol.* **65:**525–532.

Hinman, A., Kirby, C., Eddins, D., Orenstein, W., Bernier, R., Turner, P., and Bart, K., 1983, Elimination of indigenous measles from the United States, *Rev. Inf. Dis.* **5:**538–545.

Hoffmann, D. and Van Regenmortel, M. H. V., 1984, Detection of distant antigenic relationships between insect and bird lysozymes, *J. Molec. Evolut.* **21:**14–18.

Hofmann, H., Frisch-Niggemeyer, W., and Heinz, F., 1979, Rapid diagnosis of tick-borne encephalitis by means of enzyme-linked immunosorbent assay. *J. Gen. Virology* **42:**505–511.

Hogg, R., and Davidson, G., 1982, Improved specificity of ELISA for rotavirus, *Aust. Paediatr. J.* **18:**184–185.

Hovi, T., Väisänen, V., Ukkonen, P., and von Bonsdorff, C., 1982, Solid-phase enzyme immunoassay for rotavirus antigen: Faecal protease activity as a reason for false-negative results. *J. Virol. Methods* **5:**45–53.

Hoxie, J., Fitzharris, T., Youngbar, P., Matthews, D., Rackowski, J., and Rodka, S., 1987, Nonrandom association of cellular antigens with HTLF-III virions, *Human Immunol.* **18:**39–52.

Huss, B., Walter, B., Etienne, L. and Van Regenmortel, M. H. V., 1986, Grapevine fanleaf virus detection in various grapevine organs using polyclonal and monoclonal antibodies, *Vitis* **25:**178–188.

Ida, G., Siraganian, R., and Notkins, A., 1983, Cell-bound and circulating IgE antibody to herpes simplex, *J. Gen. Virol.* **64:**533–537.

Isaac, M., and Payne, R., 1982, Antibody class capture assay (ACCA) for rubella specific IgM antibody, *J. Med. Virol.* **10:**55–64.

Ivo dos Santos, J., Galvao-Castra, B., Mello, D., Pereira, H., and Pereira, M., 1987, Dot enzyme immunoassay: A simple, cheap and stable test for antibody to human immunodeficiency virus (HIV), *J. Immunol. Methods* **99:**191–194.

Jaeglé, M. and Van Regenmortel, M. H. V., 1985, Use of ELISA for measuring the extent of serological cross-reactivity between plant viruses, *J. Virol. Meth.* **11:**189–198.

Jafarpour, B., Shepherd, R. J. and Grogan, R. G., 1979, Serologic detection of bean common mosaic and lettuce mosaic viruses in seed, *Phytopathology* **69:**1125–1129.

Julkunen, I., 1984, Serological diagnosis of parainfluenza virus infections by enzyme immunoassay with special emphasis on purity of viral antigens, *J. Med. Virol.* **14:**177–187.

Kalica, A., Greenberg, H., Wyatt, R., Flores, J., Sereno, M., Kapikian, A., and Chanock, R., 1981, Genes of human (strain Wa) and bovine (strain UK) rotaviruses that code for neutralization and subgroup antigens, *Virology* **112:**385–390.

Kangro, H., 1980, Evaluation of a radioimmunoassay for IgM-class antibodies against cytomegalovirus, *Brit. J. Exptl. Pathol.* **61:**512–520.

Katz, D., Hillard, J., Eberle, R., and Lipper, S., 1986, ELISA for detection of group common and virus-specific antibodies in human and simian sera induced by herpes simplex and related simian viruses, *J. Virol. Methods* **14:**99–109.

Katz, J., and Hanson, S., 1987, Competitive and blocking enzyme-linked immunoassay for detection of fetal bovine serum antibodies to bovine viral diarrhea virus, *J. Virol. Methods* **15:**167–175.

Kendall, C., Ionescu-Matiu, I. and Dreesman, G. R., 1983, Utilisation of the biotin/avidin system to amplify the sensitivity of the enzyme-linked immunosorbent assay (ELISA), *J. Immunol. Meth.* **56:**329–339.

Keller, P., Lonergan, K., Neff, B., Morton, D., and Ellis, R., 1986, Purification of individual varicella-zoster virus (VZV) glycoproteins gpI, gpII, and gpIII, and their use in ELISA for detection of VZV glycoprotein-specific antibodies, *J. Virol. Methods* **14:**177–188.

Kennel, S. J., 1982, Binding of monoclonal antibody to protein antigen in fluid phase or bound to solid supports, *J. Immunol. Meth.* **55:**1–12.

Khan, M., Gallagher, M., Bucher, D., Cerini, C., and Kilbourne, E., 1982, Detection of influenza virus neuraminidase specific antibodies by an enzyme-linked immunosorbent assay. *Clin. Microbiol.* **16:**115.

Koenig, R., 1978, ELISA in the study of homologous and heterologous reactions of plant viruses, *J. Gen. Virol.* **40:**309–318.

Koenig, R., and Paul, H. L., 1982, Variants of ELISA in plant virus diagnosis, *J. Virol. Meth.* **5**:113–125.

Koenig, R. and Torrance, L., 1986, Antigenic analysis of potato virus X by means of monoclonal antibodies, *J. Gen. Virol.* **67**:2145–2151.

Konishi, E., and Yamaoka, M., 1982, Enzyme-linked immunosorbent assay for detection of antibodies to Japanese encephalitis virus in swine sera, *Kobe. J. Med. Sci.* **28**:7–17.

Kraaijeveld, C., Reed, S., and MacNaughton, M., 1980, Enzyme linked immunosorbent assay for detection of antibody in volunteers experimentally infected with human coronavirus strain 229E, *J. Clin. Microbiol.* **12**:493–497.

Kramer, S., and Cremer, N., 1984, Detection of IgG antibodies to measles virus using monoclonal antibodies for antigen-capture in enzyme immunoassay, *J. Virol. Methods* **8**:255–263.

Krause, P., Hyams, J., Middleton, P., Herson, V., and Flores, J., 1983, Unreliability of rotazyme ELISA test in neonates, *J. Pediat.* **103**:259–262.

Krishna, R., Meurman, O., Ziegler, T., and Krech, U., 1980, Solid phase enzyme immunoassay for determination of antibodies to cytomegalovirus, *J. Clin. Microbiol.* **12**:46–51.

Kryger, P., 1985, Significance of anti-HBc IgM in the differential diagnosis of viral hepatitis, *J. Virol. Methods* **10**:283–289.

Kunz, C., and Hofmann, H., 1971, *Zentbl. Bakteriol.* **218**:273.

Kurstak, E., 1986, *Enzyme Immunodiagnosis,* Academic Press, New York.

Kurstak, E., Tijssen, P., and Kurstak, C., 1977, Immunoperoxidase techniques in diagnostic virology and research: Principles and applications, in: *Comparative Diagnosis of Viral Diseases,* Volume II (E. Kurstak and C. Kurstak, eds.), pp. 403–448, Academic Press, New York.

Kurstak, E., Tijssen, P., van den Hurk, J., and Kurstak, C., 1980, Detection by immunoperoxidase and ELISA of arbovirus antigens and antibodies in *in vivo* and *in vitro* systems, in: *Invertebrate Systems in Vitro* (E. Kurstak, K. Maramorosch, A. Oübendorfer, eds.), pp. 365–373, Elsevier/North Holland, Amsterdam.

Kurstak, E., Tijssen, P., and Kurstak, C., 1984a, Recent advances in immunological and biochemical diagnostic virology, in: *Control of Virus Diseases* (E. Kurstak and R. Marusyk, eds.), pp. 477–500, Dekker, New York.

Kurstak, E., Tijssen, P., and Kurstak, C., 1984b, Enzyme immunoassays applied in virology: Reagents, preparation, and interpretation, *Appl. Virology* **1**:479–503.

Kurstak, E., Tijssen, P., Kurstak, C., and Morisset, R., 1986, Enzyme immunoassays and related procedures in diagnostic medical virology, *Bull. Wld. Health Org.* **64**(3):465–479.

Langer, P., Waldrop, A., and Ward, D., 1982, Enzymatic synthesis of biotin-labeled polynucleotides: Novel nucleic acid affinity probes, *Proc. Natl. Acad. Sci. USA* **78**:6633–6637.

Larke, R., 1983, Impact of measles in Canada, *Rev. Inf. Dis.* **5**:445–451.

Leslie, G. A. and Clem, L. W., 1969, Phylogeny of immunoglobulin structure and function, III. Immunoglobulins of the chicken. *J. Exp. Med.* **130**:1337–1352.

Linde, A., Hammarström, L., Persson, M., Smith, C., Sundqvist, V., and Wahren, B., 1983, Virus-specific antibody activity of different subclasses of immunoglobulins G and A in cytomegalovirus infections, *Infect. Immun.* **42**:237–244.

Lister, R. M. and Rochow, W. F., 1979, Detection of barley yellow dwarf virus by enzyme-linked immunosorbent assay, *Phytopathology* **69**:649–654.

Lister, R. M., Carroll, T. W. and Zaske, S. K., 1981, Sensitive serological detection of barley stripe mosaic virus in barley sed, *Plant Dis.* **65**:809–814.

Lommel, S. A., McCain, A. H. and Morris, T. J., 1982, Evaluation of indirect enzyme-linked immunosorbent assay for the detection of plant viruses, *Phytopathology* **72**:1018–1022.

Lommel, S. A., McCain, A. H., Mayhew, D. E. and Morris, T. J., 1983, Survey of commercial carnation cultivars for four viruses in California by indirect enzyme-linked immunosorbent assay, *Plant Dis.* **66**:167–170.

Lund, G., and Salmi, A., 1982, Radioimmunoassay of measles virus hemagglutinin protein G, *J. Virol. Methods* **4:**331–341.

Lutz, H., *et al.*, 1980, in: Feline Leukemia Viruses. (W. Hardy *et al.*, eds.) Elsevier/North Holland, Amsterdam.

Lutz, H., Pedersen, N., and Theilen, G., 1983, Course of feline leukemia virus infection and its detection by enzyme-linked immunosorbent assay and monoclonal antibodies, *Am. J. Vet. Res.* **44:**2054–2059.

Macnaughton, M., Flowers, D., and Isaacs, D., 1983, Diagnosis of human coronavirus infections in children using enzyme-linked immunosorbent assay, *J. Med. Virol.* **11:**319–325.

Madore, H., Reichman, R., and Dolin, R., 1983, Serum antibody responses in naturally occurring influenza A virus infection determined by enzyme-linked immunosorbent assay, hemagglutination inhibition, and complement fixation, *J. Clin. Microbiol.* **18:**1345–1350.

Mäkelä, M., 1987, MSc thesis, University of Alberta.

Mäkelä, M., Norrby, E., and Salmi, A., 1987, Measurement of polypeptide- and antigenic site-specific antibodies to measles virus using a competitive enzyme immunoassay, *J. Virol. Methods* **16:**65–74.

Manz, D., Wiegand, D., Behrens, F., and Ziegelmaier, R., 1981, Vergleichende serologische Untersuchungen on Blut und Milch zur Diagnostik der enzootischen Leukose des Rindes, *Zbl. Vet. Med.* **28B:**280–291.

Marco, S. and Cohen, S., 1979, Rapid detection and titer evaluation of viruses in pepper by enzyme-linked immunosorbent assay, *Phytopathology* **69:**1259–1262.

Marusyk, R., 1984, Measles: A continuing problem or eradication, in: *Appl Virology* (E. Kurstak, ed.), 211–219, Academic Press, New York.

McClurkin, A., Littledike, E., Cutlip, R., Frank, G., Coria, M., and Bolin, S., 1984, Production of cattle immunotolerant to bovine viral diarrhea virus, *Can. J. Comp. Med.* **48:**156–161.

McCullough, K. C., Crowther, J. R. and Butcher, R. N., 1985, Alteration in antibody reactivity with foot-and-mouth disease virus (FDMV) 146S antigen before and after binding to a solid phase or complexing with specific antibody. *J. Immunol. Meth.* **82:**91–100.

McKeating, J., Grundy, J., Varghese, Z., and Griffiths, P., 1986, Detection of cytomegalovirus by ELISA in urine samples is inhibited by beta 2 microglobulin, *J. Med. Virology* **18:**341–348.

McLean, B., Sonza, S., and Holmes, I., 1980, Measurement of immunoglobulin A, G, and M class rotavirus antibodies in serum and mucosal secretions, *J. Clin. Microbiol.* **12:**314–319.

McLean, D., 1979, Arbovirus vectors in Canadian arctic, in: *Arctic and Tropical Arboviruses* (E. Kurstak, ed.), pp. 7–19, Academic Press, New York.

McLaughlin, M. R., Barnett, O. W., Gibson, P. B. and Burrows, P. M., 1984, Enzyme-linked immunosorbent assay of viruses infecting forage legumes, *Phytopathology* **74:**965–969.

McMorran, J. P. and Cameron, H. R., 1983, Detection of 41 isolates of necrotic ringspot, apple mosaic, and prune dwarf viruses in *Prunus* and *Malus* by enzyme-linked immunosorbent assay, *Plant Dis.* **67:**536–538.

Meurman, O., Terho, P., and Salmi, A., 1978, Activation of rheumatoid factor during pregnancy, *Lancet* **2:**685–686.

Meurman, O., and Ziola, B., 1978, IgM class rheumatoid factor interference in the solid phase radioimmunoassay of rubella specific IgM antibodies, *J. Clin. Pathol.* **31:**483–487.

Meurman, O., 1983, Detection of antiviral IgM antibodies and its problems: A review, *Curr. Topics Microbiol. Immunol.* **104:**101–131.

Middleton, P., Holdaway, M., and Petric, M., 1977, Solid-phase radioimmunoassay for the detection of rotavirus, *Infect. Immun.* **16:**439–444.

Milan, A., and Stigbrand, T., 1981, "Sandwich" enzyme immunoassay for placental alkaline phosphatase, *Clin. Chemistry* **27:**2018.

Miller, E., Cradock-Watson, J., and Pollock, T., 1982, Consequences of confirmed maternal rubella at successive stages of pregnancy, *Lancet* **2:**781–784.

Monath, T., Nystrom, R., Bailey, R., Calisher, C., and Muth, D., 1984, Immunoglobulin M antibody capture enzyme-linked immunosorbent assay for diagnosis of St. Louis encephalitis, *J. Clin. Microbiol.* **20:**784–790.

Morell, A., Roth-Wicky, B., and Skvaril, F., 1983, Immunoglobulin G subclass restriction of antibodies against hepatitis B surface antigen, *Infect. Immun.* **39:**565–568.

Morgan, E., and Rapp, F., 1977, Measles virus and its associated diseases, *Bacteriol. Rev.* **41:**636–666.

Mowat, W. P. and Dawson, S., 1987, Detection and identification of plant viruses by ELISA using crude sap extracts and unfractionated antisera, *J. Virol. Meth.* **15:**233–247.

Muller, S., Jockers-Wretou, E., Sekaris, C. E., Van Regenmortel, M. H. V. and Bautz, F. A., 1985, Characterization of a monoclonal antibody reacting with histone H3, *FEBS Lett.* **182:**459–464.

Murphy, B., Phelan, M., Nelson, D., Yarchoan, R., Tierney, E., Alling, D., and Chanock, R., 1981, Hemagglutinin specific enzyme-linked immunosorbent assay for antibodies to influenza A and B viruses, *J. Clin. Microbiol.* **13:**554–560.

Niklasson, B., Jahrling, P., and Peters, C., 1984, Detection of Lassa virus antigens and Lassa virus-specific immunoglobulins G and M by enzyme-linked immunosorbent assay, *J. Clin. Microbiol.* **20:**239–244.

Ohmann, H., Jensen, M., Sørensen, K., and Dalsgaard, K., 1982, Experimental fetal infection with bovine viral diarrhea virus. I. Virological and serological studies, *Can. J. Comp. Med.* **46:**357–362.

Oldham, L., Moulsdale, H., Mortimer, P., Tedder, R., and Morgan-Capner, P., 1987, How sensitive are the commercial assays for anti-HTLV III/LAV? *J. Med. Virol.* **21:**75–79.

Papaevangelou, G., Roumeliotou-Karayannis, Tassopoulos, N., and Stathopoulou, P., 1984, Diagnostic value of anti-HBc IgM in high HBV prevalence areas, *J. Med. Virol.* **13:**393–399.

Pauls, F., and Dowdle, W., 1967, A serologic study of herpes virus hominis strains by microneutralization tests, *J. Immunol.* **98:**941–947.

Pedersen, N., Lowenstine, L., Marx, P., Higgins, J., Baulu, J., McGuire, M., and Gardner, M., 1986, The causes of false-positives encountered during the screening of old-world primates for antibodies to human and simian retroviruses by ELISA, *J. Virol. Methods* **14:**213–228.

Peradze, T., and Smorodintsev, A., 1983, Epidemiology and specific prophylaxis of measles, *Rev. Inf. Dis.* **5:**487–490.

Polson, A., Von Wechmar, M. B., and Van Regenmortel, M. H. V., 1980, Isolation of viral IgY antibodies from yolks of immunized hens, *Immunol. Commun.* **9:**475–493.

Polson, A., Van Heerden, D. and Van der Merwe, K. J., 1985, Coated glass beads in enzyme-linked immunosorbent assays, *Immunol. Investig.* **14:**223–231.

Portetelle, D., Bruck, C., Mammerickx, M., and Burny, A., 1983, Use of monoclonal antibody in an ELISA test for the detection of antibodies to bovine leukaemia virus, *J. Virol. Methods* **6:**19–29.

Portetelle, D., *et al.*, 1984, *Can. Top. Vet. Med. Anim. Sci.* **29:**157.

Powell, C. A. and Derr, M. A., 1983, An enzyme-linked immunosorbent blocking assay for comparing closely related virus isolates, *Phytopathology* **73:**660–664.

Quesniaux, V., Briand, J. P. and Van Regenmortel, M. H. V., 1983, Immunochemical studies of turnip yellow mosaic virus, II. Localization of a viral epitope in the N-terminal residues of the coat protein, *Molec. Immunol.* **20:**179–185.

Rapicetta, M., D'Ambrosio, E., Proietti, E., Morace, G., Donatelli, I., and Macchia, T., 1983, Antibody assay for measles virus: comparison of immune-adherence haemagglutination, single radial haemolysis, enzyme immunoassay and haemagglutination inhibition. *J. Virol. Methods* **6:** 303–310.

Rawls, W., Iwamoto, K., Adam, E., and Melnick, J., 1979, Measurement of antibodies to herpesvirus types 1 and 2 in human sera, *J. Immunol.* **104:**599–606.

Recchia, S., Rizzi, R., Acquaviva, F., Rizzetto, M., Tison, V., Bonino, F., and Verme, G., 1981, *Pathologica* **73:**773–777.

Reiner, M., and Wecker, E., 1981, A modified absorption-reduction method to detect virus-specific hemagglutination inhibiting and neutralizing IgM antibodies, *Med. Microbiol. Immunol.* **169:** 237–245.

Rice, G., Casali, P., and Oldstone, M., 1983, A new solid-phase enzyme-linked immunosorbent assay for specific antibodies to measles virus, *J. Inf. Dis.* **147:**1055–1059.

Richman, D., Schmidt, N., Plotkin, S., Yolken, R., Cherenoky, M., McIntosh, K., and Mattheis, M., 1984, Summary of a workshop on new and useful methods in rapid viral diagnosis, *J. Inf. Dis.* **150:**941–951.

Rochow, W. F., 1979, Comparative diagnosis of barley yellow dwarf by serological and aphid transmission tests, *Plant Dis. Rep.* **63:**426–430.

Roggendorf, M., Heinz, F., Deinhardt, F., and Kunz, C., 1981, Serological diagnosis of acute tick-borne encephalitis by demonstration of antibodies of the IgM class, *J. Med. Virol.* **7:**41–50.

Roggendorf, M., Wigand, R., Deinhardt, F., and Frösner, G., 1982, Enzyme-linked immunosorbent assay for acute adenovirus infection, *J. Virol. Methods* **4:**27–35.

Romaine, C. P., Newhart, S. R. and Anzola, D., 1981, Enzyme-linked immunosorbent assay for plant viruses in intact leaf tissue discs, *Phytopathology* **71:**308–312.

Rybicki, E. P. and Von Wechmar, M. B., 1981, The serology of the bromoviruses, I. Serological interrelationships of the bromoviruses, *Virology* **109:**391–402.

Sarkkinen, M., Halonen, P., Arstila, P., and Salmi, A., 1981, Detection of respiratory syncytial, parainfluenza type 2 and adenovirus antigens by radioimmunoassay and enzyme immunoassay on nasopharyngeal specimens from children with acute respiratory disease, *J. Clin. Microbiol.* **13:**258–265.

Schluederberg, A., 1965, Immune globulins in human viral infections, *Nature* **205:**1232–1233.

Schüpbach, J., Haller, O., Vogt, M., Lüthy, R., Joller, H., Oelz, O., Popovic, M., Sarngadharan, M., and Gallo, R., 1985, Antibodies to HTLV-III in Swiss patients with AIDS and pre-AIDS and in groups at risk for AIDS, *New. Engl. J. Med.* **312:**265–270.

Scotto, J., Hadchouel, M., Hery, C., Alvarez, F., Yvart, J., Tiollais, P., Bernard, O., and Brechot, C., 1983, Hepatitis B virus DNA in children's liver diseases: detection by blot hybridisation in liver and serum, *Gut* **24:**618–624.

Sejda, J., 1983, Control of measles in Czechoslovakia, *Rev. Inf. Dis.* **5:**564–567.

Shattock, A., and Morgan, B., 1984, Sensitive enzyme immunoassay for the detection of delta antigen and anti delta, using serum as the delta antigen source, *J. Med. Virol.* **13:**73–82.

Shoeman, R., Young, D., Pottathill, R., Victor, J., Conroy, R., Crowl, R., Coleman, T., Heimer, E., Lai, C., and Ganguly, K., 1987, Comparison of recombinant human immunodeficiency virus *gag* precursor and *gag/env* fusion proteins and a synthetic env peptide as diagnostic reagents, *Anal. Biochem.* **161:**370–379.

Smedile, A., Lavarini, C., Crivelli, O., Raimondo, G., Fassone, M., and Rizzetto, M., 1982, Radioimmunoassay detection of IgM antibodies to the HBV-associated delta (delta) antigen: Clinical significance in delta infection, *J. Med. Virol.* **9:**131–138.

Smith, J., Sumner, J., and Roumillat, L., 1984, Enzyme immunoassay for rabies antibody in hybridoma culture fluids and its application to differentiation of street and laboratory strains of rabies virus, *J. Clin. Microbiol.* **19:**267–272.

Soderquist, M. E. and Walton, A. G., 1980, Structural changes in proteins adsorbed on polymer surfaces, *J. Colloid and Interface Sci.* **75:**386–397.

Spratt, H., Marks, M. I., Gomersall, M., Gill, P., and Pai, C., 1978, Nosocomial infantile gastroenteritis associated with minirotavirus and calicivirus, *J. Pediatrics* **6:**922–926.

Stålhandske, P., and Pettersson, U., 1982, Identification of DNA viruses by membrane filter hybridization, *J. Clin. Microbiol.* **15:**744–747.

Stallman, N., Allan, B., and Sutherland, C., 1974, Prolonged rubella IgM antibody response, *Med. J. Austral.* **2**:629.

Stein, A., Loebenstein, G. and Koenig, R., 1979, Detection of cucumber mosaic virus and bean yellow mosaic virus in gladiolus by enzyme-linked immunosorbent assay (ELISA), *Plant Dis. Rep* **63**:185–188.

Sundquist, V-A., Linde, A., and Wahren, B., 1984, Virus-specific immunoglobulin G subclasses in herpes simplex and varicella-zoster virus infections, *J. Clin. Microbiol.* **20**:94–98.

Sureau, P., Rollin, P., and Zeller, H., 1982, Corrélations entre l'epreuve immunoenzymatique, la séroneutralisation et la réduction de foyers fluorescent pour le titrage des anticorps rabiques, *Comp. Immunol. Microbiol. Inf. Dis.* **5**:143–150.

Tamada, T. and Harrison, B. D., 1980, Application of enzyme-linked immunosorbent assay to the detection of potato leafroll virus in potato tubers, *Ann. Appl. Biol.* **96**:67–78.

Tamada, T. and Harrison, B. D., 1981, Quantitative studies on the uptake and retention of potato leafroll virus aphids in laboratory and field conditions, *Ann. Appl. Biol.* **98**:261–276.

Thomas, H., Karayiannis, P., Fowler, M., and Monjardino, J., 1985, Clinical uses of HBV-DNA assays, *J. Virol. Methods.* **10**:291–294.

Torrance, L., 1980, Use of bovine C1q to detect plant viruses in an enzyme-linked immunosorbent-type assay, *J. Gen. Virol.* **51**:229–232.

Torrance, L., 1987, Use of enzyme amplification in an ELISA to increase sensitivity of detection of barley yellow dwarf viurs in oats and in individual vector aphids, *J. Virol. Meth.* **15**:131–138.

Torrance, L. and Jones, R. A. C., 1981, Recent developments in serological methods suited for use in routine testing of plant viruses, *Pl. Pathol.* **30**:1–24.

Torrance, L. and Jones, R. A. C., 1982, Increased sensitivity ot detection of plant viruses obtained by using a fluorogenic substrate in enzyme-linked immunosorbent assay, *Ann. Appl. Biol.* **101**:501–509.

Torrance, L. and Pead, M. T., 1986, The application of monoclonal antibodies to routine tests for two plant viruses, in: Developments and Applications in Virus Testing (R. A. C. Jones and L. Torrance, eds), Association of Applied Biologists, pp. 103–118. Wellesbourne, Great Britain.

Tremaine, J. H., McKenzie, D. J. and Ronald, W. P., 1985, Monoclonal antibodies as structural probes of southern bean mosaic virus, *Virology* **144**:80–87.

Tuokko, H., 1984, Comparison of nonspecific reactivity in indirect and reverse immunoassays for measles and mumps immunoglobulin M antibodies, *J. Clin. Microbiol.* **20**:972–976.

Tuokko, H., and Salmi, A., 1983, Detection of IgM antibodies to measles virus by enzyme immunoassay, *Med. Microbiol. Immunol.* **171**:187–198.

Tuomari, D., Olsen, R., Singh, V., and Kraut, E., 1984, Detection of circulating immune complexes by a Clq/protein A-ELISA during the preneoplastic stages of feline leukemia virus infection, *Vet. Immunol. Immunopathol.* **7**:227–238.

Turner, R., Lathey, J., Van Voris, L., and Belshe, R., 1982, Serological diagnosis of influenza B virus infection: Comparison of an enzyme-linked immunosorbent assay and the hemagglutination inhibition test, *J. Clin. Microbiol.* **15**:824–829.

Ukkonen, P., Vaisänen, O., and Penttinen, K., 1980, Enzyme-linked immunosorbent assay for mumps and parainfluenza type 1 immunoglobulin G and immunoglobulin G and imunoglobulin M antibodies, *J. Clin. Microbiol.* **11**:319–323.

Vaidya, H. C., Dietzler, D. N. and Ladenson, J. H., 1985, Inadequacy of traditional ELISA for screening hybridoma supernatants for murine monoclonal antibodies, *Hybridoma* **4**:271–276.

van Drunen Littel-van den Hurk, S., and Babiuk, L., 1986, Polypeptide specificity of the antibody response after primary and recurrent infection with bovine herpesvirus 1, *J. Clin. Microbiol.* **23**: 274–282.

van Loon, A., Vander Logt, J., Heesen, F., and van der Veen, J., 1981, Quantitation of immunoglobulin E antibody to cytomegalovirus by antibody capture enzyme-linked immunosorbent assay, *J. Clin. Microbiol.* **21**:558–561.

van Loon, A., Heesen, F., and van der Logt, J., 1987, Antibody isotype response after human cytomegalovirus infection, *J. Virol. Methods* **15:**101–107.

Van Regenmortel, M. H. V., 1966, Plant virus serology, *Adv. Virus Res.* **12:**207–271.

Van Regenmortel, M. H. V., 1982, Serology and Immunochemistry of Plant Viruses, Academic Press, New York.

Van Regenmortel, M. H. V., 1986, The potential for using monoclonal antibodies in the detection of plant viruses, in: Developments and Applications in Virus Testing (R. A. C. Jones and L. Torrance, eds), pp. 89–101, Association of Applied Biologists, Wellesbourne, Great Britain.

Van Regenmortel, M. H. V. and Burckard, J., 1980, Detection of a wide spectrum of tobacco mosaic virus strains by indirect enzyme-linked immunosorbent assays (ELISA). *Virology* **106:**327–334.

Van Regenmortel, M. H. V. and Von Wechmar, M. B., 1970, A re-examination of the serological relationship between tobacco mosaic virus and cucumber virus 4. *Virology* **41:**330–338.

Vejtorp, M., 1980, The interference of IgM rheumatoid factor in enzyme-linked immunosorbent assay of rubella IgM and IgG antibodies, *J. Virol. Methods* **1:**1–9.

Vela, C., Cambra, M., Cortès, E., Moreno, P., Miguet, J. G., Pérez de San Roman, C. and Sanz, A., 1986, Production and characterization of monoclonal antibodies specific for citrus tristeza virus and their use for diagnosis, *J. Gen. Virol.* **67:**91–96.

Vestergaard, B., 1982, A prospective study of the relationship between genital herpes and carcinoma of the uterine cervix-L seroepidemiology of herpes simplex virus type 1 and type 2 in women with previous history of clinical diagnosed genital herpes, *Develop. Biol. Stand.* **52:**95–98.

Voller, A., Bartlett, A., Bidwell, D. E., Clark, M. F. and Adams, A. N., 1976, The detection of viruses by enzyme-linked immunosorbent assay (ELISA), *J. Gen. Virol.* **33:**165–167.

Walter, B., Vuittenez, A., Kuszala, A., Stocky, G., Burckard, J. and Van Regenmortel, M. H. V., 1984, Détection sérologique des virus du court-noué de la vigne par le test ELISA. *Agronomie* **4:**527–534.

Wartick, M., McCarroll, D., and Wiltbank, T., 1987, A second discriminator for biological false positive results in enzyme-linked immunosorbent assay for antibodies to human immunodeficiency virus (HTLV-III/LAV). *Transfusion* **27:**109–111.

Weber, J., Weiss, R., Roberts, C., Weller, I., Tedder, R., Clapham, P., Parker, D., Duncan, J., Carne, C., Pinching, A., and Cheingsong-Popov, R., 1978, Human immunodeficiency virus infection in two cohorts of homosexual men: Neutralising sera and association of anti-gag antibody with prognosis. *Lancet* **1:**119–121.

Wetter, C., Conti, M., Altschuh, D., Tabillion, R. and Van Regenmortel, M. H. V., 1984, Pepper mild mottle virus, a tobamovirus infecting pepper cultivars in Sicily, *Phytopathology* **74:**405–410.

Wiktor, T., *et al.*, 1973, Antigenic properties of rabies virus components, *J. Immunol.* **110:**269.

Wolff, K., Muth, D., Hudson, B., and Trent, D., 1981, Evaluation of solid-phase radioimmunoassays for diagnosis of St-Louis encephalitis infection in humans, *J. Clin. Microbiology* **14:**135–140.

Yolken, R. H., Leister, F. J., Whitcomb, L. S. and Santosham, M., 1983, Enzyme immunoassays for the detection of bacterial antigens utilizing biotin-labeled antibody and peroxidase biotin–avidin complex, *J. Immunol. Meth.* **56:**319–327.

Zrein, M., Burckard, J. and Van Regenmortel, M. H. V., 1986, Use of the biotin–avidin system for detecting a broad range of serologically related plant viruses by ELISA, *J. Virol. Meth.* **13:**121–128.

*Chapter 2*

# ISCOMs and Immunostimulation with Viral Antigens

Stefan Höglund, Kristian Dalsgaard, Karin Lövgren, Bo Sundquist, Ab Osterhaus, and Bror Morein

## 1. INTRODUCTION

With the dissection of microorganisms followed by biochemical and immunological characterization, antigens inducing protective immunity became recognized. Early attempts to use these isolated antigens as vaccines, i.e. subunit vaccines, showed that although immunogenic *in situ* as part of the microorganism, they were not immunogenic as purified antigens. Subsequent studies showed that the formation of antigen into defined multimeric forms such as protein micelles or into liposomes made them considerably more immunogenic. In a way, micelles and liposomes mimic a submicroscopic particle of a microorganism with several copies of surface antigens. By contrast, monomeric forms of antigens, e.g., envelope proteins of parainfluenza-3 virus or Semliki forest virus, not only had a low immunogenicity but had a specific suppressive effect on the immune response as well; this was shown when the monomers were given simultaneously

**Stefan Höglund** Institute of Biochemistry, Biomedical Center, Uppsala, Sweden. **Kristian Dalsgaard** State Veterinary Institute of Virus Research, Lindholm, Kalvehave, Denmark. **Karin Lövgren and Bror Morein** The National Veterinary Institute, Department of Virology, Biomedical Center, Uppsala, Sweden. **Bo Sundquist** The National Veterinary Institute, Division of Vaccine Research, Uppsala, Sweden. **Ab Osterhaus** National Institute of Public Health and Environmental Hygiene, Bilthoven, The Netherlands.

with the same antigen in a micelle (Morein *et al.*, 1982, 1983; Morein and Simons, 1985; Jennings, 1987).

In general, it seems that the immunogenicity of an antigen in a micelle or in a liposome is of the same magnitude as that of a killed microorganism. However, the immunogenicity of antigens in killed microorganisms or in micelles is often insufficient and must be enhanced by the addition of adjuvant. Unfortunately, in order to be effective, conventional adjuvants must be used in high doses, resulting in unacceptable adverse effects (Bomford, 1980).

The *i*mmuno*s*timulating *com*plex (ISCOM) was created to become a well-defined submicroscopic particle, with several copies of the antigen well exposed and with a built-in adjuvant, Quil A (Morein *et al.*, 1984). The binding of Quil A to the antigen complex was performed in an effort to reduce the amount of adjuvant required, thereby avoiding or minimizing the toxic effects connected with conventional adjuvants when used in effective doses. This chapter reviews the composition and assembly of ISCOM, as well as its function from an immunological point of view.

## 2. QUIL A: THE UNIQUE COMPOUND OF THE ISCOM

### 2.1. Origin

The adjuvant Quil A* is a component derived from the South American tree *Quillaja saponaria* Molina. Saponins are some of the most common secondary constituents of plants, found in more than 500 plant genera. Their physiological function is not clearly understood, but it has been suggested that they may in some cases protect the plant by exerting antimicrobial activity. They may also contribute to the depot nourishment of the plant by their ability to release monosaccharides. Chemically, they are glycosides with one or more mono- or oligosaccharide groups in glycosidic links with a nonpolar aglycon usually of steroid or triterpenoid structure. *Quillaja* saponins are triterpenoids. They are present in most parts of the tree, but the highest content is found in the bark of the tree. This bark, *Cortex quillajae,* is collected and sold commercially, mainly in Chile and, because of the traditional trade route, it is sometimes referred to as Panama bark. This old pharmaceutical drug is generally not produced for making immunological adjuvants. The bulk of the bark has been used in the past and is still used for making a crude extract, saponin, applied in diverse industries as film emulsions, shampoo, and food and beverages. The common property of the saponin used in these processes is its surface activity, making it useful as a mild detergent,

*Quil A is manufactured by the Danish company Superfos, Vedbaek, as a purified product of a uniform composition. It is available as a 2% sterile aqueous solution or as a lyophilized powder.

emulsifier, and froth-producing agent. The history of saponin as an immunological adjuvant dates back to early experiments with formol toxoid vaccines conducted at the beginning of the nineteenth century. It was discovered early that the formulation of vaccines and the constituents could influence their immunogenic potential quite remarkably. In a search for adjuvants, a whole variety of different substances were tested. It was then discovered that some saponins had immunological activity. The effect was quite unpredictable, however, since all batches of saponin were not active, and some of them produced intense local reactions at the site of injection. Except for a few experimental applications in laboratory animals, saponins did not have any practical use until some decades later. In 1951, R. G. Espinet reported on the application of saponin in foot-and-mouth disease vaccines. Espinet found that the saponin available in Argentina had a beneficial effect on such vaccines. Several years later, saponin was used in foot-and-mouth disease vaccines by the major manufacturers. It remained obscure why some batches of saponin had excellent adjuvant activity whereas some had no measurable effect. The approach for years was then to buy a large batch of a good one for use.

During the early 1970s, in search of the active components, Dalsgaard (1970, 1974) was able to show by thin-layer chromatography (TLC) that all the saponins with documented adjuvant activity were derived from *Quillaja saponaria Molina*. The fact that this tree is indigenous in South America evidently facilitated for Espinet (1951) the choice of this adjuvant for foot-and-mouth disease vaccines. It can by no means be excluded that saponins other than quillaja saponin may be active as immunological adjuvant. However, quillaja bark is certainly the best choice of the commercially available saponin drugs. The adjuvant component of quillaja bark, denoted Quil A (Dalsgaard, 1974), constitutes only about 1% of the bark on a weight basis. However, the supply of bark has been stable for a century, making it an attractive source of raw material for the production of Quil A.

## 2.2. Isolation

Quil A is freely soluble in water. Solutions of Quil A at a concentration of $\leq$10% can be handled readily. Therefore, isolation procedure is always started with a simple aqueous extraction of the bark. This aqueous extract contains a mixture of different plant constituents: saponins, tannins, sugars, and flavonoid compounds. The most important step in the purification of Quil A is a dialysis of this water extract. Quil A forms micelles and is retained in the dialysis bag whereas the bulk of low-molecular-weight substances will diffuse through the membrane. About 75% of the material extracted from the bark can be removed in this way without significant loss of Quil A. This is because Quil A has a relatively low critical micelle concentration (CMC) of 300 mg/liter. However, because of the hydrophobic interaction between Quil A and amphipathic molecules, several substances are trapped in the dialysis bag. Using both ion-exchange chromatography and gel filtration, most of the substances of nontriterpenoid nature will be removed (Dalsgaard, 1974). The product Quil A produces a single band in TLC silica gel, representing about five closely related compounds. These compounds were resolved in high-performance TLC (HPTLC) silica gel. The major nonpolar group of these molecules appears to be a triterpenoid with 5–10 oligosaccharide moieties. For comparison, the triterpenoid of quillajic acid is shown in Fig. 1.

## 2.3. Stability of Quil A

In the dry state, e.g. as a lyophylized powder, Quil A is very stable and can be kept for years at room temperature. In solution and after addition to vaccines stored at 5°C, no loss of Quil A can be detected after 1 year. It is recommended to keep stock solutions frozen at −20°C. The glycosidic bonds in the triterpenoid molecule may be sensitive to extreme pH values, especially low pH; autoclaving should also be avoided. The triterpenoid moiety of Quil A can be assayed by a colorimetric reaction with chlorosulfonic acid (Brieskorn and Briner, 1954). A homogeneous solution of Quil A is best suited for this method, since it is influenced by a variety of other substances. The most sensitive assay of Quil A is based on its hemolytic activity. Used in either radiodiffusion in an agarose gel containing red blood cells (RBC) or a similarly designed electrophoresis rocket hemolysis (Sundqvist *et al.*, 1983), Quil A can be detected in biological systems, including vaccines at a level above 25 μg/ml.

## 2.4. Use of Quil A as Adjuvant in Vaccines

Only during the last few years has Quil A been used for the formation of ISCOMs. However, in veterinary vaccines it has been commercially available

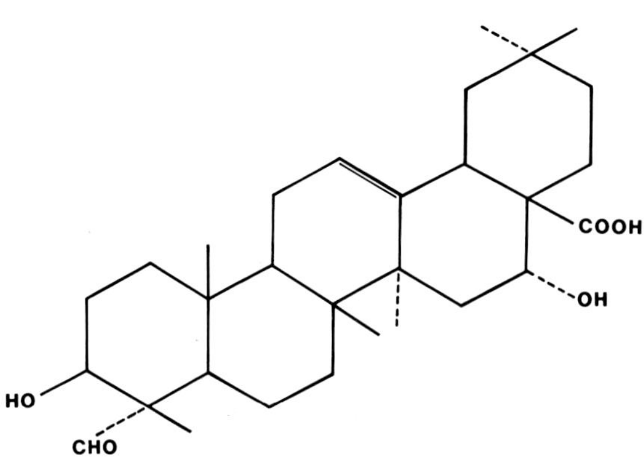


**FIGURE 1.** Triterpenoid structure of the amphipathic quillajic acid.

for several years and is now being widely used as an adjuvant for a variety of vaccines in domestic animals (Dalsgaard 1970, 1974, 1977, 1978, 1984, and references cited therein).

The assessment of the dose of Quil A as a conventional adjuvant in a given system is a balance between obtaining maximum adjuvant effect and inducing minimum local reaction at the site of injection. Being surface active and hemolytic to RBC, Quil A produces a local reaction or even tissue necrosis if applied in excessive amounts.

In large animals, i.e., cattle, pig, and sheep, a dose of 1 mg Quil A given subcutaneously (SC) is usually well tolerated. It must be emphasized that a dosage in any particular vaccine should be adjusted in each species. With 1 mg Quil A in large animals, vaccination reactions (if any) are usually limited to a small soft swelling accompanied by a short rise in body temperature, both of which disappear in 1 or 2 days, leaving no macroscopic signs at the site of injection. Formulations containing Quil A are registered additives to food and beverage, and as natural constituents of normal human diet; therefore, Quil A should not represent any hazard in the sense of meat residues.

In small animals, the optimum dose of Quil A varies from 10 μg in mice to 50–100 μg in guinea pigs and rabbits. In cats and dogs, 25 μg has been applied successfully. In dogs, this may seem a low dose, at least in the larger species, but because dogs are pet animals, there is great public concern against acceptance of any local reaction from a vaccination. $LD_{50}$ in mice is ~35 mg/kg. Toxicology studies in rats have been reported with ISCOMs containing envelope proteins of measles virus that were compared with a trivalent commercially available vaccine containing diphtheria–pertussis–tetanus–polio (DPT–polio). The ISCOM vaccine gave less local reaction than did the DPT–polio vaccine (Speijers *et al.*, 1987). The amount of Quil A included in ISCOMs may be several hundredfold lower than when used as a conventional adjuvant, provided that free Quil A is removed from the ISCOMs. With this in mind, ISCOM vaccines can be prepared with a very low toxicity; thus, toxicity should not be a serious obstacle to the use of ISCOM vaccines.

## 3. STRUCTURE AND COMPOSITION OF ISCOMs

Early ISCOM preparations were described as structures containing amphipathic proteins assembled in a multimeric complex through hydrophobic interaction with Quil A (Morein *et al.*, 1984). However, at that time there was limited knowledge about the composition of ISCOMs as well as about the minimal requirements of the components essential for formation of an ISCOM. The self-assembly of amphipathic molecules into complexes in aqueous solution is mainly due to the interaction between available hydrophobic sequences of the amphipathic molecules. Examples of such complexes are lipid vesicles, micelles of proteins or detergent molecules, and ISCOMs. It turned out that the ISCOMs, maintained by hydrophobic interaction, are energetically stable constructs that, once formed, are not easily dissociated.

A great number of ISCOMs were prepared from glycoproteins from a wide range of virus species. All ISCOMs were shown to be stable structures and moreover had a similar cagelike morphology, regardless of the properties of the protein: glycosylation, size, or tertiary structure (Morein *et al.*, 1984; Höglund *et al.*, 1986). Therefore, it was unlikely that the proteins were determining the structure of ISCOM. In a series of experiments, different lipids, with emphasis laid on cholesterol and phosphatidylcholine, were tested for ISCOM formation (Lövgren and Morein, 1988). It was then shown that an ISCOM matrix in its simplest form is composed of Quil A and cholesterol. Into this ISCOM matrix, amphipathic proteins together with an additional lipid, e.g., phosphatidylcholine, are incorporated by hydrophobic interaction to form a highly immunogenic antigen complex—the ISCOM.

## 3.1. Requirement of Lipid for the Construction of ISCOM

### 3.1.1. ISCOM Matrix

Long before ISCOMs were made, it was shown that Quil A has the ability to penetrate and bind to a number of lipids, particularly cholesterol, to form soluble lipid complexes (Bangham *et al.*, 1962; Dourmashkin *et al.*, 1962; Glauert *et al.*, 1962; Lucy *et al.*, 1964). But, to constitute the ISCOM matrix, only cholesterol was indispensable, and none of nine other lipids tested was able to interact with Quil A and replace cholesterol. By contrast, from a mixture of cholesterol and Quil A only, an ISCOM matrix was formed, with a slightly flattened cagelike morphology (Fig. 2). What, then, is the matrix? We have defined the ISCOM matrix as the cagelike structure produced through the interaction between Quil A and lipids with cholesterol being a prerequisite of its assembly. Even though the lipids tested, except cholesterol, would not alone associate with Quil A to constitute the ISCOM matrix, they were readily incorporated into the

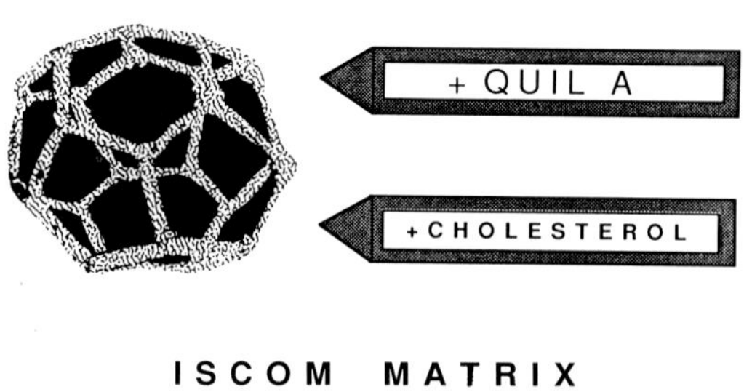


**FIGURE 2.** Model of the cagelike ISCOM matrix composed of Quil A and lipid, as has been demonstrated by electronmicroscopy.

ISCOM matrix; moreover, such a lipid was prerequisite to include a protein into the ISCOM. Although the ISCOM matrix is composed of Quil A, cholesterol, and lipid, it constitutes only a minor part of the total mass. In a typical ISCOM, 5–10% of the weight is Quil A, 1–5% is composed of cholesterol and phospholipids, and the rest is protein (i.e., the antigen). On a molar basis, the proportion of protein–Quil A–cholesterol will be approximately 1 : 1 : 1.

### 3.2. Morphology of the ISCOM Particle

The cagelike structure of the ISCOM is a characteristic feature, used for the analysis of ISCOM preparations (Morein *et al.*, 1984; Höglund *et al.*, 1986) (Fig. 3). ISCOMs prepared from viral proteins appear as a rigid isometric particle with a diameter of ~35 nm. The electron-dense negative stain readily penetrates the open structure of the ISCOM, demonstrating a regular pattern of holes in its shell (Fig. 3a,b). Structural subunits of the shell were envisaged with the aid of electronmicroscopic tomography (Skoglund *et al.*, 1986), allowing reconstruction of models of the ISCOM particle (Fig. 4). Interestingly, lyophilization of ISCOMs has proved to preserve a round shape of the ISCOM particle (Fig. 3b).

The construct of the ISCOM particle, by analogy with isometric viral capsids (Liljas, 1986), is probably energetically optimized, thus rendering the ISCOM remarkably stable. The ISCOMs have kept their morphology and immunogenicity for 3 years stored as a sterile suspension in a physiological buffer or preserved by lyophilization (Sundquist *et al.*, 1988a). By different cryopreparation techniques (Özel *et al.*, 1988) a symmetrical array of morphological subunits assembled on the shell of the ISCOM particle, exhibiting five- and six-fold symmetry (Fig. 3b) is envisaged. Interestingly, 20 morphological subunits assembled on a dodecahedral shell (Fig. 7) would mimic the ISCOM particle, as shown by electron microscopy (Höglund *et al.*, 1989; Özel *et al.*, 1989).

## 4. PREPARATION OF ISCOMs

ISCOMs have been prepared to contain a great variety of amphipathic proteins, including viral envelope proteins from about 20 different viruses, such as influenzavirus, parainfluenza-3 virus, measles virus, bovine viral diarrhea virus, bovine leukemia virus, feline leukemia virus, and simian and human immune deficiency viruses (SIV and HIV) (Morein *et al.*, 1987, and references cited therein). Membrane proteins from bacteria and protozoa were also readily integrated into ISCOMs, and in all cases tested they became highly immunogenic (Uggla *et al.*, 1989).

The detergent solubilization of a membrane involves chemical cooperation

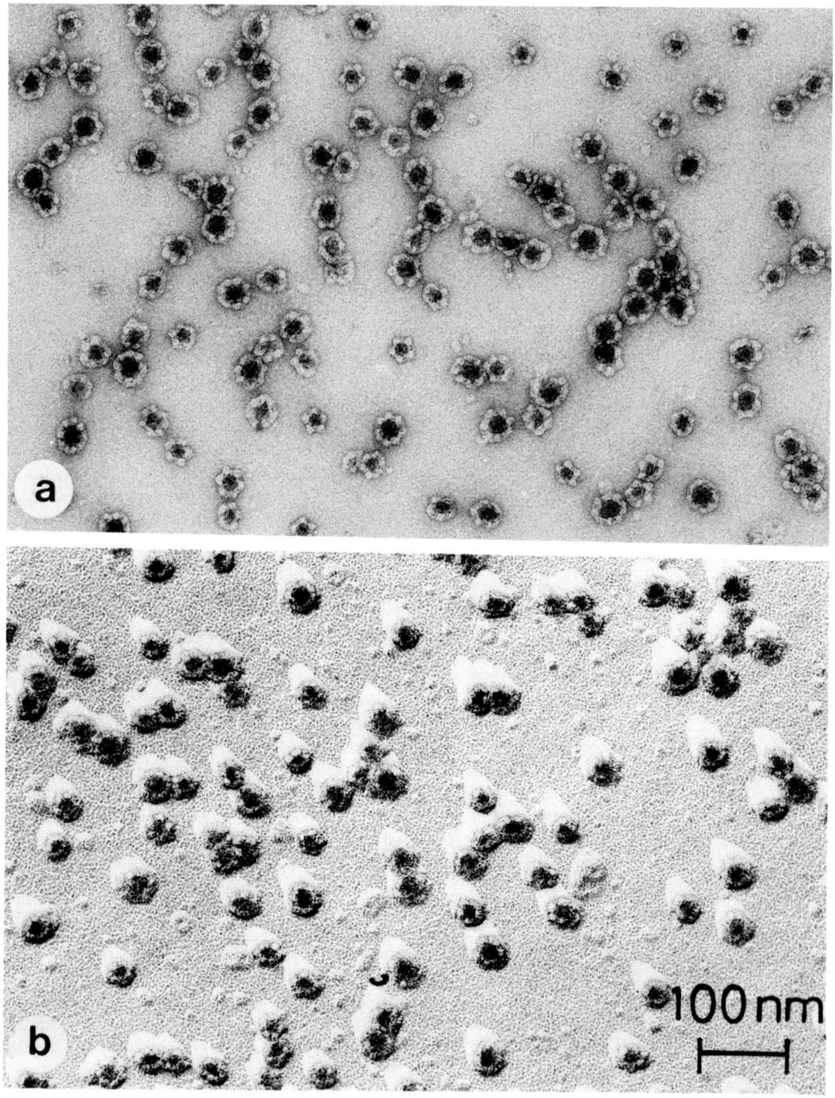


**FIGURE 3.** (a) Cagelike shell demonstrated on negatively stained ISCOM readily penetrated by the stain. The ISCOMs were preserved by freeze drying. (Courtesy of M. Özel.) (b) Specimen of ISCOMs, prepared by negative staining and shadow casting, showing a population of round ISCOM particles preserved by freeze drying. An assembly of globular morphological subunits is shown on the shell of the ISCOM particle. (Courtesy of M. Özel.)

**FIGURE 4.** Model reconstruction, produced from micrographs of an ISCOM analyzed by electron-microscopic tomography, clearly showing an open shelf composed of structural subunits. (Courtesy of L. G. Öfverstedt and Su Xiao-dong.)

between the essential components, i.e., proteins, lipids, and detergent molecules (Helenius and Simons, 1975; Tanford and Reynolds, 1976). A nonionic detergent is preferable to use, since it has a low or no denaturing effect on the hydrophilic parts of the proteins, which are likely to contain the epitopes inducing protective immunity. For certain methods, its relatively low molecular weight and the fact that it forms micelles at a high concentration, facilitate its removal by the dialysis method. Two such candidates for disintegration of viral envelopes are octylglucoside (Helenius *et al.*, 1979) and MEGA (Hildreth, 1982; Hanatani *et al.*, 1984) (Fig. 5).

### 4.1. Centrifugation Method

The centrifugation method is suitable for the preparation of ISCOMs containing viral envelope proteins (Morein *et al.*, 1984). Briefly, after harvest from the cell culture, cells and cell debris were removed by differential centrifugation at low speed. The virus is usually concentrated and purified by centrifugation through a buffered sucrose gradient at ~100,000*g* forming a zone of virus or a pellet. The purified virus is disintegrated with a detergent, preferably a nonionic detergent. The solubilized viral proteins are sedimented through a thin layer of 10% sucrose containing 0.5% detergent, e.g., Triton X-100, on top of a 20–50% sucrose gradient containing 0.1% Quil A (Fig. 6). Following rate zonal centrifugation (McEven, 1967), the ISCOMs can be obtained sedimenting at about 19 S, whereas monomer proteins will stay at the upper part of the gradient material, at approximately 4 S (Fig. 6). Protein micelles of envelope proteins generally have a higher sedimentation coefficient of about 30 S (Morein *et al.*, 1984; Helenius and Bornsdorf, 1976; Sundquist *et al.*, 1988a). The ISCOM matrix has a sedimentation coefficient of 14–18 S, depending on lipid content, which is lower than for the ISCOMs (Lövgren and Morein, 1988).

## 4.2. Dialysis Method

Basically, the formation of ISCOMs takes place while the detergent molecules are removed from the mixture of protein, lipid, detergent, and, importantly, Quil A (Fig. 6). Therefore, other methods can be used to remove the detergent. For instance, we have used a dialysis method that requires dialyzable detergents, e.g., octylglucoside or MEGA (see Fig. 5). The dialysis is performed extensively, e.g., for 3 days, against a desired buffer. In the dialysis bag, several components, not included in the ISCOMs, may be present. The ISCOMs can then be separated from contaminants by centrifugation through a cushion of 20% sucrose. A thin layer of detergent on top of the sucrose will result in retention of excess lipid and Quil A due to a slower sedimentation through the sucrose than for the ISCOMs. Nonamphipathic proteins, such as nucleoproteins of feline leukemia virus and HIV, will also be removed.

At the preparation of viral ISCOMs, there is a major concern as to how the virus is produced, e.g., what cell line is used for virus propagation, which

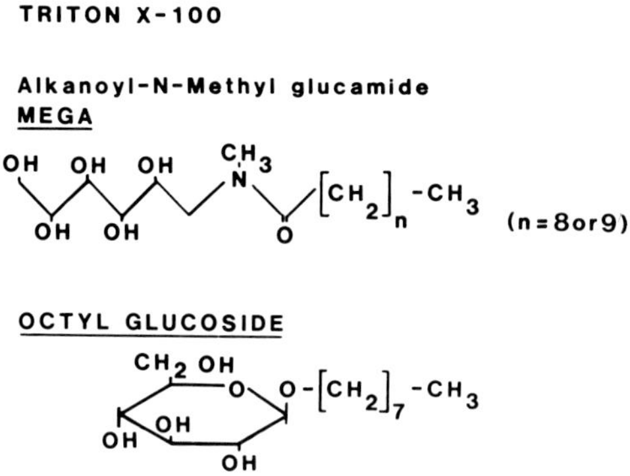


**FIGURE 5.** The nonionic detergents Triton X-100, MEGA, and octylglucoside.

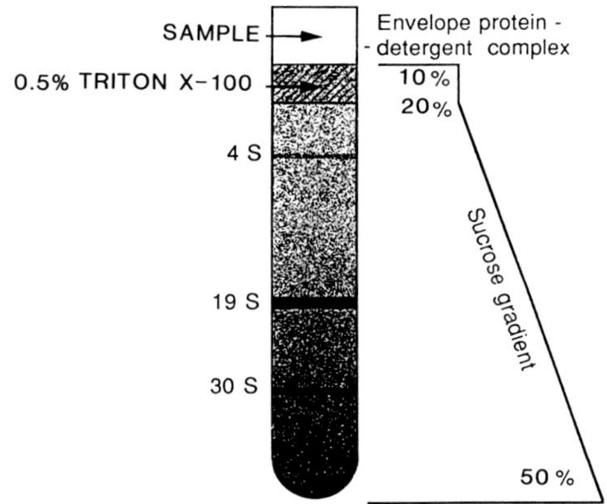


**FIGURE 6.** Preparation and isolation of ISCOMs by rate zonal centrifugation. The sample of envelope protein–detergent complex (mixed micelles of protein and detergent molecules) applied on 10% sucrose containing 0.5% Triton X-100, which will trap excess of Quil A and lipid ; below is a 20–50% sucrose gradient containing 0.1% Quil A. By centrifugation, 19 S ISCOMs are formed and separated from 4S monomer proteins and 30S micelles.

methods are employed for viral purification and subsequent viral disintegration, and what system is used for the reassembly of viral proteins into the ISCOM. An example is given below to demonstrate problems in preparation of ISCOMs from HIV proteins due to cellular components incorporated in the envelope of HIV (Gelderblom *et al.*, 1987; Henderson *et al.*, 1987; Hoxie *et al.*, 1987) and the loss of relevant antigen (Gelderblom *et al.*, 1985, 1987; Höglund *et al.*, 1986; 1988).

### 4.3. ISCOM of HIV Antigens

The HIV used was the HTLV IIIB strain grown on the human H9 cell line that expresses the class II transplantation antigen HLA-DR. Like other retroviruses, HIV has two envelope proteins, either loosely connected or completely disconnected, that originate from one intracellular precursor protein gp160. During intracellular transport, gp160 is cleaved by trypsin or a trypsin-like enzyme to the transmembraneous part gp41 and the external part gp120, both being glycosylated. During the budding process, gp120 seems to be detached from the anchoring transmembraneous gp41 (Gelderblom *et al.*, 1987).

Specific requirements have been raised in the production of an HIV vaccine, which have not yet been met in the field of viral vaccination (Barnes, 1988; Newmark, 1988). For example, it is obligatory to prepare the immunogen freed of nucleic acid because of inherent risk of the virus genome to become infectious or to be integrated into the genome of the host cell. A prospective protective antigen of HIV is most likely gp120; however, gp41 and p17 would also be required possibly to induce protective immunity (Barnes, 1988). Gp120 is not readily incorporated into ISCOMs, lacking an accessible hydrophobic region, in contrast to gp41, which has at least three sequences of hydrophobic amino acid residues (Modrow *et al.,* 1987). During purification of the virus, gp120 is efficiently separated from the virus, leaving a viral particle where more than 95% of gp120 has been detached. Such material has been used for vaccination experiments, e.g., with SIV proteins (Letvin *et al.,* 1987), obviously with a poor result as regards the immune response to gp120.

ISCOMs have been prepared from HIV proteins according to the method described above (Höglund *et al.,* 1989). The virus used was HIV, strain HTLV IIIB, grown in H9 cells. The HLA-DR protein, incorporated in the envelope of HIV, became the major constituent of the ISCOMs, whereas gp120 became a minor component hardly detectable by sodium dodecyl sulfate–polyacrylamide gel electrophoresis (SDS–PAGE) and with immunoblotting techniques. Such ISCOMs induced high-antibody titers to HLA-DR in mice, rabbits, and guinea pigs compared with the barely detectable antibody response to gp120. However, purified gp120 integrated into ISCOMs induced a high-antibody response. Antibody response was also evoked to other proteins incorporated into the ISCOMs, e.g., p24 of the nucleocapsid and the myristylated p17 of the envelope (Gelderblom *et al.,* 1987). The protein p24 can be removed from the ISCOMs by centrifugation through a detergent layer, but not the myristylated p17. It is also noteworthy to mention that HLA-DR can be removed from the viral proteins by immunosorbent affinity chromatography using specific monoclonal antibodies (Akerblom and Villacrez, 1987).

In conclusion, it should be emphasized that identification and quantification of the proteins of the ISCOM preparation is desirable to predict since the immune response to the ISCOM reflects the antigens included (Höglund *et al.,* 1989). The quality of the ISCOM preparations is also routinely analyzed by electron microscopy; the morphology and homogeneity of the ISCOM particles are then assayed.

## 5. INTEGRATION OF NONAMPHIPATHIC PROTEINS INTO ISCOM

Initially, the assembly of proteins into an ISCOM was done with native membrane proteins with a readily recognized transmembraneous region, which is surrounded by lipids *in situ* in the virus or in the cell. However, there is also of great interest to enhance the immunogenicity of a number of nonmembrane proteins. Many of these, perhaps the majority, have domains of hydrophobic amino acid residues hidden in the tertiary structure of the protein. Once made accessible, these hydrophobic amino acids may provide the necessary nonpolar character to attach the proteins to the ISCOM matrix. The possibility of enhancing the immunogenicity also encompasses a number of gene technology products (Moss *et al.,* 1988; Murray, 1988). Compared with the isolation of native membrane proteins, extraction and purification of polypeptides prepared by recombinant DNA (rDNA) techniques will often include harsh methods such as urea and SDS treatments. In particular, proteins cloned in *Escherichia coli* often contain sequences of hydrophobic amino acids, which may or may not be hidden in the tertiary structure of the protein. If not hidden, difficulties in the isolation and purification procedures will arise due to the insolubility of the protein in a polar buffer.

### 5.1. Model System

In attempts to demonstrate nonaccessible hydrophobic domains, two methods were evaluated: proteins were exposed to acidic pH or to heat treatment. In a model system, bovine serum albumin (BSA) was employed. Acid treatment of BSA at pH 2.5 in the presence of the detergent, MEGA-10 (see Fig. 5), cholesterol, phosphatidylcholine, and Quil A was followed by the removal of the detergent by dialysis. About 15% of BSA then became integrated into ISCOMs, according to the criteria described above. Alternatively, when BSA was exposed to 70°C instead of acid exposure, it also became integrated into ISCOMs. The recovery of integrated BSA ranged between 30% to 50% following heat treatment.

The immunogenicity of both acid- and heat-treated BSA incorporated into the ISCOM was tested in mice. Both preparations were more immunogenic than native and nonintegrated BSA. The serum antibody response was 10-fold higher for ISCOM preparations. Also, the dose of BSA could be reduced by 10-fold, e.g., 100 ng BSA in ISCOMs induced a serum antibody response in mice comparable to or better than 1 μg nonintegrated BSA induced (Morein *et al.*, 1989a).

### 5.2. Gp120 ISCOM

The experience with acid treatment of BSA for the preparation of ISCOMs was applied to purified gp120 of HIV. The resulting gp120 ISCOMs proved highly immunogenic in rhesus monkeys and chimpanzees. After multiple immunizations with ISCOMs containing gp120 of HIV-1B, a rhesus monkey developed a transient *in vitro* neutralizing response to the HIV-1 isolates RF and MN (Pyle *et al.*, 1989).

### 5.3. Gene-Technology Products

Proteins cloned in *E. coli* have been incorporated into ISCOMs. Examples are a *gag*-gene product from HIV, an *env*-gene product of HIV corresponding to the carboxy-terminal part (kindly supplied by S. Putney, Repligen). Both preparations were highly immunogenic in mice. Gp160 of HIV has been cloned in a vaccinia vector and produced by Vero cells, expressed as the uncleaved gp160 antigen (kindly supplied by Immuno AG, Vienna). Thus gp160 has the transmembraneous part, gp41, and does not need any treatment to expose hidden hydrophobic domains to be incorporated into ISCOMs.

### 5.4. Picornavirus

Picornaviruses do not have envelope proteins with transmembraneous hydrophobic domains. However, hydrophobic amino acid sequences exist. Fohlman *et al.* (1989) integrated three capsid proteins of coxsackievirus (CB3) by treating the virus with SDS and urea and subsequent removal of the chemicals. In a model experiment in mice, 16 ng injected three times completely protected the mice from mortality when challenge-infected with a myocarditic strain of CB3.

In conclusion, nonamphipathic proteins can also be rendered hydrophobic to be integrated into ISCOMs, and an enhanced immunogenicity can thereby be obtained. To be able to make hydrophobic domains accessible, however, partial denaturation is required that might affect crucial antigenic determinants. Since there are alternative methods for partial denaturation, a quality-control system for the important antigenic determinants should be established. Such a system is described by Merza *et al.* (1989) for gp50 of bovine leukemia virus (BLV) using monoclonal antibodies that recognize neutralizing epitopes. With the aid of this quality control, methods were worked out to make BLV ISCOMs with nondestroyed (native) neutralizing epitopes.

## 6. ISCOMS AS CARRIERS FOR SMALL MOLECULES AND PEPTIDES

The poor immunogenicity of small molecules, including synthetic peptides, is a well-known problem. To improve their immunogenicity, these molecules have been conjugated to carrier molecules, e.g., BSA, keyhole limpet hemocyanin (KLH), or random amino acid co-polymers (Erlanger, 1980; Audibert, 1982; Palfreyman *et al.*, 1984). However, in order to induce a more substantial immune response, it is generally necessary to include an adjuvant, e.g., Freund's complete adjuvant or Freund's incomplete adjuvant (Freund, 1956).

Coupled to a carrier protein, haptens may be considered as epitopes or antigenic determinants of the carrier protein. As shown by Morein and coworkers (1978, 1982, 1984, 1985), multimeric forms have several immunogenic advantages over monomeric proteins. Consequently, a hapten or a small molecule conjugated to a carrier protein included in an ISCOM should be a considerable improvement to the monomeric carrier protein.

### Hapten Model

Instead of a peptide, succinyl-activated biotin was used as a hapten model (Lövgren *et al.*, 1987). The coupling reaction is easily monitored, since $^3$H-labeled biotin is available. The conjugates produced differed only as to the number of biotin molecules coupled per carrier protein in the ISCOM, ranging from 0.1 to 20. It was concluded that at a dose of 1 μg conjugate, approximately 10 biotin molecules per protein were optimal for the antibody response. It is likely that 10 biotin molecules were needed to expose one of them as a highly immunogenic antigen determinant of a protein. In additional experiments, biotin conjugated to different carrier systems were studied as immunogens in mice. The multimeric carriers, i.e., ISCOMs and micelles, were highly immunogenic in contrast to the monomeric BSA. Ten μg of biotin-ISCOMs induced about three times higher antibiotin titers than did the same amount of biotin–micelles or biotin–BSA in Freund's complete adjuvant. Biotin–BSA inoculated without adjuvant induced only a poor antibody response.

In conclusion, these results (Lövgren *et al.*, 1987) clearly indicate that the ISCOM is an efficient carrier system for hapten presentation. This system has also been adapted for efficient presentation of synthetic peptides, including viral peptides. The dose can be kept low, $\leq$10 μg, and no additional adjuvant is required.

## 7. INDUCTION OF IMMUNE RESPONSE INCLUDING PROTECTION BY ISCOMS

Several reports showed that ISCOMs produced with various antigens induce high antibody response, e.g., against influenzavirus, paramyxovirus, togavirus, and coronavirus (Morein *et al.*, 1987; Sundquist *et al.*, 1988b). Generally, antigens included in ISCOMs induce a 10-fold higher antibody titer than do the corresponding antigen *in situ,* e.g., in the virus particle or in a protein micelle (Morein and Simons, 1985; Morein *et al.*, 1987).

### 7.1. Influenzavirus

Recently, Lövgren (1988) showed that ISCOMs containing the hemagglutinin and the neuraminidase glycoproteins from the influenzavirus H1N1, PR8 strain induced a classic serum antibody response in mice after both subcutaneous and intranasal immunization with an early immunoglobin M (IgM) response, followed by an IgG response. The antibodies were also shown to be evenly distributed in all IgG isotypes, i.e., IgG1, IgG2a and b, and IgG3. In this connection, we would like to emphasize the fact that the ISCOM efficiently induced antibody- as well as cell-mediated immunity (CMI) following local intranasal immunization. Cell-mediated immunity was demonstrated by Jones *et al.* (1988), who showed that the H1N1 influenzavirus ISCOMs after intranasal application in mice induced a cytotoxic T-cell memory response, inducing a broadened specificity with reactivity to the subserotype H3N2 of influenzavirus. In contrast to live experimental vaccine, the ISCOMs efficiently boostered the cytotoxic T-cell response.

## 7.2. Herpesvirus

For the herpesvirus, the T-cell response is considered important for protective immunity but less so for antibody-mediated immunity. Wahren *et al.* (1987) showed that ISCOMs as carriers for cytomegalovirus (CMV) antigens induced high CMI, measured with the lymphocyte-stimulation test in monkeys, in contrast to a whole-virus preparation. This CMI response was of the same magnitude, measured in counts per minute (cpm), of incorporated [$^3$H]thymidine, as that of naturally infected but healthy man. Monkeys immunized with CMV carried by monocytes also showed a CMI response provided a high dose of CMV; i.e., 150 μg was included. By contrast, when CMV antigens were presented by the ISCOMs a low dose of 15 μg was sufficient.

## 7.3. Measles Virus

In a laboratory animal model system with BALB/c mice, DeVries *et al.* (1988a) studied the activation of measles virus-specific T cells by the F protein incorporated into ISCOMs. It was demonstrated that a measles virus-specific delayed-type hypersensitivity (DTH) response was induced. The mice were immunized with 0.25 or 1.0 μg ISCOMs or micelles containing the fusion (F) protein of measles virus; they were challenged 7 days later with whole measles virus in the ear. A specific ear swelling was measured during the following three days, which was most prominent in the mice immunized with the ISCOM preparation of F protein. The F–ISCOM-immunized animals showed specific swellings of maximal 76 μm or 153 μm, respectively, whereas the F–micelle-immunized animals showed specific swellings of maximal 25 or 75 μm above background values, respectively.

The ability of F–ISCOM to induce F-specific T-cell response in mice was further shown by isolating F-specific T-cell clones from F–ISCOM-immunized BALB/c mice. The clones were shown to be specific both for ultraviolet (UV)-inactivated measles virus and for purified F protein. Most cells from each T-cell clone stained for Thy-1,2, Lyt-1, and L3T4 markers, a pattern characteristic of murine T-helper cells. The functional characterization of these clones showed that they were able to assist B-cells in the production of measles virus-specific antibodies *in vitro* (DeVries *et al.*, 1988a).

### 7.4. Rabiesvirus

Also in the rabiesvirus system, the activation of virus-specific T cells by viral proteins presented in the ISCOM was shown in a rabiesvirus-specific DTH response (Osterhaus *et al.*, 1986). BALB/c mice immunized with 0.1, 1, or 10 μg rabiesvirus ISCOMs were challenged 7 days later with inactivated rabiesvirus in the ear, and specific ear swelling was measured during the following 3 days. A specific ear swelling to a maximum of about 40% of the ear thickness was observed in the animals immunized with 1 or 10 μg of the ISCOM preparation. The ability of the rabiesvirus ISCOM to stimulate rabiesvirus specific T-cells in man was indicated by proliferative responses of rabiesvirus-specific human T-cell clones of the helper-cell phenotype upon stimulation with the ISCOM in the presence of antigen-presenting cells (Fig. 8). Importantly, CMI response was efficiently induced in different animal species by immunization with ISCOMs containing glycoproteins from cytomegalovirus, rabiesvirus, influenzavirus, and measles virus.

## 8. PROTECTION

Protection studies, if possible in the host, are the relevant tests of a vaccine. Experimental ISCOM vaccines have been tested in various animal species (Table I) and are summarized below.

### 8.1. Bovine Viral Diarrhea Virus

Bovine viral diarrhea virus (BVDV) belonging to the *Pestivirus* genus in the Togavirus family is closely related to border disease virus (BDV) of sheep and hog cholera virus (HCV) of swine. The most common form of BVDV is sub-clinical, but it can occur sporadically with severe clinical signs with low and high morbidity and mucosal disease. Recent investigations have indicated that dams may produce persistently infected offspring and that these individuals are the most important source for the spread of the infection. In susceptible dams or ewes, infection with BVDV or BDV can result in abortion, mummified fetuses, as well as persistently infected offspring. Efficient prophylaxis against BVDV and BDV is needed to protect against reproduction disturbances caused by these viruses in cattle and sheep. Consequently, such animals should be vaccinated before breeding and insemination. For that purpose, an experimental ISCOM vaccine was designed to contain the envelope proteins of a cythopathic BVDV isolate. This vaccine was tested in pregnant sheep seronegative for BVDV (Alenius *et al.*, in press, 1989). After two vaccinations spaced three weeks apart, 15 pregnant ewes responded with high serum-neutralizing antibody with a mean titer of 625 in a microplate test, while 14 nonvaccinated ewes were still

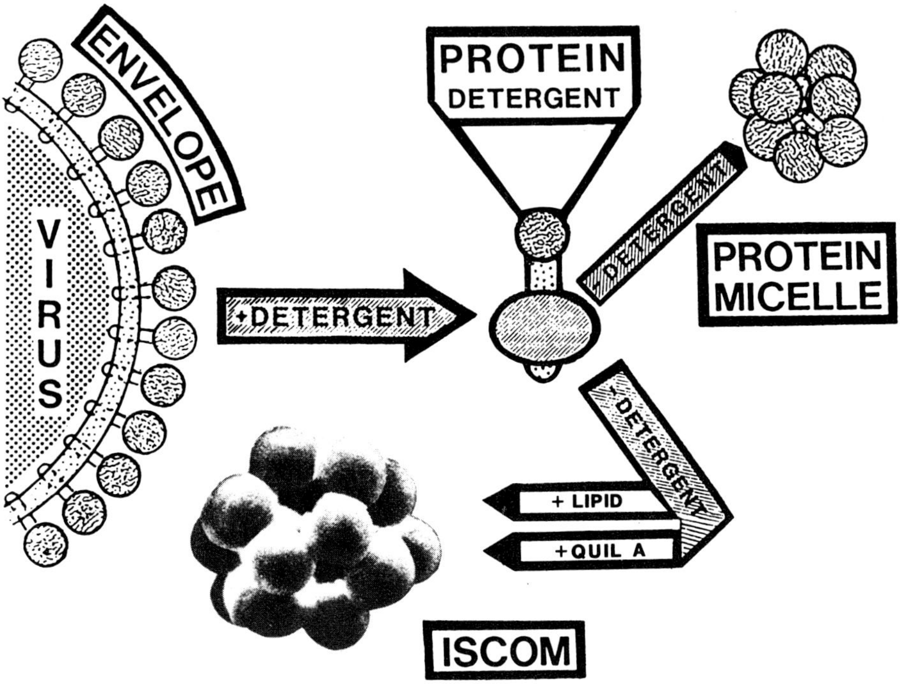


**FIGURE 7.** Schematic drawing of ISCOM preparation from viral envelope proteins. A detergent causes disintegration of the envelope, so that it will release spike protein–detergent complex. Removal of detergent (e.g., by dialysis ) will form protein micelles. Addition of Quil A and lipid, followed by the removal of detergent, will result in the assembly of Quil A, lipid and spike proteins into a symmetrical shell of globular subunits of the ISCOM particle.

seronegative. At that time, 28 and 26 fetuses were detected, respectively, in the two groups of ewes, by ultrasonic technique. The animals were challenge-infected with a recent field isolate 3 weeks after the second vaccination, i.e., between the 48th to 65th days of gestation. The 15 vaccinated ewes lambed 26 liveborn lambs, whereas the BVDV infection resulted in extensive fetal death and the 14 nonvaccinated ewes gave birth to only six liveborn lambs. This experiment demonstrates that an experimental BVDV ISCOM vaccine induced high serum-neutralizing antibody and a complete protection from abortion.

**Table I**
**Protective Immunity Induced by Various ISCOM Preparations**

| Antigen | Animal | Disease | Reference |
|---|---|---|---|
| Epstein–Barr virus | Tamarin monkey | Lymphoma | Morgan *et al.* (1988) |
| Pseudorabies virus | Swine | Lethal | Sundquist *et al.* (1988) |
| Bovine herpesvirus type 1 | Cattle | Disease | Trudel *et al.* (1988) |
| Bovine viral diarrhea virus | Sheep | Abortion | Alenius *et al.* (1989) |
| Feline leukemia virus, gp70/85 | Cat | Viremia | Osterhaus *et al.* (1985) |
| Hemagglutinin of measles virus | Mouse | Encephalitis | deVries *et al.* (1988a); Varsanyi *et al.* (1987) |
| Fusion protein of measles virus | Mouse | Encephalitis | deVries *et al.* (1988a); Varsanyi *et al.* (1987) |
| Hemagglutinin, fusion protein of canine distemper virus | Dog | Viremia | deVries *et al.* (1988b) |
| Hemagglutinin, neuraminidase (H3N8), equine influenzavirus | Horse | Pneumonia | Mumford *et al.* (1986) |
| *Toxoplasma* surface antigen | Mouse | Lethal | Uggla *et al.* (1989) |

## 8.2. Epstein–Barr Virus

There is a compelling evidence that infection with Epstein–Barr virus (EBV), belonging to the Herpesviridae family, is one of the causative events leading to the subsequent development of two important human malignancies, i.e., undifferentiated nasopharyngeal carcinoma, prevalent in China and Burkitt lymphoma, endemic in Africa. Since the former is the most common cause of cancer in men of Southern China, there is a need for a development of a vaccine to prevent or reduce the incidence of these diseases. The major envelope glycoprotein of EBV, gp340, is believed to play an important role in the induction of protective immunity and would therefore be the antigen of choice in a prospective subunit vaccine. To evaluate an EBV vaccine, an animal model has been developed using the cottontop tamarin. ISCOMs were prepared to contain purified gp340 by the dialysis method and were used at a dose level of 2 μg to vaccinate four tamarins (Morgan *et al.*, 1988). The animals were immunized three times with intervals of 2 weeks. Serum antibody was shown to neutralize EBV and EBV-induced transformation of cord blood lymphocytes. When the tamarins were infected with a challenge dose of EBV, no tumor development occurred in the vaccinated animals, although some transient lymph node enlargements were detected. However, the histologic examination showed no evidence of the presence of lymphomatous tissue reaction but of an inflammation consisting of reactive lymphomatous tissue. By contrast, in a group of four nonvaccinated tamarins, which had received the same challenge dose, the monomorphous appearance of lymphomas appeared during the third week following challenge.

These data indicate that microgram doses of gp340 incorporated into ISCOMs can protect tamarins against a complete lymphomagenic challenge of EBV, which would suggest that there is a future possibility for use of EBV vaccine in man.

## 8.3. Pseudorabies Virus

Pseudorabies or Aujezsky disease is caused by another herpesvirus, pseudorabiesvirus (PRV), which affects the respiratory tract, central nervous system, and reproduction system of swine. Morbidity and mortality rate are high and can often reach 100% in young suckling pigs. Since an eradication program of PRV employing stamping out has almost eliminated PRV in the pig populations in Denmark and England, it is likely that other countries will follow a similar policy. However, for economical reasons, many countries would like to combine an eradication program with a vaccination program. Such a program would require a diagnostic test differentiating PRV-infected from PRV-vaccinated animals. Since the ISCOM technology will selectively choose the envelope proteins, in principle it would not induce antibody to the internal proteins of the virus. In an experiment designed to explore that possibility, at least five proteins found to be present in the whole-virus preparation were lacking in the ISCOM and did not induce antibody in the vaccinated animal (Morein *et al.*, 1989b). This opened a possibility to use a competition enzyme-linked immunosorbent assay (ELISA) directed against one or more of these proteins to differentiate PRV-infected animals from vaccinated ones.

For a potency test of PRV, vaccine pigs are used. In pigs, however, the morbidity rate following experimental infection is variable and is also age restricted, which is noticeable in pigs over 6 weeks of age, at which time the animals hardly show any symptoms. In an alternative potency test model, lambs have been used for efficacy documentation (Morein *et al.*, 1989b). Lambs are extremely sensitive to PRV infection, and the infection is not age restricted, as it is in the pig. The ISCOM vaccine containing the envelope glycoproteins of PRV was tested in Hungary in lambs according to their potency test protocols (T. Soos, Control institute of Veterinary Biology, Budapest, Hungary). After a single vaccination with 3 μg PRV antigen per dose, the lambs were completely protected to challenge infection in contrast to the animals in the nonvaccinated group, in which the infection caused a 100% mortality (Morein *et al.*, 1989b).

## 8.4. Infectious Bovine Rhinotracheitis Virus

Infectious bovine rhinotracheitis (IBR) virus, or bovine herpes virus type 1, infects the respiratory tract of cattle; a variant of the virus infects the vagina and uterus. It is a major cause of economical losses in the cattle industries. Several vaccines have been tried and are available on the market, but their efficacy is questionable. Trudel *et al.* (1988) prepared ISCOMs containing the envelope proteins of IBR virus and reported that a dose of 50 μg protein induced protection to challenge infection. In contrast to commercially available vaccines, the ISCOMs conferred complete protection to disease. The animals showed no temperature rise, and there was a $10^5$ reduction in virus excretion measured in secretion, collected as nasal swabs, as compared with unvaccinated challenge-infected animals.

## 8.5. Measles Virus

ISCOMs were prepared to contain the fusion (F) protein of measles virus, isolated from virus by immunosorbent affinity chromatography. These ISCOMs induced anti-F protein antibodies in monkeys shown by hemolysis inhibition and cell-fusion inhibition assays (DeVries *et al.,* 1988a). ISCOMs prepared from a whole virus contained both the F protein as well as a small amount of hemagglutinin. In contrast to the pure F–ISCOM preparation, they also induced hemagglutination-inhibiting (HI) and virus-neutralizing serum antibodies in monkeys and in rats, besides the hemolysis-inhibiting antibodies. The BPL-inactivated measles virus preparation, from which the ISCOM preparation had been produced, was less efficient in inducing virus-neutralizing and hemagglutination-inhibiting antibodies and failed completely to induce hemolysis-inhibiting antibodies in rats (Morein *et al.,* 1984). DeVries *et al.* (1988a) could protect mice against fatal encephalopathy with ISCOMs prepared from whole virus or with F-protein ISCOMs. Varsanyi *et al.* (1987) used ISCOMs containing cell-derived measles virus envelope proteins depleted of either the H protein or the F protein by an immunosorbent technique. Both preparations induced protective immunity in mice to intracerebral challenge of a measles virus strain causing fatal encephalopathy.

In developing countries, measles virus is a major cause of mortality in young children. Therefore, it is desired to find a vaccine that can induce protective immunity, while the child still has maternal antibodies. In an attempt to mimic passive maternal immunity mice were passively given antibodies to measles virus. An antibody response could be induced in the passively immunized mice much more efficiently with a measles virus–ISCOM than with a whole measles virus preparation (DeVries *et al.,* 1988a).

## 8.6. Canine Distemper Virus

The potential of the ISCOM for the induction of protective immunity against morbillivirus infection was further shown by immunization of dogs with canine distemper virus (CDV)–ISCOMs, which contained both the F protein and the hemagglutinin of CDV. The dogs developed virus-neutralizing antibodies; in contrast to the nonvaccinated animals, they did not develop viremia or clinical signs upon intranasal challenge with the virulent Snyder–Hill strain of CDV. Immunization of the dogs with measles virus ISCOMs, prepared from either purified F protein or purified virus, resulted in partial protection against disease following challenge infection with CDV (DeVries *et al.*, 1988b).

## 8.7. Rabies Virus

Rabies virus vaccines are generally tested for efficacy in the National Institutes of Health (NIH) test (Atanasin, 1973; Seligman, 1978). In this test, mice are immunized twice 2 weeks apart. Such an immunization scheme is unfavorable for ISCOM vaccines, as we have found that a considerably longer interval is much better. For instance, with ISCOMs containing gp160 of HIV, an interval of 6 weeks was optimal, while no or hardly any booster effect was observed when the interval was only 2 weeks. In such an NIH test, rabiesvirus ISCOMs, containing the 80-kDa G protein have been tested [F. Makonnen Centers of Disease Control (CDC), Atlanta, personal communication] and by Osterhaus (1986). In a number of independent experiments with ISCOMs prepared from different rabiesvirus strains, it was found that the rabies ISCOM vaccines induced virus-neutralizing antibodies and protection in mice at least as efficient as with whole rabies virus (Osterhaus *et al.*, 1986). In an *in vitro* system for human antibody production against rabies virus, it was shown that ISCOMs were more efficient than the virus at inducing rabiesvirus-specific antibodies (UytdeHaag *et al.*, 1983; Osterhaus *et al.*, 1986). (Fig. 9).

In countries in which rabies virus is endemic, only limited resources are available for mass vaccination; consequently, vaccination often has to be given after suspected exposure to infection. Therefore, mice were pre-exposed to 50–100 $LD_{50}$ doses of rabiesvirus before being subsequently vaccinated with ISCOMs containing 3 μg G protein per dose; 90% of the ISCOM-vaccinated mice survived the challenge infection, while the same dose of commercial rabies vaccine-induced protection at a level of 30%, using the same immunization procedure (Makonnen, personal communication).

Using standard techniques for the preparation of ISCOMs from concentrated virus harvests, it was found that the ISCOM preparations contained levels of contaminating nucleic acids below 10 pg per dose. This observation would in principle make it possible to produce high yields of rabiesvirus in continuous cell lines, e.g., BHK cells, for the production of human rabies vaccines (Osterhaus, personal communication). No nucleic acid-free rabies vaccine is available on the market.

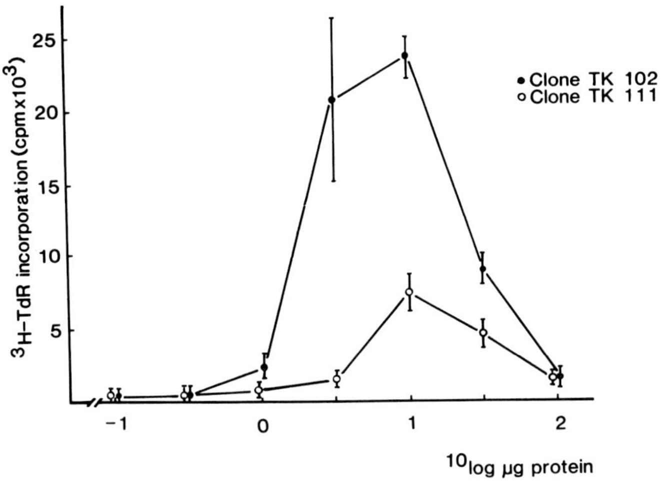


**FIGURE 8.** Proliferation of rabies virus-specific human T-cell clones upon stimulation with ISCOM.

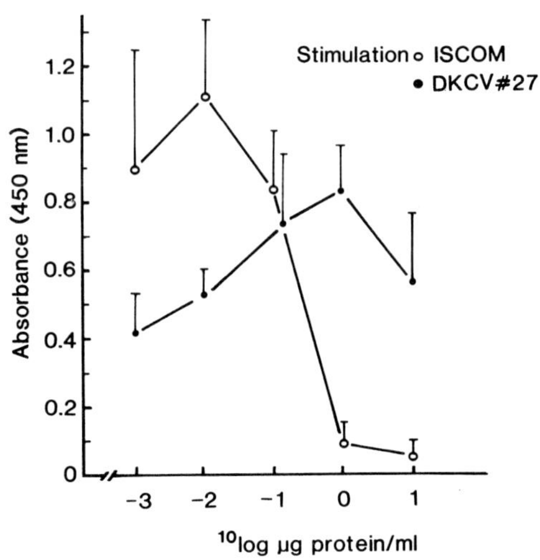


**FIGURE 9.** Secondary *in vitro* antibody (ELISA) synthesis by human PBL upon stimulation with rabies vaccine.

## 8.8. Equine Influenzavirus

Equine influenzaviruses, belonging to the Orthomyxoviridae family, are pathogens of the ciliated respiratory tract of horses, like the influenzaviruses infecting man. In susceptible horses, the inhalation and subsequent replication of virus may produce multiple epithelial wounds and disruption of the clearance mechanism of the ciliated epithelial cells. This damage of the epithelial surface might become the site for secondary infections. Two subserotypes infect horses: A/eq 1 (H7N7) and A/eq 2 (H3N8). The latter is the most pathogenic virus, causing worldwide epidemics, mostly among young, but also among old horses. Despite the fact that horses have been vaccinated with available vaccines containing A/eq 2 (H3N8) strains, they became infected, resulting in the development of clinical signs. The conventional vaccines induce a short-lived immune response with protective levels of antibodies lasting less than 6 months or even 4 months. Thus, a need for an improvement of available vaccines is obvious.

In order to study the immunogenicity of influenza virus ISCOMs, a monovalent experimental vaccine was prepared with the envelope glycoproteins hemagglutinin (HA) and neuraminidase (NA) of the equine virus strain A/eq 2/Solvalla/79, which had caused an epidemic among horses in Sweden. Six Welsh Mountain yearling ponies, seronegative for influenzavirus, were vaccinated with 15 μg HA per dose (Mumford *et al.*, 1986). The six ponies received four doses of ISCOMs administered with 8-, 27-, and 52-week intervals between doses. Six nonvaccinated ponies were included in the study as a control. The ISCOM vaccine induced comparatively high hemagglutination-inhibition (HI)-antibody response; these titers were detectable during a 50-week period after the third immunization. In a challenge experiment, a recent A/eq 2 H3N8 isolate A/eq 2/Newmarket/79 was used for challenge infection 10 weeks after the last vaccination. All nonvaccinated animals developed disease symptoms as assessed by measurement of pyrexia and coughing in contrast to all ISCOM-vaccinated animals, which were fully protected from the disease (Hannant *et al.*, 1987; Sundquist *et al.*, 1988b).

## 8.9. Feline Leukemia Virus

Oncogenic RNA viruses cause leukemia and sarcoma in several animal species. Many attempts have been made to produce vaccines against retroviruses (Morein *et al.*, 1986; Ferdinand *et al.*, 1987; Merza *et al.*, 1989; Åkerblom *et al.*, 1989). One such virus is the feline leukemia virus (FeLV), which causes a transient, persistent, latent viremia in experimentally and naturally infected cats. In a FeLV subunit vaccine, the envelope gp70/85 antigen was incorporated into ISCOMs. Immunization of specific pathogen-free (SPF) cats resulted in virus-neutralizing antibodies. A proportion of these antibodies proved to be directed to a neutralizing epitope on gp70, which was shown in competition ELISA with a virus-neutralizing monoclonal antibody, which reacts with all three subtypes of FeLV (Osterhaus *et al.*, 1985). The protective effect of FeLV ISCOM was further shown in vaccination experiments in which six 8-week-old SPF cats were immunized with FeLV ISCOMs and subsequently challenge-infected oronasally. All the cats developed virus-neutralizing antibodies, but none of the vaccinated cats developed viremia, while four of the six nonvaccinated control cats became viremic. To assess the potential of the ISCOM further as a vaccine against FeLV, a comparative trial was conducted with a commercial FeLV vaccine containing the same amount of glycoprotein gp70/85 per dose. Three groups of privately owned cats, kept under conventional conditions in single- or multiple-cat households, were vaccinated three times with either the commercial vaccine, FeLV ISCOM, or a control preparation not containing FeLV antigen. Serological responses were measured by ELISA, membrane immunofluorescence, virus neutralization, and Western blotting both before and after vaccination. Less than 6% of the cats vaccinated with the commercial vaccine developed a serological response, and there was no statistical difference from the control group. By contrast, more than 80% of the animals vaccinated with the ISCOM preparation responded in all tests (Osterhaus *et al.*, 1987; Osterhaus, *et al.*, 1989). The experiments reported clearly demonstrated that ISCOMs of viral antigens could induce protection from clinical signs, caused by viral infections, e.g., lethal disease, abortion, pneumonia, and tumorgenesis.

## 9. ISCOMS AS ANTIGENS IN IMMUNOASSAYS

A likely reason for the enhanced immunogenicity of antigens in the ISCOMs is that the antigens are well exposed on the surface. Consequently, the ISCOM should also be an effective structure for use as antigen in immunoassays. Since the membrane proteins on the surface of microorganisms, e.g., virus envelope, often are species specific, in contrast to many of the internal proteins (Maizels *et al.*, 1982); these proteins are the antigens of choice for specific immunoassays. Within this context, the ISCOM is an interesting structure for immunoassays because the procedure for preparation is selective for membrane proteins. This concept was recently tested for *Toxoplasma gondii* using the surface membrane proteins as antigen in an ELISA to measure serum antibody to this parasite in sheep and in calves. For detection of low levels of antibody to *T. gondii*, the ISCOM ELISA was found to be as sensitive as an immunofluorescence technique (IFAT) employing intact parasites or a conventional antigen ELISA, in which a whole-cell extract of the parasite is used as antigen. However, the ISCOM ELISA was found more specific than a conventional *Toxoplasma* ELISA, as exemplified by the fact that the conventional ELISA recorded antibody to the related parasite *Sarcocystis cruzi*. By contrast, no cross-reactions were observed using the IFAT or the

ISCOM ELISA (Uggla *et al.*, 1989). It was concluded that the ISCOM ELISA will combine the specificity of the IFAT with the simplicity and objectivity of the conventional ELISA. We routinely use the ISCOM technology to produce antigen for ELISA to assay for subserotype-specific response to influenza and parainfluenza viruses.

## 10. EPILOGUE

The ISCOM technology offers a way to enhance both antibody-mediated immunity (AMI) and cell-mediated immunity (CMI). This enhancement results in longer lasting immunity, thereby reducing the required number of immunizations—an important attribute for nonreplicating vaccines. As regards CMI, it is worth noting that the ISCOMS induce cytotoxic T-memory cells; this is unique for a nonreplicating vaccine form. The technology has been considered to have great significance for viral vaccines because of its importance in protective immunity. Another interesting factor is the use of ISCOMs for local application of the nonreplicating antigens that are important for evoking effective local immunity. Until now there have been few alternatives to replicating vaccines for induction of local immunity. This is partly due to a lack of adjuvants or adjuvant systems that enhance immunogenicity in local application on mucus membranes. Defined antigens, regardless of whether they originate from microorganisms propagated conventionally or by the aid of gene technology or chemical synthesis, require a system for antigenic presentation in order to optimally elicit immune response. This immunological response should in most cases include both B- and T-cell response. The construct of the ISCOM can make up the required antigen-presenting system. The first ISCOM vaccine for animal use is already on the market, and it is likely that a vaccine for humans will follow.

## 11. REFERENCES

Åkerblom, L., and Villacres, M., 1987, HIV iscoms: A vaccine model, in: *Tenth Scandinavian Virus Symposium,* Umeå, p. 72.

Åkerblom, L., Höglund, S., Strömstedt, K., Osterhaus, A., and Morein, B., 1989, Formation and characterization of FeLV iscoms, *Vaccine* (in press).

Alenius, A., Sundquist, B., and Carlsson, U., 1989, Protective effect of an iscom bovine diarrhea (BVD) virus vaccine against an experimental BVD virus infection in pregnant sheep, in preparation.

Atanasin, P., 1973, Quantitative assay and potency test of anti-rabies serum and immunoglobulin, in: *Laboratory Techniques in Rabies* (M. Kaplan and H. Koprowski, eds.), pp. 314–318, World Health Organization, Geneva.

Audibert, F., Jolivet, M., Chedid, L., Arnon, R., and Sela, M., 1982, Successful immunization with a totally synthetic diphtheria vaccine, *Proc. Natl. Acad. Sci. USA* **79:**5042–5046.

Bangham, A. D., and Horne, R. W., 1962, Action of saponin on biological cell membranes, *Nature (London)* **196**:952–953.

Barnes, D. M., 1988, Obstacles to an AIDS vaccine, *Science* **240**:719–721.

Bomford, R., 1980, The comparative selectivity of adjuvants for humoral and cell-mediated immunity. II. Effect of delayed-type hypersensitivity in the mouse and guinea pig, and cell-mediated immunity to tumour antigens in the mouse for Freund's incomplete and complete adjuvants, Alhydrogel, *Corynebacterium parvum, Bordetella pertussis,* muramyl dipeptide and saponin, *Clin. Exp. Immunol.* **39**:435–441.

Brieskorn, C. H., and Briner, M., 1954, On the analysis of triterpenes. Colorimetric determination of the total content of triterpenoid acids in labiatae. *Arch. Pharm. Detsch. Pharm. Ges.* **287**: 429–432.

Dalsgaard, K., 1970, Thin layer chromatographic fingerprinting of commercially available saponins, *Dansk. Tidskr. Farm.* **44**:327–331.

Dalsgaard, K., 1974, Saponin adjuvants. III. Isolation of a substance from *Quillaja saponaria* Molina with adjuvant activity in foot-and-mouth disease vaccines. *Arch. Ges. Virusforsch.* **44**: 243–254.

Dalsgaard, K., 1978, The application of the saponin adjuvant Quil A in foot-and-mouth disease vaccines, *Bull. Off. Int. Epiz.* **89**:963–966.

Dalsgaard, K., 1984, Assessment of the dose of the immunological adjuvant Quil A in mice and guinea pigs using sheep red blood cells as model antigens, *Zbl. Vet. Med.* **B31**:718–720.

Dalsgaard, K., Jensen, M. H., and Sörensen, K. J., 1977, Saponin adjuvants. IV. Evaluation of the adjuvant Quil A in the vaccination of cattle against foot-and-mouth disease, *Acta Vet. Scand.* **18**:349–360.

DeVries, P., Van Binnendijk, R. S., Van der Marel, P., Van Wezel, A. L., Voorma, H. O., Sundquist, B., Uytdehaag, F. G. C. M., and Osterhaus, A. D. M. E., 1988a, Measles virus fusion protein presented in an immune-stimulating complex (Iscom) induces hemolysis-inhibiting and fusion-inhibiting antibodies, virus-specific T cells and protection in mice, *J. Gen. Virol.* **69**:549–559.

De Vries, P., Uytdehaag, F. G. C. M., and Osterhaus, A. D. M. E., 1988b, Canine distemper virus (CDV) iscoms, but not measles virus iscoms, protect dogs against CDV infection, *J. Gen. Virol.* **69**:2071–2083.

Dourmashkin, R. R., Dougherty, R. M., and Harris, R. J., 1962, Electronmicroscopic observations on Rous sarcoma virus and cell membranes, *Nature (Lond.)* **194**:1116–1119.

Erlanger, B. F., 1980, The preparation of antigenic hapten-carrier conjugates; a survey, *Methods Enzymol.* **70**:85–104.

Espinet, E. G., 1951, Nuevo tipo de vacuna antiaftosa a complejo glucovirico, *Gac. Vet.* **74**:1–13.

Ferdinand, F. J., Dorner, F., and Kurth, R., 1987, Perspectives of HIV vaccine developments, *J. Virol. Methods* **17**:63–67.

Fohlman, J., Ilbäck, N. G., Friman, G., and Morein, B., 1989, Vaccination of BALB/c mice against enteroviral mediated myocarditis, *Vaccine* (submitted).

Freund, J., 1956, The mode of action of immunological adjuvants, *Adv. Tuberc. Res.* **7**:130–148.

Gelderblom, H., Reupke, H., and Pauli, G., 1985, Loss of envelope antigens of HTLV-III/LAV, a factor in AIDS pathogenesis, *Lancet* **2**:1016–1017.

Gelderblom, H. R., Hausmann, E. H. S., Özel, M., Pauli, G., and Koch, M. A. K., 1987, Fine structure of human immunodeficiency virus (HIV) and immunolocalization of structural proteins, *Virology* **156**:171–176.

Glauert, A. M., Dingle, J. T., and Lucy, J. A., 1962, Action of saponin on biological cell membranes, *Nature (Lond.)* **196**:953–955.

Hanatani, M., Nishifuji, K., Futai, M., and Tsuchia, T., 1984, Solubilization and reconstitution of membrane proteins of *Escherichia coli* using alkanoyl-*N*-methylglucamides, *J. Biochem.* **95**: 1349–1353.

Hannant, D., Jessett, D. M. O. Neill, T., Sundquist, B., and Mumford, J. A., 1987, Nasopharyngeal, tracheobronchial, and systemic immune responses to vaccination and aerosol infection with equine influenzavirus (H3N8), in: *Seventh Equine Infection Disease Conference, Kentucky.*

Helenius, A., and Simons, K., 1975, Solubilization of membranes by detergents, *Biochim. Biophys. Acta.* **415:**29–80.

Helenius, A., and von Bornsdorf, C. H., 1976, Semliki Forest virus proteins. Preparation and characterization of spike complexes soluble in detergent-free medium, *Biochim. Biophys. Acta* **436:**895–899.

Helenius, A., McCaslin, D. R., Fries, E., and Tanford, C., 1979, Properties of detergents, *Methods Enzymol.* **56:**734–749.

Henderson, L. E., Sowder, R., Copeland, T. D., Oroszlan, S., Arthur, L. O., Robey, W. G., and Fischinger, P. J., 1987, Direct identification of class II histocompatibility DR proteins in preparations of human T-cell lymphotropic virus type III, *J. Virol.* **61:**629–632.

Hildreth, J. E. K., 1982, *N*-D-Gluco-*N*-methylalkanamide compounds, a new class of non-ionic detergents for membrane biochemistry, *Biochem. J.* **207:**363–366.

Höglund, S., Åkerblom, L., Gelderblom, H., Özel, M., 1986a, Characterization of HTLV-III iscom, in: *Proceedings of the International Conference on AIDS* (J. C. Gluckman, eds.), p. 13.

Höglund, S., Morein, B., Sundquist, B., Gelderblom, H., and Özel, M., 1986b, Immunostimulating complex (iscom) of viral envelope, in: *Proceedings of the International Congress of Electron Microscopy* (T. Imura, S. Maruse, and T. Suzuki, eds.), *Kyoto, Japan,* pp. 3403–3404.

Höglund, S., Özel, M., Åkerblom, L., Villacrez, M., Gelderblom, H., and Morein, B., 1989, A construct of HIV antigens: Structural and immunological analysis, submitted.

Hoxie, J. A., Fitzharris, T. P., Youngbar, P. R., Matthews, D. M., Rackowski, J. L., and Radka, S. F., 1987, Nonrandom association of cellular antigens with HTLV-III virions, *Hum. Immunol.* **18:**39–52.

Jennings, R., Pemberton, R. M., Smith, T. L., Amin, T., and Potter, C. W., 1987, Demonstration of an immunosuppressive action of detergent-disrupted influenzavirus on the antibody response to inactivated whole virus vaccine, *J. Gen. Virol.* **68:**441–450.

Jones, P. D., Tha Hla, R., Morein, B., and Ada, G. L., 1988, Cellular immune response in the murine lung to local immunization with influenza A virus glycoproteins in micelles and iscoms, *Scand. J. Immunol.* **27:**645–652.

Letvin, L., Daniel, M. D., Kiyotaki, M., Kannagi, M., Chalifoux, L. V., Seghal, P. K., Desrosiers, R. C., Arthur, L., and Allison, A. C., 1987, AIDS-like disease in macaque monkeys induced by simian immunodeficiency virus: A vaccine trial, *Vaccine* **87:**209–213.

Liljas, L., 1986, The structure of spherical viruses, *Proc. Biophys. Mol. Biol.* **48:**1–36.

Lövgren, K., and Morein, B., 1988, The requirement of lipids for the formation of immunostimulating complexes (iscoms), *Biotechnol. Appl. Biochem.* **10:**161–172.

Lövgren, K., Lindmark, J., Pipkorn, R., and Morein, B., 1987, Antigenic presentation of small molecules and peptides conjugated to a preformed iscom as carrier, *J. Immunol. Methods* **98:** 137–143.

Lövgren, K., Uggla, A., and Morein, B., 1987b, A new approach to the preparation of a Toxoplasma gondii membrane antigen for use in ELISA, *J. Vet. Med.* **34:**274–282.

Lövgren, K., 1988, The serum antibody response distributed in subclasses and isotypes after intranasal and subcutaneous immunization with influenzavirus immunostimulating complexes, *Scand. J. Immunol.* **27:**241–245.

Lucy, J. A., and Glauert, A. M., 1964, Structure and assembly of macromolecular lipid complexes composed of globular micelles, *J. Mol. Biol.* **8:**727–748.

Maizels, R. M., Philipp, M., and Oglive, M. O., 1982, Molecules of the surface of parasitic nematodes as probes of the immune response in infection, *Immunol. Rev.* **61:**109–136.

McEven, C. R., 1967, Tables for estimating sedimentation through linear concentration gradients of sucrose solution, *Anal. Biochem.* **20:**114–149.
Merza, M. S., Linne, T., Höglund, S., Morein, B., Portetelle, D., and Burny, A., 1989, Bovine leukemia virus iscoms: Biochemical characterization, *Vaccine.*
Modrow, S., Hahn, B., Shaw, G. M., Gallo, R. C., Wong-Stahl, F., and Wolf, H., 1987, Computer-assisted analysis of envelope protein sequences of seven human immunodeficiency virus isolates: Prediction of antigenic epitopes in conserved and variable regions, *J. Virol.* **61:**570–578.
Morein, B., and Simons, K., 1985, Subunit vaccines against enveloped viruses: Virosomes, micelles and other protein complexes, *Vaccine* **3:**83–93.
Morein, B., Helenius, A., Simons, K., Pettrson, R., Kääriäinen, L., and Schirrmacher, V., 1978, Effective subunit vaccines against enveloped animal virus, *Nature (Lond.)* **276:**715–718.
Morein, B., Sundquist, B., Höglund, S., Helenius, A., and Simons, K., 1982, Protein micelles and virosomes from the surface glycoproteins of parainfluenza-3 virus, in: *Protides of the Biological Fluids* (H. Peeters, ed.), pp. 101–104, Pergamon Press, Oxford.
Morein, B., Sharp, M., Sundquist, B., and Simons, K., 1983, Protein subunit vaccines of Parainfluenza type 3 virus: Immunogenic effect in lambs and mice, *J. Gen. Virol.* **64:**1557–1569.
Morein, B., Sundquist, B., Höglund, S., Dalsgaard, K., and Osterhaus, A., 1984, Iscom, a novel structure for antigenic presentation of membrane proteins from enveloped viruses, *Nature (Lond.)* **308:**457–460.
Morein, B., Merza, M., Höglund, S., and Osterhaus, A., 1986, An approach to retrovirus vaccines, in: *Proceedings of the Retrovirus of Human AIDS and Related Animal Diseases;* "Colloque des Cent Gardes" *Paris*, pp. 168–171, (L. Montagnier), Paris.
Morein, B., Lövgren, K., Höglund, S., and Sundquist, B., 1987, The Iscom: An immunostimulating complex, *Immunol. Today* **8:**333–338.
Morein, B., Elkström, J., and Lörgren, K., 1989a, Inclusion of non-amphipathic protein (BSA) into ISCOMS increased its immunogenicity, submitted for publication.
Morein, B., Soos, T., Rusvai, M., McGwire, B. S., Bognar, K., and Belak, S., 1989b, Efficacy of immunostimulating complexes for controlling pseudorabies (Aujeszky's disease),
Morgan, A. J., Finerty, S., Lövgren, K., Scullion, F. T., and Morein, B., 1988b, Prevention of Epstein–Barr (EB) virus induced lymphoma in cottontop tamarins by vaccination with the EB virus envelope glycoprotein gp340 incorporated into iscoms, *J. Gen. Virol.* **69:**2093–2096.
Moss, B., Fuerst, T. R., Flexner, C., and Hugin, A., 1988, Roles of vaccinia virus in the development of new vaccines, *Vaccine* **6:**161–163.
Mumford, J. A., Cook, R. F., Hannant, D., Jesset, D. M., Wood, J. M., and Sundquist, B., 1986, The evaluation of antigen presentation methods and adjuvants for the enhancement of antigenicity and immunogenicity of equine influenza vaccines, in: *Proceedings of the First International Veterinary Immunology Symposium, Guelph, Canada.*
Murray, K., 1988, Application of recombinant DNA techniques in the development of viral vaccines, *Vaccine* **6:**164–174.
Newmark, P., 1988, Receding hopes of AIDS vaccines, *Nature (Lond.)* **33:**p. 699.
Osterhaus, A., 1988, Induction of anti-FeLV antibodies in cats with vaccinated virus and FeLV iscom,
Osterhaus, A., Weijer, K., Uytdehaag, F., Jarrett, O., Sundquist, B., and Morein, B., 1985, Induction of protective immune response in cats by vaccination with feline leukemia virus iscom, *J. Immunol.* **135:**591–596.
Osterhaus, A. D. M. E., Sundquist, B., Morein, B., Steenis, G., and van Uytdehaag, F., 1986, Comparison of an experimental rabies iscom subunit vaccine with inactivated dog kidney cell vaccine, in: *Proceedings of the First International Veterinary Immunology Symposium, Guelph, Canada.*
Osterhaus, A., Weijer, K., Uytdehaag, F., Knell, P., Jarrett, O., and Morein, B., 1987, Comparison

of serological responses in cats vaccinated with two different FeLV vaccine preparations, *Vet. Rec.* **121:**260.

Özel, M., Pauli, G., and Gelderblom, H., 1988, The organization of the envelope projections on the surface of HIV, *Arch. Virol.*

Özel, M., Höglund, S., Gelderblom, H., and Morein, B., 1989, The construct of iscom, in preparation.

Palfreyman, J. W., Aitcheson, T. C., and Taylor, P., 1984, Guidelines for the production of polypeptide specific antisera using small synthetic oligopeptides as immunogens, *J. Immunol. Methods* **75:**383–393.

Pyle, S. W., Morein, B., Bess, J., Nara, P. L., Nigida, S. M., Lerche, N., Fischinger, P., Arthur, L., 1989, Immune response to immunostimulatory complexes (ISCOMs) prepared from human immunodeficiency virus type 1 (HIV-1) or the HVI-1 external envelope glycoprotein (gp120) *Vaccine* (submitted).

Seligman, E. B., 1978, *Laboratory Techniques in Rabies,* 3rd ed., World Health Organization, Geneva.

Skoglund, U., Andersson, K., Strandberg, B., and Daneholt, B., 1986, Three-dimentional structure of a specific pre-messenger RNP particle established by electron microscope tomography, *Nature* **319:**560–564.

Speijers, G. J. A., Danse, L. H. J. C., Beuvery, J. J. T. W., Strik, A., and Vos, J. G., 1987, Local reactions of the Saponin Quil A and a Quil A containing iscom measles vaccine after intramuscular injection of rats: A comparison with the effect of DPT-polio vaccine, *Fund. Appl. Toxicol.* **10:**425–430.

Sundquist, B., Dalsgaard, K., and Morein, B., 1983, Assay of detergents by rocket electrophoresis in agarose gels containing red blood cells: ''Rocket hemolysis,'' *Biochim. Biophys. Res. Commun.* **114:**699–704.

Sundquist, B., Lövgren, K., Höglund, S., and Morein, B., 1988a, Influenzavirus iscoms: Biochemical characterization, *Vaccine* **6:**44–48.

Sundquist, B., Morein, B., and Jonsson, K., 1988b, Registration file for a horse influenza virus vaccine in Sweden.

Tanford, C., and Reynolds, J. A., 1976, Characterization of membrane proteins in detergent solutions, *Biochim. Biophys. Acta.* **457:**133–170.

Trudel, M., Nadon, F., Seguin, C., Simard, C., and Lussier, G., 1988, Experimental polyvalent iscom subunit vaccine based on the fusion protein induces antibodies that neutralize human and bovine respiratory syncytial virus, *Vaccine.*

Uggla, A., Araujo, F. G., Lunden, A., Lövgren, K., Remington, J. S., and Morein, B., 1989, Immunizing effects in mice of two *Toxoplasma gondii* preparations, *J. Vet. Med.* in press.

Uytdehaag, F. G. C. M., Osterhaus, A. D. M. E., Loggen, H. G., Bakker, R. H. J., Van Aasten, J. A. A. M., Kreeftenberg, J. G., Van der Marel, P., and Van Steenis, G., 1983, Induction of antigen-specific antibody response in human peripheral blood lymphocytes *in vitro* by a dog kidney cell vaccine against rabies virus (DKCV), *J. Immunol.* **131:**1234–1239.

Varsanyi, T. M., Morein, B., Löve, A., and Norrby, E., 1987, Protection against lethal Measles virus infection in mice by immune-stimulating complexes containing the hemagglutinin or fusion protein, *J. Virol.* **61:**3896–3901.

Wahren, B., Nordlund, S., Åkersson, A., Sundquist, B., and Morein, B., 1987, Monocyte and iscom enhancement of cell-mediated response to cytomegalovirus, *Med. Microbiol. Immunol.* **176:**13–17.

*Chapter 3*

# Biological Activity of Damavaricin Derivatives

Kazukiyo Onodera, Shin-ichi Ito, and Kazuya Sasaki

## 1. INTRODUCTION

The streptovaricins (Fig. 1) are a family of structurally related antibiotics, differing in the degree of oxidation or acetylation at specific carbon atoms in the common skeleton, obtained from the fermentation broth of *Streptomyces spectabiris,* an actinomycete isolated in 1957 (Siminoff *et al.*, 1957). Streptovaricins belong to the ansamycin class of antibiotics. The ansamycin antibiotics contain an ANSA group (Lüttringhaus and Grelher, 1941)—an aromatic nucleus with an aliphatic carbon chain bridging two nonadjacent carbons. The streptovaricins share this property with a number of other compounds: rifamycin (Sensi *et al.*, 1959, 1961), tolypomycin (Hartman *et al.*, 1968; Kishi *et al.*, 1969, 1972), naphthomycins (Balerna *et al.*, 1969; Williams, 1975), geldanamycin (DeBoer *et al.*, 1970), and matansinoids such as matansine (Kupchan *et al.*, 1972a), maytanprine and maytanbutine (Kupchan *et al.*, 1972b), and colubrinol (Wani *et al.*, 1973). Recently, herbimycin (Omura *et al.*, 1979), macbecin (Muroi *et al.*, 1980), ansathriens (Sugita *et al.*, 1982), and ansamitocin (Asai *et al.*, 1978) were added to this group (see Fig. 1).

**Kazukiyo Onodera** Department of Agricultural Chemistry, The University of Tokyo, Yayoi 1-1, Bunkyo-ku, Tokyo 113, Japan. **Shin-ichi Ito** Shin-Etsu Chemical Co., Ltd., Ohtemachi 2-6-1, Chiyoda-ku, Tokyo 100, Japan. **Kazuya Sasaki** Tokyo Institute of Kaken Pharmaceutical Co., Ltd., Honkomagome, Bunkyo-ku, Tokyo 113, Japan.

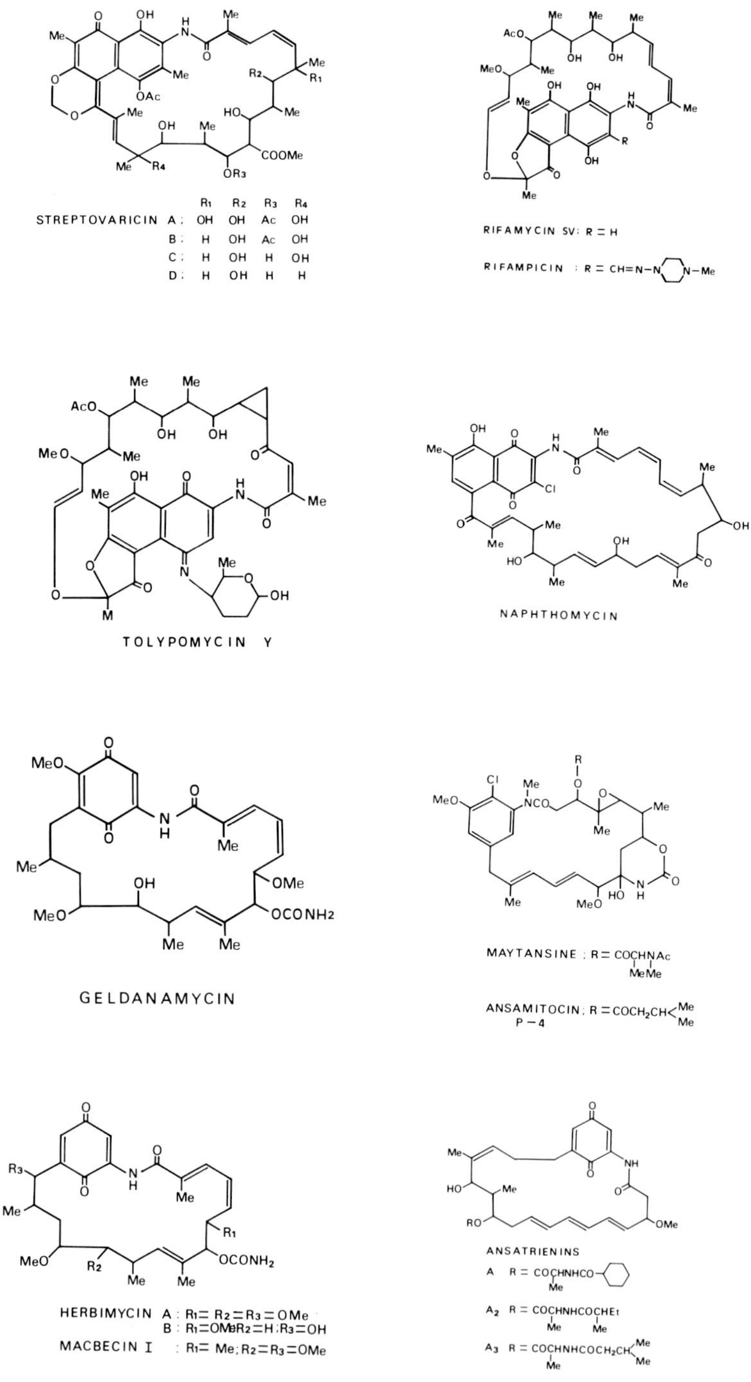


**FIGURE 1.** Structure of ansa compounds.

The streptovaricins were shown to have antibacterial activity against gram-positive and gram-negative bacteria (Siminoff *et al.*, 1957) and mycobacteria (Rhuland *et al.*, 1957) due to their ability to interact with the bacterial DNA-directed RNA polymerase (Yamazaki *et al.*, 1968a,b; Mizuno *et al.*, 1968a–c). In addition, certain members of the streptovaricin family have been shown to have antiviral activity against *Escherichia coli* infected by bacteriophage T4 (Mizuno and Nitta, 1969) or Qβ (Haruna *et al.*, 1970), poxvirus (Quintrell and McAuslan, 1970), and RNA tumor viruses (Borden *et al.*, 1971; Carter *et al.*, 1971, 1972). At least a portion of the latter antiviral activity may be attributable to the ability of the streptovaricins (Brockman *et al.*, 1971; Carter *et al.*, 1971; Rinehart, 1972) to inhibit the RNA-directed DNA polymerase (reverse transcriptase) found in RNA tumor virus (Temin and Mizutani, 1970; Baltimore, 1970).

Excellent reviews have appeared on the chemistry and biological activity of the streptovaricins and the other ansamycins (Rinehart and Shields, 1976; Milavets and Carter, 1977). This chapter describes the biological activity of damavaricin derivatives obtained from streptovaricins (Rinehart *et al.*, 1976) with special emphasis on their antitumor and antiviral activity.

## 2. CHEMISTRY AND ANTIBACTERIAL ACTIVITY OF DAMAVARICIN C DERIVATIVES

Crude streptovaricin complex was supplied by the Upjohn Company (Kalamazoo, Michigan). Streptovaricin C, the most abundant component of the streptovaricin complex, was isolated in pure crystals by a combination of silica gel chromatography and gel filtration on Sephadex LH-20 with acetone.

Upon bubbling air or oxygen through a solution of streptovaricin C, it was converted to an oxidative degradation product, damavaricin C. The nuclear magnetic resonance (NMR) spectrum of damavaricin C lacked both the methylenedioxy proton at δ5.18 (d) and δ5.83 (d) as well as the O-acetyl proton at δ1.96 present in streptovaricin C. Moreover, the infrared (IR) spectrum of damavaricin C lacked the absorption corresponding to an O-acetyl group on an aromatic ring observed at 1765 $cm^{-1}$ in streptovaricin C.

Therefore, the loss of $C_3H_2O$ during the reaction is ascribed to the hydrolysis of the methylenedioxy and the O-acetyl groups, while the loss of $H_2$ results from the autoxidation of a 1,4-dihydronaphthoquinone to a 1,4-naphthoquinone (Fig. 2).

Damavaricin C has a newly formed phenolic hydroxy group at the C-19 position of the naphthoquinone ring, thus enabling the preparation of the series of 19-O-substituted derivatives, as listed in Table I. The etherification of damavaricin C was carried out in good yield by treatment with silver oxide and a bromo- or iodo-compound. The structures of a series of 19-O-substituted deriva-

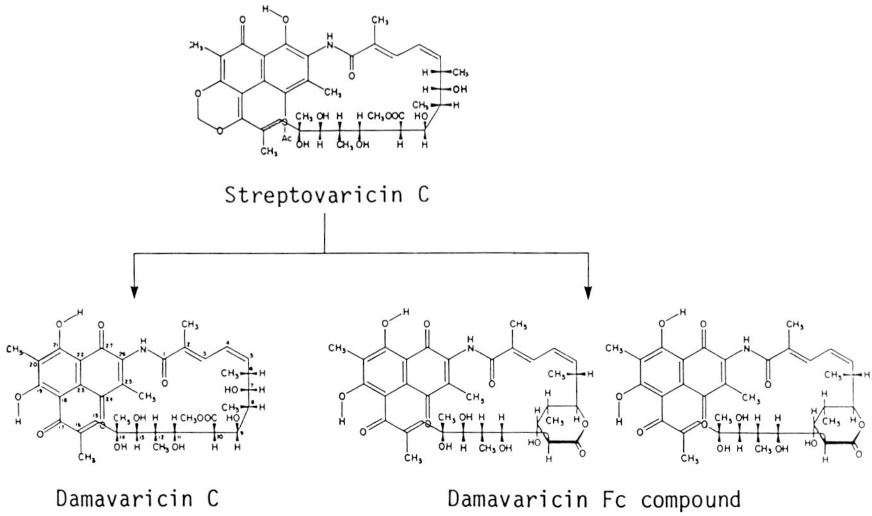


**FIGURE 2.** Conversion of streptovaricin C into damavaricin C and Fc compound.

tives were confirmed on the basis of chemical behavior and physical data, such as NMR and IR, ultraviolet (UV), and mass spectra.

Three parameters for assessing the biological activity of damavaricin C derivatives were employed: (1) antibacterial activity *in vitro,* (2) inhibitory activity against *E. coli* RNA polymerase *in vitro,* and (3) antibacterial activity against membrane mutants of *E. coli in vitro.*

Table I shows that damavaricin C has reduced antibacterial activity relative to streptovaricin C, while many of the derivatives with an ether linkage at the C-19 position of the naphthoquinone ring have equal or stronger antibacterial activity. In this series of *n*-alkyl ethers, the biological activity against *Staphylococcus aureus* and *Bacillus subtilis* seems to be related to the number of the methylene units in the alkyl chain. The antibacterial activity against *Mycobacterium smegmatis* does not closely parallel the activity against the other bacteria. It is noteworthy that none of the derivatives examined showed activity against *E. coli* or *Pseudomonas aeruginosa* at 100 μg/ml.

It is interesting to note that damavaricin C, which has low antibacterial activity, inhibited RNA polymerase almost to the same degree as streptovaricin C (data not shown). The question arises: Why are damavaricin C derivatives that block RNA polymerase unable to inhibit *E. coli* growth?

One plausible explanation is that the permeability barriers of *E. coli* and *P. aeruginosa* toward the derivatives are quite different from those of other bacteria. Hence, mutant *E. coli* with altered surface structures may be sensitive to some derivatives. Three kinds of mutants were examined. *Tol*A, B, and C are

**Table I**
**Antibacterial Activity (μg/ml) of Damavaricin C Derivatives *in Vitro*[a]**

| Compound | *S. aureus* FDA 209P | *B. subtilis* | *Mycobact.* ATCC607 | *E. coli* |
|---|---|---|---|---|
| Streptovaricin C | 2 | 10 | >100 | >100 |
| H—O—R | 50 | >100 | >100 | >100 |
| $CH_3$—O—R | 1 | 10 | 20 | >100 |
| $CH_3CH_2$—O—R | 1 | 0.5 | 20 | >100 |
| $CH_3(CH_2)_2$—O—R | 10 | 5 | 5 | >100 |
| $(CH_3)_2$—CH—O—R | 50 | 50 | 50 | >100 |
| $CH_3(CH_2)_3$—O—R | 10 | 5 | 20 | >100 |
| $CH_3(CH_2)_4$—O—R | 10 | 5 | 20 | >100 |
| $(CH_3)_2CHCH_2CH_2$—O—R | 10 | 5 | 20 | >100 |
| $CH_3(CH_2)_5$—O—R | 5 | 2 | 2 | >100 |
| $CH_3(CH_2)_6$—O—R | 2 | N.D. | 0.5 | >100 |
| $CH_3(CH_2)_7$—O—R | >100 | 50 | 10 | >100 |
| $CH_3(CH_2)_8$—O—R | >100 | N.D. | 5 | >100 |
| $CH_3(CH_2)_{10}$—O—R | >100 | 50 | N.D. | >100 |
| $CH_3(CH_2)_{11}$—O—R | >100 | >100 | 20 | >100 |
| $CH_3(CH_2)_{17}$—O—R | >100 | >100 | N.D. | >100 |
| $HOCH_2CH_2$—O—R | 50 | 50 | >100 | >100 |
| Cl—$CH_2CH_2$—O—R | 1 | 1 | N.D. | >100 |
| $CH_2$—CH—$CH_2$—O—R<br>\ /<br>O | 10 | N.D. | 2 | >100 |
| Cl—$CH_2CH_2CH_2$—O—R | 5 | 2 | 20 | >100 |
| $CH_2$=$CHCH_2$—O—R | 2 | 1 | 20 | >100 |
| $HOCH_2(CH_2)_3$—O—R | 5 | 5 | 0.5 | >100 |
| $(C_2H_5O)_2CHCH_2$—O—R | 20 | 50 | N.D. | >100 |
| $CH_2$=CH—O—$CH_2CH_2$—O—R | 10 | 20 | N.D. | >100 |

[a]R, the structure in the footnote to Table II; N.D., not determined.

membrane mutants that absorb colicins, protein antibiotics, on the cell surface but are not killed by them. These mutants were reported to show pleiotropic changes in permeability to varios antibiotics (Bernstein *et al.*, 1972; Rolfe and Onodera, 1971), and they are sensitive to varios damavaricin derivatives. However, another membrane mutant, *E. coli* deficient in phospholipase A, and the oleic and auxotropic mutants (Ohki *et al.*, 1972) are not sensitive to these derivatives (Table II). The cell membranes of colicin-tolerant mutant showed increased permeability toward those damavaricin derivatives that had no effect on the wild-type *E. coli* or the phospholipase A-deficient mutant and the oleic acid auxotropic mutant.

The results indicate that damavaricin C retains its original biological activity against the target enzyme (RNA polymerase) and that the groups introduced to the C-19 position of the molecule confer increased diffusibility to the derivatives,

**Table II**
**Antibacterial Activity (μg/ml) of Damavaricin C Derivatives against Membrane Mutants of *E. coli*[a]**

| Compound | *Tol*A, B mutants | *Tol*C mutants | Phospholipase A-deficient mutant | Oleic acid auxotroph |
|---|---|---|---|---|
| Streptovaricin C | 1 | 5 | >100 | 50 |
| H—O—R | 20 | 50 | >100 | N.D. |
| $CH_3$—O—R | 5 | 5 | >100 | 100 |
| $CH_3CH_2$—O—R | 2 | 5 | >100 | 100 |
| $CH_3CH_2CH_2$—O—R | 5 | 5 | >100 | >100 |
| $CH_3(CH_2)_5$—O—R | 10 | 100 | >100 | >100 |
| $CH_2$=$CHCH_2$—O—R | 5 | 5 | >100 | >100 |
| $CH_3COCH_2$—O—R | 5 | 10 | >100 | >100 |
| $PhCH_2$—O—R | 5 | 5 | >100 | >100 |
| $PhCH_2CH_2$—O—R | 10 | 50 | >100 | N.D. |

[a]R, represented by the following structure:

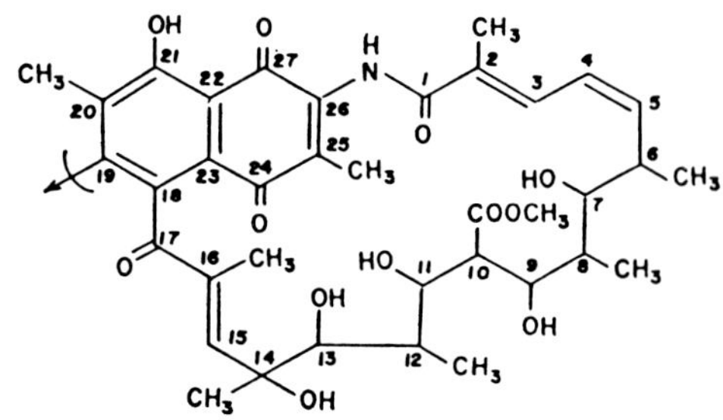

thereby facilitating cell penetration. These observations were the basis for the experiments in the following sections.

## 3. BIOLOGICAL ACTIVITY OF DAMAVARICIN C ON ANIMAL CELLS

### 3.1. Inhibition of Tumor Cell Growth *in Vitro* by Damavaricin C Derivatives

Two reports suggesting that streptovaricins have antitumor activity: first, the growth of mammary tumors was significantly inhibited by the oral administration of streptovaricin complex (Kramarski and Manthey, 1974), and second, Borden *et al.* (1971) reported that oral administration of the streptovaricin complex reduced Rausher leukemia virus (RLV)-induced splenomegaly in mice.

These results prompted us to search for derivatives that were more effective antitumor agents than the streptovaricin complex. Before assessing biological activity *in vivo,* it was necessary to obtain clues for selecting derivatives that were active *in vitro*. We employed untransformed and virus-transformed cells as an assay system.

Ethyl and *n*-undecyl derivatives of damavaricin C were the lethal to the virus-transformed cells, but corresponding untransformed cells showed less than 10% growth inhibition. Allyl, chloropropyl, *p*-phenylphenacyl, *p*-chlorophenacyl ethers, and aldol derivatives killed the virus-transformed cells preferentially, and the growth inhibition of untransformed cells was less than 20%. Derivatives such as *n*-dodecyl, acetonyl, phenethyl, and benzyl ether did not distinguish between untransformed and virus-transformed cells (Table III).

KB cells were chosen as the human-derived cancer cells. A method has been established to grow KB cells in the nude mouse. Thus, the screening of derivatives that kill KB cells *in vitro* leads to a convenient *in vitro* test system. Vinyloxyethyl, diethoxyethyl, *n*-octadecyl, and *p*-phenylphenacyl ethers showed lethal activity, and the aldol compound showed growth inhibition at 10 μg/ml. The derivatives active against KB cells differ from those active against mouse cells (Table IV).

The mode of action of those derivatives on animal cells is unclear. Two facts, are known, however, concerning the action of streptovaricin on animal

**Table III**
**Growth Inhibition of Normal and Virus-Transformed Cells *in Vitro* by Damavaricin C Derivatives (10 μg/ml)[a]**

| Derivative | Normal cells C3H-2K-C4 (× $10^6$) | Transformed cells W-2K-11 (× $10^6$) |
|---|---|---|
| Without derivative | 3.1 | 5.0 |
| Ethyl ether | 2.9 (6.5) | Lethal |
| *n*-Propyl ether | 2.4 (22.6) | Lethal |
| *n*-Butyl ether | 2.3 (25.8) | Lethal |
| *n*-Octyl ether | 2.1 (32.3) | Lethal |
| *n*-Nonyl ether | 1.0 (67.7) | Lethal |
| *n*-Undecyl ether | 2.9 (6.5) | Lethal |
| *n*-Dodecyl ether | 0.7 (77.4) | 1.7 (66) |
| Allyl ether | 2.7 (12.9) | Lethal |
| 3-chloro-*n*-propyl ether | 2.6 (16.1) | Lethal |
| Cychlohexyl ether | 2.7 (12.9) | 1.7 (66) |
| Acetonyl ether | 2.5 (19.4) | 1.8 (64) |
| Phenethyl ether | 2.4 (22.6) | 2.0 (60) |
| *p*-Phenylphenacyl ether | 2.5 (19.4) | Lethal |
| *p*-Chlorophenacyl ether | 2.5 (19.4) | Lethal |
| Aldol compound | 2.5 (19.4) | Lethal |
| Benzyl ether | 2.7 (12.9) | 1.2 (76) |

[a]Numbers in parentheses are percentages.

**Table IV**
**Growth Inhibition of KB Cells *in Vitro* by Damavaricin C Derivatives (10 μg/ml)**

| Derivative | Cell number per dish ($\times 10^6$) |
|---|---|
| Without derivative | 3.8 |
| Acetonyl ether | Lethal |
| *p*-Phenylphenacyl ether | Lethal |
| Aldol compound | 0.9 (76.3%) |
| 2-Vinyloxyethyl ether | Lethal |
| 2,2-Diethoxyethyl ether | Lethal |
| *n*-Octadecyl ether | Lethal |

cells. First, the RNA polymerase of eukaryotic cells was not inhibited by streptovaricins (Mizuno *et al.*, 1968). Second, streptovaricin D (H in place of OH at the C-14 position of streptovaricin C) selectively inhibited the nucleoside transport system in the HeLa cell membrane, whereas streptovaricin A (OAc at the C-11 position and OH at the C-6 position) and streptovaricin B (OAc at the C-11) did not show this effect (Tan and McAuslan, 1971).

The functional groups introduced at the C-19 position enhanced the ability of drugs to permeate the cell membrane of bacterial cells because of hydrophobicity of the derivatives (see Section 2). Therefore, two alternative explanations are possible for the effects of these derivatives on animal cells: (1) the functional groups introduced at the C-19 position may enhance the ability of drugs to impair membrane function, e.g., that of nucleoside transport; and (2) they confer increased permeability on compounds, enabling derivatives to reach their targets within mammalian cells, e.g., the mitochondria whose transcriptional machinery is suggested to have properties similar to those of bacterial cells in terms of antibiotic sensitivity (Wehli and Staehelin, 1971). Impairment of the exogenous nucleoside transport per se does not necessarily lead to cell death because the cells can make use of endogenous nucleoside for RNA synthesis. Therefore, we favor the second explanation.

### 3.2. Growth Inhibition of Human Lymphoid Cells *in Vitro* by *n*-Butyl Damavaricin C

Some of the damavaricin C derivatives were preferentially toxic against the virus-transformed cells *in vitro* (see Section 3.1). Then, we selected *n*-butyl damavaricin C (Bu-DvC) in a first screening toward human lymphoid cells *in vitro*.

Whereas the growth of MOLT-4B cells and ALL-6 cells was sensitive to Bu-DvC, P3HR1 cells and Raji cells were less sensitive (data not shown). Therefore, the viability of the cells was examined in the presence of the drug. It was shown that ALL-6 cells and MOLT-4B cells were killed in 2 days, but P3HR1 and Raji cells stopped cell growth and remained viable (Fig. 3). The recovery from the cessation of cell growth was observed after a 24-hr lag period, when the drug was removed by washing (data not shown).

To examine the mode of action of the drug on human cells, the effect of the drug on the macromolecular and phospholipid synthesis of lymphoid cells was studied. Inhibition of the biosynthesis of protein and RNA was observed within 4 hr and became distinct at 8 hr. DNA and phospholipid synthesis gradually stopped. The cells remained viable during these experiments (data not shown).

McAuslan and Tan (1971) reported that the transport of nucleoside into HeLa cells was inhibited by streptovaricin D within 20 min. These investigators suggested that streptovaricin D interacted with the cell membrane. Considering the results described in Section 3.1, it is conceivable that *n*-butyl damavaricin C acts on the cell membrane of human lymphoid cells.

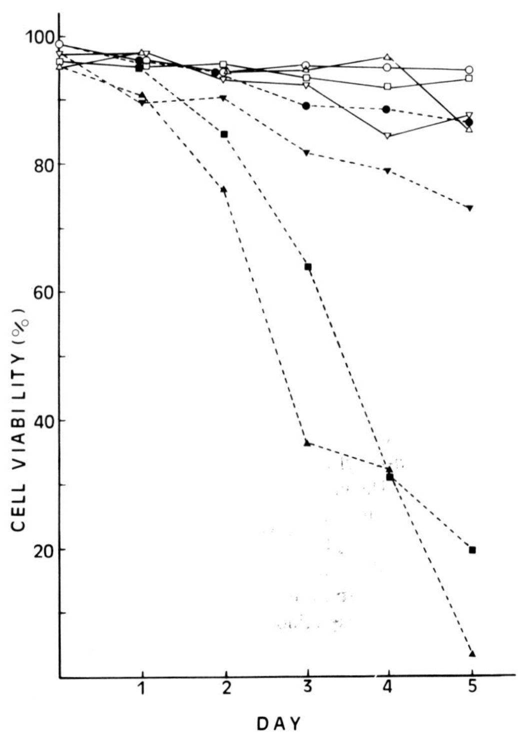


**FIGURE 3.** Effect of *n*-butyl damavaricin C (Bu-DvC) on the growth of human lymphoid cells. Human lymphoid cell lines such as Raji and P3HR1 are cell lines established from Burkitt's lymphoma, carrying Epstein–Barr virus (EBV) genome and expressing the B-cell character. MOLT-4B is a cell line derived from acute lymphoblastic leukemia (ALL), EBV genome negative, and exhibits the T-cell character. ALL-6 is a cell line established from ALL and express T-cell character. Cells were grown in RPMI-1640 medium supplemented with 10% fetal calf serum (FCS) and cultivated in a suspension culture. Viable cells were counted by the trypan blue exclusion test (unpublished data). P3HR1 cells (○) P3HR1 cells + Bu-DvC (●), MOLT-4B cells (△), MOLT-4B + BU-DvC (▲), ALL-6 cells (□), ALL-6 cells + Bu-DvC (■), Raji cells (▽), ans Raji cells + Bu-DvC (▼).

## 4. ANTITUMOR ACTIVITY *IN VIVO* OF DAMAVARICIN C DERIVATIVES

We have previously described the antitumor activity of some damavaricin C derivatives selected for their ability to inhibit preferentially the growth of virus-transformed and human-derived cancer cells *in vitro* at a relatively low concentration.

The derivatives that were active against ascites Sarcoma 180 *in vitro* were selected from the following compounds: streptovaricin A, B, C, D, and G and 19-O-substituted damavaricin C derivatives, such as ethyl ether, *n*-butyl ether, *n*-hexyl ether, 2-thenoxylmethyl ether, acetonyl ether, aldol compound, acetonyl ether oxime, acetonyl ether *O*-methyl oxime, and acetonyl ether di-*n*-pentylhydrazone. These compounds were tested using five mice per assay. The dose of each drug was 20 mg/kg per day, 24 hr after tumor implantation, and continuing for 6 days. Antitumor activity was determined by comparing the mean survival days (MSD) with that of control group. The percentage increase of life-span (%ILS) was calculated from the MSD.

The results were shown in Table V. Ethyl, *n*-butyl, 2-thenoylmethyl, acetonyl, aldol compound, hydroxyiminoacetonyl, and di-*n*-pentylhydrazonoacetonyl ether showed more than 20% ILS at the first screening. These seven compounds were retested for their antitumor activity against Sarcoma 180, using 10 mice per assay. *n*-Butyl and 2-thenoyl ether were most effective, and hydroxyiminoacetonyl ether to a lesser degree (data not shown).

**Table V**
**Effect of Damavaricin C Derivatives on Tumor Growth in Mice Bearing Ascites Sarcoma 180 *in Vivo***

| Derivative | Dose[a] (mg/kg) | MSD[b] (days ± SD) | ILS[c] (%) |
|---|---|---|---|
| Without derivative | — | 10.7 ± 1.7 | — |
| Ethyl ether | 20 | 14.2 ± 1.2 | 33 |
| *n*-Butyl ether | 20 | 12.8 ± 1.8 | 20 |
| *n*-Hexyl ether | 20 | 12.4 ± 1.4 | 16 |
| 2-Thenoylmethyl ether | 20 | 12.8 ± 2.2 | 20 |
| Acetonyl ether | 20 | 13.6 ± 1.6 | 27 |
| Aldol compound | 20 | 14.2 ± 1.2 | 33 |
| Hydroxyiminoacetonyl ether | 20 | 13.6 ± 1.6 | 27 |
| Methoxyiminoacetonyl ether | 20 | 11.6 ± 2.6 | 8 |
| Di-*n*-pentylhydrazonoacetonyl ether | 20 | 14.2 ± 1.2 | 33 |

[a]Administered intraperitoneally once daily for 6 consecutive days after injection of tumor cells.
[b]MSD, mean of survival days.
[c]Percentage increase of life span.

**Table VI**
**Effect of Damavaricin C Derivatives on Tumor Growth in Mice Bearing Solid Sarcoma 180**

| Derivatives | Dose[a] (mg/kd/day) | Average tumor weight[b] (g) | Inhibition of tumor growth (%) |
|---|---|---|---|
| Without derivative | — | 1.81 | — |
| *n*-Butyl ether | 5 | 1.51 | 17.1 |
| 2-Thenoylmethyl ether | 5 | 1.22 | 32.6 |
| Hydroxyiminoacetonyl ether | 5 | 1.02 | 43.6 |
| Di-*n*-pentylhydrazonoacetonyl ether | 5 | 1.74 | 3.9 |

[a]Administered subcutaneously once daily for 20 consecutive days beginning 24 hr after tumor implantation.
[b]Average tumor weight as measured on day 21.

Among the derivatives described above, *n*-butyl, 2-thenoylmethyl, hydroxyiminoacetonyl, and di-*n*-pentylhydrazonoacetonyl ethers were tested for their antitumor activity against solid Sarcoma 180 *in vivo*. In the case of the solid tumor, each drug was administered at a dose of 5 mg/kg per day. The tumors were excised on the 21st day, and the average tumor weight was measured. Acetonyl ether oxime and 2-thenoylmethyl ether showed 43.6% and 32.6% inhibition of tumor growth, respectively (see Table VI). *n*-Butyl ether was less effective and acetonyl ether and di-*n*-pentylhydrazone were not effective.

Some derivatives of damavaricin C were active against virus-transformed fibroblast cells, while the others were active against KB cells, the human-derived epithelial cancer cells *in vitro* (see Section 3). It was conceivable that some derivatives were active against one type of tumor and the others against a different type of tumor *in vivo*. Within this context, acetonyl ether and di-*n*-pentylhydrazone acetonyl ether were active against ascites Sarcoma 180 but not against solid Sarcoma 180. The dose used in this experiment was 20 mg/kg per day for ascites and 5 mg/kg per day for solid tumor, respectively. The dose was relatively low compared with the antitumor agents that are generally used. Moreover, *n*-butyl, 2-thenoylmethyl ether, and acetonyl ether oxime were not highly toxic to mice when administered either subcutaneously (SC) or orally (PO) (data not shown).

## 5. ANTIVIRAL ACTIVITY OF DAMAVARICIN Fc DERIVATIVES

In addition to antibacterial activity, inhibition of reverse transcriptase of RNA tumor virus was observed with streptovaricins (Rinehart *et al.*, 1974). This property has attracted much scientific interest. Damavaricin C derivatives retain their antibacterial activity but acquire cell killing activity for mammalian cells, including virus-transformed cells and human leukemia cells *in vitro*. Because of their cell-killing activity, we were unable to examine an intriguing property: the inhibition of focus formation by RNA tumor virus.

During the course of the preparation of damavaricin C by treatment of streptovaricin C with oxygenated concentrated ammonia–methanol (1 : 2), damavaricin Fc is produced. This compound was found to be an atropisomeric mixture of two compounds, damavaricin Fc and atropisodamavaricin Fc, at a ratio of 1 : 3, as determined from the peak intensities of the two phenolic hydroxy protons in the NMR spectrum (data not shown). Damavaricin Fc, in which the ansa bridge lies above the aromatic nucleus, has P helicity, as do the natural streptovaricins, whereas atropisodamavaricin Fc, in which the ansa bridge lies below the aromatic nucleus, has the M helicity (see Fig. 2). To avoid complicating nomenclature, hereafter we use damavaricin Fc. Damavaricin Fc has a δ-lactone ring between C-7 and C-10 of the ansa bridge, which has been proved to produce a change in one of its biological properties. Specifically, the loss of antibacterial activity implied that damavaricin Fc has lost its capacity to bind the RNA polymerase of prokaryotic cells. The inhibitory activity of the reverse transcriptase, however, was retained.

These observations have prompted us to prepare various derivatives of damavaricin Fc as well as damavaricin C and to examine their biological activity. It is of interest that damavaricin Fc derivatives are less toxic against animal cells *in vitro* as compared with damavaricin C derivatives (data not shown). Using these less toxic derivatives, we were able to test whether any of the compounds inhibited focus formation by mouse sarcoma virus or the mouse leukemia virus (MSV–MLV) complex.

Methyl, ethyl, and *n*-pentyl ether of damavaricin Fc was effective, whereas benzyl ether was not (see Table VII). The inhibitory activity of the derivatives for the reverse transcriptase (RDDP) *in vitro* was related to their inhibitory activity on focus formation by the MSV–MLV complex.

Next, the effect of the derivatives on the proliferation of MLV was examined. Normal rat cells were infected by MLV. After 4 days, the virions were collected from the culture fluid by centrifugation, and the proliferation of leukemia virus was compared by measuring the activity of the reverse transcriptase in the virions. The derivatives used in this experiment were not toxic against host cell growth, but the growth of the virus-infected cells was slightly inhibited. (see Table VIII). Although methyl, *n*-hexyl, *n*-undecyl ether, and damavaricin Fc itself did not inhibit viral proliferation, *n*-pentyl ether did. The hydrophobic groups introduced into the C-19 position of damavaricin Fc seem to confer the properties of penetrability and selectivity for the cells. Fortunately, the acute toxicity of the damavaricin Fc derivatives was remarkably reduced compared with streptovaricin or damavaricin C derivatives (data not shown). This result suggests that damavaricin Fc derivatives are potentially useful as antiviral agents.

**Table VII**
**Inhibition of Focus Formation by MSV/MLV and Inhibition of Reverse Transcriptase[a–c]**

| Damavaricin Fc derivative | Percent inhibition of focus formation at 5 μg/ml | | Percent inhibition of RDDP[e] at 30 μg/ml |
|---|---|---|---|
| | *N* | % | |
| Methyl ether | 76/111[d] | 28.2 | 56 |
| Ethyl ether | 39/59 | 20.0 | 34 |
| *n*-Pentyl ether | 64/111 | 42.3 | 60 |
| Benzyl ether | 48/51 | 6.0 | 0 |

[a]NRK cells were grown to $1.5 \times 10^5$ cells/dish in 20 $cm^2$ Falcon Petri dishes in Eagle's MEM supplemented with 10% fetal calf serum and tryptose phosphate broth. Cells were infected with Moloney's mouse sarcoma virus; polybrene was added to a final concentration of 2 μg/ml. At the same time, the derivatives were added at a concentration of 5 μg/ml. After 3-day incubation, the medium was removed and the fresh medium was added. After a further 3-day incubation, the cells were fixed and stained.
[b]Viral reverse transcriptase activity was measured as follows: The reaction mixture in a final volume 100 μl containing 30 mM Tris–HCl (pH 8.1), 24 mM KCl, 10 mM $MnCl_2$, 0.5 mM dATP, dCTP, dGTP, 0.025% NP40, 5 mM dithiothreithol, 7.5 mM [$^3$H]-dTTP (1 Ci/mole).
[c]Samples were incubated at 37°C for 60 min, and the reaction was terminated by the addition of 1 ml cold 10% trichloroacetic acid (TCA). The radioactivity of precipitates was determined.
[d]Foci in the presence of drug/foci in the absence of drug.
[e]Reverse transcriptase.

**Table VIII**
**Inhibition of the Growth of Mouse Leukemia Virus[a,b]**

| Experiment | Derivative | Concn. (μg/ml) | Cell number/dish (μg/ml) | Relative activity | |
|---|---|---|---|---|---|
| | | | | cpm | Percent |
| I | Without derivative | — | 2.9 | 2972 | 100 |
| | *n*-Pentyl ether | 5 | 1.6 | 87 | 2.9 |
| II | Without derivative | — | 4.8 | 3920 | 100 |
| | *n*-Pentyl ether | 3 | 4.3 | 180 | 4.6 |
| III | Without derivative | — | 4.4 | 3784 | 100 |
| | Ethyl ether | 4 | 4.2 | 3917 | 104 |
| | Methyl ether | 5 | 4.9 | 2773 | 73 |
| | *n*-Hexyl ether | 4 | 4.5 | 2689 | 71 |
| | *n*-Undecyl ether | 5 | 4.4 | 3784 | 100 |

[a]NRK cells were grown to $1.5 \times 10^5$ cells/dish in 20 $cm^2$ Petri dishes in Eagle's MEM supplemented with 10% fetal calf serum (FCS) and 10% tryptose phosphate broth. Cells were infected with Moloney mouse leukemia virus after 4-day incubation; the culture fluid was collected by centrifugation at 25,000 rpm overnight.
[b]RDDP activity of the obtained viruses was measured as described for Table VII.

## 6. SELECTIVE KILLING OF HUMAN T-CELL LYMPHOTROPIC VIRUS TYPE I-TRANSFORMED CELL LINES BY *n*-PENTYL DAMAVARICIN Fc

Several retroviruses were discovered to cause severe human disease (Poiesz *et al.*, 1980; Kalyanaraman *et al.*, 1982; Barre-sinoussi *et al.*, 1983). Human T-cell lymphotropic virus type I (HTLV-I) is the causative agent of adult T-cell leukemia (Yoshida *et al.*, 1982), one of the most aggressive human leukemias, as well as HTLV-I-associated myelopathy (HAM) (Osame *et al.*, 1986) and tropical spastic paraparesis (Roman *et al.*, 1987). However, no effective therapy for HTLV-I-related disease is known.

All HTLV-I-positive cell lines tested were highly sensitive to *n*-pentyl damavaricin Fc (*n*-pentyl DvFc) as compared with HTLV-I-negative cell lines (Table IX). Among the sensitive cell lines, MT-4 was chosen to study the mode of action of *n*-pentyl DvFc. To determine the time course of cytotoxic action of *n* -pentyl DvFc, logarithmically growing MT-4 cells were exposed to various concentrations of *n*-pentyl DvFc and the cell number was measured at various times. MT-4 cells showed no growth in the presence of *n*-pentyl DvFc at concentrations of more than 5 μg/ml under these conditions (Fig. 4). However, these MT-4 cells were viable for at least 12 hr after exposure to *n*-pentyl DvFc (5

**Table IX**
**Effect of *n*-Pentyl DvFc on Cell Growth**

| Cell lines | Minimum concentration of *n*-pentyl DvFc inhibiting cell growth by 50% (μg/ml)[a] |
|---|---|
| PBL (resting) | 44 |
| PBL (PHA-activated) | 32 |
| HTLV-I-negative T-cell lines | |
| MOLT-4 | 16 |
| SKW | 25 |
| NALM16 | 27 |
| HTLV-I-positive T-cell lines | |
| FUK | 1 |
| 467 | 3 |
| KAN | 6 |
| TLOml | 1 |
| MT-1 | 4 |
| MT-2 | 7 |
| MT-4 | 2 |

[a]The values present the concentration of *n*-pentyl DvFc showing 50% cell growth compared with DvFc-free control after incubation for 72 hr.

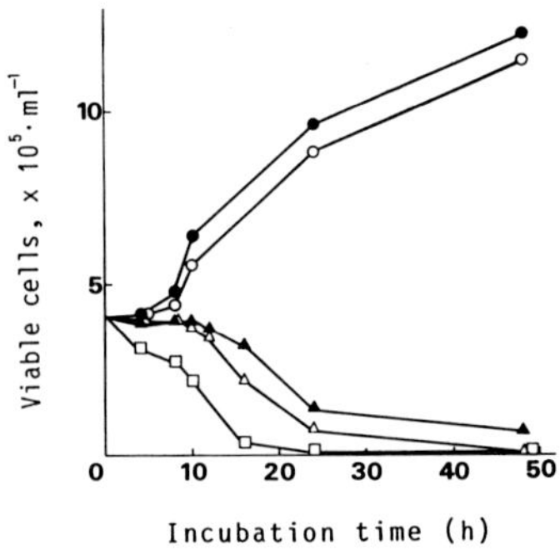


**FIGURE 4.** Effect of *n*-pentyl damavaricin Fc on the growth of MT-4 cell. MT-4 cells (4 × $10^5$ cells/ml) were incubated in RPMI-1640 supplemented with 10% FCS containing various concentrations of *n*-pentyl damavaricin Fc. Growth and viability of the cells were monitored by cell counting and by trypan blue dye exclusion. The concentrations of *n*-pentyl damavaricin Fc were 0 (●), 1 (○), 5 (▲), 10 (△), and 25 (□) μg/ml.

μg/ml), as indicated by the following experiments. MT-4 cells were incubated in the presence of *n*-pentyl DvFc (5 μg/ml) and, at various times, were washed with *n*-pentyl DvFc-free medium and incubated in the absence of the drug. The washed cells were able to resume growth in drug-free medium even after 42 hr. Similar results were obtained in experiments using 10 μg/ml of *n*-pentyl DvFc. These results imply that *n*-pentyl DvFc binds reversibly to the surface of MT-4 cells.

The activity of naphthomycin, an antibiotic of the ansamycin group, was reported to be prevented by sulfhydryl (SH) compounds (Okabe *et al.*, 1985). Therefore, we examined whether the cytotoxic action of *n*-pentyl DvFc might also be prevented by SH compounds. Both dithiothreitol and 2-mercaptoethanol diminished the action of *n*-pentyl DvFc on MT-4 in a dose-dependent manner. These results indicate that *n*-pentyl DvFc may act on the SH groups of proteins. To confirm the action of *n*-pentyl DvFc on the SH groups of proteins, we measured the inhibitory effect of *n*-pentyl DvFc on alkaline phosphodiesterase, a known SH enzyme located in the cell membrane of lymphocytes (Sugimoto *et al.*, 1981). *n*-Pentyl DvFc inhibited this enzymatic activity *in vitro* (Fig. 5); 50% inhibition was seen at a *n*-pentyl DvFc concentration of 10 μg/ml. These results suggest that *n*-pentyl DvFc may react reversibly with the SH groups of the membrane proteins of MT-4 cells.

To determine the effect of *n*-pentyl DvFc on macromolecular synthesis, we measured the incorporation of radiolabeled precursors by MT-4 cells into DNA ([$^3$H]thymidine), RNA ([$^3$H]uridine), and protein ([$^3$H]leucine). DNA synthesis was inhibited by *n*-pentyl DvFc in a dose-dependent manner. Neither RNA nor protein synthesis was significantly affected. *n*-Pentyl DvFc had no effect on the synthesis of these macromolecules in MOLT-4 cells. We wished to determine the effect of the drug on the rates of synthesis of DNA, RNA, and protein in HTLV-I-infected cells. We measured the incorporation of each macromolecular precursor into TCA-precipitable material for 30 min at various times after exposure to *n*

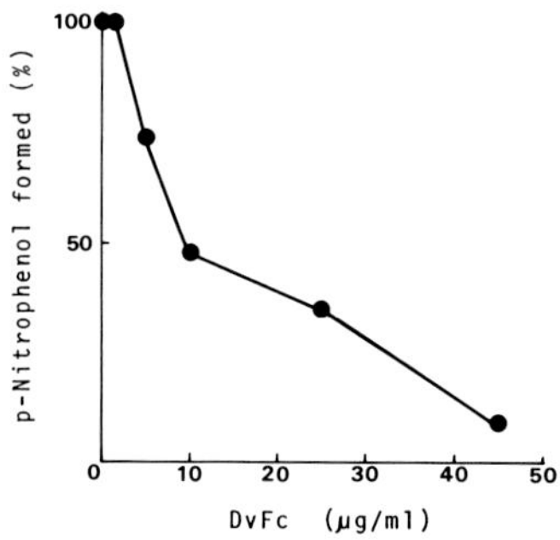


**FIGURE 5.** Inhibitory effect of *n*-pentyl damavaricin Fc on the membrane bound enzyme, alkaline phosphodiesterase of MT-4 cells. MT-4 cells were ruptured, and the extract was used as a source of SH enzyme. The reaction mixture (1 ml) contained 50 mM Tris–HCl (pH 9.0), 5 mM $MgCl_2$, 1 mM thymidine-5-monophosphate-*p*-nitrophenylester, extract from $5 \times 10^6$ MT-4 cells, and various concentrations of *n*-pentyl damavaricin Fc. After incubation at 37°C for 45 min, the enzymatic reaction was terminated by the addition of 300 μl trichloroacetic acid (TCA). The deproteinized supernatant was added to 400 μl 2.5 N NaOH, and the optical density at 400 nm was compared with standard *p*-nitrophenol.

-pentyl DvFc (5 μg/ml). The incorporation of [$^3$H]thymidine was inhibited by 50% at 4 hr, indicating that 4-hr exposure of infected cells to the drug was sufficient to depress significantly the rate of DNA synthesis. By contrast, RNA synthesis, as measured by the incorporation of [$^3$H]uridine, was increased, reaching a maximum of 150% after 4-hr exposure to *n*-pentyl DvFc, as compared with that of the cells prior to exposure. The incorporation of [$^3$H]leucine was not affected until 8 hr after exposure (data not shown).

The results described above indicate that *n*-pentyl DvFc might act directly on the cell membrane of MT-4 cells. We investigated the effect of membrane-active agents, such as quinidine and diphenylhydantoin (DPH). These drugs are a $K^+$-channel blocker and an anticonvulsant, respectively. Both types of drugs overcame the inhibitory effects of *n*-pentyl DvFc (Fig. 6). However, another $K^+$-channel blocker, 4-aminopyridine (4-AP) does not show this effect. Verapamil,

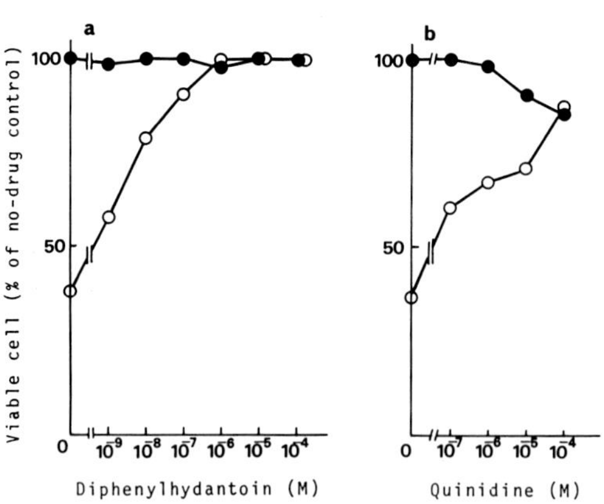


**FIGURE 6.** Effect of membrane-acting agents on the cytotoxic activity of *n*-pentyl DvFc in MT-4 cells. Logarithmically growing MT-4 cells were incubated for 24 hr in the presence of both *n*-pentyl DvFc (5 μg/ml) and membrane-acting agents (○), or of the latter only (●). (a) Effect of diphenylhydantoin; 100% corresponds $7.2 \times 10^5$ cells/ml. (b) Effect of quinidine; 100% corresponds $8.8 \times 10^5$ cells/ml.

a $Ca^{2+}$-channel blocker, was also effective at the concentration of $10^{-7}$–$10^{-4}$ M (data not shown).

To confirm whether *n*-pentyl DvFc actually caused $K^+$ efflux in MT-4 cells, $^{86}RB^+$-loaded MT-4 cells were exposed to *n*-pentyl DvFc, and the cellular $^{86}Rb$ level was examined. Efflux of 40% of cellular $^{86}Rb$ was observed within 6 min after exposure of cells to *n*-pentyl DvFc. Efflux of $^{86}Rb$ was not observed in the HTLV-I negative MOLT-4 cells under the same condition (Fig. 7). The two drugs that reversed the killing of HTLV-I-infected cells by *n*-pentyl DvFc, DPH, and verapamil are known to inhibit $Ca^{2+}$ influx into cells (Pincus and Lee, 1973; Gray *et al.*, 1987). Therefore, $Ca^{2+}$ influx into MT-4 cells might be changed by the action of *n*-pentyl DvFc. MT-4 cells were loaded with $^{45}Ca$ and then exposed to *n*-pentyl DvFc in the presence of $^{45}Ca$. A transient increase in intracellular $^{45}Ca$ ion was followed by a decrease within 10 min. Since this phenomenon was also observed in MOLT-4 cells, the transient influx of $Ca^{2+}$ could not explain the selective killing of MT-4 cells (data not shown).

We have shown the selective killing of the HTLV-I-transformed cells by *n*-pentyl DvFc and have given circumstantial evidence for the mechanism of cell killing. The results suggest that the drug acts on the cell membrane. Since $K^+$ efflux was specific for HTLV-I-infected cells, $K^+$ efflux seemed to be an essential event at an early phase in the course of the killing action of the drug. The killing action by *n*-pentyl DvFc, however, was not reversed by 4-AP, which is a blocking agent of voltage-sensitive $K^+$ channel (Hermann and Gorman, 1981). This finding suggests that the $K^+$ efflux caused by *n*-pentyl DvFc was not through the voltage-sensitive channel but through the $Ca^{2+}$-dependent one. In fact, $Ca^{2+}$ influx seems to occur in MT-4 cells immediately after the addition of DvFc, but MOLT-4 cells also showed the similar pattern of $Ca^{2+}$ flux, suggesting that drug-induced $Ca^{2+}$ influx was not specific in the infected cells. We

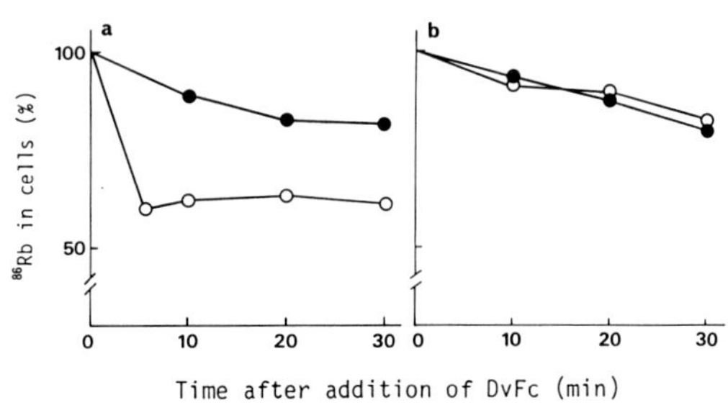


**FIGURE 7.** Effect of *n*-pentyl DvFc on the $^{86}Rb^+$ efflux from MT-4 cells. HTLV-I-infected (MT-4) cells (a) and -uninfected (MOLT-4) cells (b) ($2.4 \times 10^6$ cells/ml) were loaded overnight with 10 μCi/ml $^{86}Rb^+$ in RPMI 1640 containing 10% fetal calf serum (FCS). Cells were diluted 10-fold into nonradioactive medium, spun down (400*g*, 5 min) and resuspended in the same medium at $6 \times 10^6$ cells/ml. *n*-Pentyl damavaricin Fc (5 μg per $3 \times 10^5$ cells) was added to the culture and incubated at 37°C in a $CO_2$ incubator. At the indicated times, 100 μl cell suspension was removed and layered over 500 μl of a corn oil and dibutylphthalate (3 : 10) mixture and centrifuged in an Eppendorf microfuge (15,000 rpm, 40 sec). The tip of tube containing the cell pellet was cut off, and the radioactivity was determined. (●) *n*-Pentyl damavaricin Fc-free control; (○) *n*-pentyl damavaricin Fc treated.

speculate that if the level of intracellular $Ca^{2+}$ was similar between these two cell lines (MT-4 and MOLT-4) after exposure to DvFc, the $Ca^{2+}$-dependent $K^+$ channel of the MT-4 cell is more sensitive to the level of $Ca^{2+}$ than that of the MOLT-4 cell.

What is the target site for the action of *n*-pentyl DvFc? Cell death is prevented by the removal of *n*-pentyl DvFc or by the addition SH compounds. It seems likely that *n*-pentyl DvFc interacts with membrane proteins containing SH groups in MT-4 cells. HTLV-I antigens expressed on the cell membrane are unlikely to be the target molecules, because TLOml cells, which are also HTLV-I-infected cells but do not express HTLV-I antigen *in vitro* (Sugamura *et al.*, 1984) were also killed by *n*-pentyl DvFc. DNA synthesis in HTLV-I-infected cells was progressively inhibited with exposure to *n*-pentyl DvFc. The inhibitory activity of *n*-pentyl DvFc on SH enzymes may effect DNA synthesis, but this is probably a secondary effect.

From these results, we concluded that *n*-pentyl DvFc acts directly on the membrane, and the membranes of HTLV-I-transformed cells respond to this compound differently from the membranes of uninfected cell lines. The precise mechanism is unclear. The connections between $Ca^{2+}$ channel and $K^+$ channel are different in HTLV-I-transformed cells.

## 7. DISCUSSION

In an attempt to find the relationship between the chemical structure and function of streptovaricins, we have prepared a series of derivatives of damavaricin C. We aimed to improve the properties of the molecule so that we were able to use these compounds as antitumor or antiviral agents.

Streptovaricins have two intriguing properties. One is an inhibitory activity toward reverse transcriptase of RNA tumor viruses. This unique enzyme is involved in the proliferation and transformation of cells infected with retroviruses. Therefore, if a derivative can be identified that specifically inhibits this enzyme, we can expect to be able to prevent diseases caused by retroviruses. In addition to effects on reverse transcriptase, certain derivatives of damavaricin also impair ion transport across animal cell membranes.

During the process of chemical modification of streptovaricin C, we have obtained two compounds damavaricin C and damavaricin Fc. Both types of molecules have a free -OH group at the C-19 position, where we can introduce the hydrophobic groups by ether bond.

It is fair to say from the circumstantial evidence described in this chapter, that hydrophobic groups introduced into the C-19 position of drugs seem to related to the permeability property of the molecules and are involved in enhancing the ability of disturbing the membrane function of animal cells. Damavaricin

C and their derivatives showed the preferential killing activity of virus-transformed cells or human cancer cells *in vitro* and antitumor activity *in vivo*. Their activity as antitumor agents, however, was weak compared with that of other drugs generally used.

By contrast, the derivatives of damavaricin Fc have shown an interesting property. They have lost the antibacterial activity that streptovaricins possessed but have retained the inhibitory activity against reverse transcriptase. In addition, the toxicity of damavaricin Fc derivatives toward animal cells *in vitro* was decreased. This reduced toxicity toward cultured animal cells has encouraged us to test whether some derivatives have antiviral activity. In mouse cells infected with MSV–MLV complex, some of the derivatives of damavaricin Fc were strong inhibitors. In this case, the hydrophobic group introduced into C-19 position seemed to be involved in their activity. On the basis of these results, it appears that the inhibition of viral proliferation and viral transformation was caused by the inhibition of reverse transcriptase. Calvin's group at Berkeley, California, synthesized rifazone-$8_2$, a new rifamycin derivative, following a similar train of thought. Preferential inhibition of the growth of virus-transformed cells in culture and focus formation of chick embryo cells were observed by rifazone $8_2$ (Bissell *et al.*, 1974).

The results described in Section 6 show another aspect of the biological activity of *n*-pentyl DvFc. It was clearly demonstrated that the target of this compound in HTLV-I infected T cells was the cell membrane. We do not know the differences between the cell membranes of normal and virus-infected T cells in detail. However, it has been reported that HTLV-I-transformed cells have an altered cell surface: they expressed the IL-2 receptor (Depper *et al.*, 1984) and Ia antigen (Sugamura *et al.*, 1984). We have demonstrated that $K^+$ efflux occurred in HTLV-I-transformed cells immediately after exposure to *n*-pentyl DvFc. This phenomenon was specific for HTLV-I infected cells. Since *n*-pentyl ether was the most potent among the derivatives, the *n*-pentyl group introduced into damavaricin Fc may enhance the ability to perturb the membrane function (data not shown). Recently, adoptive immunotherapy of T cells *in vitro* has been developed (Mule *et al.*, 1984). Since *n*-pentyl DvFc can kill virus-transformed T cells preferentially, this drug can be used to improve the T-cell population of the patient with adult T-cell leukemia or to prevent the infection of human retroviruses, including acquired immune deficiency syndrome (AIDS) virus.

## 8. REFERENCES

Asai, M., Mizuta, E., Izawa, M., Haibara, K., and Kishi, T., 1978, Isolation, chemical characterization and structure of ansamitocin, a new antitumor antibiotic, *Tetrahedron* **35:**1079–1085.

Balerna, M., Keller-Schierlein, W., Martius, C., Wolf, H., and Zahner, H., 1969, Stoffwechselproduckte von Microorganismen: Napthomycin, ein Antimetabolite von Vitamin K, *Arch. Mikrobiol.* **65:**303–317.

Baltimore, D., 1970, Viral RNA-dependent DNA polymerase, *Nature (Lond.)* **226:**1209–1210.

Barre-sinoussi, F., Chermann, J. C., Rey, F., Nugeyre, M. T., Chamaret, S., Gruest, J., Dauguet, C., Axler-Blein, C., Vezine-Brun, F., Rouzioux, C., Rozeubaum, W., and Montagnier, L., 1983, Isolation of a T-lymphotropic retrovirus from a patient at risk for acquired immunodeficiency syndrome (AIDS), *Science* **220:**868–871.

Bernstein, A., Rolfe, B., and Onodera, K., 1972, Pleiotropic properties and genetic organization of the tol A, B locus of *E. coli* K12, *J. Bacteriol.* **112:**74–83.

Bissell, M. J., Hatie, C., Tischler, A. N., and Calvin, M., 1974, Preferential inhibition of the growth of virus transformed cells in culture by rifazone-$8_2$, a new rifamycin derivative, *Proc. Natl. Acad. Sci. USA* **71:**2520–2524.

Borden, E. C., Brockman, W. W., and Carter, W. A., 1971, Selective inhibition by streptovaricin of splenomegaly induced by Rauscher leukaemia virus, *Nature New Biol.* **232:**214–216.

Brockman, W. W., Carter, W. A., Li, L., Reusser, F., and Nichol, F. R., 1971, Streptovaricins inhibit RNA-dependent DNA polymerase present in an oncogenic RNA virus, *Nature (Lond.)* **230:**249–250.

Carter, W. A., Brockman, W. W., and Borden, E. C., 1971, Streptovaricins inhibit focus formation by MSV(MLV) complex, *Nature New Biol.* **232:**212–214.

Carter, W. A., Pitha, P. M., Brockman, W. W., Borden, E. C., and Marshall, L. W., 1972, Selective inhibition of viral functions: The antiviral action of interferon and streptovaricin, *Medicine (Baltimore)* **51:**181–188.

DeBoer, C., Meulman, P. A., Wnuk, R. J., and Peterson, D. H., 1970, Geldanamycin, a new antibiotic, *J. Antibiot.* **23:**442–447.

Depper, J. M., Leonard, W., Kronke, M., Waldmann, T. A., and Greene, W. C., 1984, Augmented T-cell growth factor receptor expression in HTLV-I infected human leukemic T-cell, *J. Immunol.* **133:**1691–1695.

Gray, L. S., Gnarra, J. R., Russell, J. H., and Engelhard, V. H., 1987, The role of $K^+$ in the regulation of the increase in intracellular $Ca^{2+}$ mediated by the T-lymphocyte antigen receptor, *Cell* **50:**119–127.

Hartmann, G., Behr, W., Beissner, K. A., Honikei, K., and Sippel, A., 1968, Antibiotics as inhibitors of nucleic acid and protein synthesis, *Angew. Chem. Int. Ed. Engl.* **7:**693–701.

Haruna, I., Watanabe, I., Yamada, Y., and Nagaoka, K., 1970, Specific inhibitor for RNA replicase, *Ann. NY Acad. Sci.* **173:**404–416.

Hermann, A., and Gorman, A. L. F., 1981, Effects of 4-aminopyridine on potassium currents in a molluscan neuron, *J. Gen. Physiol.* **78:**63–86.

Kalyanaraman, V. S., Sarngadharan, M. G., Robert-Guroff, M., Miyoshi, I., Blayney, D., Golde, D., and Gallo, R. C., 1982, A new subtype of human T-cell leukemia virus (HTLV-II) associated with a T-cell variant of hairy cell leukemia, *Science* **218:**571–573.

Kishi, T., Asai, M., Muroi, M., Harada, S., Mizuta, E., Terao, S., Miki, T., and Mizuno, K., 1969, Tolypomycin. I. Structure of tolypomycinone, *Tetrahedron Lett.* **2:**91.

Kishi, T., Yamana, H., Muroi, M., and Harada, S., Asai, M., Hasegawa, T., and Mizuno, K., 1972, Tolypomycin, a new antibiotic. III. Isolation and characterization of tolypomycin Y, *J. Antibiot.* **25:**11–15.

Kramarsky, B., and Manthey, W. J., 1974, Inhibition of growth of mouse mammary carcinoma by streptovaricin, *Proc. Soc. Exp. Biol. Med.* **145:**1434–1439.

Kupchan, S. M., Komoda, Y., Cout, W. A., Thomas, G. J., Smith, R. M., Karim, A., Gilmore, C. G., Haltiwanger, R. C., and Bryan, R. F., 1972a, Maytansine, a novel antileukemic ansa macrolide from Maytenus ovatus, *J. Am. Chem. Soc.* **94:**1354–1356.

Kupchan, S. M., Komoda, Y., Thomas, G. J., and Hintz, H. P. J., 1972b, Maytanprine and maytanbutine, new antileukemic ansa macrolides from *Maytenus buchananii, Chem. Commun.* **19:**1065.

Lüttringhaus, A., and Gralheer, H., 1941, Uber eine neue Ant Atropisomerer Verbindungen, *Ann. Chem.* **550:**67–98.

Milavetz, B. I., and Carter, W. A., 1977, Streptovaricins, *Pharmacol. Ther. A* **1:**289–305.

Mizuno, S., and Nitta, K., 1969, Effect of streptovaricin on RNA synthesis in phage T4-infected *Escherichia coli, Biochem. Biophys. Res. Commun.* **35:**127–130.

Mizuno, S., Yamazaki, H., Nitta, K., and Umezawa, H., 1968a, Inhibition of RNA polymerase of *Escherichia coli* by an antimicrobial substance B44P (streptovaricin), *J. Antibiot.* **21:**66–67.

Mizuno, S., Yamazaki, H., Nitta, K., and Umezawa, H., 1968b, Inhibition of DNA-dependent RNA polymerase reaction of Escherichia coli by an antimicrobial antibiotic, streptovaricin, *Biochim. Biophys. Acta* **157:**322–332.

Mizuno, S., Yamazaki, H., Nitta, K., and Umezawa, H., 1968c, Inhibition of initiation of DNA-dependent RNA-synthesis by an antibiotic B44P, *Biochem. Biophys. Res. Commun.* **30:**379–385.

Mule, J. J., Shu, S., Schwarz, S. L., and Rosenberg, S. A., 1984, Adoptive immunotherapy of established pulmonary metastases with LAK cells and recombinant interleukin-2, *Science* **225:** 1487–1489.

Muroi, M., Haibara, K., Asai, M., and Kishi, T., 1980, The structure of macbecin I and II, new antitumor antibiotics, I, *Tetrahedron Lett.* **21:**309–312.

Ohki, M., Doi, O., and Nojima, S., 1972, Mutant of *Escherichia coli* K12 deficient for detergent-resistant phospholipase A, *J. Bacteriol.* **110:**864–869.

Okabe, T., Bao, D., Isono, F., Sato, I., Fukazawa, H., Nishimura, T., and Tanaka, N., 1985, Studies on antineoplastic activity of naphthomycin, a naphthalenic ansamycin, and its mode of action, *J. Antibiot.* **38:**230–235.

Omura, S., Nakagawa, A., and Sadakane, N., 1979, Structure of herbimycin, a new ansamycin antibiotic, *Tetrahedron Lett.* **44:**4323–4326.

Osame, M., Usaki, K., Izumo, S., Ijichi, N., Amitani, H., Igata, A., Matsumoto, M., and Tara, M., 1986, HTLV-I associated myelopathy, a new clinical entity, *Lancet* **1:**1031–1032.

Pincus, J. H., and Lee, S. H., 1973, Diphenylhydantoin and calcium, *Arch. Neurol.* **29:**239–244.

Poiesz, B. J., Ruscetti, F. W., Gazder, A. F., Bunn, P. A., Minna, J. D., and Gallo, R. C., 1980, Detection and isolation of type-C retrovirus particles from fresh and cultured lymphocytes of a patient with cutaneous T-cell lymphoma, *Proc. Natl. Acad. Sci. USA* **77:**7415–7419.

Quintrell, N. A., and McAuslan, B. R., 1970, Inhibition of poxvirus replication by streptovaricin, *J. Virol.* **6:**485–491.

Rhuland, L. E., Stern, K. F., and Reames, H. R., 1957, Streptovaricin. III. In vivo studies on the tuberculous mouse, *Am. Rev. Tuberc. Pulm. Dis.* **75:**588–593.

Rinehart, K. L., 1972, Antibiotics with ansa rings, *Acc. Chem. Res.* **5:**57–64.

Rinehart, K. L., Jr., Antosz, F. J., Deshmukh, P. V., Kakinuma, K., Martin, P. K., Milavetz, B. I., Sasaki, K., and Witty, T. R., 1976, Identification and preparation of damavaricins, biologically active precursors of streptovaricins, *J. Antibiot.* **29:**201–203.

Rinehart, K. L., Jr., Antosz, F. J., Sasaki, K., Martin, P. K., Maheshwari, M. L., Reusser, F., Li, L. H., Moran, D., and Wiley, P. F., 1974, Relative biological activity of individual streptovaricins and streptovaricin acetates, *Biochemistry* **13:**861–867.

Rinehart, K. L., Jr., and Shields, L. S., 1976, Chemistry of the ansamycin antibiotics, *Fortschr. Chem. Org. Naturst.* **33:**231–307.

Rolfe, A., and Onodera, K., 1971, Demonstration of missing in a colicin-tolerant of *E. coli* K12, *Biochim. Biophys. Res. Commun.* **44:**767–773.

Roman, G. C., Spencer, P. S., Schoenberg, B. S., Madden, D., Sever, J., Hugon, J., and Ludolph,

A., 1987, Tropical spastic paraparesis: HTLV-I antibodies in patients from the Seychelles, *N. Engl. J. Med.* **316:**51.

Sensi, P., Ballotta, R., Greco, A. M., and Gallo, G. G., 1961, Rifomycin XV. Activation of rifomycin B and rifomycin O. Production and properties of rifomycin S and rifomycin SV, *Farmaco* **16:**165–180.

Sensi, P., Margalith, P., and Timbal, M. T., 1959, Rifomycin, a new antibiotic preliminary report, *Farmaco* **14:**146–147.

Siminoff, P., Smith, R. M., Sokolski, W. T., and Savage, G. M., 1957, Streptovaricin I. Discovery and bilogic activity, *Am. Rev. Tuberc. Pulm. Dis.* **75:**576–583.

Sugamura, K., Fujii, M., Kannagi, M., Sakitani, M., Takeuchi, M., and Hinuma, Y., 1984, Cell surface phenotypes and expression of viral antigens of various human cell lines carrying human T-cell leukemia virus, *Int. J. Cancer* **34:**221–228.

Sugimoto, Y., Nishimura, T., Suzuki, H., and Tanaka, N., 1981, Alteration of membrane-associated enzymes in drug-resistant sublines of mouse lymphoblastoma L5178Y cells, *J. Antibiot.* **34:**1200–1205.

Sugita, M., Furihara, K., Seto, H., Otaka, N., and Sasaki, T., 1982, The structure of mycotrienins I and II, a novel class of ansamycin antibiotic, *Agr. Biol. Chem.* **46:**1111–1113.

Tan, K. B., and McAuslan, B. R., 1971, Inhibition of nucleoside incorporation into HeLa cells by streptovaricin, *Biochim. Biophys. Res. Commun.* **42:**230–236.

Temin, H. M., and Mizutani, S., 1970, RNA-dependent DNA polymerase in virions of Rous sarcoma virus, *Nature (Lond.)* **226:**1211–1213.

Wani, M. C., Taylor, H. L., and Wall, J. F., 1973, Plant antitumor agents: Colubrinol acetate and colubrinol, antileukemic ansa macrolides from *Colubrina texensis, Chem. Commun.* **12:**390.

Wehrli, W., and Staehelin, M., 1971, Action of the rifamycins, *Bacteriol. Rev.* **35:**290–309.

Williams, T. H., 1975, Naphthomycin, a novel ansa macrocyclic antimetabolite. Proton NMR spectra and structure elucidation using lanthanide shift reagent, *J. Antibiot.* **28:**85–86.

Yamazaki, H., Mizuno, S., Nitta, K., Utahara, R., and Umezawa, H., 1968a, Biochemical mechanism of action of substance B44P, *J. Antibiot.* **21:**63–65.

Yamazaki, H., Mizuno, S., Nitta, K., Utahara, R., and Umezawa, H., 1968b, Studies on antimicrobial substance B44P (streptovaricin) produced by a strain of actinomycetes. IV, *J. Antibiot.* **21:**227–233.

Yoshida, M., Miyoshi, I., and Hinuma, Y., 1982, Isolation and characterization of human adult T-cell leukemia and its implication in the disease, *Proc. Natl. Acad. Sci. USA* **79:**2031–2035.

*Chapter 4*

# Role of Virus-like Particles in Parasitoid–Host Interaction of Insects

Otto Schmidt and Imke Schuchmann-Feddersen

## 1. INTRODUCTION

A large number of hymenopteran wasp species are endoparasitoids that lay their eggs inside lepidopteran caterpillars. A basic question in this parasitoid–host interaction is: How are the eggs of the parasitoid protected against the immune reaction of the host? At least in some systems, the protection is apparently specific, as the host is still able to react toward other implanted foreign bodies while the eggs and larvae of the parasitoid are developing and feeding in the body cavity of the living caterpillar (Salt, 1968).

It is known that wasps of braconid and ichneumonid families which are endoparasitoids, contain a secretory gland of the ovary, the calyx gland, which in many cases produces a fluid containing a particulate fraction (Bedwin, 1979; Stoltz *et al.*, 1976, Stoltz and Vinson, 1979a; Stoltz *et al.*, 1984). Because the particles are invariably observed in endoparasitoid wasp species, they appear to represent an integral prerequisite for the parasitoid life cycle (Stoltz and Vinson, 1979a). Several lines of evidence indicate that the particles are involved in the suppression of the host immune reaction (Salt, 1973; Norton and Vinson, 1977; Edson *et al.*, 1981; Feddersen *et al.*, 1986).

---

**Otto Schmidt and Imke Schuchmann-Feddersen** Institute for Biology III, University of Freiburg, D-7800 Freiburg, Federal Republic of Germany.

The biology and physiology of the parasitoid–host interaction and insect immune reactions have been reviewed previously in detail (Salt, 1968; Nappi, 1974; Ratcliffe and Rowley, 1979; Stoltz and Vinson, 1979a; Lackie, 1981; Götz and Boman, 1982). Given the multitude of insect parasitoid systems, it would be rather surprising if all were based on similar strategies in suppressing the host's immune defense. It is not our intention to cover all the findings related to the action of viruses or virus-like particles; rather, we concentrate on recent results concerning the possible role of the particles from the calyx fluid of *Venturia* in suppressing the immune reaction of the host of the parasitoid.

## 2. VIRUS-LIKE PARTICLES

In some ichneumonid parasitoids, the inclusion particles have been characterized as viruses, having fusiform nucleocapsids surrounded by envelopes (Stoltz and Vinson, 1979a). Together with particles from braconid wasps, which have cylindrical nucleocapsids of variable length surrounded either individually or as groups by a unit membrane envelope (Stoltz *et al.*, 1986), the viruses have been classified as polydnaviruses according to the possession of multipartite double-stranded circular DNA genomes (Stoltz *et al.*, 1984). However, because particles from the wasp *Venturia* appear to lack detectible nucleic acids (Bedwin, 1979; Feddersen *et al.*, 1986), we refer to these particles as virus-like particles (VLP). This implies that some characteristics, such as morphology, assembly within the nucleus of the calyx cells, and mode of secretion, have similarities to viruses (Stoltz and Vinson, 1979a; Stoltz *et al.*, 1984), whereas other typical viral properties, such as infection of host tissues and mode of transmission in the parasitoid, are absent or unknown in *Venturia* particles. In this discussion, we call particles from calyx fluid, virus-like particles, and only use the term virus in the specific context to polydnaviruses. All endoparasitic wasps invariably contain the particles, even *Venturia* where viral DNA molecules are absent. We must therefore assume that, in *Venturia* at least, the coding sequences for these particles are transmitted by the wasp genome, presumably as a provirus.

A germline transmission has been demonstrated in both braconid and ichneumonid parasitoid viruses (Fleming and Summers, 1986; Stoltz *et al.*, 1986). In *Campoletis sonorensis,* the virus DNA can be detected in low amounts in head and thorax tissues of females and in males (Fleming and Summers, 1986). Some of the DNA appears to be either extrachromosomal or integrated in *C. sonorensis* genomic DNA as judged from the characterization of offsize restriction fragments, suggesting the possibility of germline transmission of the virus DNA (Fleming and Summers, 1986). It is not clear whether the viral DNA in tissues other than the calyx gland is expressed or replicated as extrachromosomal DNA.

Recently it has been shown genetically that, in the braconid wasp *Cotesia melanoscela,* the virus DNA is transmitted through the male parasitoid germline (Stoltz *et al.*, 1986). This approach was possible by the finding of restriction fragment polymorphisms in the viral DNA of several wasp populations, which demonstrated elegantly that viral DNA can be transmitted to progeny females as a consequence of egg fertilization by male sperm cells carrying the virus (Stoltz *et al.*, 1986). It was concluded that polydnaviruses differ from typical viruses and "can just as legitimately be regarded as nuclear secretions taking a viral form, and having as a unique functional role the genetic colonization of the parasitoid's host" (Stoltz *et al.*, 1986).

In the following discussion, we describe the properties of VLP. A role of the VLP in protecting the wasp egg has been shown experimentally in a number of parasitoids (Edson *et al.*, 1981; Schuchmann-Feddersen, 1986). Therefore, mechanisms causing the suppression of the host immune reaction are probably closely associated with the functional properties of the VLP from the parasitoid.

## 3. PROPERTIES OF VLP

What properties of the VLP are relevant in this context? There are two basic concepts of VLP function, not mutually exclusive, which could account for protection of the wasp egg against cellular encapsulation by the host. The first is an inactivation of host immune functions by the viral properties of the VLP, including infection of host cells and expression of virus proteins (Stoltz and Vinson, 1979b; Edson *et al.*, 1981; Fleming *et al.*, 1983; Stoltz and Guzo, 1986). The second is suppression of the host's ability to recognize the egg as a foreign substance or to react against the egg surface. In this case, the blood cells of the host fail to recognize the parasitoid egg as a foreign body, or they recognize it but mount no reaction (Salt, 1966). This mechanism implies that the VLP contain crucial components designed to prevent either recognition of foreigness or initial reactions of encapsulation.

### 3.1. VLP Assembly in the Wasp

Virus-like particles from parasitoid wasps appear to be assembled exclusively within the nuclei of calyx gland cells (Rotheram, 1973; Stoltz and Vinson, 1979a). Particles can already be detected in calyx tissue from pupal stages by electron microscopy (Rotheram, 1973) and by immunodetection using antibodies against intact VLP (I. Schuchmann-Feddersen, unpublished data). In adult *Venturia* wasps, VLP consist of an electron-dense core surrounded by an envelope. They are densely packed within enlarged nuclei of the calyx gland cells (Fig. 1a) and are eventually secreted into the lumen of the gland. During their passage through the cytoplasm, the particles from *Venturia* carry an addi-

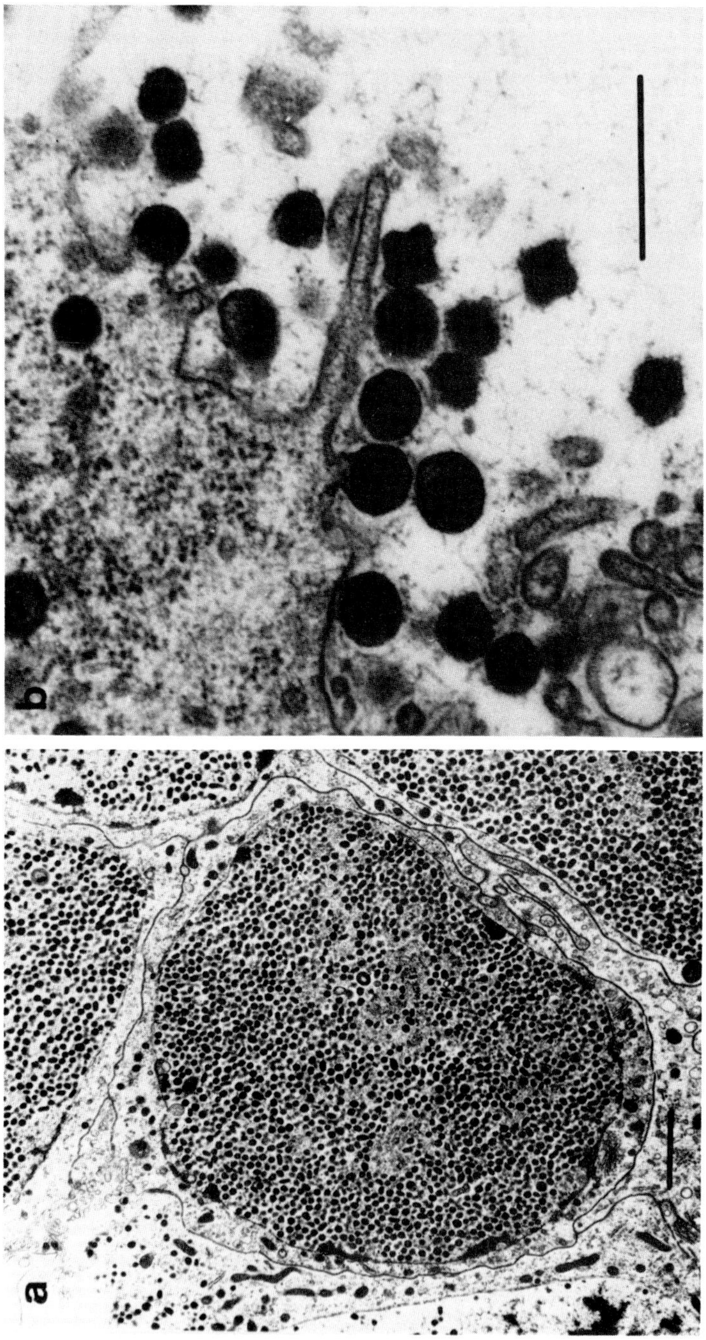

**FIGURE 1.** Virus-like particles in the calyx gland of *Venturia canescens*. Electron-dense particles are located predominantly in the nuclei of gland cells. Sections are shown from adult wasps. (a) Cells of the interior of the gland showing enlarged nuclei filled with virus-like particles. (b) Enlarged periphery of a cell with particles inside the cell and in the lumen of the gland. (After Schuchmann-Feddersen, 1986.)

tional membrane, probably acquired within the cell, which appears to be left behind when they leave the cell (Fig. 1b). In *Campoletis sonorensis,* extracellular particles have acquired a second envelope by budding through the plasma membrane of the calyx cell (Norton *et al.*, 1975). The outer membrane is then lost in the process of entering host tissues through the basement membrane (Stoltz and Vinson, 1979a). It is not clear whether this structural difference of the two types of particles reflects functional differences as well. In all cases, the VLP found on the surface of the parasitoid egg still retain the inner envelope (Fig. 2) acquired inside the nucleus, and so this structure seems to constitute an essential part of the particle.

### 3.2. Purification of VLP

Since the calyx secretions of the parasitoid play an essential part in the suppression of the host immune response (Salt, 1968, 1973; Stoltz and Vinson, 1979a; Edson *et al.*, 1981), we purified particles on CsCl gradients and asked whether they are still able to protect the egg from encapsulation. Ovaries from *Venturia* wasps were dissected, and the meiotically arrested eggs were taken from the lateral oviducts, which serve as reservoirs after passage through the calyx. Eggs were suspended in CsCl and the particles separated by density-gradient centrifugation. The particles band at a density of 1.263 g/cm$^3$, indicating that they constitute a glycoprotein fraction rather than nucleoproteins (Feddersen *et al.*, 1986).

Particles were recovered from CsCl solution by 10-fold dilution and pelleted by centrifugation. In such purified particles, no significant amounts of nucleic acids were detected after ultraviolet (UV)-spectrophotometric measurements, as reported previously on *Venturia* VLPs (Bedwin, 1979). In addition, attempted DNA and RNA purification from purified particles demonstrated no significant amounts of nucleic acids (Feddersen *et al.*, 1986). By contrast, VLP from other wasp species contain circular DNA molecules of variable size (Krell and Stoltz, 1980; Fleming *et al.*, 1983; Stoltz *et al.*, 1984). It is conceivable that the DNA was removed from *Venturia* particles during the course of the purification procedure. However, comparison of *Venturia* particle structure before and after purification failed to show any significant morphological changes (I. Schuchmann-Feddersen, unpublished observations). It is therefore likely that *Venturia* particles are being assembled within the nucleus without nucleic acids.

To test whether *Venturia* particles, purified on CsCl gradients, still function as a protective coat, we dissected eggs from the ovarioles, which are located proximal to the calyx. These eggs, which are not yet covered with VLP, were incubated for 1 hr in a suspension of purified particles and then implanted in host caterpillars. Two days later, the caterpillars were dissected and the eggs inspected. No signs of encapsulation were found in any of the 47 recovered eggs,

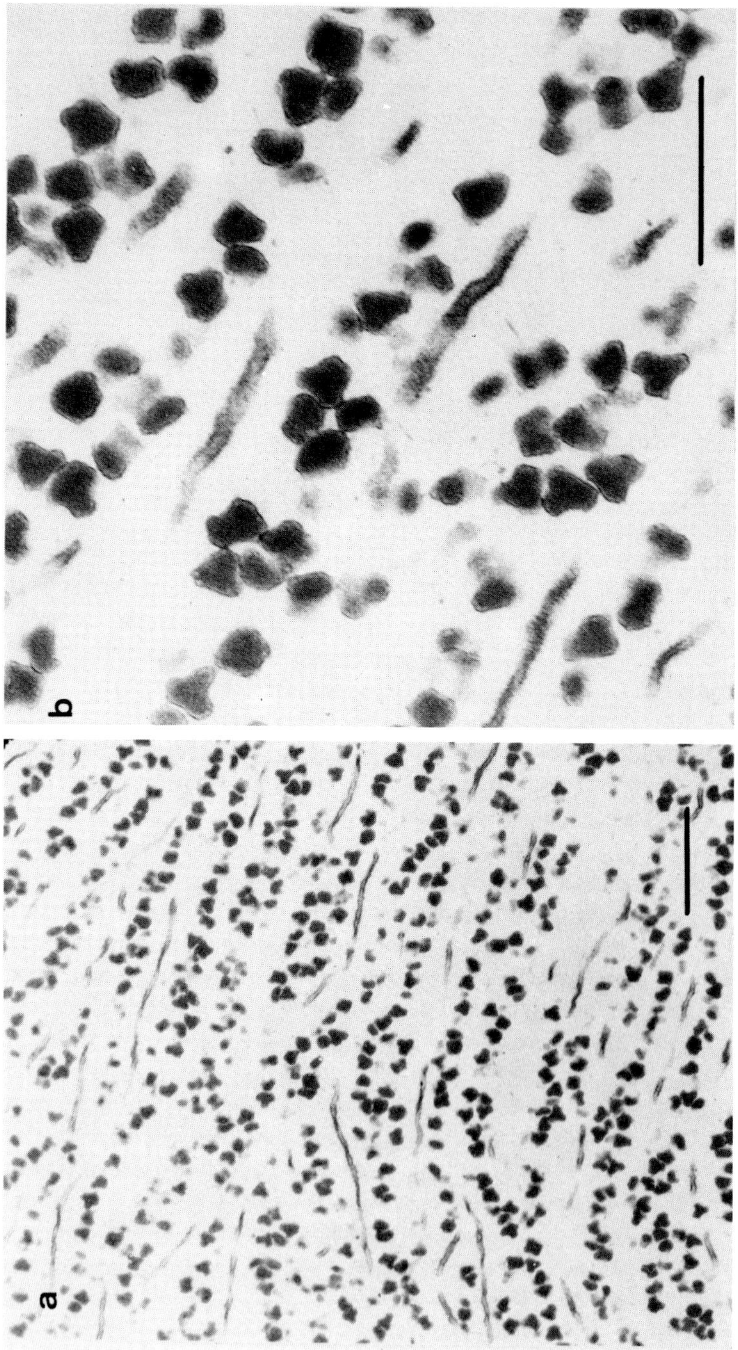


**FIGURE 2.** Virus-like particles on the chorion surface of a *Venturia* egg. Electron microscopic sections show two different enlargements of chorion folds and attached particles from the anterior part of the egg. (After Schuchmann-Feddersen, 1986.)

whereas capsules had formed on 49 of 52 ovariole eggs implanted without previous incubation in particle suspension (Feddersen *et al.*, 1986). Thus, the particles retain their protective function during the isolation procedure. However, the apparent absence of DNA in *Venturia* particles would preclude any protective function on the basis of virus-directed synthesis in the host caterpillar or upon inactivation of host hemocytes by viral infection. Since the particles coat the egg surface (Fig. 2), it is plausible that they could be involved in a suppression of the defense reaction if some of their components displayed similarities to host components. The particles could either simply mask the egg surface, preventing it from being recognized as a foreign substance, or inhibit egg encapsulation in a still unknown way.

### 3.3. Molecular Properties of VLP

In order to investigate the molecular properties of VLP, we raised antibodies against intact purified particles and used them to analyze Western blots containing cellular components from various parasitoid and host tissues. When purified VLP from *Venturia* were separated on polyacrylamide gels and stained with Coomassie blue, four main protein components of Mw 60, 40 and 35 kD were detected, as well as some minor proteins at ~80 kDa. A similar pattern is seen when purified particles are analyzed on Western blots stained with particle-specific antibodies (Fig. 3, lane 5). This finding indicates that the main protein components are recognized by the antiserum. No significant change in protein pattern and amount of proteins was observed when particles were analyzed 4 or 24 hr, respectively, after natural oviposition into the caterpillars (Fig. 4). This indicates that the particles of *Venturia* stick to the egg surface throughout most of embryogenesis.

To determine the tissue localization of the particle proteins, we tested various tissues from *Venturia* wasps. We found that only cells of the wasp ovary including the attached calyx gland are labeled: other wasp tissues are not recognized by particle-specific antibodies. This is confirmed by immunofluorescence of ovarial tissue, in which labeling is observed only within calyx gland cells (Fig. 5). This indicates that particles are synthesized in substantial amounts only in the calyx gland cells.

When Western blots containing protein extracts from tissues of unparasitized *Ephestia* host caterpillars were tested with particle-specific antibodies, a single protein component with molecular weight 42 kDa was labeled (Fig. 3, lane 6). In order to rule out coincidental cross-reaction in the polyclonal serum, we purified a particle-specific antibody reagent by eluting antibodies from cut-out regions of the Western blot filters (Towbin *et al.*, 1976; Smith and Fisher, 1984) containing the peroxidase-labeled putative antibodies of the particle components; these were then incubated to Western blots containing *Ephestia* fat-body proteins (Feddersen *et al.*, 1986).

**FIGURE 3.** Identification of *Venturia* and *Ephestia* tissue components on Western blots containing (1) *Venturia* accessory glands, (2) *Venturia* brain and thorax tissues, (3) *Venturia* ovary with calyx gland and oviduct, (4) *Venturia* ovarioles excluding calyx and oviduct, (5) purified virus-like particles, (6) *Ephestia* fat body. Left filter has been screened with rabbit preserum (collected before immunization); right filter with serum from the same rabbit obtained after immunization with purified particles. Molecular weights are given in kilodaltons. (After Feddersen, *et al.*, 1986.)

Antibodies eluted from the high-molecular-weight particle components gave a signal with the 42-kDa host caterpillar protein component (Fig. 6). The two large components of the VLP must have antigenic determinants in common, since eluted antibodies from each component bind strongly to both. No detectable cross-reaction was seen with antibodies eluted from the 40- and 35-kDa particle components.

In an additional experiment, we tested purified antibodies eluted from the host 42-kDa protein component on Western blots containing VLP proteins. As expected, these antibodies bind strongly to both VLP proteins of 52 and 60 kDa. In addition, a rather strong binding is seen to less abundant VLP proteins at ~80 kDa (Fig. 6). Apparently, antibodies against these minor components are enriched by the elution procedure. Thus, the 52-, 60-, and 80-kDa components of the VLP

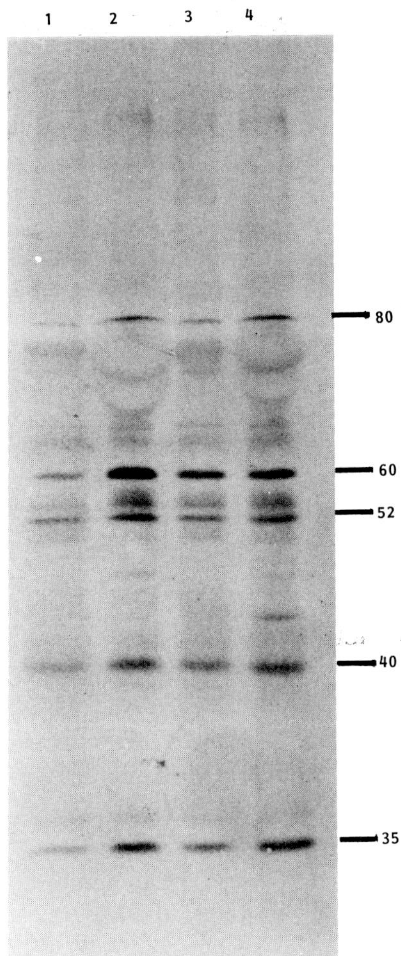


**FIGURE 4.** Identification of *Venturia* virus-like particle proteins from eggs after oviposition into *Ephestia* caterpillars (1,3) and *Laspeyresia* caterpillars (2,4). Eggs were dissected from caterpillars 4 hr (1,2) and 24 hr (3,4) after oviposition and proteins analyzed on Western blots using rabbit antiserum against purified virus-like particles from *Venturia*. Molecular weights are given in kilodaltons. (After Schuchmann-Feddersen, 1986.)

share antigenic determinants with a host component of 42 kDa, referred to as p42 protein in this discussion.

## 4. IMMUNE SUPPRESSION

### 4.1. Defense Reactions in Insects

Small substances and microorganisms entering the hemolymph are either phagocytized by hemocytes or are inclosed into nodules, or both (Ratcliffe and

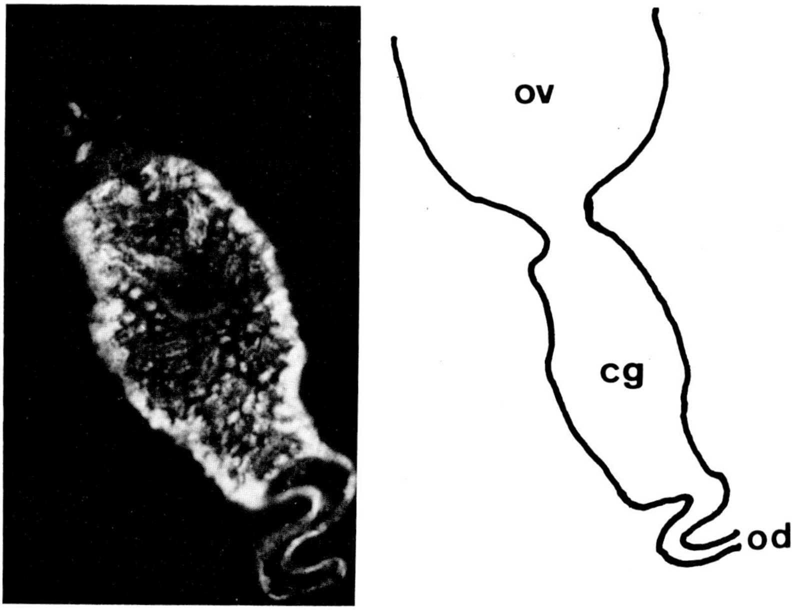


**FIGURE 5.** Immunofluorescence of ovarial tissue. The outlines of the tissues are given on the left. Only calyx gland tissues are labeled with the antibodies against the virus-like particles. ov, ovarioles; cg, calyx gland; od, oviduct. (After Schuchmann-Feddersen, 1986.)

Rowley, 1979). Nodules are formed when large amounts of small substances or microorganisms become entrapped into an extracellular coagulum which is subsequently surrounded by a multicellular sheath of hemocytes (Ratcliffe and Gagen, 1976; 1977). Foreign objects too large to be phagocytized are generally encapsulated (Salt, 1968; Grimstone *et al.*, 1967; Ratcliffe and Rowley, 1979). The encapsulation reaction is probably initiated by a plasma coagulum around the implant, which forms a dense basement lamina-like material between the adhering hemocytes (Brehelin *et al.*, 1975). Most capsules contain several structurally distinct layers of hemocytes which presumably reflect morphological changes of cells during the encapsulation process (Grimstone *et al.*, 1967; Ratcliffe and Rowley, 1979). Hemocytes close to the surface of the implant are necrotic and are surrounded by an extracellular coagulum. Subsequent layer(s) usually consist of flattened cells that form a dense aggregation, whereas hemocytes toward the periphery of the capsule are attached without undergoing significant changes in cell shape. In some dipteran species, encapsulation can occur in the absence of hemocytes by a coagulation of hemolymph components (Götz, 1969). Hemolymph clotting is also discussed as part of the coagulation process in wound healing (Bohn, 1986).

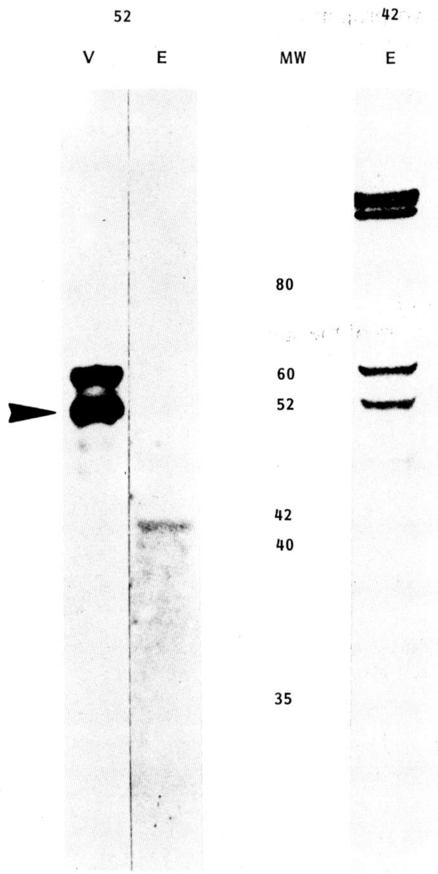


**FIGURE 6.** Western blots with ovarial tissues containing virus-like particles (V) and *Ephestia* fat body (E). Blots containing calyx cells and *Ephestia* fat body were incubated with either purified antibodies eluted from the 52-kDa particle component (52) or the 42-kDa component from the *Ephestia* fat body (42). (After Feddersen *et al.*, 1986.)

Thus, the coagulation of hemolymph components that may be secreted by hemocytes appear to play an important role in nodule formation and encapsulation. How coagulation processes are stimulated by a foreign surface or inhibited within the lining of its ''own'' hemolymph is therefore an important question of insect immunity.

## 4.2. Functional Assay of Encapsulation

The apparent similarity of VLP proteins to a host protein suggested to us a possibly important aspect of the VLP function; by potentially providing a protective coat to the egg chorion, the parasitoid egg would escape from the immune reaction of the host. The mechanism of protection could be based on either of two properties of the host-related protein component in the VLP: (1) the egg would not be recognized as a foreign body by the host because of the structural similarity to a host protein, or (2) the host-related protein in the VLP could actively prevent encapsulation in an unknown way (Salt, 1968; 1973). In both cases, the antigenic determinants of the VLP could in turn be masked experimentally by their respective antibodies, and this should either inactivate the protein or provide the egg with an additional coat consisting of vertebrate antibodies, foreign molecules, causing activation of the host immune reaction.

We tested this possibility in the following way. *Venturia* eggs covered with VLP were dissected from oviduct, incubated in serum (or as a control in preserum of the same rabbit), and then implanted into host caterpillars. After 2 days, the caterpillars were dissected and the state of encapsulation analyzed. Several degrees of encapsulation were observed as shown in Fig. 7. Serum and preserum clearly differed in their capability to induce the cellular defense reaction. No eggs completely surrounded by host cells were observed after treatment with the preserum, whereas a significant proportion was found to be encapsulated after serum treatment (Table I).

Our experiments suggest that antigenic determinants of the VLP components prevent encapsulation of the parasitoid egg and that the binding of antibodies to these determinants abolished the protective function of the particles. Thus, the experimental results appear to rule out metabolic activities of a virus to inactivate the immune response in *Ephestia* caterpillars.

One possibility is that the similarities of VLP proteins to a host component could prevent host hemocytes from recognizing the egg as a foreign body. Alternatively, components of the VLP could inhibit one of the crucial reactions of the hemolymph that normally promote the encapsulation of foreign objects. These experiments therefore provide no additional clues as to the mechanism of egg protection by the VLP. Additional information on VLP function can possibly be expected from the analysis of the molecular and physiological properties of the host p42-protein and potentially provide a key to the understanding of the mechanism by which the VLP proteins suppress the immune reaction.

## 4.3. Characterization of p42-Protein

To locate the p42-protein within the caterpillar, we analyzed protein extracts from various dissected tissues on Western blots using antibodies against VLP (Berg *et al.*, 1987). The p42-protein was found in significant amounts in the fat body. When dissected fat body was analyzed by indirect immunofluorescence, the p42-protein was found to be localized at the basal lamina surrounding the fat-body cells (Fig. 8). No labeling above background was detected in fat body cells, however, we cannot exclude that the protein is secreted in low amounts from fat-body cells and is accumulated at the basal lamina.

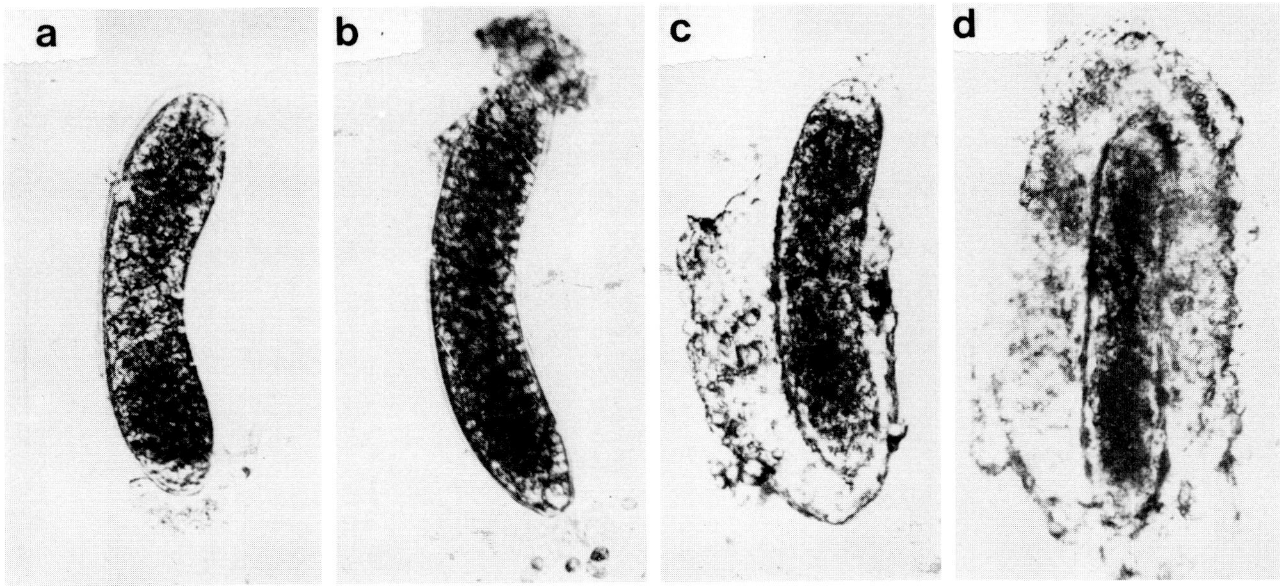


**FIGURE 7.** Encapsulation of antibody-treated *Venturia* eggs after injection into host caterpillars. (a–d) Increasing degrees of cell aggregation around the egg envelopes. Eggs ready for deposition were dissected from the reservoir of the oviduct and incubated for 30 min in a PBS solution (138 mM NaCl, 2.68 mM KCl, 7.30 nM NaHPO, 1.47 nm KHPO, 0.91 mM CaCl, 0.49 mM MgCl, pH 7.3) containing a 10-fold dilution of the rabbit antiserum against virus-like particles. The eggs were washed several times in PBS and 3–5 each were injected into a host caterpillar with a drawn-out Pasteur pipette. After 2 days, the caterpillars were dissected in PBS and the eggs recovered. (After Feddersen *et al.*, 1986.)

**Table I**
**Encapsulation of Pretreated Eggs after Injection into Host Caterpillars[a–d]**

| | Treatment of eggs | | | |
|---|---|---|---|---|
| Host reaction | Serum $N = 80$ (%) | Preserum $N = 63$ (%) | PBS $N = 51$ (%) | 0.1% SDS $N = 53$ (%) |
| No visible reaction | 13.7 | 82.5 | 100 | 0 |
| Aggregation of cells at the posterior or anterior pole | 8.1 | 15.9 | 0 | 0 |
| Localized encapsulation | 8.1 | 1.6 | 0 | 0 |
| Complete encapsulation | 70.1 | 0 | 0 | 100 |

[a]After Feddersen *et al.* (1986).
[b]*N*, number of eggs; PBS, phosphate buffer with saline; SDS, sodium dodecyl sulfate.
[c]Eggs were incubated in each of the following solutions for the indicated times: serum or preserum (30 min), PBS (2 hr), 0.1% SDS in PBS (5 min).
[d]Before injection with a drawn-out Pasteur pipette into the caterpillars, the eggs were washed several times with PBS. After 2 days, the caterpillars were dissected in PBS and the eggs examined.

In order to establish further which tissues synthesize p42-protein, we performed protein labeling experiments by injecting [$^{35}$S]methionine into caterpillars and analyzing the proteins on Western blots. The total amount of p42-protein was identified with antibodies against purified VLP. Incorporation of methionine was analyzed by exposing the same blots to X-ray film. From these experiments, it became evident that the p42-protein is one of the major labeled proteins in hemocytes and fat body (Fig. 9). More interestingly, the total amount of p42-protein, detected by labeling with antibodies, was increased in previously injected caterpillars. Perturbations caused by the injection procedure apparently induced the additional synthesis of the p42-protein.

Induction of the 42-kDa protein synthesis in *Ephestia* caterpillars is reminiscent of the induction of proteins after bacterial infection of *Hyalophora cecropia* pupae (Faye *et al.*, 1975; Boman *et al.*, 1981; Hultmark *et al.*, 1983). To explore this analogy, a suspension of bacteria together with [$^{35}$S]methionine was injected into the body cavity of *Ephestia* caterpillars (Berg *et al.*, 1987). After 3 days, the caterpillars were dissected, and protein extracts of tissues were analyzed on Western blots. The amount of antibody labeling on the p42-protein band from caterpillars after bacterial infection was increased as compared with that in uninjected larvae (Fig. 10). Similarly, in hemocytes isolated 1 day after bacterial infection, detectable amounts of the protein were seen, whereas no label was observed in hemocytes from uninfected larvae (Fig. 10). Small amounts of the p42-protein are sometimes visible in a hemocyte-free hemolymph supernatant, although a quantitative assessment of the presence of protein in isolated hemo-

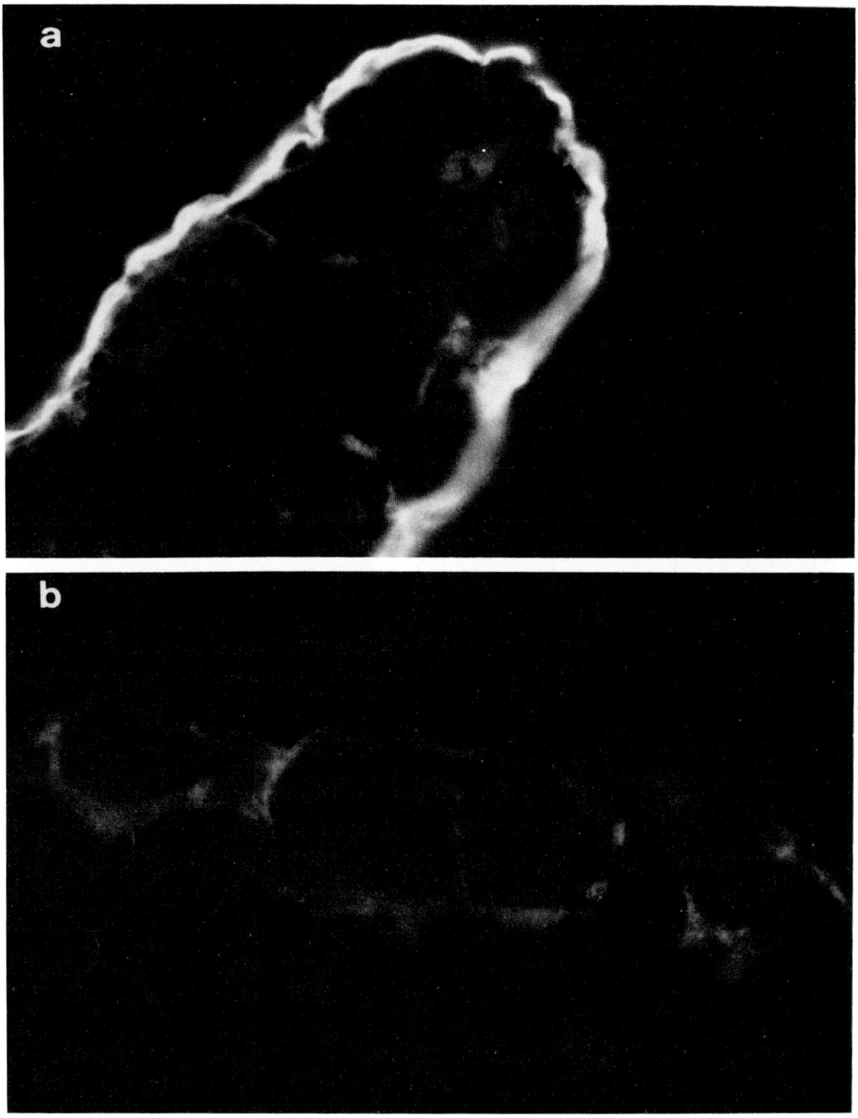


**FIGURE 8.** Immunofluorescence detection of the 42-kDa protein in isolated fat body of an *Ephestia* caterpillar. (a) Antiserum against purified VLP of *Venturia* (Feddersen *et al.*, 1986). Immunofluorescence is seen in the basal lamella of the fat body. (b) Fat body treated with preserum. (After Schuchmann-Feddersen, 1986.)

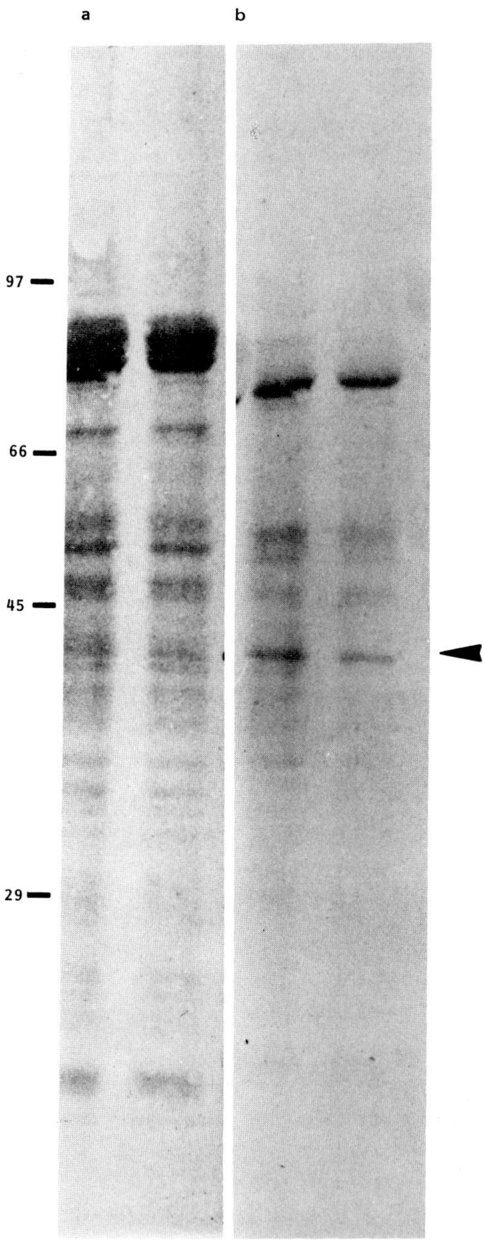


**FIGURE 9.** [$^{35}$S]Methionine-labeled proteins after bacterial infection. (Left lanes) Protein extracts from infected caterpillars separated on SDS–PAGE and transferred onto nitrocellulose filter. Proteins, visualized by amido black are shown from two individual experiments. (Right lanes) Autoradiograph of the nitrocellulose filter shown on the left. Exposure time is 4 days. A protein kDa (arrow) is among the major newly synthesized proteins. (After Berg *et al.*, 1988.)

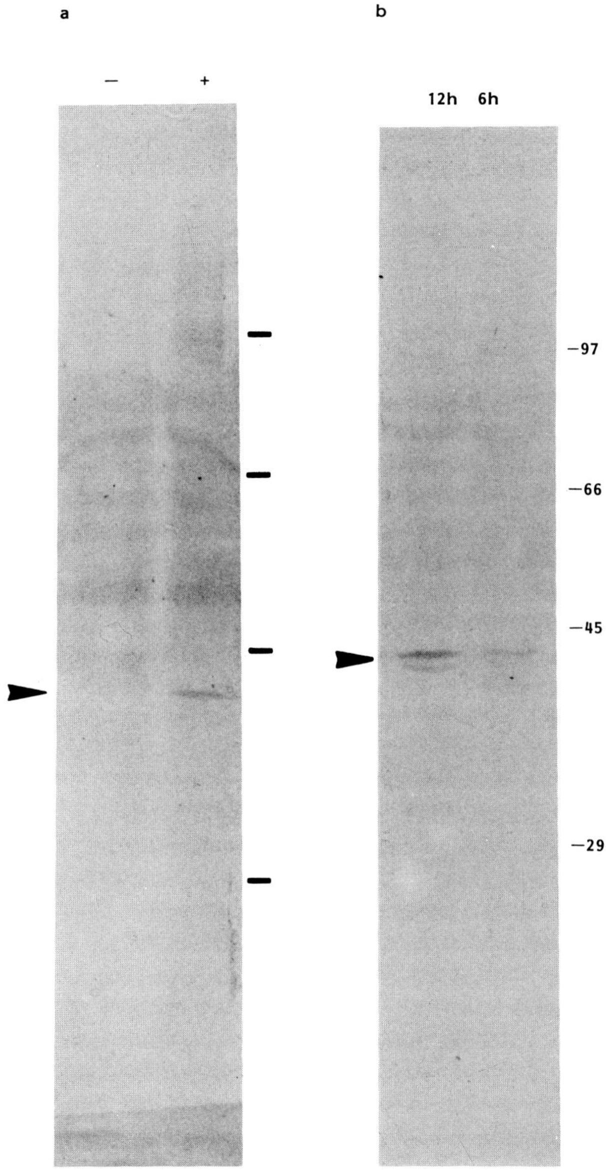


**FIGURE 10.** Induction of 42-kDa protein in *Ephestia* caterpillars infected with bacteria. (a) Protein extracts from hemocytes from uninfected (−) and from bacteria-infected caterpillars (+). (b) Protein extracts from fat body 12 hr and 6 hr after bacterial infection of caterpillars. The 42-kDa protein was visualized on Western blots by its cross-reaction to antibodies against purified VLPs of *Venturia* (Feddersen *et al.*, 1986). (After Berg *et al.*, 1988.)

lymph is difficult because of leakage artifacts or protein degradation. Nevertheless, the observations are consistent with the notion that the p42-protein is synthesized in hemocytes (or fat-body cells) and secreted into the hemolymph, where it is evidently attached to the basal lamina of fat body.

## 4.4. Antibacterial Immune Reaction

After bacterial infection, several proteins are newly synthesized in the pupae of the lepidopteran *Hyalophora cecropia* (Faye *et al.*, 1975; Boman *et al.*, 1981; Hultmark *et al.*, 1983). One of the most abundant proteins after induction appears to be a 48-kDa protein (called P4), which is apparently not directly involved in the lysis reaction of bacteria (Rasmuson and Boman, 1979; Andersson and Steiner, 1987). Our observations indicate that the p42-protein of *Ephestia* is related to the P4-protein of *Cecropia.*

The two proteins are of a similar size and are induced after physiological perturbations, in particular after bacterial infections (Boman *et al.*, 1981; Berg *et al.*, 1987). In addition, the tissue localization and the putative sites of synthesis appear to coincide in the two species. Both proteins are synthesized in hemocytes (P4-protein is actually synthesized mainly in the pupal fat body), and both proteins are found at the basal lamina of fat body (Andersson and Steiner, 1987; Berg *et al.*, 1987). Recently it was shown that anti P4-antibodies of *Cecropia* crossreact with a 42kD protein in *Ephestia* (K. Anderson and A. Will, unpublished data).

All this evidence so far obtained indicates that the p42 protein of *Ephestia* is structurally related to the P4-protein of *Cecropia.* Since the *Ephestia* p42 protein is induced in the antibacterial immune reaction, as is the P4-protein of *Cecropia,* we must ask what possible functions the two proteins might serve under any physiological condition. The answer is that so far there is no experimental evidence defining a functional role for either protein. However, the relatedness of the two proteins, which may be part of two different immune reactions, might be functionally significant. In *Cecropia,* the P4-protein is co-induced with other lysis proteins but is not directly involved in the lysing of bacteria (Andersson and Steiner, 1987; Hultmark *et al.*, 1983). In *Ephestia,* the p42 protein has antigenic similarity to VLP, which protect the parasitoid eggs from cellular encapsulation by an unknown mechanism (Salt, 1968, 1973; Feddersen *et al.*, 1986). Thus, the two proteins are part of two rather different immune reactions. Since the two proteins are both structurally and physiologically related, the question can be asked whether there is a common feature that could be assigned for a putative function. We recently suggested that the proteins are involved in the inhibition of the encapsulation during lysis of bacteria (Berg *et al.*, 1987).

### 4.5. Inhibition of Cellular Encapsulation

To test whether p42-protein synthesis in caterpillars has an influence on cellular encapsulation of foreign substances, we injected Sephadex beads into *Ephestia* caterpillars with or without bacterial infection (Berg *et al.*, 1987). When Sephadex beads are introduced into the caterpillars after the induction of a lysis reaction, they are encapsulated less frequently than in uninduced caterpillars or when co-injected with bacteria (Table II). Although there are several ways to explain this result, it is possible that after bacterial infection hemocytes secrete the p42-protein onto bacteria (and other substances in close proximity) and thereby prevent cellular encapsulation at the time when the lysis of bacteria is to occur. Thus, it appears that the p42 protein can act as a local inhibitor of cellular encapsulation.

Indirect evidence for the existence of an inhibitor of encapsulation in insects has been reported previously (Streams and Greenberg, 1969; Kitano, 1982). *Drosophila melanogaster* larvae fail to encapsulate eggs of *Pseudeucoila mellipes* if parasitized simultaneously by the resistant *Pseudeucoila bochei* (Streams and Greenberg, 1969). The fact that *P. bochei* eggs, produced after periods of continuous egg laying, are less efficient in protecting *P. mellipes* eggs from encapsulation, indicated to these authors that a quantitative factor is produced by the wasp female, which diffuses in the hemolymph of the host and provides protection to eggs of both species (Streams and Greenberg, 1969). In the parasitoid *Apanteles glomeratus*, venom of an accessory gland is co-injected together with the calyx fluid into the host *Pieris rapae crucivora* and appears to protect eggs from encapsulation (Kitano, 1986).

**Table II**
**Encapsulation of Sephadex Beads in Uninfected and Bacterium-Infected *Ephestia* Caterpillars**

| Injected material | $N$ | Encapsulated (%) | Not encapsulated (%) |
|---|---|---|---|
| *Venturia* eggs | 24 | 0 | 100 |
| Sephadex beads[b] | 100 | 82 | 18 |
| Sephadex + *E. coli*[c] | 222 | 45 | 55 |
| *E. coli*; Sephadex[d] | 166 | 17.5 | 82.5 |

[a]After Berg *et al.* (1987).
[b]Sephadex beads or *Venturia* eggs found in dissected caterpillars 3 days after injection.
[c]Sephadex suspended in *E. coli* ($10^4$/ml) was co-injected into caterpillars and encapsulation analyzed 3 days later.
[d]*Escherichia coli* ($10^4$/ml) was injected into caterpillars; after 24 hr, a suspension of Sephadex beads was injected and the encapsulation analyzed 3 days later.

## 5. VLP FUNCTION

### 5.1. Possible VLP Function in *Venturia*

If we assume that a host protein exists that has an inhibitory effect on encapsulation, the antigenically similar proteins in VLP could be related functionally to this protein. This suggests that the p42-like proteins in the VLP would inhibit the encapsulation reaction of the host hemolymph. This would clearly explain the efficiency of particle proteins in suppressing the encapsulation of the parasitoid eggs. Since VLP cover the parasitoid egg surface for at least 24 hr (see Fig. 4), a coating of VLP would provide the embryo with continuous protection against encapsulation during most of embryogenesis. The fact that antibodies against VLP inactivate the protective function of intact VLP on the egg suggests that the p42-like VLP proteins are exposed on the surface of the particles.

### 5.2. Evolutionary Aspects of VLP Function

How can we explain the emergence of p42-like proteins within VLP of the parasitoid wasp? Nothing is known about the origin and evolutionary relationship of the VLP to other insect viruses. Nevertheless, the presence of structurally related host p42 proteins in the VLP of the wasp could be logically explained in at least two ways: First, proteins of the wasp VLP might have evolved during parasitism to resemble a host protein and eventually gained the ability to mimic its function. Second, coding sequences of the lepidopteran host protein were transferred to the hymenopteran wasp by gene transfer between the two species (Syvanen, 1986), the simplest assumption being that a virus, infecting lepidopteran caterpillars, might have acquired host genome DNA sequences that code for functional domains of the cross-reacting protein. A hypothetical virus expressing the host protein at the virus coat could thus inhibit the immune response of its host. Perhaps such viruses could have a clear selective advantage against host-defense reactions, since small particles, including bacteria are normally eliminated from the hemolymph by phagocytosis (Ratcliffe and Rowley, 1979; Ratcliffe and Walters, 1983) and nodule formation (Ratcliffe and Gagen, 1976, 1977). Such viruses could have been incorporated into the life cycle of parasitoid wasps, where the antiencapsulation property of the particles was used as a "protective coat" on the egg surface of the parasitoid. Subsequently the virus became domesticated in the wasp, being expressed only in the calyx cells. Within this context, the observed differences of VLP functions in different wasp species could reflect different degrees of domestication or a process of degeneration of previously functional viruses. In some parasitoid–host systems, viruses are able to infect host cells (Stoltz and Vinson, 1979b; Stoltz and Guzo, 1986) or express virus sequences in host cells (Stoltz and Faulkner, 1978; Fleming *et al.*,

1983), whereas VLP from *Venturia* are glycoprotein particles without DNA molecules (Bedwin, 1979; Feddersen *et al.*, 1986). According to our reasoning, only the p42-like proteins of the virus coat would be relevant to the protection of the parasitoid egg. The "degeneration" of viral functions could therefore eventually have led to the DNA-lacking particles, which are nevertheless still able to function as inhibitor of encapsulation in *Venturia* (Feddersen *et al.*, 1986).

## 5.3. Virus-Related Changes in the Host

Given the vast number of hymenopteran parasitoids, it is unlikely that immune suppression in the host is achieved by an identical mechanism in all systems. The absence of DNA in *Venturia* particles is so far the exception, whereas VLP in other hymenopteran wasp species contain multipartite double-stranded circular DNA genomes belonging to the polydnavirus family (Stoltz and Vinson, 1979; Stoltz *et al.*, 1984). The introduction of polydnavirus particles into the host during egg deposition induces a number of effects related to viral activities and that are likely to weaken the host's ability to defend the parasitoid effectively (Vinson and Iwantsch, 1980).

Parasitoid viruses affect the larval prothoracic glands, causing developmental arrest of the host (Iwantsch and Smilowitz, 1975; Beckage and Templeton, 1968; Dover *et al.*, 1987). Ecdysone synthesis or secretion is abolished in the parasitized host, probably by viral infection of most tissues, leading to a drastic reduction of ecdysteroid titers in the caterpillar (Dover *et al.*, 1987).

In addition, parasitoid viruses clearly affect blood cells of the host caterpillar (Stoltz and Guzo, 1986; Davies *et al.*, 1987). In *Heliothis virescens,* the effects of parasitation on hemocyte changes can be reproduced by the injection of calyx fluid from the calyx gland of the parasitoid *Campoletis sonorensis* (Davies *et al.*, 1987). In virus-infected caterpillars, the hemolymph volume is apparently increased and less viscous. In addition, the number of plasmatocytes, when compared with other blood cell types, is significantly reduced 8–24 hr after viral infection (Davies *et al.*, 1987). However, injection of other microorganisms might also result in a depletion of plasmatocytes from insect hemolymph (Chain and Anderson, 1982; Ratcliffe and Walters, 1983), moreover, a fraction of plasmatocytes nevertheless remains in the hemolymph (Davies *et al.*, 1987). Therefore, the observation that cellular characteristics of plasmatocytes are changed after viral infection (Stoltz and Guzo, 1986; Davies *et al.*, 1987) is probably more relevant to the specificity of immune suppression than the observed reduction of hemocytes.

Plasmatocytes appear to be transformed both morphologically and functionally after injection of calyx fluid into caterpillars of *Malacosoma disstria* (Stoltz and Guzo, 1986) and *Heliothis virescens* (Davies *et al.*, 1987). In both systems, the *in vitro* aggregation, cell adhesion and spreading of plasmatocytes was significantly reduced 3–4 hr after viral infection. Plasmatocytes are known to be involved in encapsulation of foreign objects and cellular properties, like cell aggregation and spreading, are obviously necessary to accomplish the process of cellular encapsulation (Ratcliffe and Rowley, 1979). Therefore, these observations can provide an explanation for the abrogation of the immune reaction in parasitized hosts (Stoltz and Guzo, 1986; Davies *et al.,* 1987). The question is whether viral transformation of blood cells, which requires metabolic activities of VLP, can explain an immediate protection of the parasitoid egg, given the swift initial reactions toward foreign objects usually observed during the course of encapsulation (Ratcliffe and Rowley, 1979).

Interestingly, several indirect observations suggest that in addition to hemocytes some unknown molecular components of the hemolymph might also contribute to the immune response. The reduction of spreading of plasmatocytes from infected *H. virescens* caterpillars is reversed in the presence of plasma from uninfected caterpillars (Davies *et al.,* 1987). Moreover, virus-infected *M. disstria* caterpillars are able to clear yeast cells, whereas parasitized caterpillars are not (Stoltz and Guzo, 1986). Davies and co-workers conclude that an additional immune function (localized clotting?) might act independent of cellular encapsulation and might be impaired only in parasitized caterpillars, as under both conditions a cell population normally capable of *in vitro* adherence and *in vivo* encapsulation is rendered inactive (Stoltz and Guzo, 1986). From these observations, it is possible that at least part of the immune reaction toward foreign particles is performed by noncellular components of the hemolymph.

### 5.4. Immune Suppression of Parasitoid Larvae

The resistance of the larval parasite stages toward the host immune response is a puzzle. If we assume that VLP provide a protective coat to the parasitoid egg, the emerging larva must have a different mechanism of resistance; it challenges the host immune reaction with a chitinous cuticle very different from the chorion of the egg (Salt, 1968). Even if some of the particles from the egg chorion are distributed onto the larval cuticle, it would not readily provide a sufficient coating on the growing larval surface. Other properties of the parasitoid, such as larval movements, are likewise unable to explain evasion from encapsulation, as larvae in caterpillars of other species that are not habitual hosts are encapsulated even though moving vigorously.

The question is whether the presence of an inhibitor of encapsulation in the host can account for larval survival; it is conceivable, if we assume that such a protein is continuously secreted into the hemolymph and eventually attached to the growing embryo within the egg. Since parasitoid eggs are taking up substances from the hemolymph to support embryogenesis the chorion must be permeable to small components, presumably at the micropyle (Salt, 1970). Therefore, it is possible to assume that by the time the wasp larva hatches, it is covered with p42-protein. This would render the parasitoid larva resistant toward cellular encapsulation until it has grown and, by feeding on blood cells, has eventually depleted the host's hemocytes (Salt, 1968).

Such a mechanism of larval resistance toward cellular encapsulation is based on the assumption that the p42-protein is continuously synthesized from hemocytes or other tissues in low amounts. This appears to be the case in *Ephestia,* where the p42-protein is present at the basal lamina of fat body even in uninfected caterpillars (Berg *et al.*, 1987).

What is the functional significance of such a latent p42-protein production in the caterpillar? Although there is no experimental evidence, it is tempting to speculate on a general function of the protein. Since the p42-protein, which appears to inhibit cellular encapsulation, is secreted into the hemolymph, any foreign object, able to stay in the hemocoel long enough without being encapsulated, would eventually become coated by the p42-protein and thus become protected against encapsulation. A similar protective function would apply to the lining of the larval hemolymph, where the protein is found at the basal lamina of the fat body and thus provide a putative mechanism to distinguish self from foreign substances in insects.

## 5.5. p42-Protein Function and Insect Immunity

Using antibodies against purified VLP, we have tested cross-reactivity toward proteins in various lepidopteran species on Western blots (Berg *et al.*, 1987). The size of the proteins differs slightly, but the amount of antibodies bound to the corresponding proteins is not reduced in any of the species analyzed (Fig. 11). This and additional observations suggest that the proteins corresponding to the p42-protein in *Ephestia* are conserved among lepidopteran species.

The question, then, is whether eggs of *Venturia* are protected after oviposition into caterpillars of species that show strong cross-reaction of p42-like proteins. In an experiment to test this, encapsulation of *Venturia* eggs in caterpillars of *Ephestia cautella, Ephestia elutella, Sitotraga cerealella, Galleria mellonella,* and *Laspeyresia pomonella* was analyzed 2 days after oviposition. In *Laspeyresia* caterpillars, about 10% of the eggs were found to be encapsulated. In none of the other species tested was an encapsulation reaction observed against *Venturia* eggs (Schuchmann-Feddersen, 1986). This indicates that the inhibitory function of VLPs from *Venturia* is effective in other lepidopteran species, although some of these caterpillars are not habitual hosts of the parasitoid. However, a reduced protective function was observed in caterpillars from more distant species such as *Laspeyresia*.

It is important to note that some of the hosts appear to restrict normal

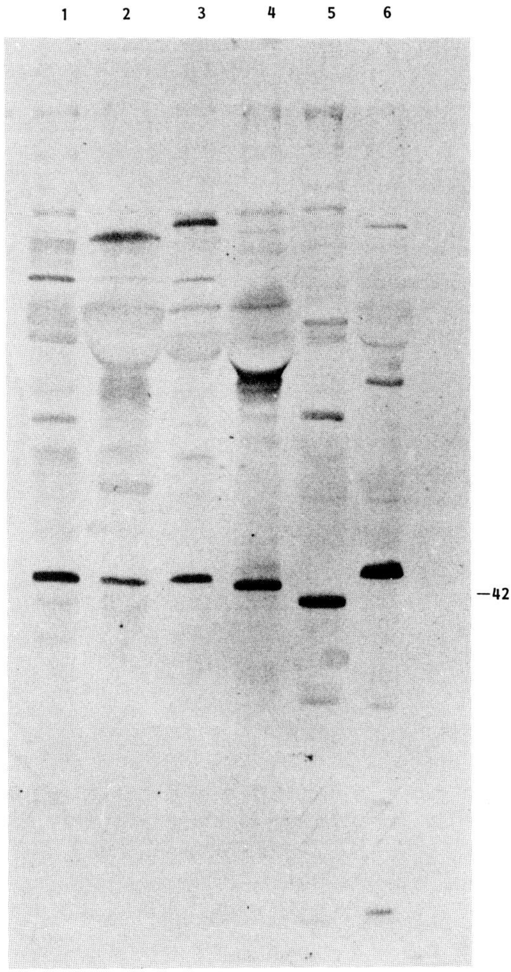


**FIGURE 11.** Immunodetection of the 42-kDa-like proteins in lepidopteran species using anti-VLP antibodies (Feddersen *et al.*, 1986). Fat body from the caterpillars was dissected and the protein extracts were analyzed on Western blots. (1) *Ephestia kühniella,* (2) *Ephestia cautella,* (3) *Ephestia elutella,* (4) *Sitotroga cerealella,* (5) *Galleria mellonella,* (6) *Laspeyresia pomonella.* (After Schuchmann-Feddersen, 1986.)

development of *Venturia* eggs at a physiological level. Although these eggs were not encapsulated, they nevertheless died at an early stage of embryogenesis (Schuchmann-Feddersen, 1986). Therefore, the host range of parasitoids is determined at least at two levels, comprising host immune reactions against the egg and physiological compatibility of parasitoid embryogenesis.

We can ask whether the existence of a p42-like protein that inhibits encapsulation would be sufficient to explain recognition of non-self within an insect.

If we assume that "hemocytes of an insect react to any surface which lacks specific properties of the surfaces which they themselves lay down or contribute to the organs and tissues of the insect of which they form a part" (Salt, 1960), it is conceivable that the p42-protein is attached to any surface structures and consequently protects it from hemolymph coagulation or hemocyte attack. Such a proposition would be a paradigm of a mechanism that enables an insect to distinguish "self" from all other surfaces lacking p42-protein. A strategy of insect immunity would be to cover one's own cell and tissue surfaces exposed to the hemolymph with a protein that inhibits encapsulation, consequently protecting these structures from coagulation or hemocyte attack (Salt, 1970).

Although a mechanism of insect immune protection that involves an inhibitor of encapsulation can explain many observations, it also raises a number of questions: What reaction is actually inhibited by the p42 protein? The inhibition could be at the cellular level by preventing cells to attach or secrete substances onto the surface. Alternatively, it could be at the humoral level by preventing the coagulation of hemolymph substances (Götz, 1969; Bohn, 1986). Can foreign substances grow resistant toward encapsulation as suggested by the apparent resistance of parasitoid larvae? Is the p42 protein involved in other functions?

Only additional experiments using a molecular approach will provide answers to these questions. Thus far, we have definitely obtained access to a general problem of insect immunity and physiology by analyzing the functions of VLP in a parasitoid–host relationship. This is reminiscent of the situation in cancer research in which the study of retroviruses in certain neoplastic transformations demonstrated the existence of cellular oncogenes that play an important part in cellular proliferation and differentiation (Bishop, 1985).

## 6. SUMMARY AND CONCLUSIONS

Insect endoparasitoids are capable of suppressing the immune reaction of their habitual hosts in a specific way. Salt (1968) characterized some of the implications:

> This seeming contradiction—that defence reactions against all kinds of foreign bodies are available to insects and that endophagous parasitoids are nevertheless able to develop in insect hosts—is resolved by recourse to one of the principles of host specificity. Although insects as a group react to every foreign body in the sense that any organism or substance evokes a reaction in most insects, each species of insect fails to make a reaction (or makes an ineffective reaction) to a small group of organisms, its habitual parasites. It is the common paradox of parasitology that defence reactions are least effective against the most noxious parasites, involving the tautology that the most noxious parasites are those against which defence reactions are least effective.

Recently, VLP of hymenopteran wasps have been shown to play a crucial part in suppressing the cellular encapsulation reaction (Stoltz and Vinson, 1979a). In some parasitoid wasps, polydnavirus particles are involved in the phenotypic transformation of hemocytes, reducing the capability of the host to mount an immune reaction towards the parasitoid egg (Stoltz and Guzo, 1986;

Davies *et al.*, 1987). However, at least in *Venturia,* the eggs are effectively protected by VLP that lack significant amounts of nucleic acids, precluding any virus expression in the host. The question was raised whether VLP could have acquired properties of the host immune system, which allows specific suppression of the immune response.

The finding of structural similarities between VLP proteins and a host component indicated that a host function is expressed in VLP (Feddersen *et al.*, 1986) and this observation has subsequently permitted the identification and characterization of a protein in caterpillars, which appears to inhibit cellular defense reactions (Berg *et al.*, 1987). On the basis of these results we continue to approach this parasitoid–host interaction, assuming that VLP have evolved in the host organism and eventually acquired the coding sequences of a host protein with properties of an inhibitor of encapsulation. Although there are several ways to explain the emergence of VLP in endophagous parasitoid wasps, a simple proposal would be that such hypothetical viruses, which were able to suppress immune reaction in lepidopteran hosts, were incorporated into a parasitoid wasp to become part of the life cycle of the parasitoid.

One of the advantages of this assumption is that it can be tested by comparing the DNA sequences of p42 protein-coding sequences in the host with the VLP-coding sequences in the wasp. DNA sequences of the *Ephestia* p42 protein and VLP proteins from *Venturia* are expected to be similar enough to be isolated from corresponding genomic libraries, using the DNA-coding sequences for the p42 protein in *Cecropia* as a probe (Lee *et al.*, 1983, Hultmark *et al.*, 1983). Moreover, the p42 protein is being purified from *Ephestia* caterpillars to test for its inhibitor function on encapsulation directly. Both approaches should enable us further to unravel insect immune reactions on a molecular level.

ACKNOWLEDGMENTS. We are indebted to Klaus Sander for initiating the project and for his continued interest. The authors thank Dieter Zissler for help with the electron microscopy and Rita Rossa for technical assistance. We thank Kerstin Andersson, Wolfgang Michalke, Günther Osche, Klaus Sander, J. B. Stoltz, and S. B. Vinson for helpful discussions and Randy Cassada and Klaus Sander for critically reading the manuscript. This work was supported by the Deutsche Forschungsgemeinschaft.

## 7. REFERENCES

Andersson, K., and Steiner, H., 1987, Structure and properties of Protein P4, the major bacterial-inducible protein in pupae of *Hyalophora cecropia, Insect. Biochem.* **17:**133–140.

Beckage, N. E., and Templeton, T. J., 1986, Physiological effects of parasitism by *Apanteles congregatus* in terminal-stage tobacco hornworm larvae, *J. Insect Physiol.* **32:**299–314.

Bedwin, O., 1979, An insect glycoprotein: A study of the particles responsible for the resistance of the parasitoid's egg to the defence reactions of its insect host, *Proc. R. Soc. [Biol.]* **205**:271–286.

Berg, R., Schuchmann-Feddersen, I., and Schmidt, O., 1988, Bacterial infection induces a moth protein which has antigenic similarity to virus-like particle proteins of a parasitoid wasp, *J. Insect Physiol.* **33**:473–480.

Blissard, G. W., Vinson, S. B., and Summers, M. D., 1968, Identification, mapping, and *in vitro* translation of *Campoletis sonorensis* virus mRNA from parasitized *Heliothis virescens* larvae, *J. Virol.* **57**:318–327.

Bishop, J. M., 1985, Viral oncogenes, *Cell* **42**:23–38.

Bohn, H., 1986, Hemolymph clotting in insects, in: *Immunity of Invertebrates* (M. Brehelin, ed.), pp. 188–207, Springer-Verlag, Berlin.

Boman, H. G., Boman, I. A., and Pigon, A., 1981, Immune and injury responses in *Cecropia* pupae. RNA isolation and comparison of protein synthesis in vivo and in vitro, *Insect Biochem.* **11**:33–42.

Brehelin, M., Hoffman, J. A., Matz, G., and Porte, A., 1975, Encapsulation of implanted foreign bodies by hemocytes in *Locusta migratoria* and *Melolontha melolontha, Cell Tissue Res.* **160:** 283–289.

Chain, B. M., and Anderson, R. S., 1982, Selective depletion of the plasmatocytes in *Galleria mellonella* following injection of bacteria, *J. Insect Physiol.* **28**:377–384.

Davies, D. H., Strand, M. R., and Vinson, S. B., 1987, Changes in differential haemocyte count and *in vitro* behaviour of plasmatocytes from host *Heliothis virescens* caused by *Campoletis sonorensis* polydnavirus, *J. Insect Physiol.* **33**:143–153.

Dover, B. A., Davies, D. H., Strand, M. R., Gray, R. S., Keeley, L. L., and Vinson, S. B., 1978, Ecdysteroid-titre reduction and developmental arrest of last-instar *Heliothis virescens* larvae by calyx fluid from the parasitoid *Campoletis sonorensis, J. Insect Physiol.* **33**:333–338.

Edson, K. M., Vinson, S. B., Stoltz, D. B., and Summers, M. D., 1981, Virus in a parasitoid wasp: Suppression of the cellular immune response in the parasitoid's host, *Science* **211**:582–583.

Faye, I., Pye, A., Rasmuson, I., Boman, H. G., and Boman, I. A., 1975, Insect immunity. II. Simultaneous induction of antibacterial activity and selective synthesis of some hemolymph proteins in diapausing pupae of *Hyalophora cecropia* and *Samia cyntia, Infect. Immun.* **12:** 1426–1438.

Feddersen, I., Sander, K., and Schmidt, O., 1986, Virus-like particles with host protein-like antigenic determinants protect an insect parasitoid from encapsulation, *Experientia* **42**:1278–1281.

Fleming, J. G. W., and Summers, M. D., 1986, *Campoletis sonorensis* endoparasitic wasps contain forms of *C. sonorensis* virus DNA suggestive of integrated and extrachromosomal polydnavirus DNAs, *J. Virol.* **57**:552–562.

Fleming, J. G. W., Blissard, G. W., Summers, M. D., and Vinson, S. B., 1983, Expression of *Campoletis sonorensis* virus in the parasitized host, *Heliotis virescens, J. Virol.* **48**:74–78.

Götz, P., 1969, Die Einkapselung von Parasiten in der Hämolymphe von Cironomus-Larven (Diptera), *Zool. Anz. Verh. Zool. Ges.* Suppl. **33**:610–617.

Götz, P., and Boman, H. G., 1982, Insect immunity, in: *Comprehensive Insect Physiology, Biochemistry and Pharmacology* (G. A. Kerkut and L. I. Gilbert, eds.), pp. 453–485, Pergamon, Oxford.

Grimstone, A. V., Rotheram, S., and Salt, G., 1967, An electron-microscopic study of capsule formation by insect blood cells, *J. Cell Sci.* **2**:281–292.

Hultmark, D., Engström, A., Andersson, K., Steiner, H., Bennich, H., and Boman, H. G., 1983, Insect immunity. Attacins, a family of antibacterial proteins from *Hyalophora cecropia, EMBO J.* **2**:571–576.

Iwantsch, G. F., and Smilowitz, Z., 1975, Relationships between the parasitoid *Hyposoter exiguae* and *Trichoplusia ni:* Prevention of host pupation at the endocrine level, *J. Insect Physiol.* **21:** 1151–1157.

Kitano, H., 1982, Effect of the venom of the gregarious parasitoid *Apanteles glomeratus* on its hemocytic encapsulation by the host, *Pieris, J. Invertebr. Pathol.* **40:**61–67.

Kitano, H., 1986, The role of *Apanteles glomeratus* venom in the defensive response of its host, *Pieris rapae crucivora, J. Insect Physiol.* **32:**369–375.

Krell, P. J., and Stoltz, D. B., 1980, Virus-like particles in the ovary of an ichneumonid wasp: Purification and preliminary characterization, *Virology* **101:**23–32.

Lackie, A. M., 1981, Immune recognition in insects, *Dev. Comp. Immun.* **5:**191–204.

Lee, J. -Y., Edlund, T., Ny, T., Faye, I., and Boman, H. G., 1983, Insect immunity. Isolation of cDNA clones corresponding to attacins and immune protein P4 from *Hyalophora cecropia, EMBO J.* **2:**577–581.

Nappi, A. J., 1974, Insect hemocytes and the problems of host recognition of foreignness, in: *Contemporary Topics in Immunology* (E. L. Cooper, ed.), pp. 207–224, Plenum, New York.

Norton, W. N., and Vinson, S. B., 1977, Encapsulation of a parasitoid egg within its habitual host: An ultrastructural investigation, *J. Invertebr. Pathol.* **30:**55–67.

Norton, W. N., Vinson, S. B., and Stoltz, D. B., 1975, Nuclear secretory particles associated with the calyx cells of the ichneumonid parasitoid *Campoletis sonorensis* (Cameron), *Cell Tissue Res.* **162:**195–208.

Rasmuson, T., and Boman, H. G., 1979, Insect immunity. V. Purification and some properties of immune protein P4 from hemolymph of *Hyalophora cecropia* pupae, *Insect Biochem.* **9:**259–264.

Ratcliffe, N. A., and Gagen, S. J., 1976, Cellular defense reactions of insect hemocytes *in vivo:* Nodule formation and development in *Galleria mellonella* and *Pieris brassicae* larvae, *J. Invertebr. Pathol.* **28:**373–382.

Ratcliffe, N. A., and Gagen, S. J., 1977, Studies on the *in vivo* cellular reactions of insects: An ultrastructural analysis of nodule formation in *Galleria mellonella, Tissue Cell* **9:**73–85.

Ratcliffe, N. A., and Rowley, A. F., 1979, Role of hemocytes in defense against biological agents, in: *Insect Hemocytes* (A. P. Gupta, ed.), pp. 331–414, Cambridge University Press, London and New York.

Ratcliffe, N. A., and Walters, J. B., 1983, Studies on the *in vivo* cellular reactions of insects: Clearance of pathogenic and non-pathogenic bacteria in *Galleria mellonella* larvae, *J. Insect Physiol.* **29:**407–415.

Ratcliffe, N. A., and Rowley, A. F., 1984, Recognition factors in insect haemolymph, *Dev. Comp. Immun.* **7:**653–656.

Rizki, R. M., and Rizki, T. M., 1984, Selective destruction of a host blood cell type by a parasitoid wasp, *Proc. Natl Acad. Sci. USA* **81:**6154–6158.

Rizki, T. M., and Rizki, R. M., 1986, Surface changes on hemocytes during encapsulation in *Drosophila melanogaster* Meigen, in: *Hemocytic and Humoral Immunity in Arthropods* (A. P. Gupta, ed.), pp. 157–190, Wiley, New York.

Rotheram, S., 1973, The surface of the egg of a parasitic insect. II. The ultrastructure of the particulate coat on the egg of *Nemeritis, Proc. R. Soc. [Biol.]* **183:**195–204.

Salt, G., 1965, Experimental studies in insect parasitism. XIII. The haemocytic reaction of a caterpillar to eggs of its habitual parasite, *Proc. R. Soc. [Biol.]* **162:**303–318.

Salt, G., 1966, Experimental studies in insect parasitism. XIV. The haemocytic reaction of a caterpillar to larvae of its habitual parasite, *Proc. R. Soc. [Biol.]* **165:**155–178.

Salt, G., 1968, The resistance of insect parasitoids to the defense reaction of their hosts, *Biol. Rev.* **43:**200–232.

Salt, G., 1970, *The Cellular Defence Reactions of Insects,* Cambridge University Press, London and New York.

Salt, G., 1973, Experimental studies in insect parasitism. XVI. The mechanism of the resistence of *Nemeritis* to defence reactions, *Proc. R. Soc. [Biol.]* **183:**337–350.

Schuchmann-Feddersen, I., 1986, Virus-artige Partikel und deren Funktion im Verlauf der parasitären Wechselbeziehung bei den Insektenarten *Venturia* und *Ephestia,* Ph.D. thesis, Freiburg.

Smith, D. E., and Fisher, P. A., 1984, Polypeptides of the *Drosophila* envelope, *J. Cell Biol.* **99:** 20–28.

Stoltz, D. B., 1986, Interactions between parasitoid-associated products and host insects: A brief overview, *J. Insect Physiol.* **32:**347–350.

Stoltz, D. B., and Cook, D. J., 1983, Inhibition of host phenoloxidase activity by parasitoid hymenoptera, *Experientia* **39:**1022–1024.

Stoltz, D. B., and Faulkner, G., 1978, Apparent replication of an unusual virus-like particle in both a parasitoid wasp and its host, *Can. J. Microbiol.* **24:**1509–1514.

Stoltz, D. B., and Guzo, D., 1986, Apparent haemocytic transformation associated with parasitoid-induced inhibition of immunity in *Malacosoma disstria* larvae, *J. Insect Physiol.* **32:**377–388.

Stoltz, D. B., and Vinson, S. B., 1979a, Viruses and parasitism in insects, *Adv. Virus Res.* **24:**125–171.

Stoltz, D. B., and Vinson, S. B., 1979b, Penetration into caterpillar cells of virus-like particles injected during oviposition by parasitoid ichneumonid wasps, *Can. J. Microbiol.* **25:**207–216.

Stoltz, D. B., Vinson, S. B., and MacKinnon, E. A., 1976, Baculovirus-like particles in the reproductive tracts of female parasitoid wasps, *Can. J. Microbiol.* **22:**1013–1023.

Stoltz, D. B., Krell, P., Summers, M. D., and Vinson, S. B., 1984, Polydnaviridae—A proposed family of insect viruses with segmented, doublestranded, circular DNA genomes, *Intervirology* **21:**1–4.

Stoltz, D. B., Guzo, D., and Cook, D., 1986, Studies on polydnavirus transmission, *Virology* **155:** 120–131.

Streams, F. A., and Greenberg, L., 1969, Inhibition of the defence reaction of *Drosophila melanogaster* parasitized simultaneously by the wasps *Pseudeucoila bochei* and *Pseudeucoila melippes, J. Invertebr. Pathol.* **13:**371–377.

Syvanen, M., 1986, Cross-species gene transfer: A major factor in evolution?, *Trends Genet.* **2:**63–66.

Towbin, H., Stähelin, T., and Gordon, J., 1979, Electrophoretic transfer of proteins from polyacrylamide gels to nitrocellulose sheets: Procedure and some applications, *Proc. Natl. Acad. Sci. USA* **76:**4350–4354.

Vinson, S. B., and Iwantsch, G. F., 1980, Host regulation by insect parasitoids, *Q. Rev. Biol.* **55:** 143–165.

Webb, B. A., and Dahlman, D. L., 1986, Ecdysteroid influence on the development of the host *Heliothis virescens* and its endoparasite *Microplitis croceipes, J. Insect Physiol.* **32:**339–345.

*Chapter 5*

# Human Leukemia and Retroviruses

Abraham Karpas

## 1. INTRODUCTION AND BACKGROUND

In 1908 the Danish veterinarians Vilhelm Ellerman and Oluf Bang showed that leukemia in chickens was transmissible by cell-free filtrates and was therefore presumably caused by a virus. This was the first virus-induced malignancy to be discovered. Then, in 1911, Peyton Rous showed that a sarcoma could be transmitted by a cell-free filtrate, which indicated that it was due to a virus; he also pointed out that the presumptive agent always caused the same type of tumor in the chick and must therefore carry specific genetic information. More recently it was shown that sarcoma viruses are essentially leukemia viruses that have transduced cellular oncogenes. Rous was so much ahead of his time that for several decades his work was kept on the shelf of Curiosa; it was not until 1966, when he was 86 years old, that he was recognized by the award of a Nobel Prize for the work he had done 55 years earlier.

The next major milestone was the discovery by Ludwik Gross in 1951 of the mouse leukemia virus. Gross's discovery changed the climate of opinion. Not only did it revive interest in chicken leukemias and sarcoma but it also started a search for possible viral involvement in other animal leukemias and sarcomas, as well as the first extensive search for viruses involved in the development of the corresponding human malignancies. During the past 20 years, viruses have been found to be involved in the development of leukemia in an increasing number of

**Abraham Karpas** Department of Haematology, University Clinical School, Cambridge, England.

animal species. Notable discoveries along the phylogenetic tree have included the finding that cat leukemia is caused by a virus, followed by the demonstration that leukemia in cattle is also caused by a virus. The first isolation and demonstration of a virus involved in the development of leukemia in subhuman primates was accomplished in the gibbon ape (reviewed previously in greater detail by Karpas, 1982).

In 1981 the first conclusive evidence of the involvement of a virus in human adult T-cell leukemia was published. This chapter reviews the work on the possible involvement of viruses in human leukemia.

## 2. RETROVIRUSES

By definition, a retrovirus contains RNA as its genetic material and the enzyme RNA-directed DNA polymerase (reverse transcriptase). Retroviruses have been classified on the basis of their biology, morphology and genomic structure. Accordingly, the retroviruses have been divided into three groups:

1. *Oncoviridae:* Further subdivided into A, B, C, and D-type particles, which are morphologically different. Many members of this group, e.g., the C types, have been shown to be oncogenic in numerous species in nature, producing hematological malignancies (leukemias/lymphoma) or mammary carcinomas (B-type).
2. *Spumaviridae:* Also called foamy viruses, often isolated in primary cell cultures, causing cytoplasmic vacuolization and cell degeneration in tissue culture. Morphologically, the foamy viruses resemble oncoviruses, but are not yet known to be associated with any disease.
3. *Lentiviridae:* Also called slow viruses, are often cytopathic to the cells in which they replicate. Viruses belonging to this group are known to cause slowly progressive disease in affected animals.

### 2.1. Transmission of Leukemia Viruses

Under natural conditions, leukemia viruses are transmitted by various modes in different animal species. In some species, such as the domestic cat, leukemia is a highly infectious disease, transmitted in much the same way that humans catch cold. A cat not previously exposed will become infected upon contact with an infected cat secreting the virus in its saliva. Some infected cats develop an immunity to the virus and remain protected from subsequent infection, but some develop leukemia and eventually die. In other species, such as the mouse, transmission is through genetic inheritance. When leukemic mice of a strain with a moderate incidence of disease are inbred, the incidence of spontaneous leukemia in their offspring increases significantly. Continuous inbreeding between such mice results in strains in which almost all the mice eventually develop leukemia. However, leukemia does not appear in leukemia-free strains of mice, even after continuous experimental exposure to the virus from the time of birth, including suckling from mice secreting the virus in their milk. In the domestic chicken, under natural conditions, both contact and genetic transmission of leukemia viruses can occur. Because there is close physical association between man and some animals in which leukemia is transmitted by contact, such as dairy cattle and the domestic cat, epidemiological studies have been undertaken to find out whether these viruses may be harmful to man. No correlation between particular animal leukemias and the disease in man has been detected.

Numerous epidemiological studies in Europe and the United States have investigated the incidence and possible transmission of leukemia in man. None of these studies provides any evidence for transmission by contact. Although the vast majority of leukemia patients are found to be the single case within their family or particular social surroundings (e.g., their school), a few observations of a higher incidence of leukemia within families have been reported—up to 14 cases, in one large family (Gunz *et al.*, 1978). One epidemiological study from the United States presents evidence that when one identical twin develops leukemia there is a 1 : 5 chance that the other twin will also develop the disease; but a similar study in Great Britain did not confirm this finding. Another study suggests that when leukemia affects a nonidentical twin, the other twin's chance of developing the disease is probably increased (Miller, 1967).

These observations do not suggest that direct genetic transmission of the disease is likely in humans, although they would not be incompatible with the existence of a transmissible predisposing agent. Further evidence for genetic predisposition is the higher incidence of the disease in certain groups of persons with genetic abnormalities. In particular, there is 20-fold higher incidence of leukemia in children with Down's syndrome, a group noted for its lower resistance to infections generally.

Suggestive evidence that leukemia in humans may be transmitted by a virus has been reported by Fialkow *et al.* (1971) and by Goh and Klemperer (1977) from studies of leukemic patients who had been treated by bone marrow transplantation. In this form of treatment both the leukemic and the normal blood-forming cells of a patient are killed by cytotoxic chemotherapy or irradiation before a marrow transplant is given. The studies were conducted on female patients who were transfused with compatible bone marrow cells (those of a brother) to replace their own leukemic cells. The transfused cells repopulated the patients' blood-forming organs, and the patients went into remission. Analysis of the chromosomes of the circulating blood cells showed that they were of donor—in this case, male—origin. When the leukemia recurred, months later, analysis of the chromosomes of the leukemic cells showed them to be the male donor cells. Whereas the donor brothers remained healthy, their transplanted white blood cells (WBC) had become leukemic in their sisters. Fialkow and co-workers suggested that an *in vivo* transmission of a virus had occurred.

In general, epidemiological studies point to a greater than average incidence of leukemia in persons affected by genetic abnormalities or exposed to radiation. However, since most of even these persons do not develop leukemia, we can assume that if proviral DNA is an essential cofactor in the development of leukemia, such genetic material is probably present only in a small percentage of the population.

Attempts to isolate a human virus by injecting leukemic material into experimental animals have proved fruitless; therefore, the scope of investigation has been largely confined to studying human leukemic cells *in vitro*. Repeated efforts by numerous research workers in the United States and Europe to isolate a retrovirus *in vitro* directly from fresh human leukemic cells have been unsuccessful with the recent exception of adult T-cell leukemia virus; the few reported isolates have turned out to be examples of laboratory contamination by various animal C-type viruses. However, the failure to isolate a virus directly from fresh leukemic cells of the most common types does not rule out the possibility of viral involvement in the development of the disease. It has been shown in some animal leukemias that leukemic cells that do not release virus *in vivo* may eventually do so, after prolonged culture, *in vitro*. The virus may be produced either spontaneously or on induction by a chemical agent such as iododeoxyuridine (IUDR), a halogenated pyrimidine nucleoside.

About 15 years ago, with the information gained from animal leukemias, and in view of the experimental failures to isolate an etiological agent directly from fresh human leukemic cells and the necessarily limited scope of experimental studies on human material, a new approach was required. One promising line of investigation appeared to be the propagation of the leukemic cells *in vitro*.

## 2.2. Human T-Cell Malignancy and Retrovirus

The most common form of T-cell malignancy in man is acute lymphoblastic leukemia (T-ALL), which affects mainly children. Mycosis fungoides and Sézary's syndrome are chronic forms which occur mainly in advanced age while T-cell chronic lymphatic leukemia is extremely rare. The above forms of leukemia are nonendemic and occur worldwide. In contrast, adult T-cell leukemia is endemic in Southwest Japan and probably in some regions of Africa. Figure 1 illustrates an approximate ratio, in percentages, of the various types of human leukemia, whereas Table I summarizes the distinction between the different types of the T-cell malignancies.

**Table I**

**Differences among Adult T-Cell Leukemia of Japan, ATL Lymphoma of Blacks, Mycosis Fungoides/Sézary Sundrome, and Acute T-Cell Leukemia**[a]

| | Epidemiology | Family clustering | Disease | Percent mortality in 12 months | Skin lesions | Retrovirus isolated | Association of virus with ATL | Percent of all leukemias worldwide |
|---|---|---|---|---|---|---|---|---|
| ALL | Most common T-ALL, nonendemic worldwide | − | Acute | 50% | − | − | − | 5 |
| MF/SS | Uncommon, nonendemic | − | Chronic | 0 | +− | ? | − | 3 |
| ATL | Endemic Southwest Japan | + | Acute | 90% | +− | ATLV | 1981 | <1 |
| ATLL | Blacks of African origin | + | Acute | 50% | +− | HTLV | 1982 | <1 |

[a]ATL, adult T-cell leukemia; ATLL, ATL lymphoma; MF/SS, mycosis fungoides/Sézary syndrome; T-ALL, T-cell acute lymphoblastic leukemia.

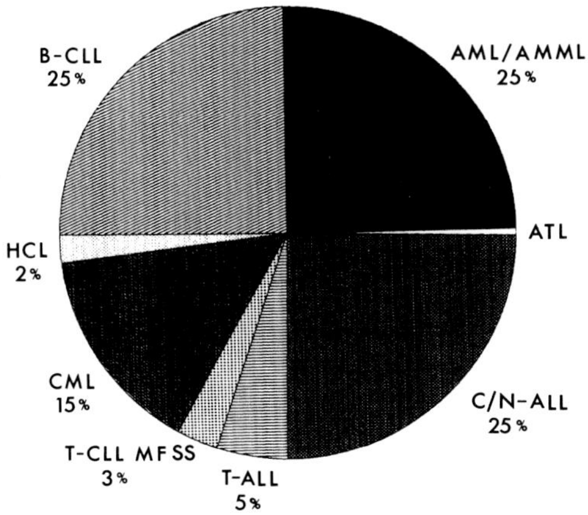


**FIGURE 1.** Outline of the various types of human leukemias and the approximate percentage which each type represents. Note that ATL, which is to date the only human leukemia proved to be caused by a retrovirus, represents probably <1% of the total number. AML, acute myeloid leukemia; AMML, acute myelomonocytic leukemia; B-CLL, B-cell chronic lymphocytic leukemia; CML, chronic myeloid leukemia; HCL, hairy cell leukemia; C/NALL, Common/null cell acute lymphoblastic leukemia; T-ALL, T-cell acute lymphoblastic leukemia; T-CLL, T-cell chronic lymphocytic leukemia; dMF, mycosis fungoides; SS, Sézary's syndrome.

### 2.2.1. Mycosis Fungoides/Sézary's Syndrome/Human T-Cell Leukemia Virus

In 1979 van der Loo and associates reported the morphological development of a C-type virus in lymph nodes of patients with mycosis fungoides and Sézary's syndrome (MF/SS) both indolent T-cell leukemia/lymphoma. In their biochemical study, they found that the particles banded at a density of 1.16 g/ml and contained reverse transcriptase activity. The second report on the presence of retroviruses in human lymphoid cells was in December 1980. Poiesz *et al.* (1980) reported the presence of a C-type virus that contained reverse transcriptase activity in a T-cell line established by Gazdar *et al.* (1980) from a patient diagnosed as having mycosis fungoides. The virus was designated HTLV. Careful study of the work on HTLV shows two distinct periods in the published studies on HTLV. The first is between December 1980 and the early part of 1982. During this period, it was repeatedly claimed that HTLV was involved with MF/SS only. The early published work on HTLV was devoted to the study of reverse transcriptase of the virus (Rho *et al.*, 1981) and to its structural proteins (Kalyanaraman *et al.*, 1981a). A separate paper dealt with the molecular hybridization studies (Reitz *et al.*, 1981). Using a wide range of retroviruses, it was shown that HTLV failed to hybridize with any of the other retroviruses tested. Likewise, the HTLV probe, complementary DNA (cDNA), was also tested against DNA preparations from a wide range of animal species but failed to show any significant level of hybridization, including human DNA preparations from six normal and one ovarian adenocarcinoma tissues. The failure to show any level of hybridization led Reitz *et al.* (1981) to conclude that HTLV is not an endogenous human virus.

However, in spite of the fact that the results of 41 hybridizations are listed, not a single hybridization study of HTLV with fresh leukemic cells from patients with T-cell malignancy is described. Although in the Introduction and Discussion parts of several other papers quoting such data are referred to, it is not in fact available. The experimental documentation of such hybridizations are essential in order to document the association between HTLV and the disease it was alleged to be involved with. The early paper on the isolation of HTLV failed to mention whether the patients who gave rise to HTLV had antibodies to the virus. A second isolate of HTLV was also reported by Poiesz *et al.* (1981) from a patient diagnosed as having Sézary's syndrome. This paper documented the hybridization studies with the patient's fresh leukemic cells; however, the paper indicated only 17% hybridization between the HTLV cDNA and leukemic cell DNA. This low percentage of hybridization could be considered negative, as controlled hybridization with DNA from normal tissue gave 3–11% (Poiesz *et al.*, 1981). In studies by Reitz *et al.* (1981), 16% hybridization between a cDNA probe of HTLV and DNA from Simian sarcoma virus was considered negative.

The report on the second isolation of HTLV by Poiesz *et al.* (1981) was followed by a separate serological study in which it was shown that the sera of only 2 of the 23 patients with T-cell malignancy and 1 of 66 normal persons had antibodies to HTLV (Kalyanaraman *et al.*, 1981b). A turning point in the studies of HTLV occurred in 1982 (Catovsky *et al.*, 1982), when it was recognized in the United Kingdom for the first time that some black patients of West Indian origin had an acute form of T-cell leukemia/lymphoma that both clinically and cytologically resembled the adult T-cell leukemia of Japanese, first described as a separate entity by Uchiyama *et al.* (1977). It also followed the arrival at the National Cancer Institute (NCI) of the Japanese T-cell lines that produced ATLV. This finding made it possible to establish that some blacks may suffer from a disease similar to ATL in Japan and that patients' sera cross-reacted with Japanese ATLV. However, in spite of the recognition in 1982 that some blacks also suffer from ATL, it was still reported up to 1983 (Popovic *et al.*, 1983) that HTLV was involved in mycosis fungoides and Sézary's syndrome (MF/SS), while it had already been established in 1981 that sera of Japanese patients with MF did not contain antibodies to ATLV (Hinuma *et al.*, 1981). Likewise our serological studies of British MF/SS sera failed to detect antibodies to ATLV (Karpas *et al.*, 1983).

### 2.2.2. HTLV-II

The first report on the isolation of HTLV-II appeared in 1982 (Kalyanaraman *et al.*, 1982). It was obtained from the Mo T-cell line, which was derived in 1976 from a spleen of a patient with a T-cell hairy cell leukemia. A second isolate of

HTLV-II was reported in 1986 from a patient with atypical hairy cell leukemia (Rosenblatt *et al.*, 1986). Hairy cell leukemia is an uncommon form of B-cell leukemia, representing approximately 1% of all leukemic cases. Few cases of hairy cell leukemia with T-cell phenotype have been reported. The serological studies with HTLV-II Mo virion in an immunoprecipitation assay showed that the sera from the two T-cell hairy cell patients that gave rise to the two isolates (HTLV-II Mo and HTLV-II NRA) reacted only with the p24 antigen, in both cases. This observation (Rosenblatt *et al.*, 1986) is unlike the reaction of sera from patients with ATL (whether Japanese or blacks) who have antibodies to the other viral structural proteins, including the envelope glycoproteins (gp61), as can also be seen in Fig. 2. False-positive reactions of human sera with p24 of HIV-1 have been reported on several occasions. Therefore, the failure of the two sera from the only two T-cell hairy cell leukemia patients who allegedly gave rise to the HTLV-II isolations is unexpected. False-positive serological studies about the presence of antibodies to HTLV-1 have been reported (Karpas *et al.*, 1985, 1986). Likewise the report from the United Kingdom on the finding of antibodies to HTLV-II in British AIDS patients from London (Tedder *et al.*, 1985) is probably also misleading, since our serological studies of more than 400 HIV-1-infected persons from London showed that only two ATLV (HTLV-1) serum-positive patients could be detected. Using the cell test we were able to determine that sera from 10 ATLV/HTLV-1-infected patients readily reacted with the Mo-T cells that produce the HTLV-II. It is reasonable to expect that HTLV-II-infected persons will readily react with ATLV/HTLV-1-infected cells. Shimotohno *et al.* (1984) cloned and sequenced the complete genome of HTLV-II Mo and demonstrated 60% homology with ATLV/HTLV-1.

In conclusion, it appears that the reported association of HTLV-II with human malignancy is extremely rare and the failure to detect antibodies to the viral envelope and to any of the other structural proteins with the exception of p24, raises serious doubt about the human origin of HTLV-II.

### 2.2.3. Adult T-Cell Leukemia Virus and Disease

In 1977 the first study of an endemic form of leukemia was reported from Japan (Uchiyama *et al.*, 1977). These workers described a form of thymus-derived (T-cell) leukemia, which occurred only in adults in a restricted area of Southwest Japan. It was therefore named adult T-cell leukemia (ATL). Miyoshi *et al.* (1979) reported the establishment of the MT-1 cell line from the leukemic cells of one of the ATL patients. This was achieved without the use of T-cell growth factor (TCGF). Hinuma *et al.* (1981) reported that the MT-1 cells produced a C-type virus and that the expression of the virus and its antigen could be amplified following the treatment of the cells with iododeoxyuridine (IUDR) (Hinuma *et al.*, 1981). Most importantly, they showed that the MT-1 contained

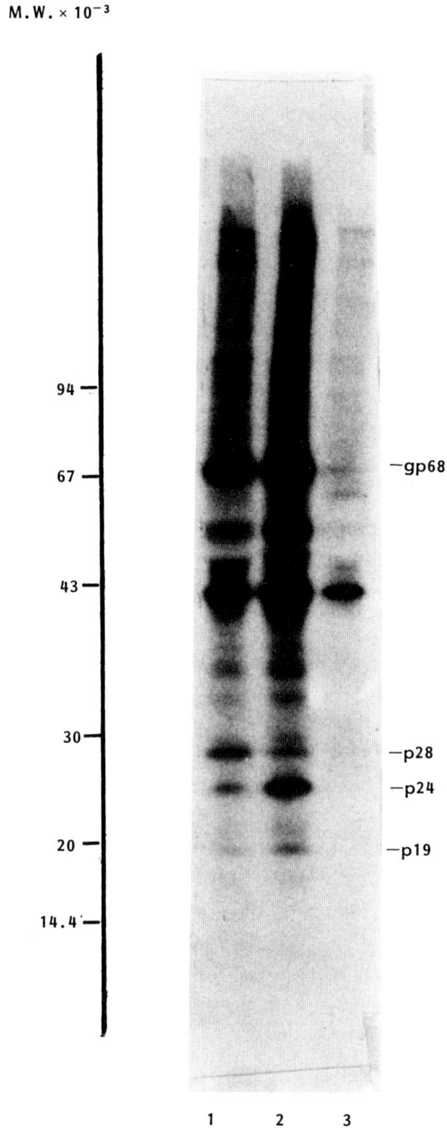


**FIGURE 2.** Immunoprecipitation of MT-2 ATLV proteins by serum from a black British patient with ATL (1) and serum from a Japanese patient (2). Both samples which had a titer of 320, precipitated all the structural proteins of the virus. The only noticeable differences were quantitative rather than qualitative. 3, normal serum.

an antigen that reacted with sera from 100% of the patients with ATL and that 26% of the healthy adults in the endemic area of Southwest Japan had antibodies to the same antigen. By contrast, only a few (2/105) healthy adults from the nonendemic areas of Japan had antibodies to the virus. That same paper also showed that the (viral) antigen that reacted with the ATL sera was not detectable in B or T human lymphoid cell lines derived from non-ATL malignancies. This single paper established a very strong case for the causal relationship of ATLV with the corresponding disease. At about the same time, it was also shown that ATLV can infect, transform and immortalize normal human T-cells following the co-cultivation of leukemic T-cells with normal cord white blood cells (Miyoshi *et al.*, 1981) again without the use of T-cell growth factor. In a subsequent paper, Yoshida *et al.* (1982) provided biochemical characterization of the ATLV. They also provided the most important confirmatory evidence that ATLV is involved with the development of ATL by documenting that the proviral DNA of ATLV was detected in fresh leukemic cells of each of the five patients with ATL used in the study. In this paper it was also shown that each patient's leukemic cells contained a single copy of ATLV and that the site of integration of the provirus within the cellular DNA differed between different individuals, suggesting that there was no specific site of integration. This was followed by a complete nucleotide sequence of ATLV proviral DNA, showing that it was made up of 9032 nucleotides and that the genome contained, in addition to the gag, pol and env genes, an additional region designated pX (Seiki *et al.*, 1983).

So far no paper provides the complete nucleotide sequence of HTLV. Our complete nucleotide sequence of ATLV/HTLV-1 isolated from blacks of Caribbean origin, revealed higher homologies between the black ATLV isolates when compared with the Japanese ATLV sequences (Malik *et al.*, 1988). Unlike the 1977 epidemiological studies concerning the occurrence of ATL in Southwest Japan, the epidemiological studies of ATL in the Western hemisphere have so far failed to provide evidence of the existence of an endemic area. A recent report of ATL in blacks has shown that seven were born in different states of North America, while two were from the Caribbean (Bunn *et al.*, 1983). Screening for antibodies to HTLV in the West Indies has not yet provided published evidence that the Caribbean is an endemic area. Only 3% (11/336) of HTLV-serum positive patients were recorded in St. Vincent (Blattner *et al.*, 1982). When ATLV was used, 1% (2/191) in Dominica were found to be serum positive, while no serum-positive case could be detected in Puerto Rico (0/104) (Hinuma *et al.*, 1983). It appears that certain blacks are likely to be carriers of the virus, since ATLV serum-positive individuals have been found in Africa (Hunsmann *et al.*, 1983). Therefore, it is possible that an endemic area(s) could be found on the African continent. Those persons whose ancestors may have come from the endemic areas in Africa to the Western hemisphere could have harbored the virus and possibly passed it to their spouses or children. Since a large number of blacks from different tribes and locations in Africa were dispersed when brought to the West, no endemic area like that in Southwest Japan is likely to be found on the American continent. The failure so far to detect an endemic area in the United States or United Kingdom, where there are some ATL patients or normal persons with ATLV/HTLV-positive sera, suggests that the virus, although exogenous, is only rarely transmitted by contact, as it is in Japan, from husband to wife (Tajima *et al.*, 1982) or from mother to child (Kumoro *et al.*, 1983). Intercontinental transfer of ATL and virus was recently documented in the case of a 49-year-old Japanese woman, who was born in the endemic area of Southwest Japan and who emigrated to the United States 26 years before developing ATL (Haynes *et al.*, 1983). Two intercontinental transfers, i.e., from Africa to the West Indies, and more recently to Britain, would explain the appearance of the disease and virus in the United Kingdom.

Seroepidemiological studies from Japan indicate that as many as 1 million Japanese may be infected with ATLV. Most of these people live in the endemic areas of Southwest Japan. These studies also indicate that approximately 1 in 1000 of ATLV-infected males and 1 in 3000 of ATLV-infected females in Japan are likely to develop the disease (Y. Hinuma, personal communication). By contrast, the distribution and incidence of ATLV/HTLV infection on the African continent have not yet been established. A recent review contains further data on ATLV/HTLV and disease (Hinuma, 1985).

**2.2.3a. Transmission of ATLV.** The natural modes of transmission are (1) sexual intercourse, (2) mother to baby via infected milk, and (3) *in utero*. The non-natural forms of transmission are in drug abusers via contaminated syringes, and recipients of blood transfusions. Retrospective study on the transmission of ATLV following blood transfusion showed that only 63.4% (26/41) seroconverted after transfusion of whole blood while none of the 14 recipients of plasma seroconverted (Okochi *et al.*, 1984).

**2.2.3b. Studies of Adult T-Cell Leukemia in the United Kingdom.** It was in the United Kingdom that adult T-cell leukemia (ATL) was first recognized to affect blacks as well (Catovsky *et al.*, 1982). The patients were of West Indian (African) origin, and to date not a single case of ATL has been diagnosed in British Caucasians. We obtained two isolates of ATLV (1010 and 1025) (Table II) following the co-cultivation of the patients' peripheral WBC with cord WBC (Karpas *et al.*, 1987). Our serological studies confirmed that ATL patients, whether Japanese or blacks, also have antibodies to all the structural proteins of the virus (see Fig. 2 and Table III). This finding is in sharp contrast to the patients' sera from whom HTLV-II was isolated whose sera reacted only with p24 of HTLV-II (Rosenblatt *et al.*, 1986). In contrast to the high prevalence of antibodies to ATLV in South West Japan none of the 70 sera samples obtained from healthy blacks of Caribbean origin in London contained antibodies to the virus (Table II).

When we screened sera from 400 HIV-positive homosexuals in London for antibodies to ATLV, using the cell test (Karpas *et al.*, 1985), only two were positive. Both were born in the United Kingdom, one was black of Caribbean origin and the other white with a black sexual partner. This very low rate of

**Table II**
**Summary of Serological Studies**[a]

| Group | ATLV/HTLV-I | HIV |
|---|---|---|
| British | | |
| Adult T-cell leukemia (blacks) | 3/3 | 0/3 |
| Mycosis fungoides/Sézary syndrome | 0/20 | 0/20 |
| Hairy cell leukemia | 0/6 | 0/6 |
| T-cell-acute lymphoid leukemia (T-ALL) | 0/15 | 0/15 |
| Italian (from Naples) | | |
| Children with T-ALL | 0/6 | 0/6 |
| Relatives of these children | 0/16 | 0/16 |
| Japanese | | |
| Adult T-cell leukemia | 5/5 | 0/5 |
| Mycosis fungoides/Sézary syndrome | 0/12 | ND |
| Israeli | | |
| Falashas | 0/56 | 0/56 |
| Americans with AIDS | 0/35 | 35/35 |
| Male homosexuals | | |
| British (London) 1985, homosexuals | 0/226 | 61/226 |
| HIV positives | 2/400 | 400/400 |
| Multiple sclerosis | 0/60 | 0/60 |

[a]The immunoperoxidase (IP) cell test was used to determine the presence of antibodies to ATLV/HTLV-I and HIV.

antibodies to ATLV is in sharp contrast to the report of 5% incidence in a similar population in London (Tedder *et al.*, 1985). Likewise our molecular hybridization studies confirmed that there was no cross-hybridization between ATLV and HIV (Fig. 3). When we screened sera from patients with other hematological malignancies as well as multiple sclerosis, none was found to possess antibodies to ATLV (Table II).

## 3. DISCORDANT CLAIMS AND CONTRIBUTIONS

### 3.1. Retroviral RNA-Directed DNA Polymerase in Human Leukemia

After Temin and Mizutani (1970) and Baltimore (1970) discovered RNA-directed DNA polymerase (reverse transcriptase) in avian and murine retroviruses, it was reported that similar reverse transcriptase (RT) activity was pres-

**Table III**
**Serological Studies of Various Cells[a,b]**

| Cells used | Japanese ATL | British ATL 1010 | British ATL 1025 | Mycosis fungoides | Sézary's syndrome | T-ALL | ATLV p19, p28 monoclonal antibodies |
|---|---|---|---|---|---|---|---|
| | Origin of sera samples | | | | | | |
| MT-2 line | + | + | + | − | − | − | + |
| Fresh ATL Patient 1025 | − | − | − | − | − | − | − |
| Karpas 1025 Umbilical cord T-cell line | + | + | + | − | − | − | − |
| Karpas 1025 Umbilical cord B-cell line | ND | ND | ND | ND | ND | ND | − |
| Karpas 1010 T-Cell line | + | + | + | − | − | − | − |
| Karpas T-ALL line | − | − | − | − | − | − | − |

[a]ATL, adult T-cell leukemia; T-ALL, acute T-cell lymphoblastic leukemia; +, present; −, absent; ND, not done.
[b]The immunofluorescence test (IF) was used in screening for the presence of ATLV/HTLV-I antigens in the various cells. As can be seen the fresh leukemia T cells of patient 1025 did not express ATLV antigen. Patient 1010 did not have circulating leukemic T-cells. The EBV-infected B-cell line of patient 1025 did not contain the p19 and p28 of ATLV; it was not tested by the indirect IF test against human Ig, since it produced Ig.

ent in fresh human leukemic cells from patients with acute lymphoid leukemia (Gallo *et al.*, 1970). This paper triggered an extensive study in numerous research centres, including our own, which attempted to confirm the presence of reverse transcriptase in human leukemic cells. These studies, which lasted several years and generated a large number of scientific publications, failed to confirm that fresh human leukemic cells, which were free of retrovirus, contained in their cytoplasm such viral reverse transcriptase activity. The polymerase activity detected in the leukemic cells was believed to be not viral but cellular. Indeed, Dr. M. Bishop in his 1980 review article on the molecular biology of retrovirus sums up the claimed discovery of RT in human leukemic cells and the large number of scientific articles that followed it by saying: "Reverse transcriptase has not yet been identified with any assurance in uninfected cells, indeed it may be unique to retrovirus. . . ." (Bishop, 1980).

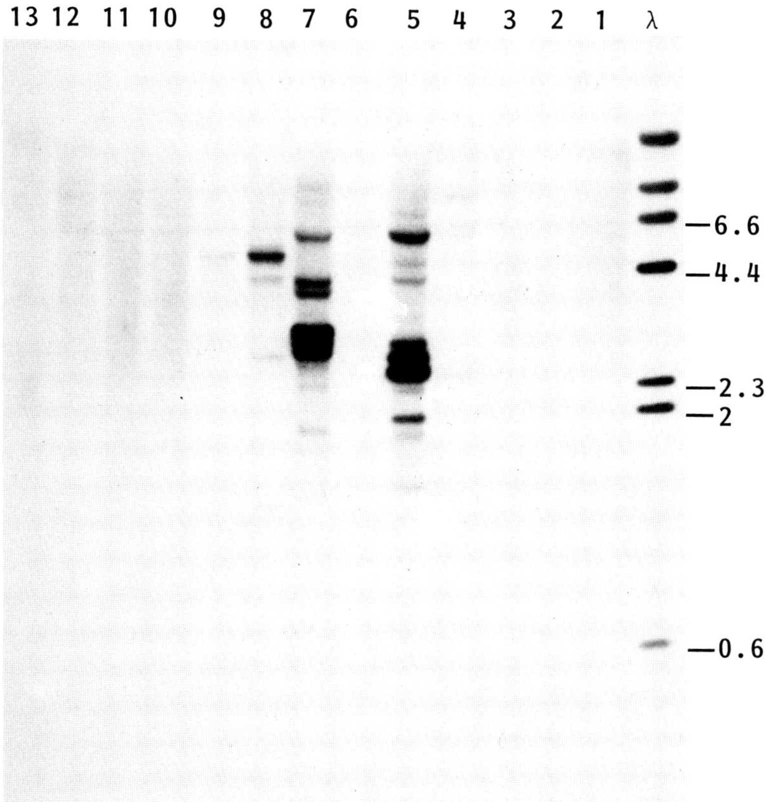


**FIGURE 3.** Southern blot hybridization for the presence of ATLV/HTLV-I (a) or HIV (b) proviral DNA sequences in cells from 1, Sézary's syndrome; 2, Karpas T-ALL cells infected with C-HIV; 3, Karpas T-ALL line; 4, Karpas 1010 T-cell line; 5, Karpas 1025 T-cell line; 6, Karpas 1025 B-cell line; 7, MT-2 cells; 8, MoB cells containing the HTLV-II genome; 9, a suspension of Karpas T-ALL and MoB cells; 10, HIV-infected CEM cells; 11, T-ALL CEM cells; 12, normal human spleen; 13, Karpas T-ALL cells co-cultivated with ATLV-producing 1010 cells. *Hind*III digested bacteriophage was used as a size marker. ATLV/HTLV-I related proviral sequences (a) could be detected only in the ATLV/HTLV-I infected cells (MT-2, 1010, 1025) and HTLV-II-infected MoB cell DNA, while HIV proviral sequences (b) could be found only in the DNA of the CEM cells infected with a French strain of HIV and in the DNA of the Karpas T cells infected with the Cambridge HIV (From Karpas, Malik, and Lida, 1987).

## 3.2. Retroviral Isolation from Human Leukemic Cells

### 3.2.1. HL23 Virus

Following the discovery of RNA-directed DNA polymerase, several reports on the isolation of retroviruses from fresh human leukemic cells turned out to be

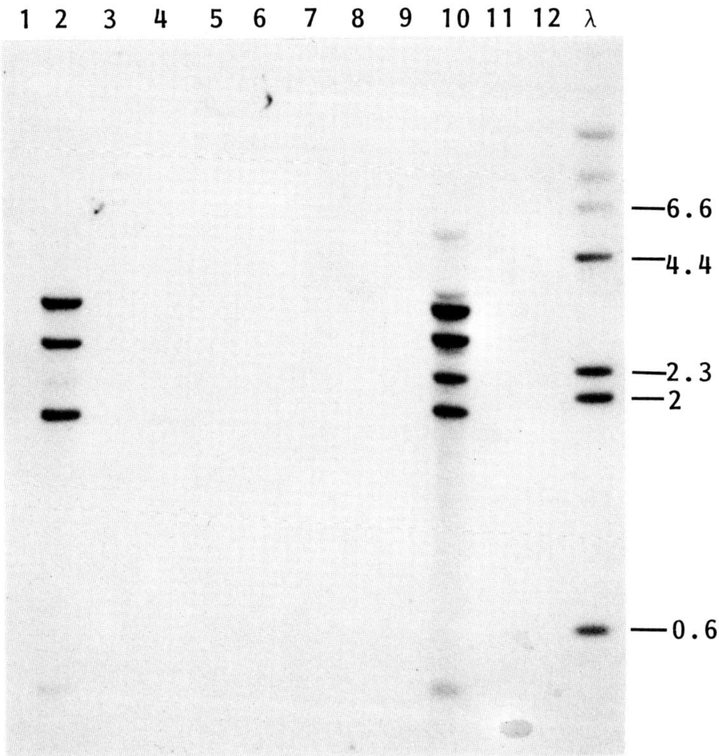


**FIGURE 3.** (*Continued*)

laboratory contaminations by animal retroviruses. However, once the virus in question was identified as an animal rather than a human retrovirus, it was usually accepted as an accidental contamination. However, the reported isolation of a human retrovirus (Gallagher and Gallo, 1975) designated as HL23 virus from fresh leukemic cells stands out. It was claimed that "although closely related, the HL23V-1 virus and SSV are not identical . . . " and that "genomic differences do exist between HL23-1 (A-204) and SSV. . . ." (Teich *et al.*, 1975). Two groups of investigators from the NCI (Okabe *et al.*, 1976) and Columbia University (Ohino *et al.*, 1976), who studied the HL23 virus isolate, established that the virus in question was a mixture of monkey viruses, the gibbon ape virus and its sarcoma counterpart (SSV) as well as the baboon endogenous virus (BaEV), which is a distinct primate retrovirus. In these two papers, published in *Nature* (March 13, 1976), both groups made the point that by using immunological and molecular hybridization studies, the HL23 virus was a mixture of those primate retroviruses. Despite conclusive evidence on the properties of HL23 virus isolates, a further major article appeared in *Nature* in which it was claimed that fresh human acute myeloid leukemia cells contained baboon endogenous virus sequences (Wong-Staal and Gallo, 1976).

This was followed by a serological study that reported that "of 39 sera, 100% showed reactivity against baboon endogenous virus, whereas only 49% showed reactivity against simian sarcoma-associated virus. . . ." (Kurth *et al.*, 1977). This claim prompted further work by two different teams that showed that the tests employed were in fact an uncontrolled artifact (Snyder and Fleisner, 1980; Barbacid *et al.*, 1980).

### 3.2.2. HTLV and ATLV

For approximately 2 years, it was repeatedly claimed that HTLV was involved in the development of mycosis fungoides (MF) and Sézary's syndrome (SS) (Gallo and Reitz, 1982; Popovic *et al.*, 1983). It may not be unreasonable to suggest that the HTLV that preceded the arrival of ATLV at the NCI was distinct from the HTLV that was subsequently isolated from black patients with ATL. We, like other groups, spent considerable time to verify these early claims about the involvement of ATLV/HTLV in MF/SS but were unable to confirm them (Karpas *et al.*, 1983, 1987). Only after it was recognized in the United Kingdom that some blacks may also suffer from a disease similar to the ATL of Japan was it accepted that HTLV/ATLV is not involved in MF and SS. The early reported studies of ATLV of Japanese and HTLV of blacks already showed some biological differences between the black and Japanese strains of the virus involved in the development of ATL:

1. The Japanese ATLV infected not only T but also B (EBV-infected) cells (Yamamoto *et al.*, 1982), while HTLV infected only T cells (Gallo *et al.*, 1983). Indeed, our studies of ATL from a British black subject confirm that the malignant T cells contained the provirus while the EBV-infected B lymphoblasts from black ATL patients were negative for the virus (Karpas *et al.*, 1987).
2. The Japanese ATLV can relatively easily infect, transform, and immortalize T cells, which also grow very well *in vitro* without T-cell growth factor (TCGF). By contrast, the initial report about HTLV indicated that only T cells from relatives could be infected and that their growth was dependent on the addition of TCGF (Ruscetti *et al.*, 1983). Although we succeeded in infecting cord T cells with two independent isolates from British blacks with ATL, the growth of the infected T cells depended for approximately 1 year on the periodic addition of fresh cord cells.

3. None of the epidemiological studies of blacks shows the high degree of family clustering shown in Japan (Tajima *et al.* 1982). This might reflect the lower level of infectivity by the African viral strain of ATL.
4. Our molecular cloning and complete nucleotide sequence of the ATLV isolated from British black show higher homologies between isolates originating from similar geographical areas—approximately 1% divergence as compared with 2.3% divergence (Malik *et al.*, 1988).

Other facets of the studies of HTLV led to publication of what we know as incorrect data:

1. The first report on the infective nature of HTLV claimed that only cells from relatives of patients can be infected by the virus (Ruscetti *et al.*, 1983). It also stated that "all HTLV-1 infected cultured cells always remained dependent on the addition of TCGF for growth." This was in sharp contrast to the work from Japan, which showed that ATLV can transform and immortalize T cells of unrelated donors and that the infected cells can grow without TCGF, i.e., interleukin-2 (IL-2).
2. Although several papers claimed that HTLV provirus sequences were present in fresh leukemic T cells of acute lymphoblastic leukemia (ALL) and Sézary's syndrome, experimental evidence to support such claims was not provided (Kalyanaraman *et al.*, 1981a, 1982; Gallo and Reitz, 1982).
3. The paper that claimed that HTLV provirus had a common site of integration in the malignant ATL cells had to be retracted (Hahn *et al.*, 1983).
4. The paper that claimed homology between class 1 HLA and HTLV (Clarke *et al.*, 1983) also turned out to be untenable (Jennings and Even, 1984).
5. It was also claimed that ATL was prevalent in Southern Italy (Pandolfi *et al.*, 1985). However, our studies of sera from patients with T-cell leukemia and their relatives did not reveal any positive cases (see Table IV).

### 3.3. Adult T-Cell Leukemia Virus ATLV/HTLV in AIDS

In May 1983 three papers that appeared in *Science* claimed that HTLV-1 was involved in AIDS (Gallo *et al.*, 1983, Gelmann *et al.*, 1983; Essex *et al.*, 1983). Subsequently, five more papers about the involvement of HTLV-1 in AIDS were published (Schupbach, Sarngadharan and Gallo, 1984; Essex *et al.*, 1983; Evatt *et al.*, 1983; Mathiz *et al.*, 1984; Essex *et al.*, 1984). Once Gallo's group succeeded in propagating the AIDS virus isolate ($LAV_{BRU}$) they had received from France (Montagnier, 1984), they changed its name to HTLV-IIIB. This was probably done to avoid a retraction of their earlier publication on the involvement of HTLV-I in AIDS. In order to justify the naming of the causative agent of AIDS as HTLV-III, they produced data on the cross-hybridization between the leukemia virus and the AIDS virus (Hahn *et al.*, 1984; Arya *et al.*, 1984). No other group, including ourselves (see Fig. 3), was able to reproduce this hybridization because, unlike the ATLV/HTLV-I, which belong to the Oncoviridae, the AIDS virus belongs to the lentiviridae (Karpas *et al.*, 1987).

### 3.4. HTLV in Multiple Sclerosis

Reports by the groups of Koprowski and Gallo that human T-cell lymphotropic retroviruses may be involved in the development of multiple sclerosis (MS) has aroused much interest (Kaprowski *et al.*, 1985). The report was based largely on immunological evidence, using enzyme-linked immunosorbent assays (ELISA) with viral antigens or disrupted virions, and we have accordingly sought confirmation by screening sera and cerebrospinal fluid (CSF) from MS patients against cell lines infected, respectively, with adult T-cell leukemia virus (ATLV/HTLV-I) of Japanese cells (MT-1 and MT-2 lines) and separately with our own isolate from British black patients with ATL. We also used the MoT cell line, which produces HTLV-II and our own T-cell line containing a local isolate of lymphadenopathy AIDS virus (C-LAV). We have used the cell-based immunoperoxidase (IP) method, which we developed and found to be a highly sensitive test (Karpas *et al.*, 1985), both for antibodies to the AIDS virus as well as to ATLV.

As can be seen in Table II, we have tested altogether 46 samples of sera that originated from MS patients from the United Kingdom, Sweden, and Germany; in addition to the sera, we also had CSF samples from 15 of these patients. The CSF was also tested for the presence of antibodies to the same retroviruses. In addition, we tested 20 paired sera and CSF from patients with other neurological disorders. In order to control the specificity of the reaction, we have used sera both from Japanese and black British patients with ATL, as well as sera from healthy persons found to be infected with the AIDS virus. Table II, which summarizes our findings, indicates that none of the MS patients had circulating or CSF antibodies to ATLV/HTLV-I, to HTLV-II, or to the AIDS virus, either in their sera or in their CSF.

In our preliminary studies, we tried to isolate a retrovirus from the peripheral WBC of two patients with MS. The WBC cells were separated on Ficoll before being divided into two parts. One part was co-cultivated with fresh human cord WBC and the other part was cultured with the Karpas T cells. If ATLV/HTLV-I or HIV-I were involved in MS, one could expect the cord WBC to become infected and express the viral antigen. By contrast, if an AIDS-like lentivirus was involved, one would have expected cell lysis of both the cord T-cells as well as of our T-cell line, or at least the development of an antigen that will react with sera containing antibodies to the AIDS virus, or ATLV.

The two cultures of cord cells, as well as the two cultures of our T-cell line, were tested for the expression of ATLV/HTLV-I and HIV-1 after 2 and 4 weeks in culture, using the IP method. Neither the cord WBC cultures nor our T-cell line express any antigens that cross-reacted with human sera that contained antibodies to ATLV nor to LAV sera. Since we succeeded in isolating ATLV following the co-cultivation of a peripheral blood that did not contain detectable levels of ATL cells, one could have expected that if a similar virus is involved in MS, at least one of the cord WBC should have become infected. Likewise, since sera of patients with ATL contained antibodies which cross-reacted with HTLV-II, one could have expected the sera and CSF of such patients to contain antibodies which would cross-react with ATLV/HTLV-I or HTLV-II-infected cells if a related virus is involved in MS.

As can be seen in Table II, we failed to detect antibodies to the oncoviruses (ATLV/HTLV-I, HTLV-II) or to the lentiviruses (HIV-1) in any of the 46 sera and 15 CSF from MS patients or in the sera/CSF of 20 other patients with neurological disorders. In addition to our failure to demonstrate an association between ATLV and HIV in MS, an independent group of investigators failed as well (Karpas *et al.*, 1986; Hauser *et al.*, 1986).

The claimed presence of antibodies that cross-reacted with the known human retroviruses is based on ELISA tests. We now know that some of the ELISA tests give a high rate of false-positive results. The report of 37% of ATLV/HTLV-I positive in Israeli Falashas turns out to be misleading, probably because of the deficiency of the ELISA used (Karpas *et al.*, 1986). Likewise, it has been repeatedly reported that some of the ELISA AIDS test kits as well as Western blots used for AIDS screening give a high rate of false-positive results (Barnes, 1987). The use of such ELISA methods could probably explain the premature claims to the involvement of HTLV in MS. It appears therefore that the possible viral involvement of MS is still a challenging research target. Should a retrovirus eventually be isolated from MS, it is likely to be distinct from the known human retroviruses.

### 3.5. HTLV-III

Human T-cell leukemia virus type III was the original name given by Dr. Gallo's group to the virus that the French isolated from a patient at risk for AIDS. Subsequently the meaning of the L in HTLV was changed from leukemia to lymphotropic and the justification for naming the AIDS virus also HTLV was the claim that it cross-hybridized with ATLV/HTLV-1 (Hahn *et al.*, 1984; Arya *et al.*, 1984). However, these studies could not be confirmed (see Fig. 3). Thereafter, the International Committee on the Nomenclature of Viruses decided to rename the virus human immunodeficiency virus (HIV).

### 3.6. HTLV-IV

After Daniel *et al.* (1985) gave a sample of the simian immunodeficiency virus they isolated from monkeys with AIDS-like symptoms to Dr. M. Essex, his group reported the isolation of a new human immunodeficiency virus from West Africa, which they named HTLV-IV (Kanki *et al.*, 1986). More recently it was established that Essex HTLV-IV was in fact a laboratory contamination of the human cells with the SIV strain they had received from Daniel (Mulder, 1988).

### 3.7. HTLV-V

Recently Manzari *et al.* (1987) reported from Italy the isolation of a new human retrovirus from T-cell lymphoma/leukemia. In spite of the fact that it has been established that ATLV/HTLV-1 belongs to the Oncoviridae, while HIV belongs to the Lentiviridae. Manzari and associates claimed that HTLV-V is related to both HTLV-1 and HIV. However, although they claim that the electron micrographs are of C-type morphology, in fact those illustrated are distinct from any known oncovirus. It also appears that the cells that proliferate *in vitro* are not the patient's malignant cells. More data will be needed to establish whether HTLV-V is indeed a new human retrovirus.

## 4. HUMAN LEUKEMIA *IN VITRO*

Unlike normal and malignant human fibroblasts and epithelial cells, which have been grown successfully in numerous laboratories for many years, human hemic cell lines from leukemic patients have been established by only a few investigators. Hemic cell lines have also been established from the peripheral blood and bone marrow of normal patients. However, until recently it was not possible to learn with certainty the nature of the cell lines, i.e., to determine whether those derived from leukemic patients represented an *in vitro* proliferation of the patient's malignant or normal cells. Cytological and cytochemical characteristics have generally been sufficient to establish the nature of the fresh cells but are often insufficient for cells that have undergone long-term *in vitro* growth, in which they may lose their characteristic features.

An important development in the study of cultured human hemic cells, and of leukemic cells in particular, was the discovery by Reedman and Klein (1973) of a simple test to detect cells that are latently infected with Epstein–Barr virus (EBV), the causative agent of infectious mononucleosis. Approximately 90% of a given population will be infected at some time with EBV; however, most affected persons do not develop a clinical illness but continue to carry latent virus in some lymphoid cells. During *in vitro* culture of hemic cells from a leukemic patient previously infected with EBV, even a few EBV-positive lymphoblasts derived from normal circulating B lymphocytes are sufficient to outgrow the leukemic cells. Fresh leukemic cells are usually free of EBV; for this reason, the examination of cultured cells for EBV is one of the first tests to determine whether a newly established cell line from a leukemic blood or bone marrow represents a proliferation of the patient's malignant cells or his normal, i.e., EBV-infected, cells.

## 4.1. Virus-like Particles

Epstein–Barr virus-negative hemic cells have been established only from patients with hematological malignancies, and all are thought to represent an *in vitro* proliferation of the patients' malignant cells. To the best of my knowledge, no one has ever succeeded in establishing *in vitro* an EBV-negative hemic cell line from a normal patient. In our laboratory, repeated electron microscopic examination of 15 human leukemic cell lines established *in vitro* has not shown any morphologically detectable known virus particles. However, I have been able to detect in three of the cell lines the spontaneous formation of virus-like particles (VLP) within the cytoplasm of the leukemic cells (Karpas *et al.*, 1980; Karpas and Fischer, 1980). Similar VLP aggregated within so-called inclusion bodies have been reported over the past 20 years by several investigators in the cytoplasm of fresh human leukemic lymphoblasts and myeloblasts. A systematic search for inclusion bodies in children with acute leukemia demonstrated that such VLP-containing inclusions could be found in the fresh leukemic cells of 12 of the 28 patients studied (Smith, 1978).

It is impossible at this stage to be certain how human VLP relate to the animal leukemia viruses; however, the ability of IUDR or of bromodeoxyuridine (BUDR) to induce the expression of cell-integrated viral information is well documented. A 40- to 50-fold increase in the number of VLP-containing cytoplasmic inclusion bodies (following culture of three leukemic cell lines in the presence of BUDR) therefore constitutes support for the hypothesis that these bodies represent the synthesis of viral structural proteins (Karpas and Fischer, 1980). The reason for the expression of what can be interpreted as incomplete particles in the human leukemic cells is not yet known. Particle formation may be incomplete because of incomplete proviral information in the cells, or there may be a block in particle assembly.

Any discussion of the nature of VLP and their possible relationship to leukemia in humans must first take into account two major sources of potential confusion and error. One is the possibility of a laboratory-borne contamination of the human cells with one or more extraneous viruses. The second concerns the distinction between VLP and normal cellular organelles, which can have a certain morphological resemblance to viruses. The possibility of cross-contamination is unlikely, mainly because the cells do not contain any cytopathic agent, nor has any cross-reaction been observed between antisera against the major groups of animal leukemia viruses and the human leukemic cell line established in our laboratory. Furthermore, vesicular stomatitis virus has been used by Zavada in a most sensitive test to confirm that the VLP-producing cell lines are not contaminated with any of the known animal leukemia viruses.

A careful ultrastructural comparison with cellular organelles such as multivesicular bodies permits easy discrimination from VLP. Another point of distinction is that no known cellular organelle increases drastically in number in cells dying from IUDR treatment.

Because a conventional experimental model, i.e., an animal susceptible to human leukemia, is not available, different approaches must be explored in order to show whether VLP are involved in the development of the leukemic transformation. Experimental studies have recently been initiated in two directions. One is to find out whether the proteins of VLP are also present in fresh leukemic cells but not in normal white blood cells. This will be done by the preparation of antibodies specific to the purified VLP proteins, which will be used in a test against both leukemic and normal cells. The second is to determine whether the VLP-associated nucleic acid is responsible for the malignant state of the cells. This would be the case if the leukemic cell genome contained copies of viral nucleic acid in an amount several orders of magnitude higher than that found in normal cell genomes. If the answer to one or both of these questions turns out to be positive, the association of VLP with human leukemia will have been established.

### 4.2. Production of Complete Retrovirus in Cultured Leukemic Cells

One of the leukemic cell lines that produced virus-like particles (Karpas 241 cells) was kept in liquid nitrogen for about 8 years. A vial of cells that was recently thawed was grown *in vitro* for about 3 months before re-examination by electron microscopy. In addition to the VLP reported previously, the cytoplasm of the cells also contained membrane-bound inclusion bodies. The contents of the inclusion bodies were not uniform in appearance; some contained distinct forms of enveloped virus particles in an electron-dense matrix (Fig. 4a). The envelope of the distinct virus particles was denser than that of the matrix and varied in diameter within the inclusion. An electron-dense core could be seen in most particles. The ultrathin sections of some of the particles showed that the core was surrounded by a clear limiting membrane (Fig. 4b), which appeared cylindrical in shape. Some of the larger enveloped virus forms contained two distinct cores (Fig. 4c). Likewise, in some of the enveloped forms, the core appeared to be acentric. Occasionally, budding forms from the inclusion membrane into the center of the sac could also be observed.

Many of the cytoplasmic inclusions contained what appeared to be a range of morphologically distinct particles, some of which were similar to the VLP reported earlier, as well as a wide range of intermediate forms and what seemed to be mature (enveloped) particles. Many inclusions nearer to the cell membrane appeared to contain mainly morphologically complete particles, i.e., virus with an electron-dense core surrounded by an electron-lucent matrix within an outer envelope. The cytoplasm also contained small electron-dense particles, which lay in clusters but not within a limiting membrane (Fig. 5a). These electron-dense particles could be seen at the site of extracellular budding near the cell membrane (Fig. 5b). Although typical budding forms could be found, most of the extracellular particles appeared to be localized in cluster forms, probably due to the emptying off of the virus containing sacs into the outside (Fig. 6a–c).

In the cells that grew in the presence of BUDR, numerous empty loop formations could be seen at the cell membrane, consisting of a double membrane with numerous complete forms of virus around. High-power magnification of individual particles showed that the membrane contained knoblike projections. These individual complete particles could be distinguished morphologically both from ATLV and HIV (Fig. 7). Although ATLV was more similar to the particles produced by the 241 cells, each ATLV particle contained an electron-dense core that occupied most of the space within the envelope with a very narrow electron-lucent rim. However, none of our ATLV appeared to contain a distinct limiting membrane around the core, nor did any of the ATLV particles described in Japanese ATL cells.

The retrovirus produced in the 241 cell line appears to belong on morphological grounds to the oncoviridae subgroup. The particles are essentially spherical, which in cross section appear to differ in diameter. The individual particles are made of an electron-dense core surrounded by an intermediate inner cylindrical membrane (Fig. 8) enclosed within an external envelope. This outer envelope exhibits spikelike projections (Fig. 8). Although most enveloped forms had a diameter of approximately 100 nm, there were also numerous particles with a significantly larger diameter. This was not due so much to variation in the size of the cores, but mainly to a larger envelope. In many sections, virus particles with a double nucleotide core could be seen (Figs. 4c and 8d). Similar forms have been reported in B-type, i.e., mouse mammary tumour virus (MMTV), rather than in C-type virus (Sarkar *et al.*, 1973).

Lying free within the cytoplasm of the cells were clusters of electron-dense structures suggestive of viral cores, partially surrounded by a less dense material resembling the membrane of the virus particles. Further support that these cytoplasmic particles represent viral cores can be seen in Fig. 5b, which shows the assembly of such a core into a budding formation.

Again, it is important to emphasize that such electron-dense particles within the cytoplasm are often seen in MMTV-infected cells. Where the complete 241 particles differ greatly from ATLV is in the detailed anatomy of the virus. In ATLV the electron-dense core fills most of the space within the envelope, regardless of the diameter of the particle (Fig. 7c). By contrast, the core even in

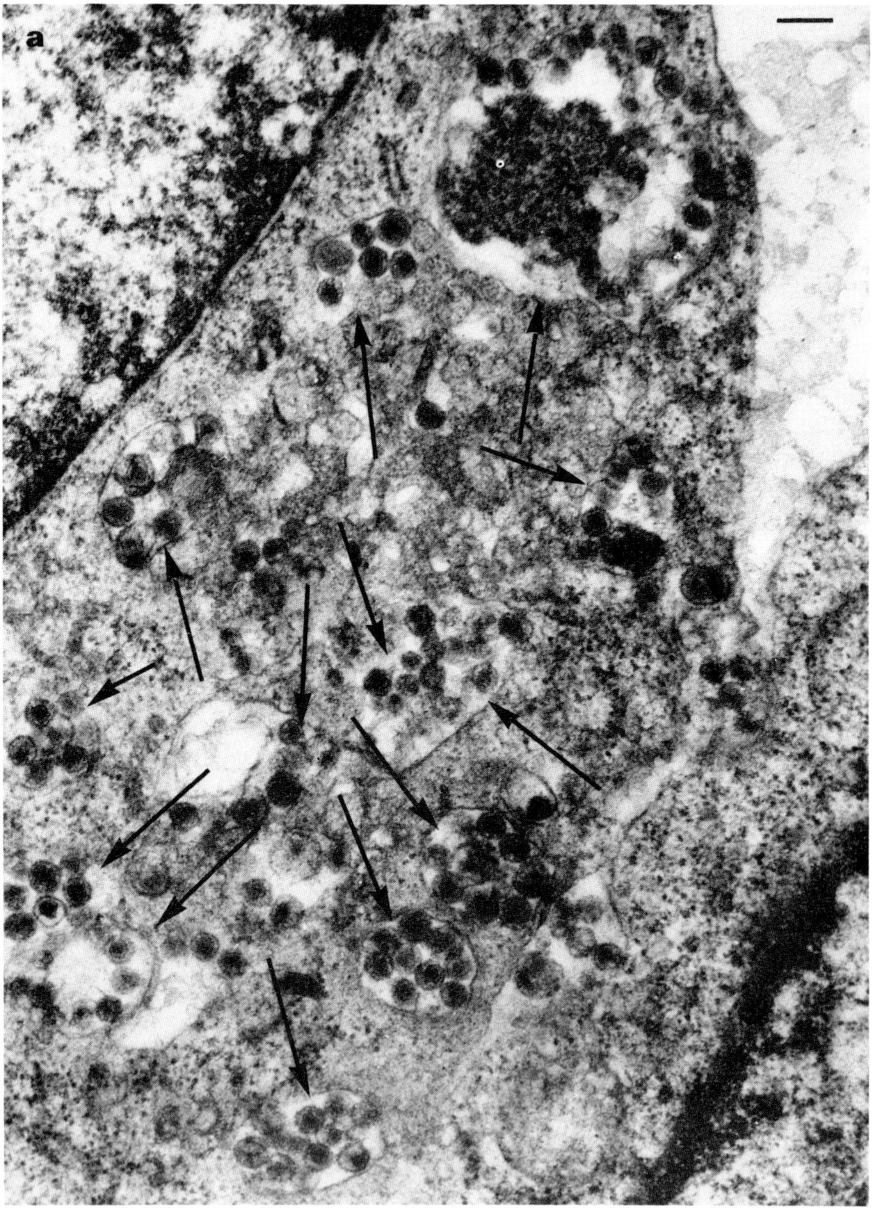


**FIGURE 4.** (a) Ultrathin section through the cytoplasm of Karpas 241 cells cultured in the presence of BUDR, showing numerous membrane-bound inclusions (→) containing a large number of virus particles. (b) High magnification of two such inclusions showing distinct enveloped forms of virus particles lying within an electron dense matrix (→). In some of the virions, the core appears

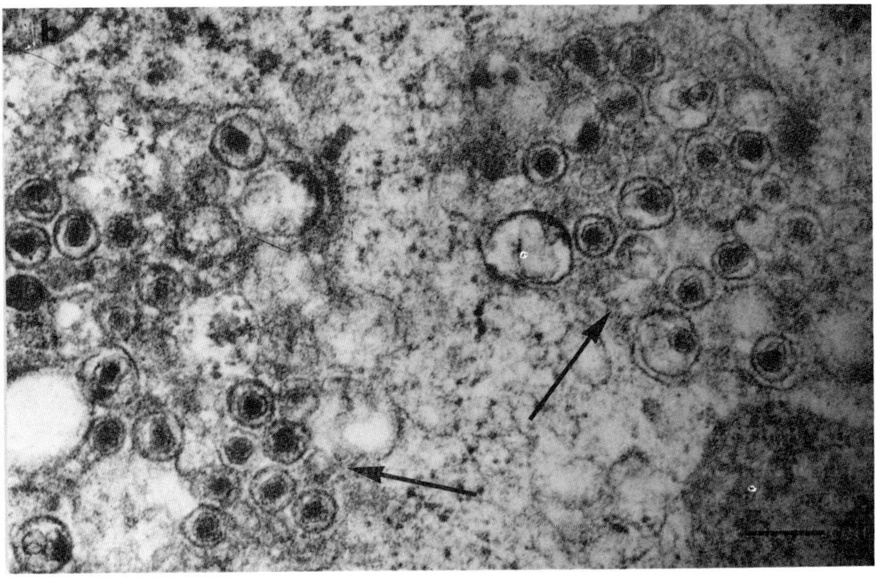

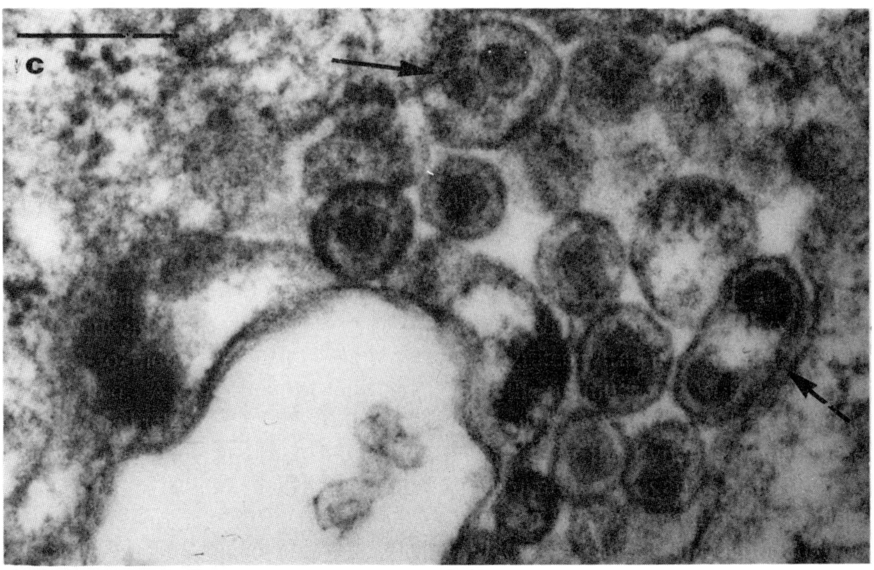

excentric. (c) Section through an inclusion body that also contains two large particles, each containing two distinct cores (→). Bar = 200 nm.

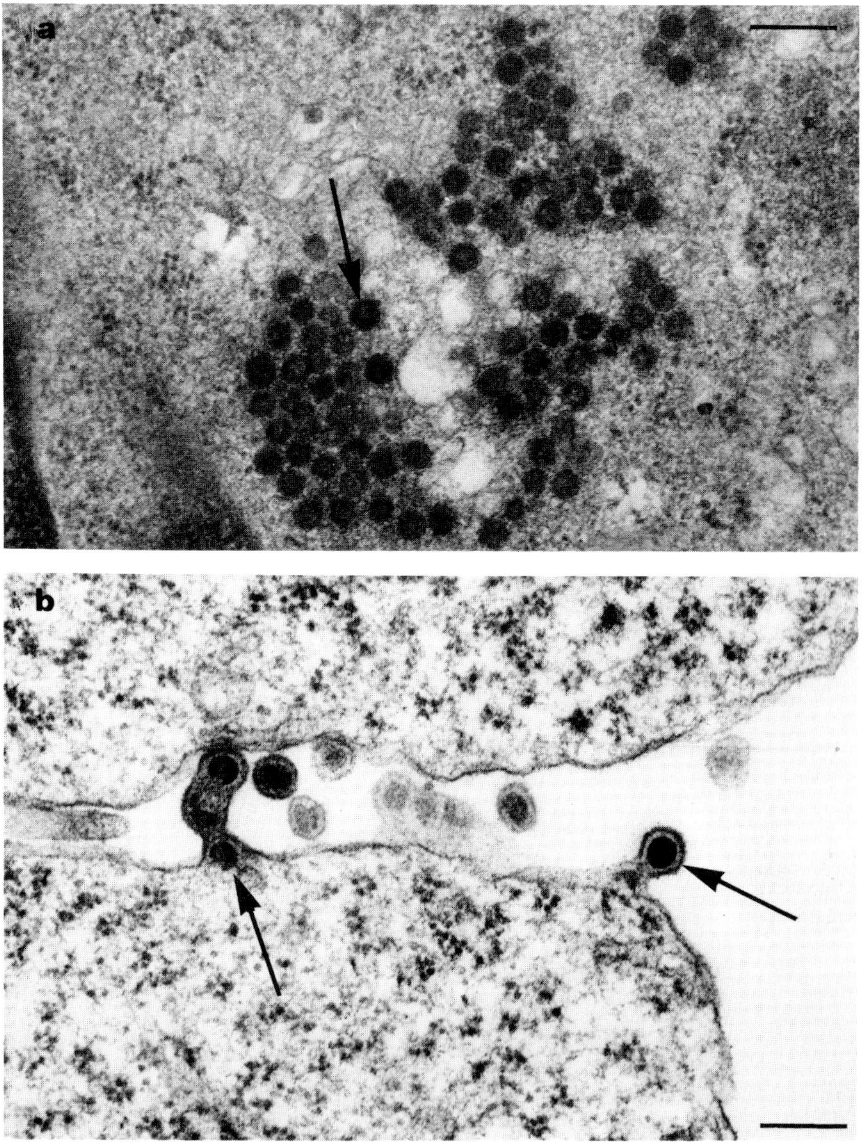


**FIGURE 5.** (a) Cluster of cytoplasmic electron-dense particles resembling those seen in human cells that produce B-type particles. (b) Assembly of such electron-dense particles into a budding formation. This ultrathin section also shows typical oncovirus in budding form (→). Bar = 200 nm.

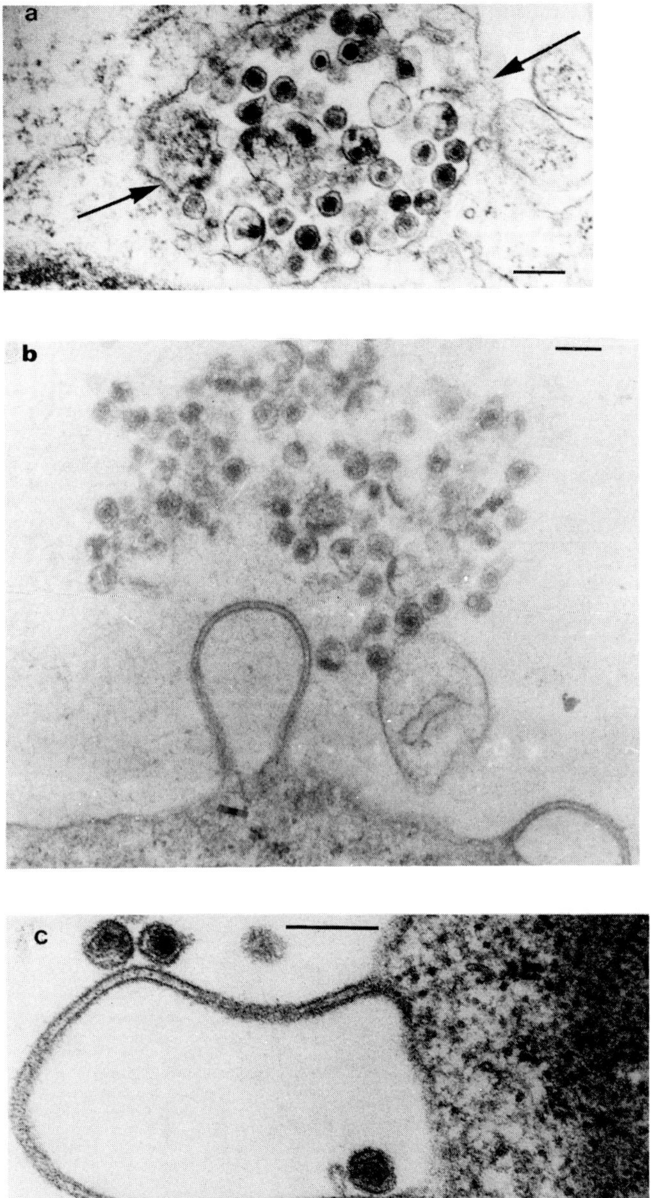


**FIGURE 6.** (a) Thin section showing a membrane-bound inclusion body full of distinct virus particles, the cell membrane probably about to empty the particles out to the extracellular space (→). (b) Empty loop formation projecting at the periphery of the cells, surrounded by numerous virus particles which it probably released. (c) Double-membraned loop formation with a few distinct enveloped virions which clearly show a core surrounded by a limiting membrane. Bar = 200 nm.

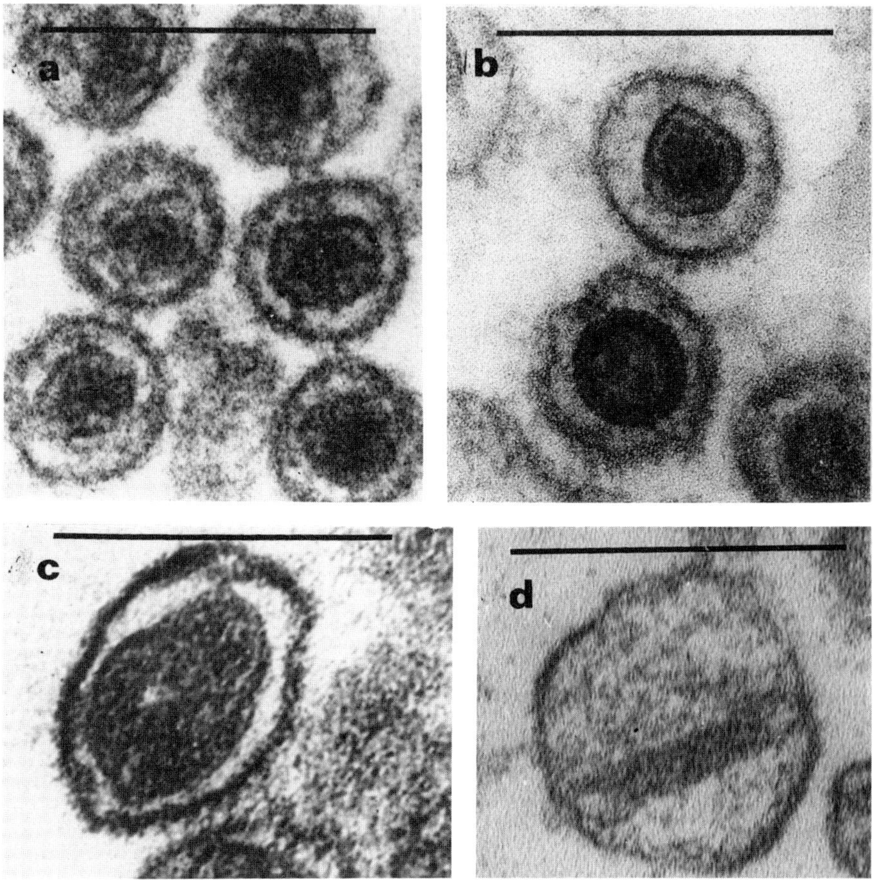


**FIGURE 7.** (a,b) Several extracellular 241 particles showing an electron-dense core surrounded by a limiting membrane, with electron-lucent space, all within an outer envelope. All particles contain knoblike projections. (c) Adult T-cell leukemia virus (ATLV/HTLV-IO, which is made up of a large electron-dense core, surrounded by a narrow electron-lucent space within an outer envelope. (d) Human immunodeficiency virus (HIV) showing a rod-shaped electron-dense core surrounded by a matrix of a lower density within a distinct envelope. Bar = 200 nm.

the very large 241 particles does not appear to be any larger than the core in smaller particles, and the space between the core and membrane appears to be electron-lucent or containing material of lesser density.

A most significant difference between the virus produced by the 241 cells and other retroviruses is the formation of most mature forms of virus within the cytoplasm, i.e., the formation of complete enveloped particles within membrane-bound electron-dense cytoplasmic inclusions. The presence of complete retro-

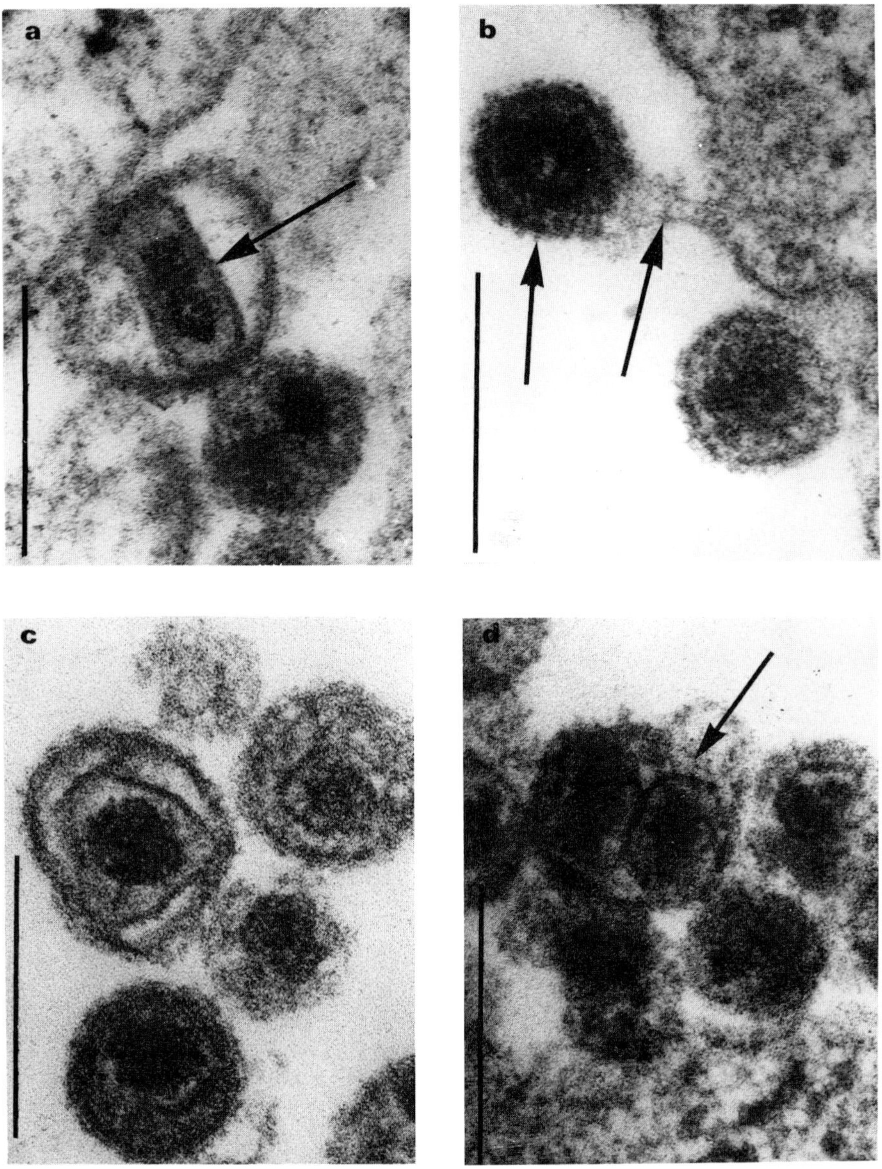

**FIGURE 8.** Range of particles produced by 241 cells. (a) Particles within the electron-dense core surrounded by an intermediate cylindrical membrane (→) within an outer envelope. (b) Budding particle with knoblike projections (→). (c) Cross section of a particle showing a distinct anatomy of the virus, i.e., an electron-dense core, surrounded by a distinct membrane and an outer envelope. (d) Particle with two distinct cores and surrounded by a limiting membrane (→). Bar = 200 nm.

viruses within the cytoplasm of fresh human leukemic cells was documented by van der Loo (1979).

The electron microscopic studies also suggest that virus full inclusion bodies expel themselves through the cell membrane (Fig. 6b,c). Preliminary enzymatic studies indicate that the 241 particles contain reverse transcriptase activity. This, together with the morphological characteristics, supports the notion that the virus produced by the 241 cells is a retrovirus. Our serological studies also confirm that the virus produced by the 241 cells is distinct from ATLV and HIV. We are now trying to clone the proviral DNA of this virus, as a prelude to examining fresh human leukemic cell DNA for viral sequences.

## 5. CONCLUSION

A critical overview of the history of human lymphotropic retroviruses became necessary because of the large amount of misleading and conflicting information published on this subject. Before going over the main highlights of the research, as well as sorting out the various claims and counter claims, it is important to emphasize that retroviral involvement in human leukemia is still a challenging research target. There is still no evidence, as yet, for retroviral involvement in any of the forms of human leukemia (see Fig. 1), with the exception of adult T-cell leukemia of Japanese origin and some blacks of African origin.

The earlier claims to the discovery of retroviral reverse transcriptase activity in the cytoplasm of virus-free human leukemic cells (Gallo *et al.*, 1970), and the subsequent claims to the isolation of a human retrovirus, HL23 virus (Gallagher and Gallo, 1975), turned out to be misleading. The cytoplasmic polymerase was probably cellular rather than retroviral (Bishop, 1980) and the HL23 virus turned out to be a laboratory contamination by three distinct monkey viruses (Woolly monkey, gibbon ape, and baboon viruses) (Okabi *et al.*, 1976; Ohino *et al.*, 1976).

The first report on retroviral involvement in human T-cell malignancy (mycosis fungoides and Sézary's syndrome) was by a Dutch group in 1979 (van der Loo, 1979). However, these investigators failed to grow the virus *in vitro* and therefore could not substantiate their observation. The reported discovery and subsequent studies of a C-type virus by Poiesz *et al.* (1980), named HTLV in the United States from cultured T cells can be divided into two periods. During the first period, 1980–1982, there was no positive evidence that the original HTLV was involved in any form of human malignancy or that it was a human virus. In fact, the only hybridization studies with fresh leukemic cells of a black patient with ATL were negative (17% with leukemic DNA 3–11% with normal DNA) (Poiesz *et al.*, 1981). The second period followed the arrival of the Japanese ATLV at the laboratory of Dr. Gallo. Thereafter, the hybridization studies as well as other studies became positive. The 1982 reported observation that blacks may also have ATL with clinical manifestations resembling Japanese ATL (Catovsky *et al.*, 1982), and the demonstration of antibodies to ATLV/HTLV in such patients indicated that certain blacks of African origin are likely to be carriers of the same or very closely related virus. The finding of ATLV serum-positive individuals in Africa would suggest that the infection of blacks in the West is of African origin. By contrast, the studies reported from Japan were novel and accurate from the outset, i.e. (1) the 1977 discovery of the endemic adult T-cell leukemia as a newly recognized disease in man (Uchiyama *et al.*, 1977), (2) the 1979 report on the establishment of an ATL cell line (Miyoshi *et al.*, 1979), and (3) the 1981 report that it produced a C-type virus (Hinuma *et al.*, 1981). This 1981 report on the association of the virus (ATLV) with the disease indicated from the outset that ATLV is endemic in Southwest Japan, that 100% (44/44) of ATL patients and 26% (26/100) of normal adults in the endemic area are serum positive, while serum positivity outside the endemic area is uncommon (Hinuma *et al.*, 1981). It was also in 1981 that the first report of the malignant transformation of normal human T cells by ATLV were published (Miyoshi *et al.*, 1981). All subsequent studies from Japan have confirmed and extended the evidence of close association between ATLV and ATL alone. Like the serological studies, the molecular biological investigations from Japan were accurate from the outset, i.e., the frequent presence of one ATLV proviral copy per cell, the monoclonality of the disease and the random site of integration (Yoshida *et al.*, 1982). It was also in Japan that the first complete ATLV sequence was achieved and where the pX region was discovered (Seiki *et al.*, 1983). We reported the first complete nucleotide sequence of ATLV which was isolated from a black ATL patient. Our sequence comparisons showed strong sequence conservation between isolates from blacks and Japanese (2.3% divergences). However, we also found higher homologies between isolates originating from similar geographical areas (approximately 1% divergence) (Malik *et al.*, 1987). Although ATLV/HTLV are exogenous viruses, they appear to be of low-level infectivity, but with familial clustering (Tajima *et al.*, 1982).

The observation that ATLV is probably transmitted via mother's milk provides an opportunity to reduce the spread of the virus, which illustrates the importance of studies that follow the isolation of the causative agent. The largest published epidemiological survey for antibodies to ATLV/HTLV outside Japan was done in Germany, where more than 4000 sera from leukemic and cancer patients and normal patients were screened for antibodies (Dolken *et al.*, 1983). The failure to find any positive cases is in agreement with our observations thus far that ATLV/HTLV are involved only with ATL of Japanese and some blacks of African origin. Intercontinental transfer of ATLV by Japanese migrating to the United States has been documented recently (Haynes *et al.*, 1983). A similar migration of virus-carrying African blacks to the United States and Britain could possibly be responsible for the random appearance of the disease in the United

Kingdom and the United States. Since the virus is not involved in the most common forms of human T-cell leukemia, it is more reasonable to designate it as ATLV, which also indicates the unique and rare form of T-cell leukemia in which it is involved. In addition the claim of Koprowski *et al.* (1985) that the virus is involved in multiple sclerosis is misleading (Hauser *et al.*, 1986; Karpas *et al.*, 1986). Likewise, the claims that ATLV cross-reacts and cross-hybridizes with the causative agent of AIDS, HIV (Hahn *et al.*, 1985) are also misleading (Karpas *et al.*, 1987). Therefore, to date it appears that ATLV is involved only with the development of ATL, while the possible viral etiology of all the other types of human leukemias and lymphomas remains a challenging research target.

## 6. REFERENCES


Arya, S. K., Gallo, R. C., Hahn, B. H., Shaw, G. M., Popovic, M., Salahudin, S. Z., and Wong-Staal, F., 1984, Homology of genome of AIDS-associated virus with genomes of human T-cell leukemia viruses, *Science* **225**:927–930.

Baltimore, D., 1970, RNA-dependent DNA polymerase in virions of RNA tumour viruses, *Nature (Lond.)* **226**:1209–1211.

Barbacid, M. O., Bolognesi, D., and Aaronson, S. A., 1980, Humans have antibodies capable of recognizing onco-viral glycoproteins. Demonstration that these antibodies are formed in response to cellular modification of glycoproteins rather than as a consequence of exposure to virus, *Proc. Natl. Acad. Sci. USA* **77**:1617–1621.

Barnes, D. H., 1987, New questions about AIDS test accuracy, *Science* **238**:884–885.

Barré-Sinoussi, F., Chermann, J. C., Rey, F., Nugeyre, M. T., Chamaret, S., Gruest, J., Dauget, C., Axler-Blin, C., Vezinet-Brun, F., Rouzioux, C., Rozenbaum, W., and Montagnier, L., 1983, Isolation of a T-lymphotropic retrovirus from a patient at risk for acquired immune deficiency syndrome (AIDS), *Science* **220**:868–871.

Bishop, J. M., 1980, The molecular biology of RNA tumor viruses: A physician's guide, *N. Engl. J. Med.* **303**:675–682.

Blattner, W. A., Kalyanaraman, V. S., Robert-Guroff, M., Lister, T. A., Galton, D. A. G., Sarin, P. S., Crawford, M. H., Catovsky, D., Greaves, M., and Gallo, R. C., 1982, The human type-C retrovirus, HTLV, in Blacks from the Caribbean region, and relationship to adult T-cell leukemia/lymphoma, *Int. J. Cancer* **30**:257–264.

Bunn, P. A., Schechter, G. P., Jaffe, E., Blayney, D., Young, R. C., Matthews, M. J., Blattner, W., Broder, S., Robert-Guroff, M., and Gallo, R. C., 1983, Clinical course of retroviruses associated adult T-cell lymphoma in the United States, *N. Engl. J. Med.* **309**:257–264.

Catovsky, D., Rose, M., Goolden, A. W. G., White, J. M., Bourikas, G., Brownell, A. I., Blattner, W. A., Greaves, M. F., Galton, D. A. G., McCluskey, D. R., Lampert, I., Ireland, R., Bridges, J. M., and Gallo, R. C., 1982, Adult T-cell lymphoma-leukaemia in Blacks from the West Indies, *Lancet* **1**:639–643.

Clarke, M. F., Gelmann, E. P., and Reitz, M. S., Jr., 1983, Homology of human T-cell leukaemia virus envelope gene with class I HLA gene, *Nature (Lond.)* **305**:60–62.

Daniel, M. D., Letvin, N. L., King, N. W., Kannagi, N., Sehgal, P. K., Hunt, R. D., Kanki, P. J., Essex, E., and Desrosiers, R. C., 1985, Isolation of T-cell tropic HTLV-III like retrovirus from macaques, *Science* **228**:1201–1204.

Dolken, G., Bross, K. H., Chosa, T., Schneider, J., Bayer, H., and Hunsmann, G., 1983, No evidence of HTLV infection among leukaemic patients in Germany, *Lancet* **2**:1495.

Essex, M., McLane, M. F., Lee, T. H., Falk, L., Howe, C. W. S., Mullins, J. I., Cabradilla, C., and Francis, D. P., 1983a, Antibodies to cell membrane antigens associated with human T-cell leukemia virus in patients with AIDS, *Science* **220:**859–862.

Essex, M., McLane, M. F., Lee, T. H., Tachibana, N., Mullins, J. I., Kreiss, J., Kasper, C. K., Poon, M-C., Landay, A., Stein, S. F., Francis, D. P., Cabradilla, C., Lawrence, D. N., and Evatt, B., 1983b, Antibodies to human T-cell leukemia virus membrane antigens (HTLV-MA) in hemophiliacs, *Science* **221:**1061–1064.

Essex, M., McLane, M. F., Lee, T. H., Falk, L., Howe, C. W. S., Mullins, J. I., Cabradilla, C., and Francis, D. P., 1983c, Antibodies to cell membrane antigens associated with human T-cell leukemia virus in patients with AIDS, *Science* **220:**859–862.

Evatt, B. L., Stein, S. F., Francis, D. P., Lawrence, D. N., McLane, M. F., McDougal, J. S., Lee, T. H., Spira, T. J., Cavradilla, C., Mullens, J. I., and Essex, M., 1983, Antibodies to human T-cell leukaemia virus-associated membrane antigens in haemophiliacs; evidence for infection before 1980, *Lancet* **2:**698–700.

Fialkow, P. J., Thomas, E. D., Bryant, J. I., and Neiman, P. E., 1971, Leukaemic transformation of engrafted human marrow cells in vivo, *Lancet* **1:**251–255.

Gallagher, N. S., and Gallo, R. C., 1975, Type C RNA tumor virus isolated from cultured human acute myelogenous leukemic cells, *Science* **187:**350–353.

Gallo, R. C., and Reitz, M. S., Jr., 1982, Human retroviruses and adult T-cell leukemia-lymphoma, *J. Natl. Cancer Inst.* **69:**1209–1214.

Gallo, R. C., Yang, S. S., and Ting, R. C., 1970, RNA dependent polymerase in human acute leukaemia cells, *Nature (Lond.)* **228:**927–929.

Gallo, R. C., Mann, D., Broder, S., Ruscetti, F. W., Maeda, M., Kalyanaraman, V. S., Robert-Guroff, M., and Reitz, M. S., Jr., 1982, Human T-cell leukemia/lymphoma virus (HTLV) is in T but not B lymphocytes from a patient with cutaneous T-cell lymphoma, *Proc. Natl. Acad. Sci. USA* **79:**4780–4782.

Gallo, R. C., Sarin, P. S., Gelmann, E. P., Robert-Guroff, M., Richardson, E., Kalyanaraman, V. S., Mann, D., Sidhu, G. D., Stahl, R. E., Zolla-Pazner, S., Leibowitch, J., and Popovic, M., 1983, Isolation of human T-cell leukemia virus in acquired immune deficiency syndrome (AIDS), *Science* **220:**865–867.

Gazdar, A. F., Carney, D. N., Bunn, P. A., Russell, E. K., Jaffe, E. S., Schechter, G. P., and Guccion, J. G., 1980, Mitogen requirements for the in vitro propagation of cutaneous T-cell lymphomas, *Blood* **55:**409–417.

Gelmann, E. P., Popovic, M., Blayney, D., Masur, H., Sidhu, G., Stahl, R. E., and Gallo, R. C., 1983, Proviral DNA of a retrovirus, human T-cell leukemia virus, in two patients with AIDS, *Science* **220:**862–865.

Goh, K., and Klemperer, M. R., 1977, In vivo leukemic transformation of engrafted marrow cells, *Am. J. Hematol.* **2:**283–290.

Gunz, F. W., Gunz, J. P., Vincent, P. C., Bergin, M., Johnson, F. L., Bashir, H., and Kirk, R. L., 1978, Thirteen cases of leukemia in a family, *J. Natl. Cancer Inst.* **60:**1243–1250.

Hahn, B., Manzari, V., Colombini, S., Franchini, G., Gallo, R. C., and Wong-Staal, F., 1983, Common site of integration of HTLV in cells of three patients with mature T-cell leukaemia-lymphoma: A retraction, *Nature (Lond.)* **305:**340.

Hahn, B. H., Shaw, G. M., Arya, S. U., Popovic, M., Gallo, R. C., and Wong-Staal, F., 1984, Molecular cloning and characterization of the HTLV-III virus associated with AIDS, *Nature (Lond.)* **312:**166–169.

Hauser, S. L., Aubert, C., Burks, J. S., Kerr, C., Lyon-Caen, O., de The, G., and Brahic, M., 1986, Analysis of human T-lymphotropic virus sequences in multiple sclerosis tissue, *Nature (Lond.)* **322:**176–177.

Haynes, B. F., Miller, S. E., Moore, J. O., Dunn, P. H., Bolognesi, D. P., and Metzgar, R. S., 1983, Identification of human T-cell leukemia virus in a Japanese patient with adult T-cell leukemia and cutaneous lymphomatous vasculitis, *Proc. Natl. Acad. Sci. USA* **80:**2054–2059.

Hinuma, Y., 1985, Adult T-cell leukemia virus. The present picture, in: *The Cytobiology of Leukaemias and Lymphomas* (D. Quaglino and F. G. J. Hayhoe, eds.), Twentieth Serono Symposia, pp. 435–450, Raven, New York.

Hinuma, Y., Nagata, K., Hanaoka, M., Nakai, M., Matsumoto, T., Kinoshita, K., Shirakawa, S., and Miyoshi, I., 1981, Adult T-cell leukemia: antigen in an ATL cell line and detection of antibodies to the antigen in human sera, *Proc. Natl. Acad. Sci. USA* **78:**6476–6480.

Hinuma, Y., Chosa, T., Komoda, H., Mori, I., Suzuki, M., Tajima, K., Hung-pan, I., and Lee, M., 1983, Sporadic retrovirus (ATLV)-seropositive individuals outside Japan, *Lancet* **1:**824–825.

Hunsmann, G., Schneider, J., Schmitt, J., and Yamamoto, N., 1983, Detection of serum antibodies to adult T-cell leukemia virus in nonhuman primates and in people from Africa, *Int. J. Cancer* **32:**329–332.

Jennings, P. A., and Even, J., 1984, No real homology of human T-cell leukaemia virus envelope with class I HLA, *Nature (Lond.)* **308:**85.

Kalyanaraman, V. S., Sarngadharan, M. G., Bunn, P. A., Minna, J. D., and Gallo, R. C., 1981, Antibodies in human sera reactive against an internal structural protein of human T-cell lymphoma virus, *Nature (Lond.)* **294:**271–273.

Kalyanaraman, V. S., Sarngadharan, M. G., Poiesz, B., Ruscetti, F. W., and Gallo, R. C., 1981, Immunological properties of a type C retrovirus isolated from cultured human T-lymphoma cells and comparison to other mammalian retroviruses, *J. Virol.* **38:**906–915.

Kalyanaraman, V. S., Sarngadharan, M. G., Robert-Guroff, M., Miyoshi, I., Blayney, D., Golde, D., and Gallo, R. C., 1982, A new subtype of human T-cell leukemia virus (HTLV-II) associated with a T-cell variant of hairy cell leukemia, *Science* **218:**571–573.

Kanki, P. J., Barin, F., M'Boup, S., Allan, J. S., Romet-Lemonne, J. L., Marlink, R., McLane, M. F., Lee, T-H., Arbeille, B., Denis, F., and Essex, M., 1986, New human T-lymphotropic retrovirus related to Simian T-lymphotropic virus type III (STLV-$III_{AGM}$), *Science* **232:**238–243.

Karpas, A., 1982, Viruses and leukemia, *Am. Sci.* **70:**277–285.

Karpas, A., and Fischer, P., 1980, Ultrastructural expression of virus-like particles in human leukemic cells growing in vitro, *Leukemia Res.* **4:**315–329.

Karpas, A., Worman, C. P., Khalid, G., Neumann, H., Hayhoe, F. G. J., Newell, D., and Stewart, J. W., 1980, Phenotypic characterization of a virus producing line established from a patient with leukaemia/lymphoma: Comparison with T-lymphoblastic and myeloblastic lines, *Br. J. Haematol.* **44:**415–424.

Karpas, A., Hatanaka, M., and Hinuma, Y., 1983, The virus of adult T-cell leukaemia of Japan (ATLV) is not involved in mycosis fungoides/Sezary syndrome, in: *Leukaemia Reviews International* (M. A. Rich, ed.), pp. 86–87, Marcel, New York.

Karpas, A., Gillson, W., Bevan, P. C., and Oates, J. K., 1985, Lytic infection by British AIDS virus and development of rapid cell test for antiviral antibodies, *Lancet* **2:**695–697.

Karpas, A., Maayan, S., and Raz, R., 1986, Lack of antibodies to adult T-cell leukaemia virus and to AIDS virus in Israeli Falashas, *Nature (Lond.)* **319:**794.

Karpas, A., Kämpf, U., Sidèn, Å., Koch, M., and Poser, S., 1986, Lack of evidence for involvement of known human retroviruses in multiple sclerosis, *Nature (Lond.)* **322:**177–178.

Karpas, A., Malik, K., and Lida, J., 1987, Studies of human retroviruses in relation to adult T-cell leukaemia, acquired immune deficiency syndrome and multiple sclerosis, *Arch. Virol.* **95:**237–249.

Koprowski, H., De Freitas, E. C., Harper, M. E., Sandberg-Willheim, M., Sheremata, W. A., Robert-Guroff, M., Saxinger, C. W., Feinberg, M. B., Wong-Staal, F., and Gallo, R. C., 1985, Multiple sclerosis and human T-cell lymphotropic retroviruses, *Nature (Lond.)* **318:**154–160.

Kumoro, A., Hayami, M., Fuji, H., Miyahara, S., and Hirayam, M., 1983, Vertical transmission of adult T-cell leukaemia virus, *Lancet* **1:**240.

Kurth, R., Teich, N. M., Weiss, R., and Oliver, R. T., 1977, Natural antibodies reactive with primate type-C viral antigens, *Proc. Natl. Acad. Sci. USA* **74:**1237–1241.

Malik, K. T. A., Even, J., and Karpas, A., 1988, Molecular cloning and complete nucleotide sequence of an adult T cell leukaemia virus/human T cell leukaemia virus type I (ATLV/HTLV-I) isolate of Caribbean origin: relationship to other members of the ATLV/HTLV-I subgroup, *J. Gen. Virol.* **69:**1695–1710.

Manzari, V., Gismondi, A., Barillari, G., Morrone, S., Modesti, A., Albonici, L., De Marchis, L., Fazio, V., Gradilone, A., Zani, M., Frati, L., and Santoni, A., 1987, HTLV-V: a new human retrovirus isolated in a Tac-negative T cell lymphoma/leukemia, *Science* **238:**1581–1583.

Mathez, D., Leibowitch, J., Catalan, P., Essex, M., and Zaguri, D., 1984, HTLV and AIDS in France, *Lancet* **1:**799.

Montagnier, L., 1984, AIDS priority, *Nature (Lond.)* **310:**446.

Miller, R. W., 1967, Persons with exceptionally high risk of leukemia, *Cancer Res.* **27:**2420–2423.

Miyoshi, I., Kubonishi, I., Sumida, M., Yoshimoto, S., Hiraki, S., Tsubota, T., Kobashi, H., Lai, M., Yanaka, T., Kimura, I., Miyamoto, K., and Sato, J., 1979, Characteristics of a leukemic T-cell line derived from adult T-cell leukemia, *Jpn. J. Clin. Oncol.* **9:**485–494.

Miyoshi, I., Kubonishi, I., Yoshimoto, S., Akagi, T., Ohtsuki, Y., Shiraishi, Y., Nagata, K., and Hinuma, Y., 1981, Type C virus particles in a cord T-cell line derived by co-cultivating normal human cord leukocytes and human leukaemic T-cells, *Nature (Lond.)* **294:**770–771.

Mulder, C., 1988, A case of mistaken non-identity, *Nature* **331:**562–563.

Ohino, T., Kife, D. W., Spiegelman, S., Gallo, R. C., and Gallagher, R. E., 1975, Characterization of a virus (HL23V) isolated from cultured acute myelogenous leukaemic cells, *Nature (Lond.)* **260:**266–268.

Okabe, H., Gilden, R. V., Hatanaka, M., Stephenson, J. R., Gallagher, R. E., Gallo, R. C., Tronick, S. R., and Aronson, S. A., 1976, Immunological and biochemical characterization of type C viruses isolated from cultured human AML cells, *Nature (Lond.)* **260:**264–266.

Okochi, K., Sato, H., and Hinuma, Y., 1984, A retrospective study on transmission of adult T cell leukemia virus by blood transfusion: seroconversion in recipients, *Vox Sang.* **46:**245–253.

Pandolfi, F., De Rossi, G., Lauria, F., Ranucci, A., Barillari, G., Manzari, V., Semenzato, G., Liso, V., Pizzolo, G., and Aiuti, F., 1985, T-helper phenotype chronic lymphocytic leukaemia and "adult T-cell leukaemia" in Italy, 1985, *Lancet* **2:**633–636.

Poiesz, B. J., Ruscetti, F. W., Gazdar, A. F., Bunn, P. A., Minna, J. D., and Gallo, R. C., 1980, Detection and isolation of type C retrovirus particles from fresh and cultured lymphocytes of a patient with cutaneous T-cell lymphoma, *Proc. Natl. Acad. Sci. USA* **77:**7415–7419.

Poiesz, B. J., Ruscetti, F. W., Reitz, M. S., Kalyanaraman, V. S., and Gallo, R. C., 1981, Isolation of a new type C retrovirus (HTLV) in primary uncultured cells of a patient with Sézary T-cell leukaemia, *Nature (Lond.)* **294:**268–271.

Popovic, M., Sarin, P. S., Robert-Guroff, M., Kalyanaraman, V. S., Mann, D., Minowada, J., and Gallo, R. C., 1983, Isolation and transmission of human retrovirus (human T-cell leukemia virus), *Science* **29:**856–859.

Reedman, B. M., and Klein, G., 1973, Cellular localisation of an Epstein-Barr virus (EBV)-associated complement fixing antigen in producer and non-producer lymphoblastoid cell lines, *Int. J. Cancer* **11:**499–520.

Reitz, M. S., Poiesz, B. J., Ruscetti, F. W., and Gallo, R. C., 1981, Characterisation and distribution of nucleic acid sequences of a novel type C retrovirus isolated from neoplastic human T lymphocytes, *Proc. Natl. Acad. Sci. USA* **78:**1887–1891.

Rho, H. M., Poiesz, B. J., Ruscetti, F. W., and Gallo, R. C., 1981, Characterisation of the reverse transcriptase from a new retrovirus (HTLV) produced by a human cutaneous T-cell lymphoma cell line, *Virology* **112:**355–360.

Rosenblatt, J. D., Golde, D. W., Wachsman, W., Giorgi, J. V., Jacobs, A., Schmidt, G. M., Quan,

S., Gasson, J. C., and Chen, I. S. Y., 1986, A second isolate of HTLV-II associated with atypical hairy-cell leukemia, *N. Engl. J. Med.* **315:**372–377.

Ruscetti, F. W., Robert-Guroff, M., Ceccherini-Nelli, L., Minowada, J., Popovic, M., and Gallo, R. C., 1983, Persistent in vitro infection by human T-cell leukemia–lymphoma virus (HTLV) of normal human T-lymphocytes from blood relatives of patients with HTLV-associated mature T-cell neoplasms, *Int. J. Cancer* **31:**171–180.

Sarkar, N. H., Moore, D. H., Kramarsky, B., and Chopra, H. C., 1973, Oncornaviruses. 3. The mammary tumor virus, in: *Ultrastructure of Animal Viruses and Bacteriophages: An Atlas* (A. J. Dalton and F. Haguenau, eds.), pp. 307–311, Academic, New York.

Schüpbach, J., Sarngadharan, M. G., and Gallo, R. C., 1984, Antigen in HTLV infected cells recognised by leukemia and AIDS sera are related to HTLV viral glycoprotein, *Science* **224:** 607–610.

Seiki, M., Hattori, S., Hirayama, Y., and Yoshida, M., 1983, Human adult T-cell leukemia virus: Complete nucleotide sequence of the provirus genome integrated in leukemia cell DNA, *Proc. Natl. Acad. Sci. USA* **80:**3618–3622.

Shimotohno, K., Wachsman, W., Takahashi, Y., Golde, D. W., Miwa, M., Sugimura, T., and Chen, I. S. Y., 1984, Nucleotide sequence of the 3′ region of the infectious human T-cell leukemia virus type II genome, *Proc. Natl. Acad. Sci. USA* **81:**6657–6661.

Smith, H., 1978, Inclusion bodies containing virus-like particles in acute lymphoblastic leukemia of childhood, *Leukemia Res.* **2:**133–140.

Snyder, H. W., and Flassner, E., 1980, Specificity of human antibodies to oncovirus glycoproteins. Recognition of antigen by natural antibodies directed against carbohydrate structures, *Proc. Natl. Acad. Sci. USA* **77:**1622–1626.

Tedder, R. S., Shanson, D. C., Jeffries, D. J., Cheinsong-Popov, R., Clapham, P., Dalgleish, A., Nagy, K., and Weiss, R. A., 1984, Low prevalence in the UK of HTLV-I and HTLV-II infection in subjects with AIDS, with extended lymphadenopathy, and at risk of AIDS, *Lancet* **2:**125–127.

Teich, N. M., Weiss, R. A., Salahuddin, S. Z., Gallagher, R. E., Gillespie, D. H., and Gallo, R. C., 1975, Infective transmission and characterization of a C-type virus released by cultured human myeloid cells, *Nature (Lond.)* **256:**551–555.

Temin, H. M., and Mizutani, S., 1970, RNA directed DNA synthesis and RNA tumour viruses, *Nature (Lond.)* **226:**1211–1213.

Uchiyama, T., Yodoi, J., Sagawa, K., Takatsuki, K., and Uchino, H., 1977, Adult T-cell leukemia, clinical and hematological features of 16 cases, *Blood* **50:**481–492.

van der Loo, E. M., van Muijen, G. N. P., van Vloten, W. A., Beens, W., Scheffer, E., and Meijer, C. J. L. M., 1979, C-type virus-like particles specifically localized in Langerhans cells and related cells of skin and lymph nodes of patients with mycosis fungoides and Sézary's syndrome, *Virchows Arch.* **31:**193–203.

Wong-Staal, F., Gillespie, D., and Gallo, R. C., 1976, Proviral sequences of baboon endogenous type C-RNA virus in DNA of human leukaemic tissues, *Nature (Lond.)* **262:**190–195.

Yamamoto, N., Matsumoto, T., Koyanagi, Y., Tanaka, Y., and Hinuma, Y., 1982, Unique cell lines harbouring both Epstein–Barr virus and adult T-cell leukaemia virus, established from leukaemia patients, *Nature (Lond.)* **299:**367–369.

Yoshida, M., Miyoshi, I., and Hinuma, Y., 1982, Isolation and characterization of retrovirus from cell lines of human adult T-cell leukemia and its implication in the disease, *Proc. Natl. Acad. Sci. USA* **79:**2031–2035.

*Chapter 6*

# Cell-Activation Responses to Cytomegalovirus Infection

## Relationship to the Phasing of CMV Replication and to the Induction of Cellular Damage

Thomas Albrecht, I. Boldogh, M. Fons, C. H. Lee, S. AbuBakar, J. M. Russell, and W. W. Au

## 1. INTRODUCTION

Evidence for the presence of cytomegaloviruses (CMV) as pathogenic agents in man was first obtained by examination of tissues derived from still-born infants that demonstrated severe developmental abnormalities (Jesionek and Kiolemeno-

**Thomas Albrecht, I. Boldogh, M. Fons, C. H. Lee, and S. AbuBakar** Department of Microbiology, University of Texas Medical Branch, Galveston, Texas 77550. **J. M. Russell** Department of Physiology and Biophysics, University of Texas Medical Branch, Galveston, Texas 77550. **W. W. Au** Department of Preventive Medicine and Community Health, University of Texas Medical Branch, Galveston, Texas 77550. *Permanent address of I.B.*: Department of Microbiology, Medical University of Debrecen, Debrecen, H-4012 Hungary. *Present address of C.H.L.*: Department of Microbiology, College of Natural Sciences, Chung Buk University, Chong Ju 360-763, Korea.

glou, 1904; Ribbert, 1904; Lowenstein, 1907). The salivary glands and other tissues of these infants contained enlarged cells with prominent nuclear inclusions (NI) that were eventually recognized as pathognomonic for this disease, i.e., cytomegalic inclusion disease (CID). It was not until about half a century after the initial descriptions of the cytomegalic cells that the virus responsible for these manifestations was isolated and partially characterized (Smith, 1956; Rowe *et al.*, 1956, Weller *et al.*, 1957). Cell cultures infected with freshly isolated or laboratory-adapted CMV demonstrated the characteristic enlarged cells with eosinophilic NI (Weller *et al.*, 1957). Noting the unique and characteristic features of the cytopathology induced by these viruses, Weller *et al.* (1960) suggested that this group of viruses be renamed cytomegaloviruses (CMV).

Using stocks prepared from these early isolates as a source of antigenic material, it was quickly shown that infection with CMV was much more common than could have been predicted by the frequency of clinically apparent CID (Rowe *et al.*, 1956). Seroepidemiological studies suggest that CMV infections are among the most common of all viral infections of man; however, these infections are most often inapparent, possibly latent (Weller, 1971). Clinical disease resulting from primary infection or from reactivation of latent CMV infection occurs only infrequently. Notwithstanding the relatively small percentage of otherwise healthy immunologically competent persons who experience clinical disease, generalized CMV infection is the bane of the developing fetus and individuals whose immune system is compromised, e.g., organ transplant recipients (Hill *et al.*, 1964; Hedley-Whyte and Craighead, 1965; Rifkind, 1965). Perhaps the most dramatic demonstration of the potentially devastating nature of CMV infections in the immunocompromised adult is observed in individuals with acquired immune deficiency syndrome (AIDS), where systemic CMV infection is often the immediate cause of death (Macher *et al.*, 1983).

The discovery that CMV has the capacity to form persistent infection (chronic and latent) is not particularly surprising, given its taxonomic classification. CMV is a member of the herpesvirus family and, along with other members of this group, shares a number of biophysical characteristics and a proclivity for forming persistent infections. Infectious CMV particles, like those of other herpesviruses, contain a relatively large double stranded DNA genome (Kilpatrick and Huang, 1977) surrounded by a protein capsid (Wright and Goodheart, 1964) and an envelope obtained from modified cellular membranes (McGavran and Smith, 1965; Severi *et al.*, 1988). CMV nucleocapsids are assembled in the nucleus (Ruebner *et al.*, 1965; McGavran and Smith, 1965) following at least three periods of transcription (DeMarchi *et al.*, 1980; Wathen *et al.*, 1981; Chua *et al.*, 1981; DeMarchi, 1983a) and translation (Stinski, 1978), and replication of the DNA genome (Goodheart *et al.*, 1964), features that are also common to other herpesviruses.

Cytomegaloviruses, however, are different from other members of the herpesvirus group in several unique and important characteristics. For example, the replication cycle for CMV is unusually long, with progeny virus first detected at about 49 hr postexposure (PE*) and maximum infectious yields obtained at 97 hr PE or later (Albrecht *et al.*, 1980a; Cavallo *et al.*, 1981). Also, productive† infection with CMV is more phylogenetically restricted than with other herpesviruses. For CMV, productive infection is restricted to cells of the same or of a closely related species to that from which the virus was isolated. Furthermore, although infection *in vivo* is characteristically apparent in epithelial and endothelial cells, infection resulting in high yields of infectious virus *in vitro* is usually restricted to relatively undifferentiated fibroblasts or myometrial cells (Weller, 1971).

Studies of CMV-directed transcription and translation indicate an orderly progression of macromolecular synthesis that generally appears to be similar to that for herpes simplex viruses (HSV), except for their protracted time course (Stinski, 1978; DeMarchi *et al.*, 1980; Wathen *et al.*, 1981; Chua *et al.*, 1981; DeMarchi, 1983a). Thus, molecular approaches have confirmed the slow replication cycle of CMV and have shown an orderly, if protracted, pattern for the synthesis of CMV-specific macromolecules. The available data do not explain why CMV transcription and translation in particular and CMV replication in general are substantially slower than those for HSV.

Accordingly, a previous attempt to explain the slow replication cycle of CMV (St. Jeor and Hutt, 1977) could be quite important in elucidating the molecular mechanisms of CMV replication. In this report, it was suggested that replication of CMV was dependent on cellular DNA synthesis, since a temporal coincidence between cellular and CMV DNA replication was observed for several cycles. Thus, the slow replication cycle for CMV might be a result of the time required for the cells to begin each round of cellular DNA synthesis. Other researchers, however, have been unable to confirm these findings and have concluded that CMV DNA synthesis is independent of cellular DNA synthesis (DeMarchi and Kaplan, 1976; Boldogh *et al.*, 1978b). Our studies of CMV DNA

*Although it has been the general practice to measure the time of infection for CMV after allowing 1 hr for adsorption at 37°C and replacement of the infecting fluids with maintenance medium (i.e., 0 hr postinfection), the recent findings of Valyi-Nagy *et al.* (1988) suggest that this practice is no longer acceptable. These investigators demonstrated that important events occur within the infected cell by 20 min after exposing the cells to the virus at 37°C (PE). Accordingly, we have converted all of our earlier data and those of others to hr PE by adding 1 hr to the time PI, with the exception of those few investigators who have adsorbed CMV at 4°C, shifted to 37°C at a later time, and measured time beginning at the shift to 37°C. Although these latter investigators labeled their time PI, virus-induced synthetic events in these studies should occur at about the same time as hr PE and should be about 1 hr behind other early studies measured in hr PI.

†A productive infection may be defined as an infection from which infectious progeny virus is derived. It should be noted that the efficiency of such infections varies widely.

synthesis suggest that the bulk of CMV DNA replication occurs at a time when cellular DNA synthesis is undetectable (Albrecht, 1973; St. Jeor *et al.*, 1974; Albrecht *et al.*, 1985). Therefore, considering these latter reports, there seems to be little support for the concept that CMV DNA synthesis is dependent on cellular DNA synthesis. Indeed, the suppression of cellular DNA synthesis by nucleoside analogues is reported to increase the CMV yield (St. Jeor and Rapp, 1973a) or to induce conversion of nonpermissive* cells to a permissive† state for CMV replication (St. Jeor and Rapp, 1973b).

It may be useful to consider additional characteristics of CMV infection in examining other approaches for providing an adequate explanation for and an understanding of the protracted replication cycle of CMV and for the cellular and molecular pathogenesis of CMV infections. CMV has a tropism for glandular tissues (Smith and Rasmussen, 1963). Such tissues have been postulated to be activated to work (e.g., secretion) through a signal pathway that is similar to those proposed for stimulation of cells to contract or to proliferate (Berridge, 1987). CMV infection is associated with stimulation of cellular macromolecular synthesis and, in some circumstances, increased mitotic activity. Early cellular responses to CMV infection resemble, to some extent, those associated with smooth muscle contraction. Accordingly, it might be reasonable to postulate that the cellular response to CMV infection is related to cell activation.

This chapter considers the evidence for activation of cells by CMV; examines the possibility that CMV infection stimulates and perhaps disrupts the controls for a cellular signal pathway that may be related to those pathways involved in cell activation by growth factors, hormones, and serum factors; and, evaluates the possible relationship between cell activation and CMV replication with regard to both biochemical and morphological events in the CMV replication cycle. Furthermore, since the cellular and molecular pathogenesis of CMV infections may be related to these activation phenomena, it will be useful to consider also how these phenomena may be related to the induction of cellular damage and CMV disease.

## 2. CELL ACTIVATION BY GROWTH FACTORS

In recent years, cell biologists and physiologists have made considerable progress in improving our understanding of two of the fundamental questions of biology: How are cells stimulated to work (e.g., contract, secrete) and to prolife-

*Nonpermissive cells are those cells that fail to produce yields of infectious virus. These cells, however, may produce various virus-specified products and demonstrate cytopathic effects.
†Permissive cells are those cells that are capable of producing infectious virus.

rate? As described in recent reviews (e.g., Michell, 1986; Rozengurt, 1986; Berridge, 1987; Alkon and Rasmussen, 1988) and as briefly summarized here, substantial parts of the signal pathway involved in cell activation by growth factors and hormones have been elucidated. Therefore, we summarize only the major points related to the proposed signal pathway for activation of cells by growth factors. Our recent work and that of others suggest that CMV uses this signal pathway to modify the intracellular environment to one that is more permissive for efficient CMV replication. The time course for the CMV-induced biosynthetic processes, as well as morphological responses to CMV replication, may be related to the development of CMV-induced pathophysiological responses that, in turn, may be dependent on disruption of the cellular controls for cell activation leading to protracted activation signals.

Growth factors such as platelet-derived growth factor (PDGF) bind to specific receptors on the cell surface and, in so doing, initiate a signal across the cell membrane that activates phospholipase C. Phospholipase C cleaves the membrane-associated inositol lipid phosphatidylinositol 4,5-bisphospate ($PIP_2$) to yield two secondary messengers: 1,2-diacylglycerol (DG) and inositol 1,4,5-trisphosphate ($IP_3$). DG remains within the plane of the membrane and stimulates the activity of protein kinase C. Protein kinase C then activates $Na^+/H^+$ exchange, which leads to $Na^+$ entry and possibly to alkalinization of the intracellular fluid. $IP_3$ induces the release of stored intracellular $Ca^{2+}$. The release of the $Ca^{2+}$ stores then triggers $Ca^{2+}$ influx, which leads to an increase of the cytosolic free $Ca^{2+}$ concentration ($[Ca^{2+}]_i$). The combined signals of increased cellular pH and $[Ca^{2+}]_i$ are proposed to serve as signals to stimulate the cell to work or proliferate (Berridge, 1987). The obvious problem of avoiding stimulation of the cell to proliferate each time that it is induced to work may be accomplished by requiring more intense and prolonged signals for initiation of cell proliferation (Berridge, 1987). Other secondary messengers, such as cyclic adenosine monophosphate, (cAMP) and cyclic guanosine monophosphate (cGMP) are involved also in these activation phenomena (e.g., Alkon and Rasmussen, 1988). For example, through the action of a kinase or a lipase, DG can enter the arachidonic acid cascade, producing prostaglandins (PG) that can affect the cellular levels of cAMP and cGMP. Increased cellular levels of cAMP and cGMP can result in the activation of protein kinases that can modulate the activity of a variety of cell functions including $Ca^{2+}$ by regulating the $[Ca^{2+}]_i$.

## 3. CELL ACTIVATION BY CYTOMEGALOVIRUS

A number of early findings suggest, or are consistent with the view, that CMV infection results in some degree of activation of the infected cell. Infection of arrested permissive and nonpermissive (i.e., abortively* infected) cells results in stimulation of cellular DNA synthesis (St. Jeor *et al.*, 1974; Furukawa *et al.*, 1975; Albrecht *et al.*, 1976; DeMarchi and Kaplan, 1976; DeMarchi and Kaplan, 1977a,b; Hirai *et al.*, 1977; Boldogh *et al.*, 1978a; Yamanishi and Rapp, 1979; Radsak and Schmitz, 1980; DeMarchi, 1983b). Consistent with this observation, increased synthesis or activity of a number of host cell macromolecules that are associated with the transition from the $G_0/G_1$ phase to the S phase of the cell cycle is observed. CMV infection also causes infected cells to undergo a contraction-like response. Thus, rounding CMV-infected cells demonstrate a decrease in their diameter, the presence of actin bundles proximate to the plasmalemma, a highly invaginated nuclear membrane profile, and a sensitivity to inhibition by smooth muscle relaxing agents, providing further evidence for some sort of cell activation in response to CMV infection (Albrecht *et al.*, 1983, 1984). The dependence of CMV expression and replication on the physiological state of the host cell (Albrecht *et al.*, 1976; DeMarchi and Kaplan, 1977a; Boom *et al.*, 1988), while not necessarily demonstrating a requirement for cell activation, is consistent with this view (see Section 3.3).

### 3.1. Stimulation of Cellular DNA Synthesis and Cell Proliferation

The first indications that CMV infection induces a proliferative signal were provided in the initial observations by Albrecht and Rapp (1973), Lang *et al.* (1974), and St. Jeor *et al.* (1974) that CMV infection induces growth-, agarose-, IUdR (5-iodo-2′-deoxyuridine)-, or serum-arrested cells to initiate replication or DNA synthesis. The latter study demonstrated that two strains of CMV induced an increase in the rate of DNA synthesis both in cells that sustain productive infections (human embryo lung fibroblasts) and in cells that sustain inefficient infections—Vero cells, an established cell line derived from African green monkey kidney tissue. CMV preparations inactivated by heat or by high doses of ultraviolet (UV) light, i.e., 300 mean lethal hits per plaque-forming unit (PFU), failed to stimulate cellular DNA synthesis, which was interpreted to suggest that the presence of the virion alone is insufficient to stimulate cellular DNA synthesis. It is still controversial, however, whether expression of virus information is necessary for the induction of cellular DNA synthesis.

It is known that the synthetic capacity of the entire virus genome is not necessary to initiate arrested cells, as stimulation of cellular DNA synthesis is observed in nonpermissive guinea pig, hamster, mouse, and rabbit cells that are

*An abortive infection is an infection from which infectious progeny virus is not obtained, although some viral expression may occur. Abortive infections may result from infection of nonpermissive cells or from maintenance of infected, permissive cells under restrictive conditions (e.g., in the presence of nucleoside analogues or at supraoptimal temperatures).

unable to express late CMV genes (Furukawa *et al.*, 1975; Albrecht *et al.*, 1976; Boldogh *et al.*, 1978a; DeMarchi, 1983b). Furthermore, Hirai *et al.* (1977) and Yamanishi and Rapp (1979) observed that neither phosphonoacetic acid nor restrictive temperatures blocked stimulation of cellular DNA synthesis, confirming that CMV DNA synthesis, hence the expression of late genes, are not required for the induction of cellular DNA synthesis.

Stocks of CMV are composed of a very large proportion of defective particles, many of which contain partial DNA cores (Smith and Rasmussen, 1963). These defective virions display biological activity (Albrecht *et al.*, 1974) and apparently are involved in the initiation of cellular DNA synthesis, since stocks of CMV exposed to low doses of UV light (Albrecht *et al.*, 1976; DeMarchi and Kaplan, 1977b; Boldogh *et al.*, 1978a) or several cycles of undiluted passage (DeMarchi and Kaplan, 1977b) demonstrate increased ability to stimulate cellular DNA synthesis. It is tempting to raise the point that the expression of immediate early and early genes of CMV would be favored under these conditions, since the late genes with a larger target size would be more susceptible to damage than immediate early or early genes. DeMarchi and Kaplan (1976, 1977a), however, were unable to detect CMV antigens in human embryo lung cells stimulated to initiate DNA synthesis and Boldogh *et al.* (1978a) could not correlate the stimulation of cellular DNA synthesis with expression of early nuclear and cytoplasmic antigens. It may be premature, however, to conclude that the CMV virion is sufficient to induce cellular DNA synthesis, since the antibody preparations used in these studies may not have detected low levels of expressed CMV proteins. Additional work is needed to determine definitively whether expression of CMV information is required for the stimulation of cellular DNA synthesis.

Since CMV infection stimulates semiconservative synthesis of cellular DNA (St. Jeor *et al.*, 1974; DeMarchi, 1983b), this event should be associated with stimulation of other cellular activities and with an increase in cell division. Increased synthesis or activity of a number of host cell macromolecules has been reported, including thymidine kinase (Zavada *et al.*, 1976; Estes and Huang, 1977; DeMarchi, 1983b), $\alpha$-type DNA polymerase (Hirai and Watanabe, 1976; Hirai *et al.*, 1976; Yamanishi and Rapp, 1979), RNA synthesis (Tanaka *et al.*, 1975), DNA-dependent RNA polymerase (Tanaka *et al.*, 1978), ornithine decarboxylase (Isom, 1979), putrescine uptake (Tyms and Williamson, 1980), plasminogen activator (Yamanishi and Rapp, 1979), exonuclease and topoisomerases (Huang *et al.*, 1984), mitochondrial DNA synthesis (Furukawa *et al.*, 1976), glucose uptake (Landini, 1984), and alterations in the pattern of glycosphingolipid synthesis (Radsak and Wiegandt, 1984). The increased synthesis or activity of most of these host cell macromolecules is associated with the entry of cells into S phase and the initiation of cellular DNA synthesis, though the synthesis of cellular histones has been reported as unchanged (Radsak and

Schmitz, 1980). On the basis of these findings, it might seem reasonable to suggest that CMV infection results in a general nonspecific stimulation of host cell metabolism, including stimulation of host cell DNA synthesis (Kamata *et al.*, 1978, 1979; DeMarchi, 1983b). It may be noteworthy, however, that stimulation of a number of these activities is observed for cells cultured in conditions in which stimulation of cellular DNA synthesis is not observed (Furukawa *et al.*, 1975, 1976; Hirai and Watanabe, 1976; Hirai *et al.*, 1976; Zavada *et al.*, 1976; Estes and Huang, 1977).

Increased mitotic activity in CMV-infected growth-arrested hamster cells and in agarose-arrested human embryo skin muscle cells was observed in early work (Albrecht and Rapp, 1973; Lang *et al.*, 1974), although a quantitative assessment of this increase was not made. Later studies (Albrecht *et al.*, 1976; Nachtigal and Nachtigal, 1978; Boldogh *et al.*, 1978a) provided a quantitative evaluation and demonstrated a substantial increase in the mitotic index of CMV-infected serum-arrested hamster embryo cells and human embryo lung cells. These findings indicate that the initiation of cellular DNA synthesis by CMV can result in duplication of the cellular genome and cell division.

In summary, CMV infection of arrested, permissive, and nonpermissive cells results in a proliferative signal that may lead to the initiation of cellular DNA synthesis and cell division. There is substantial evidence to suggest that defective CMV particles may participate in initiating this proliferative signal. There are no convincing data to support the necessity of CMV gene expression for induction of this proliferative signal; additional work is needed to clarify this matter.

### 3.2. Contraction of CMV-Infected Cells

Activation of cells is also indicated by the induction of cells to initiate work such as contraction or secretion (e.g., Berridge, 1987). CMV-infected cells undergo a characteristic sequence of morphological responses, as illustrated in Fig. 1 (Albrecht *et al.*, 1984). Although it has been recognized for some time that CMV-infected fibroblasts assume a rounded shape, beginning about 6 hr PE (e.g., Furukawa *et al.*, 1973), it was not appreciated until relatively recently that the diameter of these rounded cells decreases by about 13–25 hr PE, depending on the intensity of the infection (Albrecht *et al.*, 1983, 1984; Nokta *et al.*, 1988). These rounded cells demonstrate features that resemble those observed in contracting smooth muscle cells. The nuclear membrane profiles are highly convoluted (Fig. 2), a characteristic that has been considered an indication of contraction for myofibroblasts (Majno, 1979). Further ultrastructural evaluation of these cells reveals structures proximate to the plasmalemma that resemble actin bundles, which is consistent with a possible relationship between CMV-induced cell rounding and cell contraction (Fig. 2).

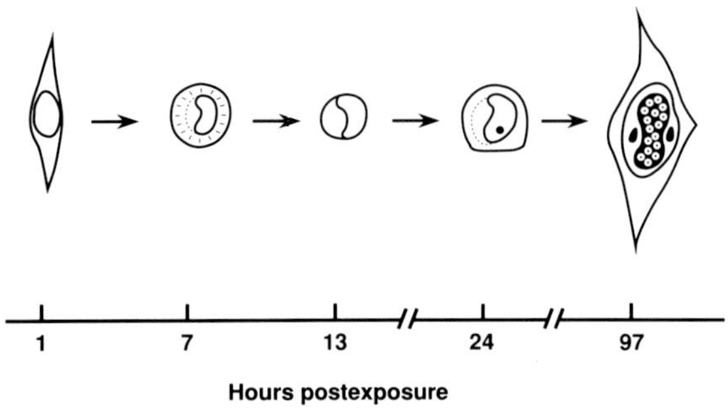


**FIGURE 1.** Diagrammatic representation of changes in cell shape and size in response to CMV infection. All cells are drawn to scale (~1000 : 1).

Since CMV-induced cell rounding is an early virus-induced event, occurring in a variety of conditions restrictive for CMV replication (Fioretti *et al.*, 1973; Waner and Weller, 1974; Albrecht *et al.*, 1976, 1980b; Hirai *et al.*, 1977), we were interested in determining the mechanism of cell rounding and in examining what role, if any, this phenomenon might play in the replication of CMV and in the pathogenesis of CMV infection. Accordingly, we examined the effects of smooth muscle relaxing agents on the rounding of CMV-infected cells. If there was a physiological basis for the observed similarities between contraction of smooth muscle cells and CMV-induced cell rounding, these agents might also be expected to interfere with the rounding of CMV-infected cells. Indeed, pa-

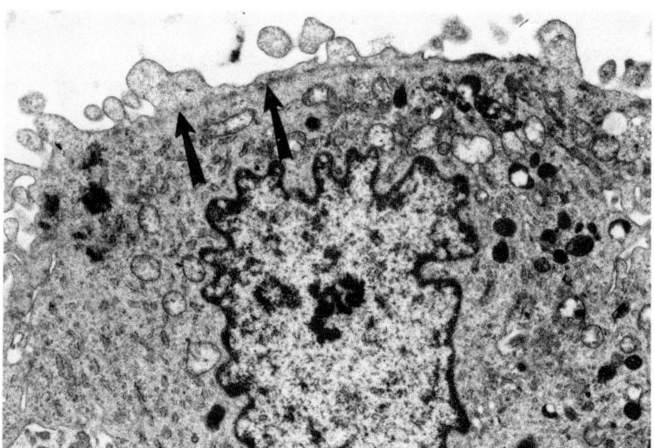

**FIGURE 2.** Cell rounding at 6 hr PE. The nuclear membrane demonstrates numerous invaginations; actin bundles (arrows). (×15,400)

paverine, sodium nitroprusside, and verapamil inhibited the rounding of CMV-infected cells at 6 hr PE (Albrecht *et al.*, 1983), supporting the view that a contractile-like response is involved in the rounding and decrease in size of CMV-infected cells. While this response could be provoked by a number of mechanisms, the data for verapamil suggest that $Ca^{2+}$ influx is involved in the development of these cytopathic effects.

### 3.3. Effect of the Physiological State of the Cell on the Expression and Replication of CMV

Stimulation of cellular DNA synthesis, the expression of CMV antigens, and the replication of CMV are related to the physiological state of the host cells. This is true both for cells that sustain abortive infections (Albrecht *et al.*, 1976; DeMarchi, 1983b; Boom *et al.*, 1988) and for those that sustain productive infections (St. Jeor *et al.*, 1974; DeMarchi and Kaplan, 1977a; DeMarchi, 1983b).

For abortively infected hamster embryo cells, Albrecht and colleagues found that the expression of CMV antigens and the development of cytopathic effects are substantially more efficient in serum-arrested cells that demonstrate stimulation of cellular DNA synthesis than in nonarrested, confluent cells in which cellular DNA synthesis is not initiated following CMV infection. Similar findings were obtained with abortively infected rabbit kidney cells in which maximum increases in the percentage of cells stimulated to initiate DNA synthesis are 9.9- and 2.4-fold in serum-arrested and in nonarrested cells, respectively (calculated from DeMarchi, 1983b). One explanation for these findings is that there may be a relationship between the induction of a cellular proliferative signal and the expression of CMV information and that CMV infection can initiate cellular DNA synthesis in the arrested, but not the nonarrested, confluent, abortively infected cells. Alternatively, it is possible that an inhibitor of CMV expression is present in sufficient quantity in the nonarrested cells, but not in the serum-arrested cells. The latter explanation seems unlikely, though, since preliminary experiments failed to detect a soluble inhibitor of CMV expression in the nonarrested hamster cells (T. Albrecht and F. Rapp, unpublished studies).

For permissive human embryo lung cells, the results, at least at first appearance, seem contradictory. St. Jeor *et al.* (1974) demonstrated that infection of serum-arrested cells resulted in a significant increase in cellular DNA synthesis while infection of nonarrested, confluent cells resulted in inhibition of cellular DNA synthesis which was apparent within the first 25 hr after exposure to CMV (Fig. 3). DeMarchi and Kaplan (1977a) obtained similar results for serum-arrested and nonarrested confluent cells; however, they did not detect the synthesis of virus antigens at 49 hr PE in the initiated cells. Furthermore, these investigators observed that subconfluent, actively growing cells were more sus-

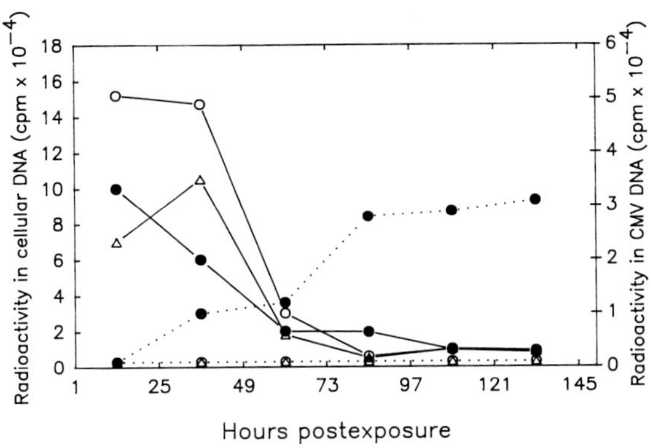


**FIGURE 3.** Effect of CMV infection on DNA synthesis in nonarrested, confluent human embryo lung cells. Total [$^3$H]methyl thymidine radioactivity incorporated into cellular DNA (———) or into virus DNA (·····) in CMV-infected cells (●), ultraviolet light-irradiated (50 lethal hits per plaque forming unit) CMV-infected cells (△), or mock-infected cells (○) is plotted at the midpoint of each 24-hr pulse period.

ceptible to viral infection than confluent contact-inhibited cells, as demonstrated by a greater percentage of cells developing CMV-specific antigens and by a greater yield of infectious virus. Stimulation of cellular DNA synthesis was not detected in the subconfluent actively replicating cells. It is possible, however, that the high level of cellular DNA synthesis present in the actively replicating cells prevented the detection of any stimulatory effect the virus may have had. Although the increase in permissiveness of the actively replicating cells relative to the stationary cells tends to support the findings obtained with abortively infected cells, the differences in the responses for serum-arrested and nonarrested confluent cells require closer examination.

In the DeMarchi and Kaplan study (1977a), stimulation of cellular DNA synthesis was observed only in serum-arrested confluent cells; virus antigens were not detected in those cells that initiated cellular DNA synthesis. In nonarrested confluent cells in which stimulation of cellular DNA synthesis was not observed, virus antigens were detected in 25–50% of confluent cells, even though the multiplicity was varied from 1 to 10 PFU/cell. Considering these observations with those from their earlier study (1976), DeMarchi and Kaplan (1977a) concluded that (1) "cellular DNA synthesis is probably not required for viral DNA or antigen synthesis"; (2) "detectable virus-induced stimulation of cellular DNA synthesis and the synthesis of virus antigens appear to be mutually exclusive events"; and, (3) "although the physiological state of the cells affects their ability to respond to infection with CMV, the functions required for the growth of CMV are not those provided by cells in S phase."

While our findings are in agreement with most of those of DeMarchi and Kaplan (1976, 1977a), the implications from their latter two conclusions are troubling in that they may suggest that CMV cannot replicate in an activated cell because CMV antigens are not synthesized and that cell activation is not related to the replication of CMV. Given this interpretation, these results appear to be contradictory to those for nonpermissive or actively replicating cells. It is possible, however, that cell activation is induced in both the subconfluent actively replicating and in the confluent nonarrested cells, which do not demonstrate initiation of cellular DNA synthesis, as well as in those confluent serum-arrested cells that demonstrate initiation of cellular DNA synthesis. This seemingly contradictory circumstance could arise in the permissively infected cells if CMV infection and/or activation of the cells would lead to synthesis of a virus- (DeMarchi, 1983b) or cell-specified product that inhibited cellular DNA synthesis after the activation process had been initiated. It could also arise if the controls for cellular activation were perturbed by CMV infection (Table I), so that after initiating the activation process, an intracellular environment inconsistent with cellular DNA synthesis would result. Accordingly, those permissive cells that demonstrated enhancement of cellular DNA synthesis would have produced insufficient quantities of the inhibitor or would have maintained reasonable control over the cell-activation process. The lack of virus antigens at 48 hr in many of the cells in the confluent cultures (i.e., those that are induced to initiate DNA synthesis) could be related to the inability of the virus to prolong the activation signals in these cells sufficiently, as described in Sections 4 and 5. The greater permissivity of the actively growing cells tends to support this view, as does the finding that cellular thymidine kinase is stimulated even when cellular DNA synthesis is not (DeMarchi, 1983b). In the future, the acquisition of data for the effect of CMV infection on early cellular activation responses rather than stimulation of cellular DNA synthesis, which is usually maximum between 24 and 48 hr after infection, should help clarify this issue.

## 4. CELLULAR PATHOPHYSIOLOGICAL RESPONSE TO CMV INFECTION

If CMV infection is inducing some sort of cell activation phenomenon, it should be possible to test this hypothesis by determining whether it is supported by physiological cellular responses consistent with those induced by other activation phenomena. Infection of permissive cells with CMV results in sequentially ordered physiological responses, including hydrolysis of inositol lipids, $Ca^{2+}$ influx, the intracellular sequestering of $Ca^{2+}$, a significant increase in the $[Ca^{2+}]_i$ $Na^+$ entry (possibly through activation of $Na^+/H^+$ exchange), and increases in the cellular levels of cAMP and cGMP. The limited studies available suggest

**Table I**
**Relationship between the Duration of Cell Activation and the Cellular Effects of CMV**

| Stimulus | Duration of activation signals | Effect on cellular work | Effect on cellular DNA | References |
|---|---|---|---|---|
| Acetylcholine | [→ | Contraction | None | Moummi *et al.* (1988) |
| Cholecystokinin | [→ | Secretion | None | Tsunoda *et al.* (1988) |
| PDGF | [———→ | Contraction, ? | Stimulation | Heldin *et al.* (1979) |
| CMV + papaverine | [→ | Contraction | None | Albrecht *et al.* (1984) |
| CMV + amiloride | [——→ | Contraction, relaxation | None | Nokta *et al.* (1988); Fons *et al.* (1987) |
| CMV (abortive infection) | [———→? | Contraction, ? | Stimulation | Albrecht *et al.* (1976) |
| CMV (productive infection) | [——————→ | Contraction, relaxation, enlargement | Inhibition | Albrecht *et al.* (1980b) |

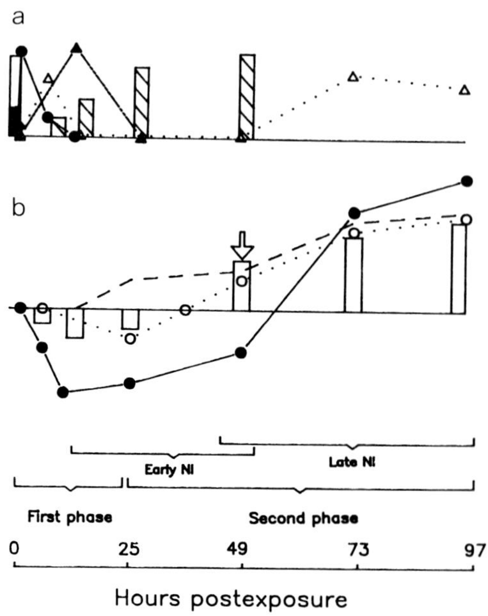


**FIGURE 4.** Relationship between CMV-induced physiological cellular responses, CMV DNA synthesis, development and progression of NIS, formation of nucleocapsids, and yields of infectious CMV. (a) Increases in $IP_3$ (open bar), DG (solid bar), $Ca^{2+}$ influx (●———●), $[Ca^{2+}]_i$ (hatched bar), cAMP (△···△), cGMP (▲———▲) are represented. (b) Increases and decreases in $Na^+/K^+$ ATPase (open bars), cell size (○···○), CMV DNA synthesis (dashed line without symbols), first detection of CMV nucleocapsids (open arrow), and CMV infectious yields (●———●) are represented.

The limited studies available suggest that these physiological responses to CMV infection can be divided into two broad phases (Fig. 4). The first phase appears to be related to $Ca^{2+}$ influx, rising $[Ca^{2+}]_i$, and events secondary to them. This first phase is associated with the morphological responses of cell rounding and contraction (Albrecht *et al.*, 1984). The second phase may be mediated by $Na^+$ entry. This second phase is associated with the morphological response of cytomegaly (Nokta *et al.*, 1988). Although these physiological responses (Fig. 5) are those expected in cellular activation, the time course for these responses is protracted (see Table I and Fig. 4), in contrast to the transient responses observed in other cell-activation phenomena (Berridge, 1987).

## 4.1. Inositol Lipid Hydrolysis

Hydrolysis of $PIP_2$, yielding the secondary messengers $IP_3$ and DG, appears to serve as a signal for cell activation by a number of growth factors and other stimulators of cell activity. In CMV-infected cells, increases in the cellular levels of $IP_3$ and DG are maximum at 20 min after exposing the cells to CMV stock (Valyi-Nagy *et al.*, 1988) (Table II). Significant increases in $IP_3$ and DG are not detected after inactivation of virus stock with heat (56°C for 30 min) or by neutralization with CMV-specific serum. Even though active CMV appears to be required to stimulate the increases in cellular $IP_3$ and DG, it is unlikely that expression of CMV immediate early genes is responsible for this cellular response to CMV infection. CMV-specific antigen synthesis is not detectable by

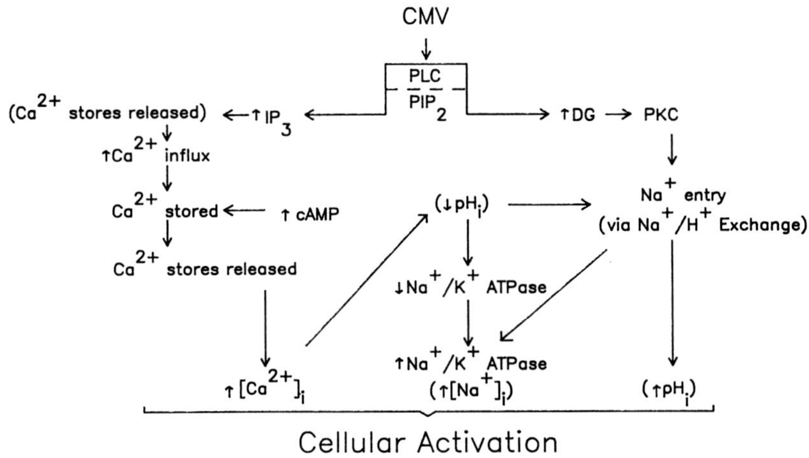


**FIGURE 5.** Proposed scheme indicating the known and possible (parentheses) physiological responses to CMV infection and their interactions based on findings for CMV infection and for other cell activation phenomena. PLC, phospholipase C; $PIP_2$, phosphatidylinositol bisphosphate; ↑, increased intracellular level or activity, $IP_3$, inositol trisphosphate; DG, diacylglycerol; PKC, protein kinase C; ↓, decreased intracellular level or activity.

anticomplement immunofluorescence until 40 min PE. At the time when $IP_3$ and DG reached maximum levels as a result of $PIP_2$ hydrolysis (20 min PE), only membrane fluorescence is detectable, which represents the presence of adsorbed virus components on the membrane of infected cells (Valyi-Nagy *et al.*, 1988). Thus, it seems possible that the adsorption of CMV to its membrane receptor stimulates the hydrolysis of $PIP_2$, initiating a cascade of physiologic responses (Fig. 5).

## 4.2. $Ca^{2+}$ Influx and Increased $[Ca^{2+}]_i$

Following CMV infection, there is a substantial alteration in intracellular $Ca^{2+}$ homeostasis, as demonstrated by changes in $Ca^{2+}$ influx, $Ca^{2+}$ sequestering, and $[Ca^{2+}]_i$ (Fig. 6). $Ca^{2+}$ uptake by CMV-infected cells is observed within the first 2 hr PE, at which time it is at its maximum rate. $Ca^{2+}$ influx blockers, such as verapamil and nifedipine, inhibit the $Ca^{2+}$ influx (Nokta *et al.*, 1987), although diltiazem has no measurable effect (Table III). In CMV-infected cells, total intracellular $Ca^{2+}$ is increased at 9 and 13 hr PE but is decreased from 13 to 25 hr PE (Fig. 6). A second increase in total intracellular $Ca^{2+}$ is observed from 49 to 97 hr PE. An increase in $[Ca^{2+}]_i$ is apparent by 6 hr PE and by 13 hr, the $[Ca^{2+}]_i$ in CMV-infected cells is doubled relative to the concentration at 1 hr PE. The $[Ca^{2+}]_i$ was measured with the Quin 2 fluorescence procedure (Tsien *et al.*, 1982), which in CMV-infected cells is likely to

**Table II**
**Stimulation of Inositol Lipid Hydrolysis by Human Cytomegalovirus**

| Cells | Increase in radioactivity relative to mock-infected cells[a] (%) | | | |
|---|---|---|---|---|
| | 20 min PE[b] | | 40 min PE | |
| | $IP_3$[c] | DG[d] | $IP_3$ | DG |
| Mock-infected[e] | 0 | 0 | 0 | 0 |
| CMV-infected | | | | |
| MOI[f] = 5 | 93.5 | 29.0 | 2.1 | 0.8 |
| MOI = 10 | 92.0 | 32.0 | 3.2 | 1.5 |
| CMV stock[g] neutralized with CMV-specific serum | 0.9 | 0.1 | 0.0 | 0.0 |
| CMV stock[g] neutralized with VZV-specific serum | 92.0 | 28.5 | 2.1 | 0.6 |
| CMV stock[g] heated to 56°C for 30 min | 2.1 | 0.2 | NT | 0.2 |
| CMV stock[h] and papaverine (80 μM) | −36.0 | −63.0 | NT | NT |

[a]At 0 hr PE.
[b]PE: postexposure.
[c]$IP_3$: inositol 1,4,5-trisphosphate.
[d]DG: 1,2-diacylglycerol.
[e]With virus-free supernatant.
[f]MOI: multiplicity of infection.
[g]MOI prior to inactivation was 5 PFU/cell.
[h]MOI was 5 PFU/cell, papaverine was added to the culture fluids at 0 hr PE.

provide minimal estimates of any increase in $[Ca^{2+}]_i$ (Nokta *et al.*, 1987). The increase in $[Ca^{2+}]_i$ can be inhibited by verapamil and nifedipine but is insensitive to inhibition by diltiazem, suggesting that the apparent increase in $[Ca^{2+}]_i$ is related to an influx of $Ca^{2+}$ mediated by mechanisms that are blocked by verapamil or nifedipine. Furthermore, since the total cellular $Ca^{2+}$ increases substantially as $[Ca^{2+}]_i$ increases, these findings imply that during this early phase of CMV infection most of the $Ca^{2+}$ that enters the cells is captured into intracellular stores. At later times (13–25 hr PE) after the initial influx of $Ca^{2+}$, both the influx rate and total cellular $Ca^{2+}$ decrease, while $[Ca^{2+}]_i$ increases significantly. This implies that a major source of the free cytosolic $Ca^{2+}$ at this stage is likely to be $Ca^{2+}$ released from intracellular stores.

The necessity for active CMV for stimulation of increased $[Ca^{2+}]_i$ was tested by inactivating CMV stock with heat, specific antisera, or UV light (Nokta *et al.*, 1987). Increases in Quin 2 fluorescence were significantly reduced for

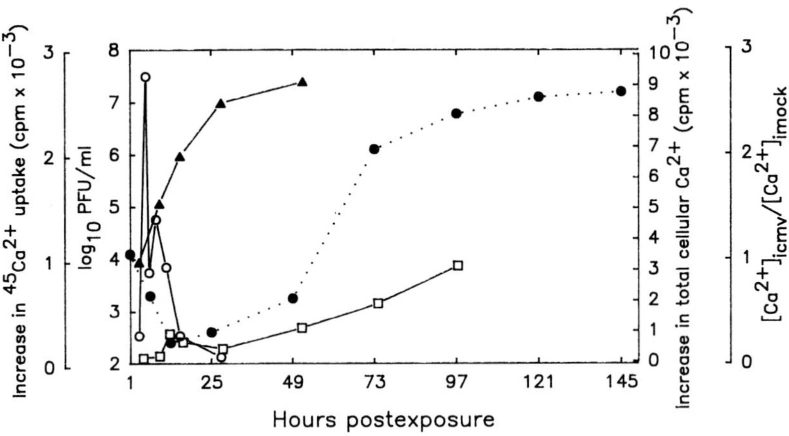


**FIGURE 6.** $Ca^{2+}$ responses to CMV infection in permissive human embryo fibroblasts. The increase in $^{45}Ca^{2+}$ uptake (○——○) is the difference between the radioactivity taken up by CMV- and mock-infected cells in 30-min pulses at the indicated times. The increase in total cellular $^{45}Ca^{2+}$ (□——□) is the difference between the radioactivity accumulated by CMV- and mock-infected cells, corrected for the change in cell size, at the indicated times for a continuous label beginning 1 hr PE. The fold increase in $[Ca^{2+}]_i$ (▲——▲) is indicated by the quotient of the $[Ca^{2+}]_i$ for CMV-infected cells divided by the $[Ca^{2+}]_i$ for mock-infected cells (Nokta *et al.*, 1987). Data are superimposed on the replication curve for CMV (●···●). PFU, plaque-forming units.

cells infected with inactivated CMV, indicating that active CMV virions are required for the increase in $[Ca^{2+}]_i$. Inhibitors of the formation of competent messenger RNA (mRNA) (3′-deoxyadenosine) or of translation (cycloheximide) also interfere with the increase in Quin 2 fluorescence, suggesting that biosynthetic events within the CMV-infected cell are necessary for the increase in $[Ca^{2+}]_i$. Thus, infectious CMV appears to be required for the sustained increase

**Table III**
**Sensitivity of CMV-Induced $Ca^{2+}$ Influx and $[Ca^{2+}]_i$ Increase to Inhibition by $Ca^{2+}$ Influx Blockers**

| Drug (100 μM) | Inhibition: $Ca^{2+}$ influx | Inhibition: $[Ca^{2+}]_i$ increase |
|---|---|---|
| Verapamil | + | + |
| Nifedipine | + | + |
| Diltiazem | − | − |

in $[Ca^{2+}]_i$. It is not at all clear, however, whether CMV virions are sufficient or whether early CMV-directed synthetic events are also required for $Ca^{2+}$ influx and the early increase in $[Ca^{2+}]_i$.

There appears to be a relationship between the $Ca^{2+}$ responses to CMV infection and the development and progression of CMV-induced cytopathologies on the one hand (Albrecht *et al.*, 1983, 1984), and the replication of infectious CMV on the other (Albrecht *et al.*, 1984, 1987). It has also been suggested that there is a relationship between the rate of $Ca^{2+}$ influx and the rate of development of CMV-induced cytopathic effects (Garnett, 1981). Even though we have reported that $Ca^{2+}$ influx blockers interfere with the development of CMV-induced cytopathic effects (Albrecht *et al.*, 1981, 1983, 1984), it seems unlikely that $Ca^{2+}$ influx *per se* is the trigger. The time course for the rate of increase in $Ca^{2+}$ influx and the rate of development of cytopathic effects are not the same in our data. The increase in cytosolic $Ca^{2+}$ activity is a more likely triggering event. It has a better temporal correlation with the development of CMV-induced cytopathologies and can be prevented by $Ca^{2+}$ influx blockers. It should be recognized, however, that other secondary messengers may also be involved in the development of physiological and morphological responses to CMV infection (Lee and Albrecht, 1987).

### 4.3. Increased Cellular Levels of Cyclic Nucleotides

Since many of the cell-activation phenomena (e.g., contraction, secretion) are also related to the intracellular levels of cyclic nucleotide secondary messengers, we have examined the relationship of the cellular levels of cAMP and cGMP to cell activation in CMV-infected cells (Lee and Albrecht, 1987). CMV infection of permissive human fibroblasts results in sequentially ordered increases in cAMP and cGMP (Fig. 7). The apparent intracellular level of cAMP is substantially, but transiently, increased during the early phase of CMV replication (at 7 hr PE), returns to levels similar to the mock-infected controls between 13 and 49 hr PE, and is elevated during the late phase of CMV replication. Closer examination of the time course for the early cAMP increase indicates that it is first apparent by 3 hr PE and that it attains a maximum level at 7 hr PE. The intracellular level of cGMP is increased significantly relative to the mock-infected controls from 6 to 25 hr PE, with the maximum increase observed at 13 hr PE. CMV- and mock-infected cells have similar levels of cGMP from 49 to 97 hr PE.

The finding that the cAMP increase, beginning at 3 hr PE, follows the time of maximum $Ca^{2+}$ influx within 1 hr suggests that there may be a relationship between these physiologic responses, as has been suggested for other cell activation phenomena (Rasmussen, 1981). It is possible that the $Ca^{2+}$ influx leads to the induction of the cAMP increase through activation of the $Ca^{2+}$-dependent adenylate cyclase (Blumenthal and Stull, 1980; Rasmussen and Barrett, 1984),

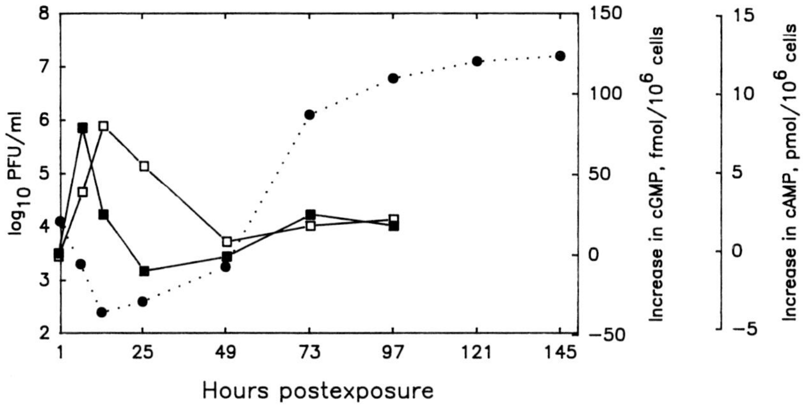


**FIGURE 7.** Increases in the cellular levels of cAMP (■——■) and cGMP (□——□) following CMV infection in permissive human embryo fibroblasts. The increases are the differences between of level for each cyclic nucleotide in CMV- and mock-infected cells, corrected for the change in cell size, at the times indicated (Lee and Albrecht, 1987). The data are superimposed on the replication curve for CMV (●···●). PFU, plaque-forming units.

but other factors may contribute to the cAMP increase. For example, CMV infection results in increased cellular levels of $IP_3$ and DG. Through the activity of a kinase or a lipase, DG can stimulate arachidonic acid metabolism and the formation of prostaglandins. Since prostaglandins have been shown to stimulate adenylate cyclase activity in other tissues (Sha'afi and Naccache, 1985), it is possible that DG-mediated stimulation is also involved in the CMV-induced increase in cAMP. These two pathways leading to increased cAMP levels are not mutually exclusive and it is possible that both contribute to the increased levels of cAMP in CMV-infected cells. It is not possible from the available data, however, to determine the relative contribution of $[Ca^{2+}]_i$ and prostaglandins to the CMV-induced cAMP increases.

How the cyclic nucleotide responses are involved in the cellular and molecular pathogenesis of CMV infections is unknown; however, by comparison with other cell-activation phenomena, some possibilities may be considered. cAMP is reportedly involved in the regulation of the cell cycle (Haddox *et al.*, 1980), and there is considerable evidence that cAMP triggers a cascade of $G_1$ and late $S$–$G_2$ events by activating cAMP-dependent protein kinases (Haddox *et al.*, 1980; Boynton and Whitfield, 1983). Activation of one of these kinases, holoenzyme I, has been related to diminished $Ca^{2+}$ signals and their effects; to increased mRNA, transfer RNA (tRNA), and polyamine synthesis; and to stimulation of glucose and amino acid uptake. These phenomena have been well documented in the $G_1$ phase of the cell cycle (Sand *et al.*, 1981; Boynton and Whitfield, 1983), and it may be more than coincidental that some of these phenomena are observed during the early phase of CMV gene expression. In CMV-infected cells, the increase in cAMP is related temporally to a decrease in $Ca^{2+}$ influx and to an increase in the intracellular sequestering of $Ca^{2+}$ (Nokta *et al.*, 1987). In this connection, cAMP has been associated with the energy-dependent sequestering of intracellular $Ca^{2+}$ in other tissues (Bhalla *et al.*, 1978; Kattenburg and Daniel, 1984). It is not possible from the present findings, however, to determine the relative contributions of increased $[Ca^{2+}]_i$ and of increased DG levels to the cAMP increases, nor is it possible to determine whether the early cAMP increase is involved in shunting cytosolic $Ca^{2+}$ to stores. It may be noteworthy, though, to observe that, in papaverine-treated CMV-infected cells in which the cAMP increase has been blocked, the CMV-induced $[Ca^{2+}]_i$ increase is also inhibited, particularly at later times in CMV infection (Albrecht *et al.*, 1985).

A strong correlation between the increased cAMP and the morphological cellular responses is not apparent in our data. There is a striking temporal relationship, however, between the increased cGMP and the development of morphological responses. The contraction of infected cells is maximum or near maximum at the time, i.e., 13 hr PE, (Albrecht *et al.*, 1984) that cGMP levels are highest in infected cells. Furthermore, the relaxation and subsequent enlargement of CMV-infected cells (Albrecht *et al.*, 1980a) is temporally related to low cGMP levels, similar to those for mock-infected cells. Papaverine inhibits the decrease in cGMP content in CMV-infected cells and blocks the progression of morphological responses in the contracted stage (Albrecht *et al.*, 1984; Lee and Albrecht, 1987), consistent with a relationship between cGMP and at least the progression of the CMV-induced morphological responses. The relationship between cGMP levels and contraction–relaxation, however, is controversial for noninfected smooth muscle cells. Although early work suggested that increased cellular levels of cGMP were associated with contraction of smooth muscle cells (Lee *et al.*, 1972; Dunham *et al.*, 1974), further studies failed to show a good quantitative correlation between cGMP levels and contraction (Schultz *et al.*, 1978). Furthermore, more recent work has indicated that several drugs (e.g., nitroprusside, nitroglycerine) that are recognized for their relaxing effect on smooth muscle stimulate increased levels of cGMP (Diamond and Blisard, 1976; Katsuki and Murad, 1977). In CMV-infected cells, however, treatment with these drugs is associated with arrest of the development of virus-induced morphological changes at the contracted cell stage. Thus, in CMV infection of these permissive cells, maintenance of high levels of cGMP is related to a general failure of the cells to relax and enlarge.

### 4.4. $Na^{+}$ Entry and the Development of Cytomegaly

The phenomenon of cytomegaly is one of the most intriguing and often overlooked aspects of CMV research. Our laboratory has viewed cytomegaly as an indication of the establishment of molecular alterations in cell physiology and morphology that may be necessary for efficient CMV replication. Support for this view can be found in the observations that (1) virus progeny are not detected until cell enlargement begins, even though CMV DNA synthesis is initiated at a much earlier time (see Fig. 4); (2) substantial increases in the rate of CMV DNA synthesis and in yields of infectious virus coincide with cell enlargement (see Fig. 4); and (3) the most potent anti-CMV compounds we have studied (e.g., papaverine) prevent the development of cytomegaly (Albrecht *et al.*, 1984, 1987; Nokta *et al.*, 1988). Considering the possible relationship between the development of cytomegaly and CMV replication, it seems prudent to evaluate the molecular and cellular mechanisms controlling the development of cytomegaly and to determine whether they are related to the development of an intracellular environment providing for efficient CMV replication. The development of cytomegaly appears to be related to $Na^+$ entry, which in turn may be related to the protracted cell-activation signals associated with CMV infection.

Entry of $Na^+$ appears to be a general feature of cells stimulated to progress through the cell cycle (Rozengurt and Mendoza, 1980) including stimulation of cellular DNA synthesis by growth factors and tumor promoters (Moolenaar *et al.*, 1983; Grinstein *et al.*, 1985; L'Allemain *et al.*, 1984, 1985). There are several distinct mechanisms through which $Na^+$ may enter a cell. $Na^+$ may enter cells through $Na^+$ channels due to the gradient that results from the high extracellular $[Na^+]$ and relatively low intracellular $[Na^+]$. A large part of the energy expenditure of a cell is used to maintain this gradient by extrusion of $Na^+$ through the ouabain-sensitive $Na^+/K^+$ ATPase (Skou, 1965). $Na^+$ may also enter cells through exchange processes, such as the amiloride-sensitive $Na^+/H^+$ antiport (Moolenaar *et al.*, 1983; L'Allemain *et al.*, 1984), bumetanide-sensitive $Na^+/K^+/Cl^-$ co-transport (Owen and Prastein, 1985), or the 4,4′-diisothiocyanatostilbene-2,2′-disulfonic acid (DIDS)- and 4-acetamido-4′-isothiocyanatostilbene-2,2′-disulfonic acid (SITS)-sensitive, $Na^+$-dependent $Cl^-/HCO_3^-$ exchange (L'Allemain 1985).

Our results show a temporal relationship between $Na^+$ entry and cell enlargement. The time course for the CMV-induced change in cell size is shown in Fig. 8. These data indicate that there is an approximately linear increase in cell size from 25 to 73 hr PE, at which time infected cells are about 1.9-fold larger than their preinfection size. Reducing cellular $Na^+$ uptake by lowering the $[Na^+]$ in the culture fluids blocks the development of cytomegaly (Nokta *et al.*, 1988). In medium containing a reduced $Na^+$ concentration (4 mM) and *N*-methyl-D-glucamine (NMDG) (139 mM), CMV-infected cells fail to enlarge and at 73 hr PE are flattened, similar in size and shape to the relaxed infected cells (at about 25 hr PE) in physiological $[Na^+]$ (143 mM). Pharmacological inhibition of $Na^+$ entry with amiloride produces similar results (Nokta *et al.*, 1988). Since amiloride is an inhibitor of $Na^+/H^+$ exchange (Paris and Pouyssegur, 1983) at the concentrations (150–200 μM) used in these experiments, our findings suggest that $Na^+$

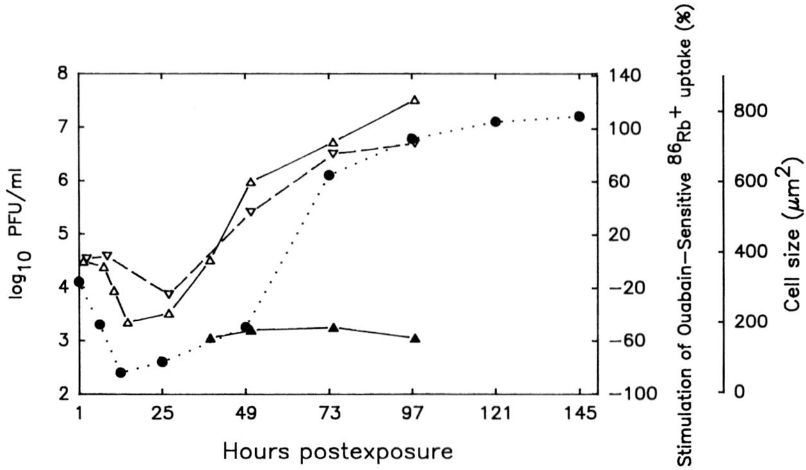


**FIGURE 8.** $Na^+$ responses to CMV infection in permissive human embryo fibroblasts and the change in cell size. Stimulation of $Na^+/K^+$ ATPase activity is indicated by the percent increase or decrease in ouabain-sensitive $^{86}Rb^+$ uptake (△———△). Amiloride (150 μM), an inhibitor of $Na^+/H^+$ exchange, blocked the increase in ouabain-sensitive $^{86}Rb^+$ uptake (▲———▲). Cell size (▽———▽) was measured in the horizontal plane and, since data are not available for the increase in the vertical plane, these should be considered minimal estimates of the increase in cell size. Nevertheless, they should be a good indication of the rate of change in cell size. The data are superimposed on the replication curve for CMV (●···●). PFU, plaque-forming units.

entry in CMV-infected cells is through amiloride-sensitive pathways and possibly through $Na^+/H^+$ exchange.

If cytomegaly results as a consequence of increased $Na^+$ entry (together with an anion, presumably $Cl^-$), this osmotic burden should result in an increased rate of $Na^+$ efflux and possibly an increase in the $Na^+/K^+$ pump activity, as shown in Fig. 8. The activity of the $Na^+/K^+$ ATPase is depressed in CMV-infected human fibroblasts at 13 and 25 hr PE, as indicated by a decrease in ouabain-sensitive $^{86}Rb^+$-uptake. By 36 hr, the $Na^+/K^+$ pump regains its preinfection level of activity. From 49 to 97 hr PE, the activity of the $Na^+/K^+$ pump demonstrates a nearly linear increase in activity, coinciding with enlargement of the infected cells. Presumably, this increase results either from increased $[Na^+]_i$ or an increased number of pump sites within the plasmalemma (Nokta *et al.*, 1988). In the presence of amiloride or 4 mM $Na^+$ (and 139 mM NMDG), the enhancement of $Na^+/K^+$ pump activity is not observed (Nokta *et al.*, 1988). Measurements of the concentration effect of amiloride on the development of cytomegaly and on the increase in $Na^+/K^+$ pump activity suggest that they have similar sensitivities with half-saturation constants in the range of 12–15 μM, consistent with the entry of $Na^+$ through $Na^+/H^+$ exchange.

These results show the development of cytomegaly and $Na^+$ movements to be closely related. Events that enhance $Na^+$ entry may lead to an increased intracellular $Na^+$ concentration, which in turn would stimulate the activity of the $Na^+/K^+$ pump (Smith and Rozengurt, 1978). The increase in the plasma membrane potential observed following CMV infection (Landini and Rugolo, 1984) may be a reflection of increased $Na^+$ pump activity. Indeed, the total $TPP^+$ (tetraphenylphosphonium) accumulation, a measure of membrane potential, increased steadily throughout infection in Landini and Rugolo's study and seems to correlate with the increase in activity of the $Na^+/K^+$ pump observed by Nokta *et al.* (1988). A similar, but transient, increase in $Na^+/K^+$ pump activity has been noted in fibroblasts after stimulation with serum, growth factors, or tumor promoters (Rozengurt and Mendoza, 1980). Although $Na^+$ homeostasis in many tissues and cells is controlled through several pathways involved in $Na^+$ entry (Paris and Pouyssegur, 1983; Moolenaar *et al.*, 1983; Cassel *et al.*, 1985), $Na^+/H^+$ exchange is one of the most active $Na^+$ pathways in fibroblasts and the one proposed to be involved in $Na^+$ entry in activation of cells to work or proliferate. For CMV-infected cells, both cell enlargement and the increase in $Na^+/K^+$ pump activity are sensitive to amiloride at concentrations consistent with the entry of $Na^+$ through $Na^+/H^+$ exchange. $Na^+$ entry through other mechanisms does not seem important in the development of cytomegaly, since concentrations of amiloride in the range that block $Na^+$ channels (0.1 μM) (Reuss *et al.*, 1986) and since inhibitors of $Na^+$-dependent $Cl^-/HCO_3^-$ and $Na^+/K^+/Cl^-$ exchange do not appear to affect CMV-induced cell enlargement (Table IV). In CMV-infected cells, the increase in $Na^+$ entry may result from DG stimulation of protein kinase C (Berridge and Irvine, 1984) and from protracted cell activation signals and/or intracellular acidification related to the CMV-induced increases in metabolic activity. Available data are not sufficient to determine the relative contribution of

**Table IV**
**Effect of $Na^+$ Entry Blockers on the Development of Cytomegaly**

| Drug | Inhibition of cytomegaly |
|---|---|
| Amiloride (30–150 μM) | + |
| SITS (30–300 μM)[a] | – |
| DIDS (3–30 μM)[b] | – |
| Bumetanide (1–30 μM) | – |

[a]SITS: 4-acetamido-4′-isothiocyanatostilbene-2,2′-disulfonic acid.
[b]DIDS: 4,4′-diisothiocyanatostilbene-2,2′-disulfonic acid.

these two pathways, which are not mutually exclusive and are each capable of provoking increased $Na^+/H^+$ exchange.

## 5. PHASING OF CMV DNA SYNTHESIS, DEVELOPMENT OF NIS, AND FORMATION OF NUCLEOCAPSIDS

### 5.1. Pattern of CMV DNA Synthesis

The time course for the rate of CMV DNA synthesis in nonarrested confluent cells appears to be biphasic (see Fig. 3). CMV DNA synthesis (as measured by the rate of incorporation of radioactive thymidine) begins about 17 hr PE (Albrecht, 1973), increases through 25 hr PE, and levels off soon afterward. A second phase of CMV DNA synthesis beings at about 37 hr PE and reaches a maximum rate between 73 and 97 hr PE. Since synthesis of CMV DNA during the second phase is at a greater rate and for a substantially longer period of time, the bulk of CMV DNA is made during this interval. The biphasic pattern observed for CMV DNA synthesis does not appear to be an artifact of the experimental procedures. It has been observed with plaque-purified virus; with several laboratory strains of CMV; after infection of cloned cells; after infection of cells derived from several tissues; and using radioactive hypoxanthine, as well as radioactive thymidine, to measure the rate of CMV DNA synthesis (T. Albrecht, M. Fons and S. Cerruti-Sola, unpublished studies). Stinski (1978) used molecular hybridization techniques to demonstrate a similar pattern of DNA synthesis for CMV strain Towne, which he attributed to the dependence of CMV DNA synthesis on cellular DNA synthesis (St. Jeor and Hutt, 1977).

In their study, St. Jeor and Hutt observed repeated cycles of CMV and cellular DNA synthesis, which were coincident with about a 24-hr periodicity. They concluded that there was a relationship between CMV and cellular DNA synthesis and that the multiphasic nature of CMV DNA synthesis was a result of this relationship. Other investigators have been unable to confirm this relationship. We were unable to correlate CMV and cellular DNA synthesis in early studies (Albrecht, 1973; St. Jeor *et al.*, 1974) (see Fig. 3). In these experiments, cellular DNA synthesis was substantially inhibited within the first 25 hr PE and, when the bulk of CMV DNA synthesis occurred, cellular DNA synthesis was either at a very low level or was undetectable. DeMarchi and Kaplan (1976, 1977a) noted that cellular DNA synthesis is not required for the synthesis of CMV antigens measured at 49 hr PE or for the synthesis of CMV DNA, since they observed the synthesis of both in cells treated with 5-fluorouracil (5-FU) and thymidine, which demonstrated very low levels of cellular DNA synthesis. Although the relationship between cellular and CMV DNA synthesis may still be controversial, there seems to be little evidence to support the view that CMV

DNA synthesis is dependent on cellular DNA synthesis. Such a view, however, does not necessarily imply that cell activation is not involved in CMV DNA synthesis (see Fig. 4).

If the biphasic pattern measured with the incorporation of radioactive precursors or by molecular hybridization techniques accurately reflects the time course for CMV DNA synthesis, some explanation for this interesting pattern and for the slow replication cycle for CMV is warranted. One explanation that merits consideration is that one or more precursors or enzymes needed for CMV replication are in short supply during the early phase of CMV DNA synthesis. Although data are not available for all the components needed for the replication of CMV, for those for which we do have data, there is no indication of insufficient quantity or activity. For example, the CMV DNA polymerase is first detected at 13 hr PE and attains nearly maximum activity by 30 hr PE (Hirai *et al.*, 1977). Thus, the time course for the increase in CMV DNA polymerase activity precedes that for CMV DNA synthesis and, since the increase in activity of the CMV DNA polymerase is monophasic, it does not explain either the biphasic pattern of CMV DNA synthesis or the slow replication cycle for CMV.

### 5.2. Development and Progression of Early and Late NIs

Morphological and ultrastructural analysis of NIs in CMV-infected cells lends some support for the biphasic nature of CMV DNA synthesis. CMV DNA synthesis takes place within NIs (Goodheart *et al.*, 1964); and, accordingly, their size and distribution provides some insight into the characteristics of CMV DNA synthesis. At 25 hr PE, CMV-infected cells demonstrate small ringlike structures (Fig. 9A,D), early NIs (Albrecht *et al.*, 1980a), composed of granular and fibrillar material (Cavallo *et al.*, 1981). Virus particles are not detected in the nucleus at this time. By 49 hr, larger NIs are evident in most cells (Fig. 9E). Ultrastructural analysis of cells at this time routinely demonstrates the presence of NIs in various stages of development, denoted by their size and organization (Fig. 9B). Even though early NIs are still present at 49 hr, larger NIs made up of a network of electron-dense branching material, often enclosing, or partially enclosing, electron-lucent areas are evident. We have adopted the term *cellula* for each structure consisting of an electron-lucent area enclosed by an electron-dense network (Cavallo *et al.*, 1981). Nucleocapsids are present in most nuclei at this time and, when present in relatively small numbers, are located preferentially in the vicinity of the NI. At 73 hr PE and later, large well-developed NIs are present in most cells (Fig. 9C). These NIs are composed of large cellulae that contain a punctiform central area by light microscopic examination (Fig. 9F). At this time, the areas of grainy material within the cellulae are larger and better defined; and, relatively large numbers of nucleocapsids are associated with the NI (Cavallo *et al.*, 1981; Fons *et al.*, 1986).

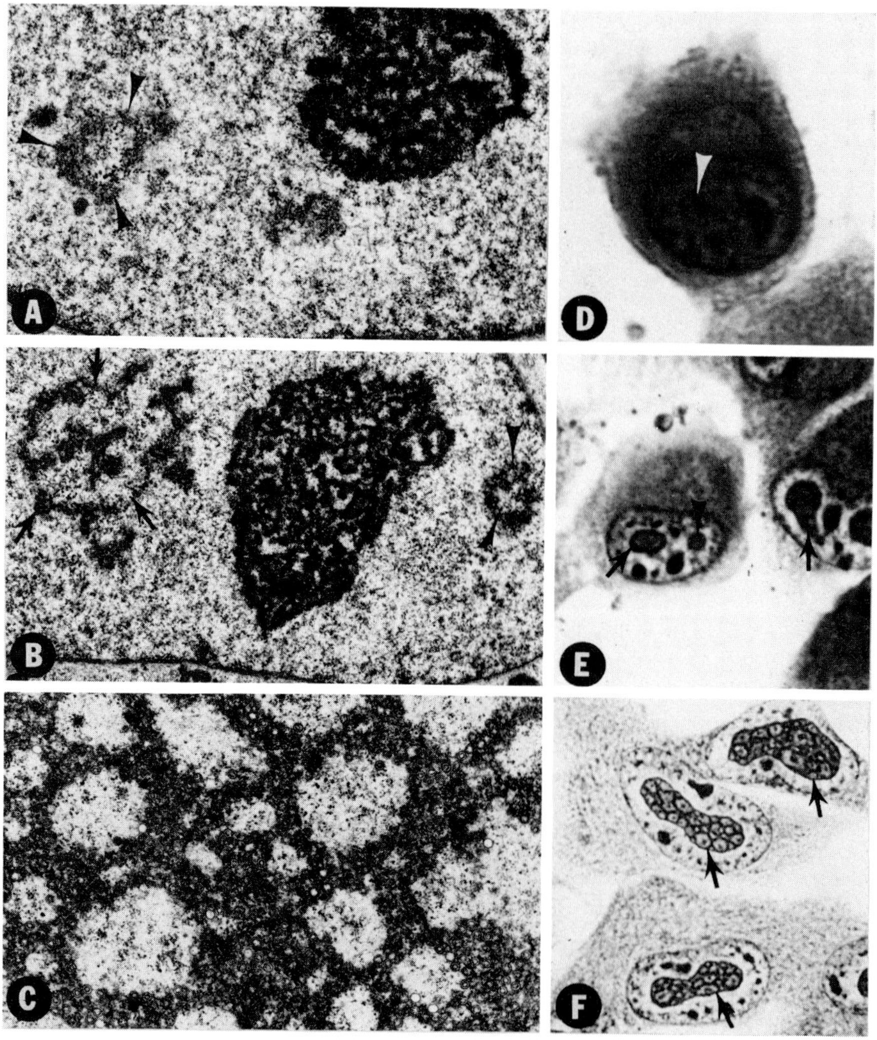


**FIGURE 9.** Human embryo fibroblasts infected with CMV for 25 (A,D), 49 (B,E), or 97 (C,F) hr PE. An early NI (arrowheads) is evident at 25 hr. At 49 hr, an early (arrowheads) and a developing late NI (arrows) constituted of a branching electron-dense network are demonstrated (B). Virus particles are present in the vicinity of the NI. A well-developed late NI constituted of numerous cellulae is evident at 97 hr (F). The electron-dense network is packed with capsids and nucleocapsids at this time (C). (A–C) ×9954; (D) ×1575; (E,F) ×491.

It is apparent from these observations that the increase in size and complexity of the NI closely parallels the rate of CMV DNA synthesis (see Fig. 3). The early NIs demonstrate a relatively small size and only a limited increase in size, while the late NIs demonstrate larger and progressively increasing size. The morphological characteristics of the early and late NIs are distinct. The formation of cellulae and nucleocapsids is associated with the late NI, evident only during the late phase of CMV DNA synthesis. These observations have led us to propose that the first phase of CMV DNA synthesis takes place within the early NI and that it is not associated with the formation of nucleocapsids. Accordingly, the daughter DNA strands formed during this phase of CMV DNA synthesis may be conserved and, at a later time, coinciding with the initiation of the second phase of CMV DNA synthesis, these daughter strands initiate replication, which is associated with the formation cellulae and nucleocapsids as described in Section 5.3.

## 5.3. Organization of the Cellulae and Nucleocapsid Assembly

The organization of the CMV NI into cellulae may result from the unique patterns of molecular regulation associated with CMV infection. Analysis of the fine structure of cellulae reveals a consistent organizational arrangement and suggests that there are three zones (Fig. 10) based on the distribution of virus particles (Fons and Albrecht, 1986). Within the central electron-lucent area (zone 1), there are many coarse electron-dense granules and small usually fine fibers. Capsid and nucleocapsid structures are rarely observed in this area. Zone 2 is defined as the interface between the electron-lucent zone (1) and the electron-dense zone (3). Virus structures that appear to be developing capsids and nucleocapsids, such as those shown in Fig. 11, are noted predominantly in zone 2. Most of these particles appear to be incomplete capsids with fibrous material similar to that observed in zone 1 in ·arious degrees of condensation. The band of electron-dense material enclosing each cellula (zone 3) demonstrates abundant well-developed capsids and nucleocapsids.

We interpret these observations to suggest that zone 1 may be the site of CMV DNA synthesis. Although other investigators (Fong, 1982) have suggested that CMV DNA synthesis takes place in the electron-dense network (our zone 3), this seems inconsistent with our observations since the free fibers which may contain CMV DNA are found in zone 1 and since the most complete particles are found within zone 3. Zone 2 appears to be the site in which nucleocapsid assembly and maturation begin. These observations are consistent with the view that when the DNA replicated during the first phase of CMV DNA synthesis begins replication, the newly replicated DNA (from the second-phase) forms the cellulae as it diffuses from its site of synthesis and encounters structural components for the virus particles forming capsids. Accordingly, the structure and

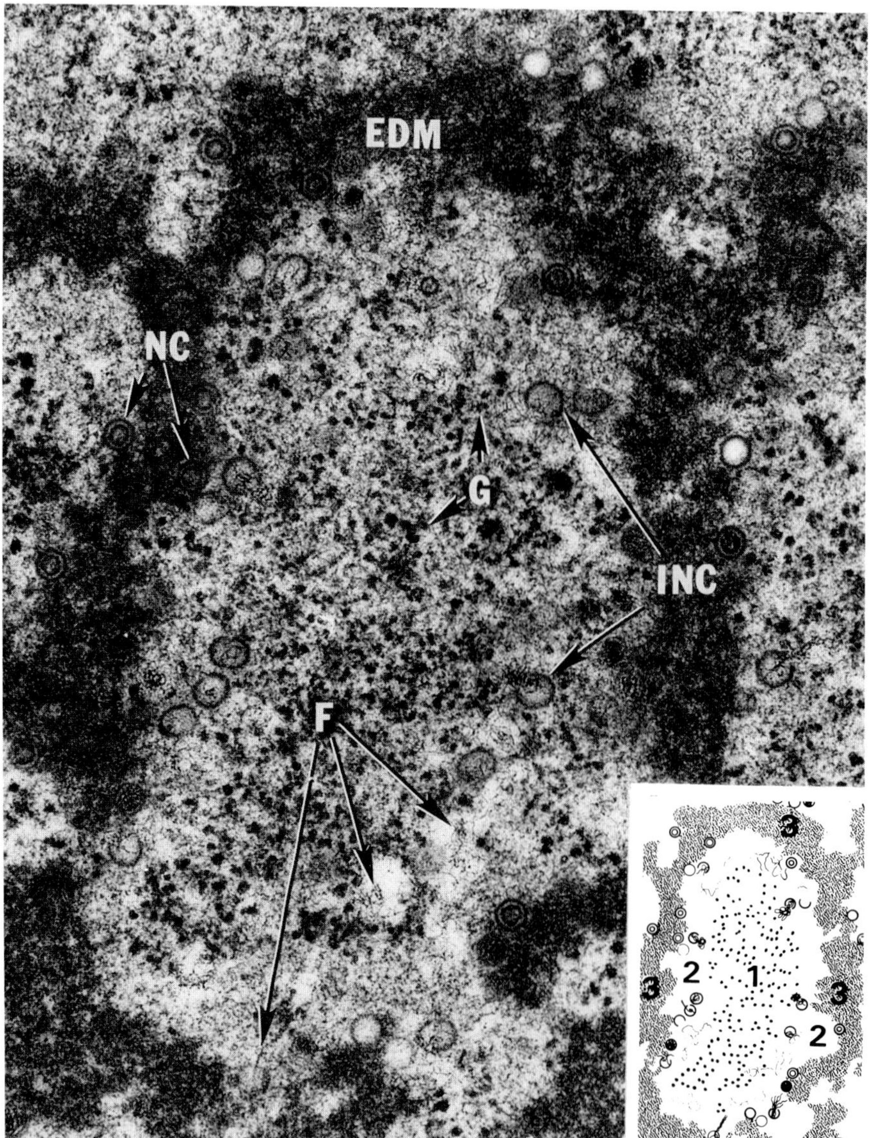


**FIGURE 10.** Cellulae demonstrating three zones based on the distribution of virus particles. In zone 1 fibers (F) and coarse granules (G) are evident within the electron-lucent area. Incomplete capsids and nucleocapsids (INC) are demonstrated in zone 2 at the interface of the electron-lucent area and the electron-dense material. Mature capsids and nucleocapsids (NC) are present in the electron-dense material (EDM), zone 3. (×40,810) Insert, diagrammatic representation of the three zones within the cellulae.

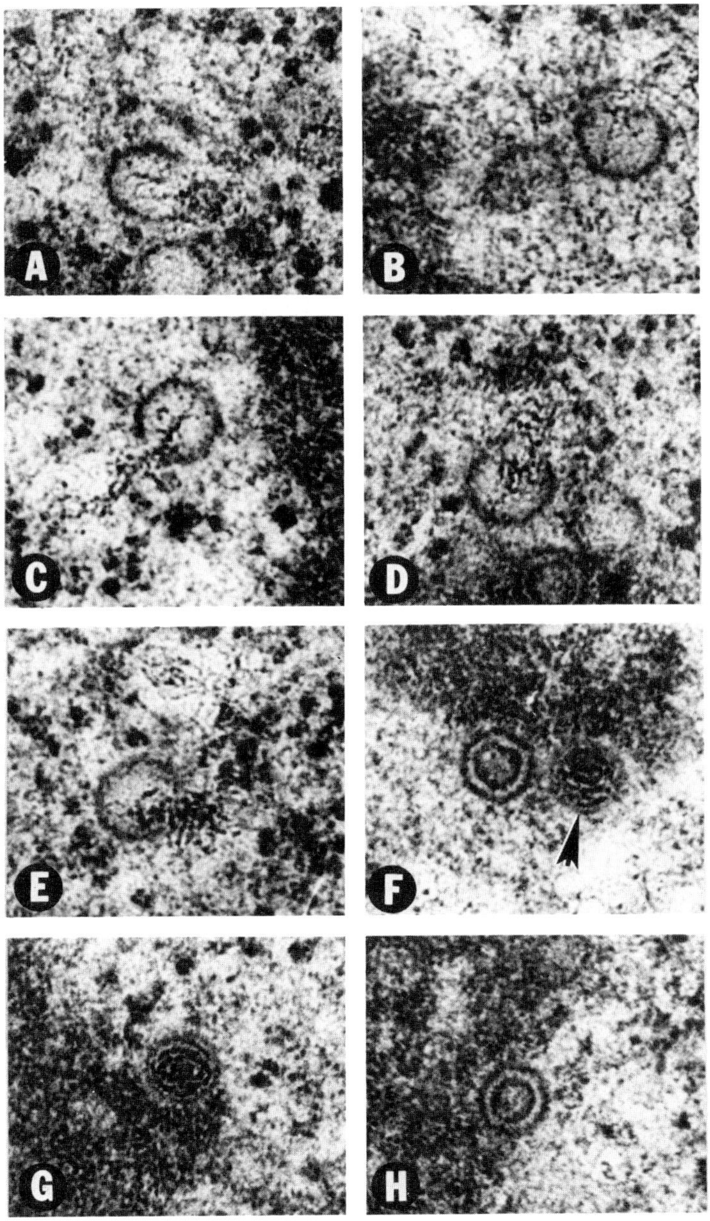


**FIGURE 11.** Incomplete CMV nucleocapsids from zone 2 (Fig. 10) arranged in a sequence to depict the apparent increased condensation of the fibrous material into incomplete capsids. A exhibits loose fibrous material in association with an incomplete capsid. B through E demonstrate increasingly condensed fibers in association with incomplete nucleocapsids. F (arrowhead) and G demonstrate the fibrous material within capsids. An apparently mature nucleocapsid similar to those observed adjacent to the nuclear membrane is evident in H. (×53,570)

organization of the cellulae and the late NI are related to the biphasic pattern for CMV DNA synthesis, with the second phase of CMV DNA synthesis possibly occurring in the cellulae.

Although, these observations provide an explanation for the relationship of the early and late NIs to the two phases of CMV DNA synthesis, they do not provide an understanding of why the daughter strands synthesized during the first phase of CMV DNA synthesis do not immediately, *in toto,* serve as templates for new rounds of CMV DNA synthesis, yielding a logarithmic increase instead of a biphasic pattern. It is now well established that the activity of CMV DNA polymerase is enhanced at salt concentrations (e.g., $(NH_4)_2SO_4$ (100 mM), KCl (200 mM)) that inhibit the activity of the cellular DNA polymerases. In fact, this characteristic is used to distinguish the activity of viral and cellular enzymes *in vitro* (Huang, 1975; Nishiyama *et al.*, 1983). This finding suggests that the viral and cellular enzymes may have differing optimal conditions *in vivo* as well. The cellular responses to CMV infection can be divided into two broad phases. The early phase is associated with the physiological responses of $Ca^{2+}$ entry and rising $[Ca^{2+}]_i$ and with the morphological responses of cell rounding and contraction. The late phase is characterized by $Na^+$ entry and the development of cytomegaly. If the second phase of the CMV-induced cellular physiologic response is aborted, stimulation of cellular DNA synthesis is observed (e.g., in hamster embryo fibroblasts); however, if the second phase is allowed to proceed, the cellular environment becomes intolerant for cellular DNA synthesis (e.g., in human embryo fibroblasts). On the other hand, although the intracellular environment resulting from the first phase of the physiological response is associated with low levels of CMV DNA synthesis, the continuing modification of the intracellular environment with the second phase of the physiological response (e.g., $Na^+$ entry) is associated with much higher levels of CMV DNA synthesis, the formation of the late NI consisting of cellulae, and the formation of virus progeny. It is not known whether there are any structural or functional differences in the CMV DNA synthesized during the first and second phases of CMV DNA synthesis that might assist in explaining the two phase pattern for CMV DNA synthesis and the failure to detect progeny virus during the first phase.

### 5.4. Effect of Inhibitors of the CMV-Induced Physiological Responses on CMV Replication

Consideration of the series of physiological cellular responses (see Fig. 4) identified in CMV-infected cells and of their temporal relationship with morphologic responses to CMV infection, particularly the development and progression of NI and cytomegaly, suggests that these physiological responses may be involved in the replication of CMV. Table V summarizes data for the effects of

**Table V**
**Effects of Inhibitors of CMV-Induced Cellular Physiological Responses on CMV DNA Synthesis, Development of NIs, and Yields of Infectious CMV**

| | Inhibition of | | | | |
|---|---|---|---|---|---|
| | | | NIs | | |
| Drug | Cytomegaly | CMV DNA synthesis | Early | Late | CMV yields |
| Papaverine (100 μM) | 5+[a] | 3+ | 3+ | 5+ | 5+ |
| Nitroprusside (100 μM) | 3+ | NT[b] | NT | NT | 3+ |
| Forskolin (30 μM) | — | NT | — | — | — |
| Verapamil (100 μM) | 1+ | NT | — | 1+ | 1+ |
| Nifedipine (100 μM) | 1+ | NT | — | 1+ | 1+ |
| Diltiazem (100 μM) | — | NT | — | — | — |
| Amiloride (150 μM) | 4+ | 3+ | — | 3+ | 4+ |
| SITS (300 μM)[c] | — | NT | — | — | — |
| Bumetanide (30 μM) | — | NT | — | — | — |

[a]For all parameters except inhibition of infectious yields: —, no inhibition; 1+, 1–24% inhibition; 2+, 35–49% inhibition; 3+, 50–74%; 4+, 75–99% inhibition; 5+, >99% inhibition; for inhibition of infectious yields, each + equals 1 $\log_{10}$-fold reduction in infectious yields relative to the infectious yield in the absence of drugs.
[b]NT: not tested
[c]SITS: 4-acetamido-4′-isothiocyanatostilbene-2,2′-disulfonic acid.

drugs that inhibit physiological responses to CMV infection and also interfere with the replication of CMV. Evaluation of these data suggest several points that should be considered. The drug (papaverine), which interferes with the earliest physiological responses, increased DG and $IP_3$ levels (Valyi-Nagy *et al.*, 1988), inhibits other physiological responses as well (e.g., the increases in $[Ca^{2+}]_i$ and $Na^+/K^+$ pump activity, the development of cytomegaly) and is the most potent of these drugs in reducing infectious yields of CMV. Drugs that inhibit $Ca^{2+}$ responses also interfere with $Na^+$ responses but are not as potent as papaverine. The first phase of CMV DNA synthesis and development of early NIs is relatively more resistant to the effects of these inhibitors, particularly inhibitors of $Na^+$ entry than are the second phase of CMV DNA synthesis and development of late NIs (Fig. 12). These findings suggest that there is a cascade of physiological responses in CMV-infected cells as proposed in Fig. 5 and that these physiological responses may be related to transformation of the intracellular environment to one consistent with the second phase of CMV DNA synthesis, the formation of late NIs, and production of progeny virus. More detailed information on the effects of these drugs on physiological responses to CMV infection and on CMV replication is available elsewhere (Albrecht *et al.*, 1987; Nokta *et al.*, 1987, 1988; Fons *et al.*, 1987; Lee and Albrecht, 1987).

$Na^+$ entry and cytomegaly seem important for efficient CMV replication.

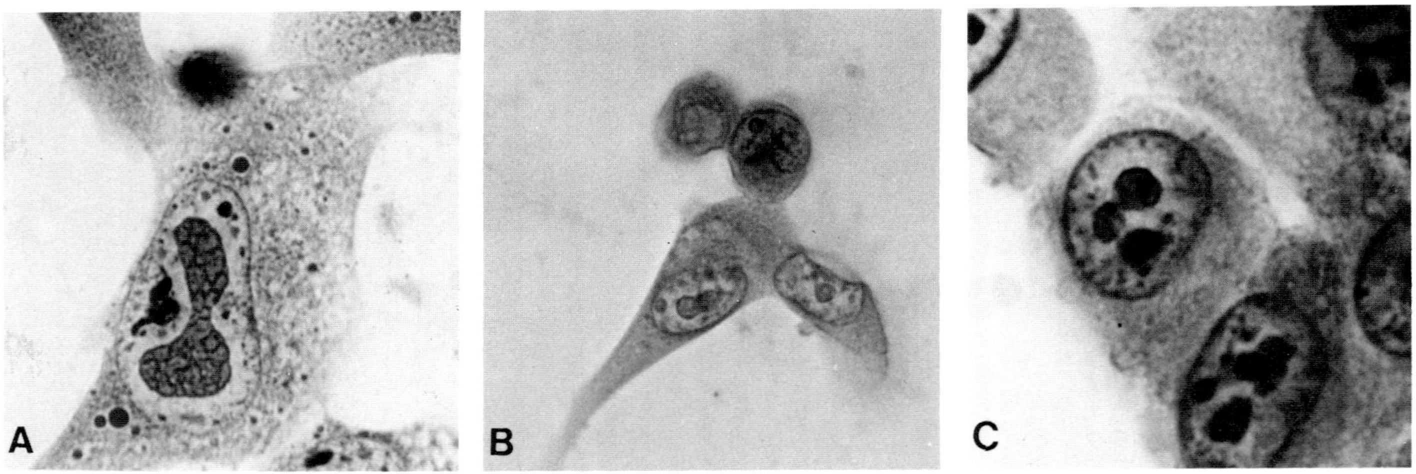


**FIGURE 12.** Effect of inhibition of the CMV-induced physiological changes on cellular morphological responses and on the development and progression of NIs. CMV-infected human fibroblasts at 97 hr PE: A, untreated; B, treated at 1 hr PE with papaverine (100 μM); C, treated at 1 hr PE with amiloride (150 μM). (A, C, ×722.5; B, ×365.5)

Therefore, we have focused our recent investigations on the relationship between $Na^+$ entry and cell enlargement on the one hand and on CMV DNA synthesis, development and progression of the NIs, and formation of infectious virus on the other. The necessity of $Na^+$ entry for efficient CMV replication is demonstrated by the sensitivity of CMV infectious yields to inhibition by amiloride (Fons *et al.*, 1987), low $Na^+$-(NMDG) medium, and papaverine (Albrecht *et al.*, 1987). Amiloride reduces CMV yields in a concentration-dependent manner between 30 and 150 μM (Fons *et al.*, 1987). Substitution of NMDG (139 mM) for $Na^+$ (4 mM) in the culture fluids also inhibits CMV replication, providing further evidence for the involvement of $Na^+$ entry in CMV replication. How $Na^+$ entry is involved in CMV replication is not altogether clear, but it appears that this relationship involves inhibition of the second phase of CMV DNA synthesis and the formation of late NIs. Amiloride (150 μM)-treated CMV-infected cells contain very small NIs (Fig. 12) without the numerous characteristic cellulae seen in untreated infected cells (Cavallo *et al.*, 1981; Fons and Albrecht, 1986). As in growth factor or serum stimulation of cellular DNA synthesis, amiloride is effective in inhibiting CMV DNA synthesis (Fons *et al.*, 1987). We have observed an approximately 30% inhibition of CMV DNA synthesis before 37 hr PE, which increases to an approximately 70% inhibition at later times PE in the presence of amiloride (150 μM). Cellular DNA synthesis is also appreciably inhibited at this concentration of amiloride. Consistent with the data for cell enlargement, SITS, DIDS, and bumetanide have negligible effects on CMV DNA synthesis, formation of NIs, and CMV replication (Table V).

Since CMV DNA synthesis may be associated with increased $[Na^+]_i$ and a concomitant rise in pH, the reported inhibition of CMV DNA synthesis and replication by monensin, a carboxylic $Na^+$ ionophore (Kaiser and Radsak, 1987) would seem surprising and at first glance contradictory to our findings. Monensin, however, also affects the transport of glycoproteins within the golgi apparatus (Tartakoff, 1983). Therefore, as Kaiser and Radsak point out, CMV DNA synthesis may be dependent on an early viral glycoprotein that is inhibited in the Golgi apparatus by monensin. Consistent with this view are the observations that other inhibitors of glycosylation, 2-deoxy-D-glucose (Radsak and Weder, 1981) and tunicamycin (Kaiser and Radsak, 1987), also inhibit CMV DNA synthesis. It is also possible that continued cellular glycoprotein synthesis could be required for maintenance of the CMV-induced physiologic changes (e.g., as part of the $Na^+$ and/or $Ca^{2+}$ regulatory systems).

## 6. GENOTOXIC EFFECTS: INDUCTION OF CHROMOSOMAL DAMAGE AND CELL-CYCLE PERTURBATION BY CMV

The concern that CMV can cause genomic changes that can lead to long-term biological problems is an appropriate one, especially when a substantial portion of the population carries persistent CMV infection. In at least three reports, CMV infection has been shown to induce chromosome aberrations in human fibroblasts (Nachtigal and Nachtigal, 1978; AbuBakar *et al.*, 1988) and in human fetal cells (Luleci *et al.*, 1980). It was recently shown (AbuBakar *et al.*, 1988) that CMV infection of permissive human fibroblasts induces profound chromosome damage, such as chromosome breakage, pulverization resembling premature chromosome condensation (PCC) (Johnson and Rao, 1970; Hittelman *et al.*, 1981), prolonged arrest of chromosomes in metaphase (C-mitosis), and cell-cycle inhibition as indicated by much reduced mitotic indices. In this study, confluent $G_0$ cells were infected with UV-irradiated or with nonirradiated CMV, subcultured at 25 hr PE, and harvested 48 hr later for cytogenetic analysis. Cytogenetic analysis of these cells showed that there was a significant inverse relationship between UV dose and induction of chromosome aberrations and pulverization (e.g., Fig. 13A). When actively growing cells were infected with UV-irradiated CMV, similar inverse UV-dose-dependent relationships were observed. The abnormalities seen, however, were chromosome breakage and C-mitosis. Consistent with this observation, Kamiya *et al.* (1986) observed an increased number of cells arrested in mitosis after CMV infection of both permissive and nonpermissive cells. These observed chromosomal abnormalities suggest that CMV infection can induce the production of cells with unbalanced genomic material, with the potential to induce cellular dysfunction or cause cell death.

The cell-culture conditions for these cytogenetic and cell-cycle studies were slightly different from those used for the pathophysiological studies in order to obtain sufficient mitotic cells. In particular, $G_0$-infected cells were subcultured to permit cell proliferation, and UV-irradiated CMV was used predominantly to provide for abortive infection. The existing pathophysiological data are not yet sufficient to explain adequately the cytogenetic changes or vice versa, although future experiments will almost certainly be conducted to determine whether there is a causal relationship.

Many herpesviruses have been shown to induce chromosome damage in infected cells (Stich and Yohn, 1970; Harnden, 1974; Yoon *et al.*, 1976; Webber and Whalley, 1977); however, the mechanisms for induction of the abnormalities remain obscure. The CMV-induced increased $[Ca^{2+}]_i$, $Na^+$ entry, and development of cytomegaly (Albrecht *et al.*, 1984; Nokta *et al.*, 1987, 1988) indicate an altered intracellular ionic environment that may contribute to chromosomal alterations. Since externally manipulated ionic imbalances and pH changes can cause chromosome damage in cells in culture (Cipollaro *et al.*, 1986; Brusik, 1986), it would not be surprising that intracellular physiological changes induced by CMV could also elicit similar chromosomal alterations. Indeed, the elevated $[Ca^{2+}]_i$ could conceivably affect the formation of the mitotic spindle and thereby inhibit chromosome segregation. It is also possible that the CMV-induced patho-

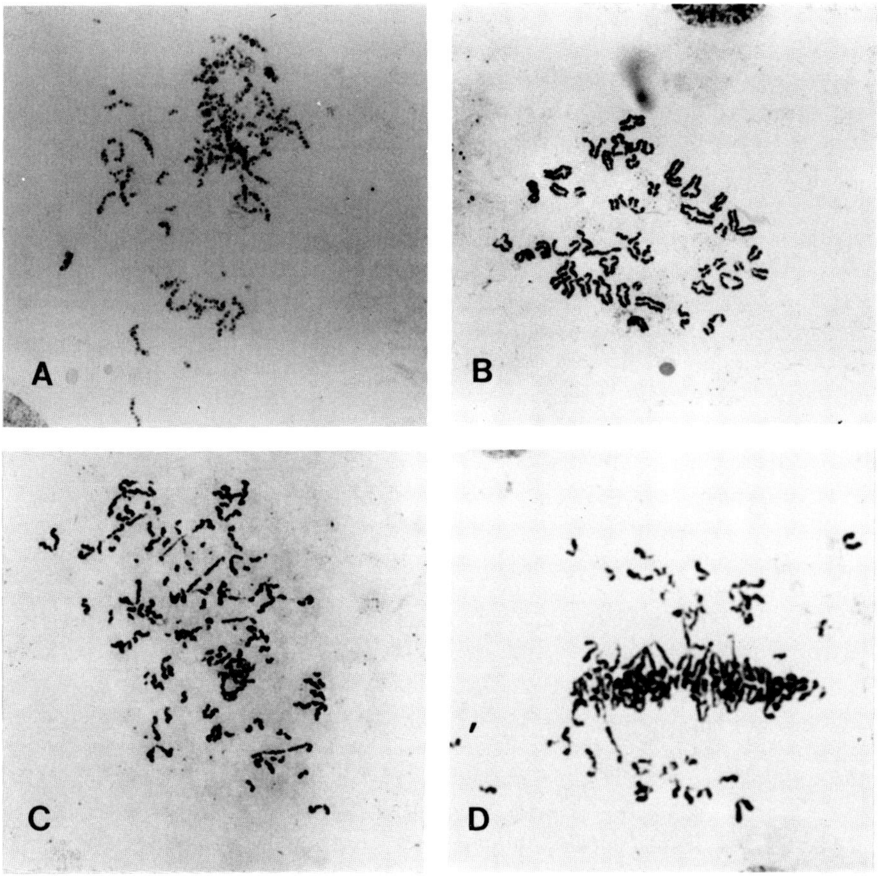


**FIGURE 13.** Various types of chromosome and mitotic abnormalities induced by CMV in human lung fibroblasts. A, a cell showing prematurely condensed chromosomes; B, a cell demonstrating problems in chromatid segregation. The phenomenon suggests that the kinetochore proteins may be abnormal. C, an abnormal anaphase cell with chromatid breaks and apparently random segregation of the chromatids; D, a metaphase/anaphase cell demonstrating an asynchronous and, therefore abnormal, segregation process. (×876)

physiological responses lead to the release of lysosomal enzymes, the abnormal function of nucleases, and other processes. The precise cause of chromosomal damage induced by CMV remains to be determined.

Our data show that CMV infection can cause interference with cell replication in both serum-arrested and proliferating cells. In $G_0$-infected cells, cellular DNA replication is perturbed, leading to the appearance of PCC chromosomes. In actively growing cells, cellular DNA synthesis does not appear to be as affected by CMV infection, but the segregation of chromatids during mitosis is blocked, leading to the appearance of C-mitotic chromosomes. PCC and C-mitosis may be alternative expressions of cell cycle perturbation by CMV. The observed effect is thus dependent on the stage of the cell cycle ($G_0$ or $G_1$, respectively) at the time of infection.

Two hypotheses can be suggested to explain the genotoxic damage induced by CMV. According to the first hypothesis, CMV may cause perturbation of cellular protein synthesis, possibly by dissociation of histone synthesis from cellular DNA synthesis (Radsak and Schmitz, 1980). Under conditions in which critical cellular proteins, such as histones, are unavailable in $G_0$ infection, cellular DNA replication may be aberrant or interrupted. Therefore, the chromatin may be unable to organize into normal condensed chromosomes and may appear as $G_1$- and S-PCC. In $G_1$ infection, the alteration of protein synthesis or function caused by CMV infection may occur during the latter part of the cell cycle. For example, the synthesis of kinetochore proteins may be affected. Figure 13B shows a mitotic preparation in which the kinetochores of the chromosomes appear to be disrupted. Insufficient or faulty kinetochore proteins may prevent segregation of chromatids during mitosis and may induce other types of cell-division anomalies.

The second hypothesis suggests that the cellular physiological changes induced by CMV infection generate activation signals that may be out of phase with the remaining cellular activities and therefore perturb the normal regulatory control of cell replication. If these cell-activation signals are present too early in the cell cycle, they may function like mitotic factors, which have been shown by Rao (1982) to be capable of inducing PCC. Ionic imbalance brought on by CMV infection may eventually create an inappropriate environment (perhaps osmotic and/or pH) for cellular DNA polymerase to function properly (see Sections 3.3 and 5.3). This and other physiological changes may cause the premature termination of cellular DNA synthesis and produce $G_1$- and S-PCC in $G_0$-infected cells.

In $G_1$ infection, the CMV-induced cellular changes may not occur early enough to affect the already ongoing cellular DNA replication process. The CMV-induced cellular physiological changes, however, may affect mitotic segregation of the chromatids by disrupting the polymerization of mitotic spindles. Figure 13C,D shows aberrant anaphase cells prepared from CMV-exposed cell cultures.

The two hypotheses presented above are not mutually exclusive. Further work is needed to understand the contribution of these proposed mechanisms to the induction of genomic damage and cell-cycle perturbation by CMV infection. It may be noteworthy that the genotoxic capacity demonstrated by CMV in our studies is consistent with its oncogenic potential, as recently reviewed by MacNab (1987).

## 7. SUMMARY AND PERSPECTIVES

It has been recognized for more than three decades that viruses bind to specific receptors (e.g., Levine and Sagik, 1956). In part, the pathogenesis of viral diseases is a reflection of the distribution of these receptors among the various cells composing an organism. Apparently, it has not been appreciated that the binding of a virus to its receptor could accomplish more than the specific adsorption of the virus to the cell. Our studies suggest that the interaction of at least one virus with its receptor may be used to initiate changes in the intracellular environment that put the synthetic needs of the virus at a selective advantage over those of the cell. The cellular receptors identified for several other viruses include: acetylcholine receptor (AChR), rabies virus; epidermal growth factor receptor (EGFR), vaccinia virus; β-adrenergic hormone receptor, reovirus T3; and complement receptor 2, Epstein–Barr virus (EBV) (Crowell, 1987). Thus, it would not be particularly surprising to find some degree of cell activation resulting from the attachment and binding of at least some other viruses to their receptors.

The present findings suggest that CMV uses an activation pathway present in fibroblasts that may be similar to those for some growth factors. The time course for the activation signals, however, appears to be substantially protracted for CMV infection of permissive cells relative to the duration of activation signals proposed for hormones or growth factors (see Table I and Fig. 4). We interpret this finding to suggest that CMV infection may be disturbing, in some way, the cellular controls regulating the duration of the activation signals. The work describing the CMV effect on cell activation has been done, almost exclusively, in fibroblasts. It is likely, however, that similar activation responses are initiated in CMV infection of other cell types (e.g., glandular epithelial cells, endothelial cells), since these cells have differentiated to perform work that is stimulated through similar activation pathways. The studies relating cell-activation signals to CMV replication are relatively recent, however, and much additional work is needed to determine the specific molecular mechanisms regulating these responses and their relationship to the pathogenesis of CMV disease.

The progressively developing physiological responses induced by the interaction of CMV with its receptor and by subsequent CMV-induced events may be related to the pathogenesis of CMV disease in several important ways. Adaptation of CMV to this mode of replication may assist it in successfully infecting and replicating in the wide range of tissues observed to support CMV infection *in vivo,* which allows for efficient viral spread through its presence in a relatively large number of secretions (Weller, 1971). The prolonged replication cycle for CMV and the phasing of CMV-induced synthetic events may provide the virus with the opportunity to form persistent infections. It would appear from our data that some CMV DNA synthesis may occur before the cell is committed to the production of infectious virus and the lysis of the cell. If this hypothesis is correct, any modification of the cultural conditions that interferes with development of the second phase of the physiological response (and development of cytomegaly) would have a tendency to select for persistent infections, as was demonstrated with supraoptimal temperature (Albrecht *et al.*, 1980b; Li and Albrecht, 1982). It may be important in future work to consider the possibility that the effects of CMV on the physiology, morphology, and the genome of infected cells suggest that CMV may alter cellular functions in subtle ways that could be related to slowly progressive disease. The findings suggesting the possible involvement of CMV in some cancers (Rapp *et al.*, 1975; Huang and Roche, 1978; Melnick *et al.*, 1978; Huang *et al.*, 1983, 1984; Rapp and Robbins, 1984), atherosclerosis (Adam *et al.*, 1987), and schizophrenia (Albrecht *et al.*, 1980c) may be examples that suggest of such a relationship.

When the expression of CMV information is restricted, and presumably when the signal pathway initiated by CMV infection is of limited duration (e.g., in abortive infection), stimulation of cellular DNA synthesis may be observed. The induction of this proliferative signal by CMV is consistent with its demonstrated oncogenic potential *in vitro* (Albrecht and Rapp, 1973; Geder *et al.*, 1976; Boldogh *et al.*, 1978c) and with its suspected relationship with neoplastic processes *in vivo* (Rapp *et al.*, 1975; Huang and Roche, 1978; Melnick *et al.*, 1978; Huang *et al.*, 1983, 1984; Rapp and Robbins, 1984). Although the biophysical and biochemical characteristics for the three groups of DNA viruses (herpesviruses, adenoviruses, papovaviruses) with demonstrated oncogenic potential differ widely, they share the ability to initiate cellular DNA synthesis in arrested cells (Dulbecco *et al.*, 1965; Gershon *et al.*, 1966; Takahashi *et al.*, 1966; Gerber and Hoyer, 1971; St. Jeor *et al.*, 1974). Levine and Burger (1972) postulated that the ability of these viruses to stimulate cellular DNA synthesis is related to their capacity to transform cells. Future experiments should determine whether the induction of genomic damage, as well as the stimulation of cellular DNA synthesis, by CMV may be related to malignant transformation of infected cells, and whether the CMV-induced proliferative signal is related mechanistically to that for activation by growth factors. Such a finding would not be altogether new, however, since a relationship has been observed between polyoma middle T antigen, the hydrolysis of inositol lipids, and cell transformation (Whitman *et al.*, 1985; Kaplan *et al.*, 1986).

In conclusion, CMV induces a cascade of physiological responses similar to those observed for factors that induce cells to work or to proliferate. The time course for this signal pathway in CMV-infected cells is protracted, which may be related to the phasing of the CMV-induced pathophysiological responses and may lead to the phasing of CMV DNA synthesis and the development and progression of NIs. The phasing of the physiological responses and the expression of CMV may provide for important advantages in developing a mode of pathogenesis based on infection of specialized tissues and persistence. The initial $Ca^{2+}$-related events appear to be conducive to cell activation, to stimulation of cellular macromolecular synthesis, and to early CMV functions, while the later phase demonstrating prominent $Na^{+}$-related events appears to favor CMV DNA replication and late CMV functions. A detailed understanding of the cellular and molecular mechanisms involved in cell activation by CMV could provide information for development of novel therapeutic approaches, as demonstrated by the remarkable potency of papaverine *in vitro* (Albrecht *et al.*, 1987). Furthermore, the protracted nature of cell activation as induced by CMV may be of interest to cell biologists and physiologists, as it may provide a model of pathological regulation of these phenomena well suited for detailed analysis.

ACKNOWLEDGMENTS. The authors thank Dr. R. Pollard, Dr. W. R. Fleischmann, Jr., and Dr. R. McKendall for critically reading the manuscript. Research in the authors' laboratory was supported by USPHS research contract NO1 AI 42557 (TA) and by USEPA research grant R812086 (TA). M. Fons and C. H. Lee are recipients of J. W. McLaughlin Fellowships.

## 8. REFERENCES

AbuBakar, S., Au, W. W., Legator, M. S., and Albrecht, T., 1988, Induction of chromosome aberrations and mitotic arrest by cytomegalovirus in human cells, *Environ. Mutagen.* **12:**409–420.

Adam, E., Melnick, J. L., Probtsfield, J. L., Petrie, B. L., Burek, J., Bailey, K. R., McCollum, C. H., DeBakey, M. E., 1987, High levels of cytomegalovirus antibody in patients requiring vascular surgery for atherosclerosis, *Lancet* **2:**291–293.

Albrecht, T. B., 1973, Studies on the oncogenic potential of human cytomegalovirus, Ph.D. thesis, Pennsylvania State University, University Park, Pennsylvania.

Albrecht, T., and Rapp, F., 1973, Malignant transformation of hamster embryo fibroblasts following exposure to ultraviolet-irradiated human cytomegalovirus, *Virology* **55:**53–61.

Albrecht, T., St. Jeor, S. C., Funk, F. D., and Rapp, F., 1974, Multiplicity reactivation of human cytomegalovirus inactivated by ultra-violet light, *Int. J. Radiat. Biol.* **26:**445–454.

Albrecht, T., Nachtigal, M., St. Jeor, S. C., and Rapp, F., 1976, Induction of cellular DNA synthesis and increased mitotic activity in Syrian hamster embryo cells abortively infected with human cytomegalovirus, *J. Gen. Virol.* **30:**167–177.

Albrecht, T., Cavallo, T., Cole, N. L., and Graves, K., 1980a, Cytomegalovirus: Development and progression of cytopathic effects in human cell culture, *Lab. Invest.* **42:**1–7.

Albrecht, T., Li, M.-L., Cole, N., Downing, E., and Funk, F. D., 1980b, Replication of human cytomegalovirus at supraoptimal temperatures is dependent on the virus strain, multiplicity of infection and phase of virus replication, *J. Gen. Virol.* **51:**83–97.

Albrecht, P., Torrey, E. F., Boone, E., Hicks, J. T., and Daniel, N., 1980c, Raised cytomegalovirus antibody level in cerebrospinal fluid of schizophrenic patients, *Lancet* **2:**769–772.

Albrecht, T., Speelman, D. J., and Steinsland, O., 1981, Similarities between human cytomegalovirus induced cell rounding and contraction of myofibroblasts or smooth muscle cells, in: *Abstracts of the Sixth International Workshop on Herpesviruses, July 27–31, Bologna, Italy,* p. 71.

Albrecht, T., Speelman, D. J., and Steinsland, O. S., 1983, Similarities between cytomegalovirus-induced cell rounding and contraction of smooth muscle cells, *Life Sci.* **32:**2273–2278.

Albrecht, T., Li, J.-L., Speelman, D., Ball, R., Nokta, M., Fons, M., Lee, C. H., Steinsland, O., Thompson, W. C., and Carney, D. H., 1984, Cellular responses to human cytomegalovirus infection, in: *CMV: Pathogenesis and Prevention of Human Infection* (S. A. Plotkin, S. Michelson, J. S. Pagano, and F. Rapp, eds.) [Birth Defects Orig. Art. Ser. **20:**21–34], Liss, New York.

Albrecht, T., Nokta, M., Lee, C. H., and Fons, M., 1985, Possible correlations between the phasing of cellular responses to cytomegalovirus (CMV) infection, development of nuclear inclusions (NI), and the pattern of CMV DNA synthesis, in: *Abstracts of the Tenth International Herpesvirus Workshop, August 11–16, Ann Arbor, Michigan,* p. 58.

Albrecht, T., Lee, C. H., Speelman, D. J., and Steinsland, O. S., 1987, Inhibition of cytomegalovirus replication by smooth-muscle relaxing agents, *Proc. Soc. Exp. Biol. Med.* **186:**41–46.

Alkon, D. L., and Rasmussen, H., 1988, A spatial-temporal model of cell activation, *Science* **239:** 998–1004.

Berridge, M. J., 1987, Inositol lipids and cell proliferation, *Biochim. Biophys. Acta* **907:**33–45.

Berridge, M. J., and Irvine, R. F., 1984, Inositol trisphosphate, a novel second messenger in cellular signal transduction, *Nature (Lond.)* **312:**315–320.

Bhalla, R. C., Webb, R. C., Singh, D., and Brock, T., 1978, Role of cAMP in rat aortic microsomal phosphorylation and calcium uptake, *Am. J. Physiol.* **234:**H508–H514.

Blumenthal, D. K., and Stull, J. T., 1980, Activation of skeletal muscle myosin light chain kinase by calcium (2+) and calmodulin, *Biochemistry* **19:**5608–5614.

Boldogh, I., Gonczol, E., Gartner, L., and Vaczi, L., 1978a, Stimulation of host DNA synthesis and induction of early antigens by ultraviolet light irradiated human cytomegalovirus, *Arch. Virol.* **58:**289–299.

Boldogh, I., Gonczol, E., and Vaczi, L., 1978b, Lack of correlation between HCMV- and cellular DNA synthesis, in: *Abstracts of the International Congress for the Hungarian Society for Microbiology, Budapest, Hungary,* p. 91.

Boldogh, I., Gonczol, E., and Vaczi, L., 1978c, Transformation of hamster embryonic fibroblast cells by UV-irradiated human cytomegalovirus, *Acta Microbiol. Acad. Sci. Hung.* **25:**269–275.

Boom, R., Sol, C. J. A., Minnaar, R. P., Geelen, J. L. M. C., Raap, A. K., and van der Noordaa, J., 1988, Induction of gene expression under human cytomegalovirus immediate early enhancer-promoter control by inhibition of protein synthesis is cell cycle-dependent, *J. Gen. Virol.* **69:** 1179–1193.

Boynton, A. L., and Whitfield, J. F., 1983, The role of cyclic AMP in cell proliferation: A critical assessment of the evidence, *Adv. Cyclic Nucl. Res.* **15:**193–294.

Brusik, D., 1986, Genotoxic effects in cultured mammalian cells produced by low pH treatment conditions and increased ion concentrations, *Environ. Mutagen.* **8:**879–886.

Cassel, D., Whiteley, B., Zhuang, Y. X., and Glaser, L., 1985, Mitogen-independent activation of $Na^+/H^+$ exchange in human epidermoid carcinoma A431 cells: Regulation by medium osmolarity, *J. Cell. Physiol.* **122:**178–186.

Cavallo, T., Graves, K., Cole, N. L., and Albrecht, T., 1981, Cytomegalovirus: An ultrastructural study of the morphogenesis of nuclear inclusions in human cell culture, *J. Gen. Virol.* **56:**97–104.

Chua, C. C., Carter, T. H., and St. Jeor, S. C., 1981, Transcription of the human cytomegalovirus genome in productively infected cells, *J. Gen. Virol.* **56:**1–11.

Cipollaro, M., Corsale, G., Esposito, A., Ragucci, E., Staino, N., Giordano, G. G., and Pagano, G., 1986, Sublethal pH decrease may cause genetic damage to eucaryotic cells: A study on sea urchins and *Salmonella typhimurium, Teratogen. Carcinogen. Mutagen.* **6:**275–287.

Crowell, R. L., 1987, Cellular receptors in virus infections, *ASM News* **53:**422–425.

DeMarchi, J. M., 1983a, Post-transcriptional control of human cytomegalovirus gene expression, *Virology* **124:**390–402.

DeMarchi, J. M., 1983b, Correlation between stimulation of host cell DNA synthesis by human cytomegalovirus and lack of expression of a subset of early virus genes, *Virology* **129:**274–286.

DeMarchi, J. M., and Kaplan, A. S., 1976, Replication of human cytomegalovirus DNA: Lack of dependence on cell DNA synthesis, *J. Virol.* **18:**1063–1070.

DeMarchi, J. M., and Kaplan, A. S., 1977a, Physiological state of human embryonic lung cells affects their response to human cytomegalovirus, *J. Virol.* **23:**126–132.

DeMarchi, J. M., and Kaplan, A. S., 1977b, The role of defective cytomegalovirus particles in the induction of host cell DNA synthesis, *Virology* **82:**93–99.

DeMarchi, J. M., Schmidt, C. A., and Kaplan, A. S., 1980, Patterns of transcription of human cytomegalovirus in permissively infected cells, *J. Virol.* **35:**277–286.

Diamond, J., and Blisard, K. S., 1976, Effects of stimulant and relaxant drugs on tension and cyclic nucleotide levels in canine femoral artery, *Mol. Pharmacol.* **12:**688–692.

Dulbecco, R. L., Hartwell, H., and Vogt, M., 1965, Induction of cellular DNA synthesis by polyoma virus, *Proc. Natl. Acad. Sci. USA* **53:**403–410.

Dunham, E. W., Haddox, M. K., and Goldberg, N. D., 1974, Alteration of vein cyclic 3′:5′ nucleotide concentrations during changes in contractility, *Proc. Natl. Acad. Sci. USA* **71:**815–819.

Estes, J. E., and Huang, E.-S., 1977, Stimulation of cellular thymidine kinases by human cytomegalovirus, *J. Virol.* **24:**13–21.

Fioretti, A., Furukawa, T., Santoli, D., and Plotkin, S. A., 1973, Nonproductive infection of guinea pig cells with human cytomegalovirus, *J. Virol.* **11:**998–1003.

Fong, C. K. Y., 1982, Ultrastructural localization of cytomegalovirus DNA synthesis in infected guinea-pig cells, *J. Gen. Virol.* **60:**235–245.

Fons, M., and Albrecht, T., 1986, Cytomegalovirus: The relationship of nucleocapsid assembly to the organization of cellulae, *Arch. Virol.* **91:**351–356.

Fons, M., Cerruti-Sola, S., Hunt, P., Nokta, M., and Albrecht, T., 1987, The effect of amiloride on CMV-induced $Na^+/K^+$ ATPase activity, cytomegaly and replication, in: *Abstracts of the Twelfth International Herpesvirus Workshop, July 30–August 4, Philadelphia,* p. 381.

Fons, M. P., Graves, K., Cavallo, T., Pollard, R., and Albrecht, T., 1986, Human cytomegalovirus: Development and progression of nuclear inclusions by primary clinical isolates and laboratory-adapted strains, *Proc. Soc. Exp. Biol. Med.* **181:**416–422.

Furukawa, T., Fioretti, A., and Plotkin, S. A., 1973, Growth characteristics of cytomegalovirus in human fibroblasts with demonstration of protein synthesis early in viral replication, *J. Virol.* **11:** 991–997.

Furukawa, T., Tanaka, S., and Plotkin, S. A., 1975, Stimulation of macromolecular synthesis in guinea pig cells by human CMV, *Proc. Soc. Exp. Biol. Med.* **148:**211–214.

Furukawa, T., Sakuma, S., and Plotkin, S. A., 1976, Human cytomegalovirus infection of WI-38 cells stimulates mitochondrial DNA synthesis, *Nature (Lond.)* **262:**414–416.

Garnett, H. M., 1981, Increased ability of human embryonic fibroblasts to accumulate $Ca^{2+}$ due to cytomegalovirus infection, *Cytobios* **31:**107–116.

Geder, L., Lausch, R., O'Neill, F., and Rapp, F., 1976, Oncogenic transformation of human embryo lung cells by human cytomegalovirus, *Science* **192:**1134–1137.

Gerber, P., and Hoyer, B. H., 1971, Induction of cellular DNA synthesis in human leucocytes by Epstein–Barr virus, *Nature (Lond.)* **231:**46–47.

Gershon, D., Sachs, L., and Winocour, E., 1966, The induction of cellular DNA synthesis by simian virus 40 in contact-inhibited and X-irradiated cells, *Proc. Natl. Acad. Sci. USA* **56:**918–925.

Goodheart, C. R., McAllister, R. M., and Filbert, J. E., 1964, Human Cytomegaolvirus: DNA synthesis and migration in infected cells studied autoradiographically, *Virology* **23:**603–608.

Grinstein, S., Cohen, S., Goetz, J. D. and Rothstein, A. 1985. Osmotic and phorbol ester-induced activation of $Na^+/H^+$ exchange: Possible role of protein phosphorylation in lymphocyte volume regulation. *J. Cell. Biol.* **101:**269–276.

Haddox, M. K., Magun, B. E., and Russell, D. H., 1980, Differential expression of type I and type II cyclic AMP-dependent protein kinases during cell cycle and cyclic AMP-induced growth arrest, *Proc. Natl. Acad. Sci. USA* **77:**3445–3449.

Harnden, D. G., 1974, Viruses, chromosome and tumors: The interaction between viruses and chromosomes, in: *Chromosomes and Cancer* (J. German, ed.), pp. 151–191, Wiley, New York.

Hedley-Whyte, E. T., and Craighead, J. E., 1965, Generalized cytomegalic inclusion disease after renal homotransplantation, *N. Engl. J. Med.* **272:**473–475.

Heldin, C.-H., Westermark, B., and Westeson, A., 1979, Platelet-derived growth factor: Purification and partial characterization, *Proc. Natl. Acad. Sci. USA* **76:**3722–3726.

Hill, R. B., Jr., Rowlands, D. T., Jr., and Rifkind, D., 1964, Infectious pulmonary disease in patients receiving immunosuppressive therapy for organ transplantation, *N. Engl. J. Med.* **271:** 1021–1027.

Hirai, K., and Watanabe, Y., 1976, Induction of α-type DNA polymerase in human cytomegalovirus-infected WI-38 cells, *Biochim. Biophys. Acta* **447:**328–339.

Hirai, K., Furukawa, T., and Plotkin, S. A., 1976, Induction of DNA polymerase in WI-38 and guinea pig cells infected with human cytomegalovirus (HCMV), *Virology* **70:**251–255.

Hirai, K., Maeda, F., and Watanabe, Y., 1977, Expression of early virus functions in human cytomegalovirus infected HEL cells: Effect of ultraviolet light-irradiation of the virus, *J. Gen. Virol.* **38:**121–133.

Hittelman, W. W., Rao, P. N., and McCredie, K. B., 1981, Premature chromosome condensation studies in human leukemia, in: *Genes, Chromosomes and Neoplasia* (F. E. Arrighi, P. W. Rao, and E. Stubblefield, eds.), pp. 379–403, Raven, New York.

Huang, E.-S., 1975, Human cytomegalovirus. III. Virus-induced DNA polymerase, *J. Virol.* **16:** 298–310.

Huang, E.-S., and Roche, J. K., 1978, Cytomegalovirus DNA and adenocarcinoma of the colon: Evidence for latent viral infection, *Lancet* **1:**957–960.

Huang, E.-S., Boldogh, I., and Mar, E.-C., 1983, Human cytomegaloviruses: Evidence for possible association with human cancer, in: *Viruses Associated with Human Cancer* (L. A. Phillips, ed.), pp. 161–194, Dekker, New York.

Huang, E.-S., Mar, E.-C., Boldogh, I., and Baskar, J., 1984, The oncogenicity of human cytomegalovirus, in: *CMV: Pathogenesis and Prevention of Human Infection* (S. A. Plotkin, S. Michelson, J. S. Pagano, and F. Rapp, eds.) [*Birth Defects Orig. Art. Ser.* **20:**193–211], Liss, New York.

Isom, H. J., 1979, Stimulation of ornithine decarboxylase by human cytomegalovirus, *J. Gen. Virol.* **42:**265–278.

Jesionek, A., and Kiolemenoglou, B., 1904, Ueber einen Befund von protozoenartigen Gebilden in den Organen eines hereditar-leutischen Fotus, *MMW* **51:**1905–1907.

Johnson, R. T., and Rao, P. N., 1970, Mammalian cell fusion: Induction of premature chromosome condensation in interphase nuclei, *Nature (Lond.)* **226:**717–722.

Kaiser, C.-J., and Radsak, K., 1987, Inhibition by monensin of human cytomegalovirus DNA replication, *Arch. Virol.* **94:**229–245.

Kamata, T., Tanaka, S., and Watanabe, Y., 1978, Human cytomegalovirus-induced chromatin factors responsible for changes in template activity and structure of infected cell chromatin, *Virology* **90:**197–208.

Kamata, T., Tanaka, S., and Watanabe, Y., 1979, Characterization of the human cytomegalovirus-induced chromatin factor responsible for activation of host cell chromatin template, *Virology* **97:** 224–228.

Kamiya, S., Tanaka, J., Ogura, T., Ogura, H., Sato, H., and Hatano, M., 1986, Rabbit kidney cells abortively infected with human cytomegalovirus are arrested in mitotic phase, *Arch. Virol.* **89:** 131–144.

Kaplan, D. R., Whitman, M., Schaffhausen, B., Raptis, L., Garcea, R. L., Pallas, D., Roberts, T. M., and Cantley, L., 1986, Phosphatidylinositol metabolism and polyoma-mediated transformation, *Proc. Natl. Acad. Sci. USA* **83:**3624–3628.

Katsuki, S., and Murad, F., 1977, Regulation of adenosine cyclic 3′,5′-monophosphate and guanosine cyclic 3′,5′-monophosphate levels and contractility in bovine tracheal smooth muscle, *Mol. Pharmacol.* **13:**330–341.

Kattenburg, D. M., and Daniel, E. E., 1984, Effects of an endogenous cyclic AMP-dependent protein kinase catalytic subunit on Ca-uptake by plasma membrane vesicles from rat mesenteric artery, *Blood Vessels* **21:**257–266.

Kilpatrick, B. A., and Huang, E.-S., 1977, Human cytomegalovirus genome: Partial denaturation map and organization of genome sequences, *J. Virol.* **24:**261–276.

L'Allemain, G., Paris, S., and Pouyssegur, J., 1984, Growth factor action and intracellular pH regulation in fibroblasts: Evidence for a major role of the Na/H antiport, *J. Biol. Chem.* **259:** 5809–5815.

L'Allemain, G., Paris, S., and Pouyssegur, J., 1985, Role of a $Na^+$-dependent $Cl^-/HCO_3^-$ exchange in regulation of intracellular pH in fibroblasts, *J. Biol. Chem.* **260:**4877–4883.

Landini, M. P., 1984, Early enhanced glucose uptake in human cytomegalovirus-infected cells, *J. Gen. Virol.* **65:**1229–1232.

Landini, M. P., and Rugolo, M., 1984, Increased accumulation of a lipophilic cation (tetraphenylphosphonium) in human embryo fibroblasts after infection with cytomegalovirus, *J. Gen. Virol.* **65:**2269–2272.

Lang, D. J., Montagnier, L., and Latarjet, R., 1974, Growth in agarose of human cells infected with cytomegalovirus, *J. Virol.* **14:**327–332.

Lee, C. H., and Albrecht, T., 1987, Cyclic nucleotide responses to cytomegalovirus (CMV) infection: Partial correlation with inhibition of CMV yields by papaverine, in: *Abstracts of the Twelfth International Herpesvirus Workshop, July 30–August 4, Philadelphia,* p. 382.

Lee, T.-P., Kuo, J. F., and Greengard, P., 1972, Role of muscarinic cholinergic receptors in regulation of guanosine 3′ : 5′-cyclic monophosphate content in mammalian brain, heart muscle and intestinal smooth muscle, *Proc. Natl. Acad. Sci. USA* **69:**3287–3291.

Levine, A. J., and Burger, M. M., 1972, A working hypothesis explaining the maintenance of the transformed state by SV40 and polyoma, *J. Theor. Biol.* **37:**435–446.

Levine, S., and Sagik, B. P., 1956, The interactions of Newcastle disease virus (NDV) with chick embryo tissue culture cells: Attachment and growth, *Virology* **2:**57–68.

Li, J.-L. H., and Albrecht, T., 1982, Characterization of human cells persistently infected with cytomegalovirus and exposed to a chemical carcinogen, *Int. J. Cancer* **29:**49–55.

Lowenstein, C., 1907, Uber protozoenartige Gebilde in den Organen von Kindern, *Zentralbl. Allg. Pathol. Anat.* **18:**513–518.

Luleci, G., Skizli, M., and Gunalp, A., 1980, Selective chromosomal damage caused by human cytomegalovirus, *Acta Virol.* **24:**341–345.

Macher, A. M., Reichert, C. M., Straus, S. E., Longo, D. L., Parrillo, J., Lane, H. C., Fauci, A. S., Rook, A. H., Manischewitz, J. F., and Quinnan, G. V., Jr., 1983, Death in the AIDS patient: Role of cytomegalovirus, *N. Engl. J. Med.* **309:**1454.

MacNab, J. C. M., 1987, Herpes simplex virus and human cytomegalovirus: Their role in morphological transformation and genital cancers, *J. Gen. Virol.* **68:**2525–2550.

Majno, G., 1979, The story of the myofibroblasts, *Am. J. Surg. Pathol.* **3:**535–542.

McGavran, M. H., and Smith, M. G., 1965, Ultrastructural, cytochemical, and microchemical observations on cytomegalovirus (salivary gland virus) infection of human cells in tissue culture, *Exp. Mol. Pathol.* **4:**1–10.

Melnick, J. L., Lewis, R., Wimberly, I., Kaufman, R. H., and Adam, E., 1978, Association of cytomegalovirus (CMV) infection with cervical cancer: Isolation of CMV from cell cultures derived from cervical biopsy, *Intervirology* **10:**115–119.

Michell, R. H., 1986, Inositol lipids and their role in receptor function: History and general principles, in: *Receptor Biochemistry and Methodology: Phosphoinositides and Receptor Mechanisms* (J. W. Putney, Jr., ed.), Vol. 7, pp. 1–24, Liss, New York.

Moolenaar, W. H., Tsien, R. Y., van der Saag, P. T., and de Laat, S. W., 1983, $Na^+/H^+$ exchange and cytoplasmic pH in the action of growth factors in human fibroblasts, *Nature (Lond.)* **304:** 645–648.

Moummi, C., Magous, R., Strosberg, D., and Bali, J. P., 1988, Muscarinic receptors in isolated smooth muscle cells from gastric antrum, *Biochem. Pharmacol.* **37:**1363–1369.

Nachtigal, M., and Nachtigal, S., 1978, Interactions between human herpesviruses and host cell chromosomes, *Arch. Roum. Pathol. Exp. Microbiol.* **37:**223–249.

Nishiyama, Y., Maeno, K., and Yoshida, S., 1983, Characterization of human cytomegalovirus-induced DNA polymerase and the associated 3′-to-5′, exonuclease, *Virology* **124:**221–231.

Nokta, M., Eaton, D., Steinsland, O. S., and Albrecht, T., 1987, $Ca^{2+}$ responses in cytomegalovirus-infected fibroblasts of human origin, *Virology* **157:**259–267.

Nokta, M., Fons, M. P., Eaton, D. C., and Albrecht, T., 1988, Cytomegalovirus: Sodium entry and development of cytomegaly in human fibroblasts, *Virology* **164:**411–419.

Owen, N. E., and Prastein, M. L., 1985, Na/K/Cl cotransport in cultured human fibroblasts, *J. Biol. Chem.* **260:**1445–1451.

Paris, S., and Pouyssegur, J., 1983, Biochemical characterization of the amiloride-sensitive $Na^+/H^+$ antiport in Chinese hamster lung fibroblasts, *J. Biol. Chem.* **258:**3503–3508.

Radsak, K., and Schmitz, B., 1980, Unimpaired histone synthesis in human fibroblasts infected by human cytomegalovirus, *Med. Microbiol. Immunol.* **168:**63–72.

Radsak, K. D., and Weder, D., 1981, Effect of 2-deoxy-D-glucose on cytomegalovirus-induced DNA synthesis in human fibroblasts, *J. Gen. Virol.* **57:**33–42.

Radsak, K., and Wiegandt, H., 1984, Glycosphingolipid synthesis in human fibroblasts infected by cytomegalovirus, *Virology* **138:**300–309.

Rao, P. N., 1982, The phenomenon of premature chromosome condensation, in: *Premature Chromosome Condensation* (P. N. Rao, R. T. Johnson, and K. Sperling, eds.), pp. 1–41, Academic, New York.

Rapp, F., and Robbins, D., 1984, Cytomegalovirus and human cancer, in: *CMV: Pathogenesis and Prevention of Human Infection* (S. A. Plotkin, S. Michelson, J. S. Pagano, and F. Rapp, eds.) [*Birth Defects Orig. Art. Ser.* **20:**175–192], Liss, New York.

Rapp, F., Geder, L., Murasko, D., Lausch, R., Ladda, R., Huang, E.-S., and Webber, M. J., 1975, Long-term persistence of cytomegalovirus genome in cultured human cells of prostatic origin, *J. Urol.* **16:**982–990.

Rasmussen, H., 1981, *Calcium and cAMP as Synarchic Messengers,* Wiley, New York.

Rasmussen, H., and Barrett, P. Q., 1984, Calcium messenger system: An integrated view, *Physiol. Rev.* **64:**938–984.

Reuss, L., Cassel, D., Rothenberg, P., Whiteley, B., Mancuso, D., and Glaser, L., 1986, Mitogens and ion fluxes, in: *Current Topics in Membranes and Transport* (L. J. Mandel and D. J. Benos, eds.), pp. 3–54, Academic, New York.

Ribbert, H., 1904, Uber protozoenartige Zellen in der Niere eines syphilitischen Neugeboren und in der Parotis von Kindern, *Zentralbl. Allg. Pathol. Anat.* **15:**945–948.

Rifkind, D., 1965, Cytomegalovirus infection after renal transplantation, *Arch, Intern. Med.* **116:** 554–558.

Rowe, W. P., Hartley, J. W., Waterman, S., Turner, H. C., and Heubner, R. J., 1956, Cytopathogenic agent resembling human salivary gland virus recovered from tissue cultures of human adenoids, *Proc. Soc. Exp. Biol. Med.* **92:**418–424.

Rozengurt, E., 1986, Early signals in the mitogenic response, *Science* **234:**161–166.

Rozengurt, E., and Mendoza, S., 1980, Monovalent ion fluxes and the control of cell proliferation in cultured fibroblasts, *Ann. NY Acad. Sci.* **339:**175–190.

Ruebner, B. H., Hirano, T., Slusser, R. J., and Medearis, D. N., Jr., 1965, Human cytomegalovirus infection: Electron microscopic and histochemical changes in cultures of human fibroblasts, *Am. J. Pathol.* **46:**477–496.

Sand, T. E., Bronstad, G. O., Digernes, V., and Christoffersen, T., 1981, Evidence for stimulatory and inhibitory effects of cAMP on DNA synthesis in primary cultures of adult hat hepatocytes, *Adv. Cyclic Nucl. Res.* **14:**683.

Schultz, G., Schultz, K.-D., Bohme, E., and Kreye, V. A. W., 1978, The possible role of cyclic GMP in the actions of hormones and drugs on smooth muscle tone: Effects of exogenous cyclic GMP derivatives, in: *Advances in Pharmacology and Therapeutics: Ions—Cyclic Nucleotides—Cholinergy* (J. C. Stoclet, ed.), Vol. 3, pp. 113–122, Pergamon, Oxford.

Severi, B., Landini, M. P., and Govoni, E., 1988, Human cytomegalovirus morphogenesis: An ultrastructural study of the late cytoplasmic phases, *Arch. Virol.* **98:**51–64.

Sha'afi, R. I., and Naccache, P. H., 1985, Relationship between calcium, arachidonic acid metabolites and neutrophil activation, in: *Calcium in Biological Systems* (R. P. Rubin, G. B. Weiss, and J. W. Putney, eds.), pp. 137–146, Plemem, New York.

Skou, J. C., 1965, Enzymatic basis for active transport of $Na^+$ and $K^+$ across cell membrane, *Physiol. Rev.* **45:**596–617.

Smith, J. B., and Rozengurt, E., 1978, Serum stimulates the $Na^+$, $K^+$ pump in quiescent fibroblasts by increasing $Na^+$ entry, *Proc. Natl. Acad. Sci. USA* **75:**5560–5564.

Smith, K. O., and Rasmussen, L., 1963, Morphology of cytomegalovirus (salivary gland virus), *J. Bacteriol.* **85:**1315–1325.

Smith, M. G., 1956, Propagation in tissue cultures of a cytopathogenic virus from human salivary gland virus (SGV) disease, *Proc. Soc. Exp. Biol. Med.* **92:**424–430.

Stich, H. F., and Yohn, D. S., 1970, Viruses and chromosomes, *Prog. Med. Virol.* **12:**78–127.

Stinski, M. F., 1978, Sequence of protein synthesis in cells infected by human cytomegalovirus: Early and late virus-induced polypeptides, *J. Virol.* **26:**686–701.

St. Jeor, S. C., and Hutt, R., 1977, Cell DNA replication as a function in the synthesis of human cytomegalovirus, *J. Gen. Virol.* **37:**65–73.

St. Jeor, S. C., and Rapp, F., 1973a, Cytomegalovirus replication in cells pretreated with 5-iodo-2′-deoxyuridine, *J. Virol.* **11:**983–990.

St. Jeor, S. C., and Rapp, F., 1973b, Cytomegalovirus: Conversion of nonpermissive cells to a permissive state for virus replication, *Science* **181:**1060–1061.

St. Jeor, S. C., Albrecht, T. B., Funk, F. D., and Rapp, F., 1974, Stimulation of cellular DNA synthesis by human cytomegalovirus, *J. Virol.* **13:**353–362.

Takahashi, M., Van Hoosier, G. L., and Trentin, J. J., 1966, Stimulation of DNA synthesis in human and hamster cells by human adenovirus types 12 and 5, *Proc. Soc. Exp. Biol. Med.* **122:** 740–749.

Tanaka, S., Furukawa, T., and Plotkin, S. A., 1975, Human cytomegalovirus stimulates host cell RNA synthesis, *J. Virol.* **15:**297–304.

Tanaka, S., Ihara, S., and Watanabe, Y., 1978, Human cytomegalovirus induces DNA-dependent RNA polymerases in human diploid cells, *Virology* **89:**179–185.

Tartakoff, A. M., 1983, Perturbation of the structure and function of the Golgi complex by monovalent carboxylic ionophores, *Methods Enzymol.* **98:**47–59.

Tsien, R. Y., Pozzan, T., and Rink, T. J., 1982, T-cell mitogens cause early changes in cytoplasmic free $Ca^{2+}$ and membrane potential in lymphocytes, *Nature (Lond.)* **295:**68–71.

Tsunoda, Y., Takeda, H., Otaki, T., Asaka, M., Nakagaki, I., and Sasaki, S., 1988, A role for $Ca^{2+}$ in mediating hormone-induced biphasic pepsinogen secretion from the chief cell determined by luminescent and fluorescent probes and X-ray microprobe, *Biochim. Biophys. Acta* **941:**83–101.

Tyms, A. S., and Williamson, J. D., 1980, Polyamine metabolism in MRC-5 cells infected with human cytomegalovirus, *J. Gen. Virol.* **48:**183–191.

Valyi-Nagy, T., Bandi, Z., Boldogh, I., and Albrecht, T., 1988, Hydrolysis of inositol lipids: An early signal of human cytomegalovirus infection, *Arch. Virol.* **101:**199–207.

Waner, J. L., and Weller, T. H., 1974, Behavior of human cytomegaloviruses in cell cultures of bovine and simian origin, *Proc. Soc. Exp. Biol. Med.* **145:**379–384.

Wathen, M. W., Thomsen, D. R., and Stinski, M. F., 1981, Temporal regulation of human cytomegalovirus transcription at immediate early and early times after infection, *J. Virol.* **38:** 446–459.

Webber, C. E., and Whalley, J. M., 1977, Chromosome aberrations induced by a marsupial herpes virus in cultured marsupial mouse cells, *Exp. Cell. Res.* **106:**437–440.

Weller, T. H., 1971, The cytomegaloviruses: Ubiquitous agents with protean clinical manifestations, *N. Engl. J. Med.* **285:**203–214, 267–274.

Weller, T. H., Macauley, J. C., Craig, J. M., and Wirth, P., 1957, Isolation of intranuclear inclusion producing agents from infants with illnesses resembling cytomegalic inclusion disease, *Proc. Soc. Exp. Biol. Med.* **94:**4–12.

Weller, T. H., Hanshaw, J. B., and Scott, D. E., 1960, Serologic differentiation of viruses responsible for cytomegalic inclusion disease, *Virology* **12:**130–132.

Whitman, M., Kaplan, D. R., Schaffhausen, B., Cantley, L., and Roberts, T. M., 1985, Association of phosphatidylinositol kinase activity with polyoma middle-T competent for transformation, *Nature (Lond.)* **315:**239–242.

Wright, H. T., Jr., and Goodheart, C. R., 1964, Human cytomegalovirus. Morphology by negative staining, *Virology* **23:**419–424.

Yamanishi, K., and Rapp, F., 1979, Production of plasminogen activator by human and hamster cells infected with human cytomegalovirus, *J. Virol.* **31:**315–419.

Yoon, J. W., Kim, S. M., Hahn, E. C., and Kenyon, A. J., 1976, Lymphoproliferative diseases of fowl: Chromosome breaks caused in lymphocytes by JM-V herpesvirus, *J. Natl. Cancer Inst.* **56:**757–762.

Zavada, V., Erban, V., Rezacova, D., and Vonka, V., 1976, Thymidine kinase in cytomegalovirus infected cells, *Arch. Virol.* **52:**333–339.

*Chapter 7*

# Rubella Virus Products and Their Distribution in Infected Cells

D. S. Bowden and E. G. Westaway

## 1. INTRODUCTION

### 1.1. Taxonomy

The family *Togaviridae* is presently composed of four genera, *Alphavirus, Rubivirus, Pestivirus,* and *Arterivirus,* of which rubella virus (RV) is the only member of the *Rubivirus* genus (Westaway *et al.,* 1985a). The family name *Togaviridae* (from the Latin *toga,* for "cloak" or "shroud," referring to the virus envelope) was initially adopted by the Arbovirus Study Group to describe only two genera, *Alphavirus* and *Flavivirus,* formerly the group A and group B arboviruses, respectively (Fenner *et al.,* 1974). The arbovirus groups comprised a large and diverse set of arthropod-borne (arbo) animal viruses that shared the characteristic of having replicative cycles in both arthropod and vertebrate hosts (Casals, 1957). This system of grouping proved to be taxonomically unacceptable; therefore, a new family, the *Togaviridae,* was defined using more concise parameters (Fenner *et al.,* 1974). This definition allowed for the inclusion of several nonarboviruses of similar morphology and size that were classified into two new genera, the *Rubivirus* genus and *Pestivirus* genus (Fenner *et al.,* 1975, 1976).

---

**D. S. Bowden** MacFarlane Burnet Centre for Medical Research, Fairfield Hospital, Fairfield, Victoria 3078, Australia. **E. G. Westaway** Department of Microbiology, Monash University, Clayton, Victoria 3168, Australia.

With recent characterization of the molecular nature of viruses, there was a further taxonomic revision of the family, with the *Flavivirus* genus being elevated to family status (Westaway *et al.*, 1985b) and the addition of a new genus, the *Arterivirus*. The *Togaviridae* possess a single strand of positive-sense RNA of $\sim 4 \times 10^6$ $M_r$. Translation of structural proteins is from at least one subgenomic RNA corresponding to the 3′-end of the genome. The virus particles comprise an icosahedral nucleocapsid surrounded by an envelope of host lipid and two or three virus-specified polypeptides, usually glycosylated. Morphogenesis of the 50- to 70-nm virus particles occurs by budding through cytoplasmic or plasma membranes (Westaway *et al.*, 1985a).

## 1.2. Historical Perspective

Rubella virus infections postnatally are usually mild diseases in which complications are rare. The virus is transmitted by direct contact via respiratory tract secretions, usually by droplet infection. Contagiousness is generally only moderate. The incubation period is 14–18 days; clinical manifestations include a typical rash accompanied by tender and enlarged lymph nodes of the head and neck and a low-grade fever (Heggie and Robbins, 1969; Herrmann, 1979). The disease may be complicated by the presence of joint pain in the form of polyarthritis and polyarthralgia (Simpson, 1940; Herrmann, 1979). These have now become frequent manifestations associated with immunization by present attenuated vaccines (Cooper *et al.*, 1969; Ogra and Herd, 1971).

The disease was originally described by two German physicians, de Bergen in 1752 and Orlow in 1758, and was known as Rotheln (Wesselhoeft, 1947). The measles-like symptoms of the disease and the early German influence led to the misnomer German measles, a term still used today even though rubella has been shown to be a distinct entity from measles. The present name for the disease, rubella, was introduced by Veale in 1866 (Wesselhoeft, 1947). The viral nature of rubella was established in 1938 by Hiro and Tasaka (Forbes, 1969) after transmission of the disease using filtrates of infected nasal washings. The belief that rubella was an inconsequential and transient disease was dissipated when Gregg, a Sydney ophthalmologist, published observations linking congenital malformations in infants to the occurrence of rubella early in the mother's pregnancy (Gregg, 1941). Gregg's attention had been aroused by a sharp rise in the incidence of congenital cataracts after the rubella epidemic of 1940–1941. Many of the children suffered from other defects, such as heart abnormalities, deafness, microencephaly, cerebral palsy, and mental retardation. This report was largely overlooked; it was published after the outbreak of World War II and was also viewed with a degree of skepticism. In the following years, many papers appeared confirming Gregg's observations (Forbes, 1969), stimulating renewed interest in the disease and in measures that might lead to its control.

It was not until 1962 that recovery of the virus was accomplished independently by Weller and Neva (1962) at the Harvard School of Public Health and by Parkman *et al.* (1962) at the Walter Reed Army Institute of Research. Both groups employed different techniques. Weller and Neva (1962) detected cytopathic effects (CPE) in human amnion cell cultures infected with RV from blood and urine specimens, whereas Parkman *et al.* (1962) demonstrated the presence of RV indirectly by resistance to challenge with an echovirus in African green monkey kidney cells previously infected with RV obtained from throat washings. Subsequently, it has been shown that RV can be detected by CPE and plaque assay in a wide variety of cell systems (Herrmann, 1979; Horzinek, 1981).

The first attenuated vaccine strain was developed by Parkman *et al.* (1966) by serially passing a strain of RV 77 times (HPV77) in vervet monkey kidney cells. Further vaccines were developed by passage of the original vaccine in duck embryo fibroblasts (HPV77·DE5) and by attenuation of the virus in primary rabbit kidney cultures (Cendehill vaccine), human diploid (W138) cells (RA27/3 vaccine), and primary guinea pig cultures (To-336 vaccine) (Banatvala, 1977).

## 2. VIRUS STRUCTURE

### 2.1. Proteins

The identification of RV structural proteins has been hindered by the inherently low virus yields obtained from infected tissue culture fluids, by the fragility of the virus particle, and by contamination with host proteins. A number of diverse techniques have been applied to concentrate and purify RV; this diversity has probably contributed to some of the discrepant results obtained. Different systems of nomenclature have been used to describe RV structural proteins; the current nomenclature follows that used for alphavirus structural proteins, *viz.*, E1, E2, and C (Baltimore *et al.*, 1976). Results published before 1980 (Table I) appear to be incompatible based on each investigator's own interpretation; closer examination shows that they share several features and are consistent with many later reports (Table II). For example, Liebhaber and Gross (1972) identified eight RV polypeptides (Table I), but these fall into three size groups corresponding to E1 (60–63kDa), a broad E2 (47–55kDa), and C (30–32kDa).

The more definitive recent studies of RV structural proteins (Table II) established that RV contains a core protein, C (30–38kDa) and two glycosylated envelope proteins, E1 (55–62kDa) and a heterogeneous E2 (42–54kDa) (Fig. 1); minor differences in reported molecular-weight ($M_r$) values of the individual proteins are probably a reflection of the gel systems used rather than of strain variation. E1, E2, and C must be produced in equimolar amounts from the same

**Table I**
**Early Reports of Rubella Virus Structural Proteins**

| Reference | Structural proteins ($M_r \times 10^{-3}$) | |
|---|---|---|
| | Major | Minor |
| Liebhaber and Gross (1972) | 63 (GP)[a], 60 (GP), 32, 30 | 56 (GP), 54 (GP), 52 (GP), 47 (GP) |
| Vaheri and Hovi (1972) | 62.5 (GP), 45–50 (GP), 35 | |
| Bardeletti *et al.* (1975) | 58[b], 48, 31, 16.5 | 75, 40, 26, 10 |
| | 63[c], 42, 30 | 88, 76, 54, 36, 24 |
| Payment *et al.* (1975a) | 63 (GP), 50 (GP), 35 | |

[a]GP, shown to be glycosylated.
[b]High concentration of denaturing agents.
[c]Low concentration of denaturing agents.

polycistronic messenger (see Section 3) and are present in virions in similar proportions (Bowden and Westaway, 1984). On this basis, the much greater incorporation of [$^{35}$S]methionine into E2 relative to E1 (Fig. 1) is in accord with the number of methionine residues in the deduced translation sequences, six or seven (E2) and three (E1) (Clarke *et al.*, 1987; Vidgren *et al.*, 1987). However, two reports have claimed anomalous molar proportions of 5 : 1 : 5 for E1 : E2 : C (Vaheri and Hovi, 1972; Waxham and Wolinsky, 1985a).

Several investigators have resolved the heterogeneous E2 into two or more glycoproteins by gel electrophoresis (Ho-Terry and Cohen, 1982; Oker-Blom *et al.*, 1983; Toivonen *et al.*, 1983; Waxham and Wolinsky, 1983; Bowden and Westaway, 1984; Kalkkinen *et al.*, 1984). Using isoelectric focusing (IEF) in two-dimensional gels, Waxham and Wolinsky (1985a) defined at least 15 different

**Table II**
**Recent Reports of Rubella Virus Structural Proteins**

| Reference | Structural proteins ($M_r \times 10^{-3}$) |
|---|---|
| Ho-Terry and Cohen (1980) | 55 (GP)[a], 46 (GP), 33 |
| Van Alstyne *et al.* (1981) | 44, 41, 24, 19 |
| Trudel *et al.* (1982) | 59, 45, 33 |
| Oker-Blom *et al.* (1983) | 58 (GP), 47 (GP), 45 (GP), 33 |
| Toivonen *et al.* (1983) | 62 (GP), 44–51 (GP), 35 |
| Waxham and Wolinsky (1983) | 62 (GP), 47–54 (GP), 38 |
| Bowden and Westaway (1984) | 59 (GP), 43–48 (GP), 34 |
| Dorsett *et al.* (1985) | 63 (GP), 45–48 (GP), 30 |

[a]GP, shown to be glycosylated.

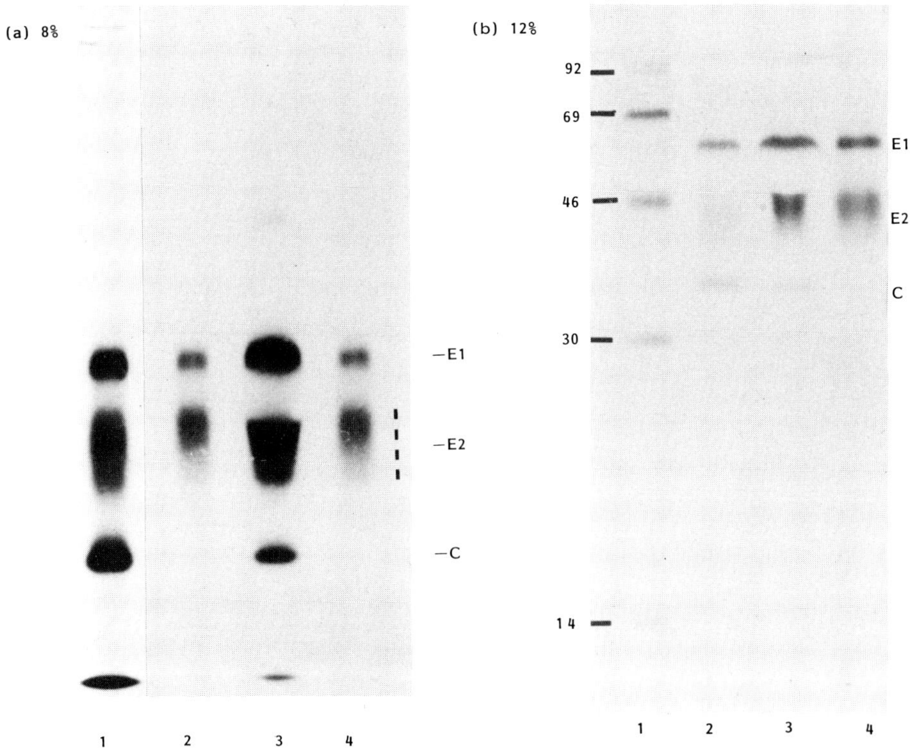


**FIGURE 1.** Fluorograms of polyacrylamide slab gels with rubella virus (RV) proteins from purified virus labeled in amino acids and carbohydrate. Virus was concentrated from infected cell culture fluids by precipitation with polyethylene glycol and purified by rate-zonal sedimentation. E1 and E2 designate the envelope glycoproteins and C the core protein. (a) 8% gel. Lane 1, [$^{35}$S]methionine-labeled RV structural proteins; lane 2, [$^{3}$H]glucosamine-labeled RV proteins; lane 3, [$^{35}$S]methionine-labeled RV proteins that have been immunoprecipitated; lane 4, [$^{3}$H]glucosamine-labeled RV proteins from virus purified by two sequential rate-zonal sedimentations. (From Bowden and Westaway, 1984.) (b) 12% gel. Lane 1, $^{14}$C-labeled molecular-weight markers; lanes 2, 3, and 4 correspond to lanes 1, 3, and 4, respectively, of part (a).

isoelectric species of E2. The polypeptide backbones of the heterogeneous E2 glycoprotein were shown to be essentially identical by two dimensional tryptic peptide analysis (Oker-Blom *et al.*, 1983) and by limited-digest peptide maps (Ho-Terry and Cohen, 1982; Waxham and Wolinsky, 1983). Kalkkinen *et al.* (1984) demonstrated that the two glycoproteins within E2 described by Oker-Blom *et al.*

(1983) also have very similar amino acid compositions and identical amino-terminal and carboxy-terminal sequences. These results attribute the migration differences in gels of the E2 components to some variation in the degree of glycosylation. Evidence for this variation was obtained by Oker-Blom *et al.* (1983) and Bowden and Westaway (1984), who found that the distribution of radioactively labeled sugars in the E2 region of gels was indicative of the presence of complex glycans in the slower migrating fraction of E2 and high-mannose or simple glycans in the faster migrating fraction (Fig. 2). Recent nucleotide sequence data support the notion of variation in glycosylation, as it is evident that only one gene codes for E2 and there are probably four potential glycosylation sites (Clarke *et al.*, 1987; Vidgren *et al.*, 1987). E1 and C were not detected by Van Alstyne *et al.* (1981) probably because their technique of *in vitro* labeling of gradient-purified virus prior to immunoprecipitation was relatively inefficient compared with labeling *in vivo*.

In addition to E1, E2, and C, components of ~100,000 $M_r$ have been observed in purified virus preparations (Vaheri and Hovi, 1972; Payment *et al.*, 1975a; Ho-Terry and Cohen, 1980; Oker-Blom *et al.*, 1983; Toivonen *et al.*, 1983; Waxham and Wolinsky, 1983; Dorsett *et al.*, 1985). One of these consists of a dimeric form of E1 and the other a mixed dimer of E1 and E2 (Waxham and Wolinsky, 1983; Dorsett *et al.*, 1985). It is apparent that E1 and E2 are closely associated in the mature virion, probably linked by disulfide bonds; both have been shown to be located externally by digestion with trypsin or pronase, by radiolabeling *in vitro,* and by precipitation of whole virus with individual monoclonal antibodies (Ho-Terry and Cohen, 1980; Oker-Blom *et al.*, 1983; Waxham and Wolinsky, 1983; Green and Dorsett, 1986). Some recent data suggest that E2 is not as prominently displayed on the surface as E1 because of its relative

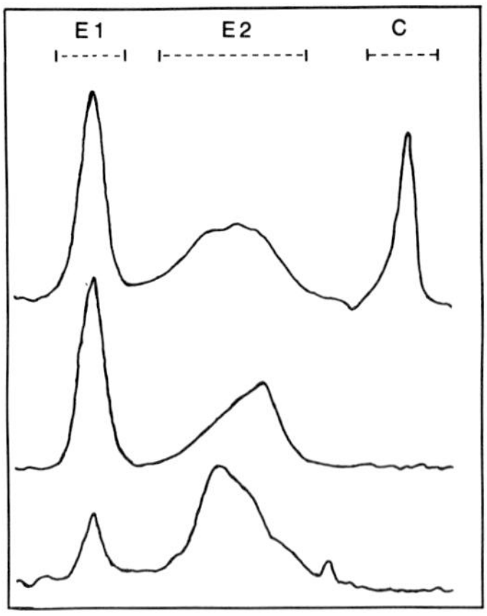


**FIGURE 2.** Densitometer traces of fluorograms of 8% slab gels after electrophoresis of RV proteins E1, E2, and C from purified virus labeled with [$^{35}$S]methionine (top), [$^{3}$H]mannose (middle), and [$^{3}$H]galactose (bottom). (From Bowden and Westaway, 1984.)

inaccessibility to glycosidase treatment (Ho-Terry and Cohen, 1984), to neuraminidase (Pettersson *et al.*, 1985), and to specific antibody (Waxham and Wolinsky, 1985b). Studies with monoclonal antibodies indicate that E1 contains multiple epitopes involved in hemagglutination inhibition and neutralization (Waxham and Wolinsky, 1985b; Green and Dorsett, 1986; Ho-Terry and Cohen, 1986) of which three have been mapped to an antigenic domain between amino acids 245 to 285 of the predicted amino acid sequence (Terry *et al.*, 1988). Strain-specific antigenic differences have been located in E2 (Dorsett *et al.*, 1985). Monoclonal antibodies have also demonstrated that E2 has at least one epitope involved in neutralization (Green and Dorsett, 1986).

Functionally, the core protein C is associated with the RNA forming the nucleocapsid (Vaheri and Hovi, 1972; Payment *et al.*, 1975a; Oker-Blom *et al.*, 1983; Waxham and Wolinsky, 1983). When unreduced, C migrates in gels as a dimer of 78,000 $M_r$ (Waxham and Wolinsky, 1983). Sedimentation studies on detergent disrupted virus suggest that the capsid is composed of two disulfide-linked dimers of C (Dorsett *et al.*, 1985). The amino acid sequence deduced from the nucleotide sequence includes two cysteine residues and a high proportion of proline (14.1%), and of arginine (14.4%) and lysine (1.1%) residues (Clarke *et al.*, 1987; Takkinen *et al.*, 1988); the basic amino acids should provide positive charges for binding of C to virion RNA.

## 2.2. RNA

Direct evidence of the RNA nature of the RV genome was first shown by Brodersen and Thomssen (1969) and Hovi and Vaheri (1970a), both groups detecting [$^3$H]uridine in cores of virions labeled during growth. Several reports described the 38–40S RNA as being single-stranded, polyadenylated and infectious (Hovi and Vaheri, 1970b; Sedwick and Sokol, 1970; Katow and Matsuno, 1980; Oker-Blom *et al.*, 1984). The estimated molecular weight by gel electrophoresis was $3.8 \times 10^6$ (Oker-Blom *et al.*, 1984). After RNase digestion of virion RNA, two putative 5′-cap structures remained (Oker-Blom *et al.*, 1984). The base composition of [$^{32}$P]phosphate-labeled RV RNA was shown by Katow and Matsuno (1980) to be guanine, 31.1; adenine, 21.9; cytosine, 34.3; and uracil, 12.8 per 100 nucleotides, significantly different from the base composition of alphavirus RNA. In some virion RNA preparations, an additional but minor single-stranded component was detected (Sedwick and Sokol, 1970; Norval, 1979; Oker-Blom *et al.*, 1984) which may represent a degradation product of the genomic RNA, a host cell RNA contaminant or the genome of an incomplete or defective-interfering (DI) particle. Serial undiluted passage of RV in Vero cells has been shown to generate DI articles (Terry *et al.*, 1985), even when plaque-purified virus was used to produce the infecting stocks (Frey and Hemphill, 1988). One of the predominant RNA species encapsidated appeared to be the subgenomic RNA (Terry *et al.*, 1985; Frey and Hemphill, 1988). Similar passages of RV in L cells produced high interference virus stocks of heterogeneous density in which the virion RNA was also predominantly subgenomic (Bohn and Van Alstyne, 1981).

### 2.3. Lipids

Rubella virus contains lipid as a structural component of the viral envelope similar to other members of the *Togaviridae*. The proportion of lipid in RV purified from BHK-21 cells is 18.8% of the dry weight of the virus (Voiland and Bardeletti, 1980) or 0.245 mg per mg viral protein (Bardeletti *et al.*, 1975). The lipid content is composed of approximately 70% phospholipid, of which 4.4% is cardiolipin, a phospholipid relatively specific to the inner mitochondrial membrane (Bardeletti and Gautheron, 1976). The differences in lipid content between the virion and the BHK-21 plasma membrane together with the presence of cardiolipin in RV suggested to Bardeletti and Gautheron (1976) that RV maturation in BHK-21 cells takes place predominantly at an internal site, possibly through mitochondrial membranes.

Labeling experiments with [$^3$H]palmitic acid during viral growth demonstrated that the fatty acid was associated with E1 and E2 (Waxham and Wolinsky, 1985a). Palmitic acid may help anchor the glycoproteins within the lipid bilayer after being covalently attached during transport through the endoplasmic reticulum (ER), as has been described for alphaviruses (Schmidt and Schlesinger, 1980; Berger and Schmidt, 1985).

## 3. VIRAL REPLICATION

### 3.1. Virus-Specific RNA

Rubella virus RNA synthesis in RV-infected cells is detectable by labeling in the presence of the DNA transcription inhibitor, actinomycin D. Reduction of virus yields occurred only when actinomycin D was added to cells at infection or before infection, possibly because of toxic effects during the latent period of 8–12 hr (Maes *et al.*, 1966; Woods and Robbins, 1968; Wong *et al.*, 1969; Hovi and Vaheri, 1970a; Sedwick and Sokol, 1970; Payment *et al.*, 1975b; Bowden and Westaway, 1984; Hemphill *et al.*, 1988). However, Nakhasi *et al.* (1988) provided evidence that actinomycin D added early in infection reduced the synthesis of viral RNA and viral proteins; they suggested that this reduction in yield by actinomycin D was by an effect on the host cell DNA which directs the synthesis of a factor necessary for replication of viral RNA. A 38–40S virus-specific RNA was identified in RV-infected BHK-21 cells and some years later in Vero cells which became the preferred cell line for molecular studies of RV (Wong *et al.*, 1969; Hovi and Vaheri, 1970a,b; Sedwick and Sokol, 1970; Oker-Blom *et al.*, 1984; Vidgren *et al.*, 1987; Hemphill *et al.*, 1988). The intracellular 38–40S RNA could be detected as early as 12–15 hr after infection (Hovi and Vaheri, 1970a; Hemphill *et al.*, 1988); its synthesis reached a peak by 26 hr, about the same time as the peak of infectious virus, and then declined.

An additional virus-specified single-stranded RNA sedimenting at 24–26S has also been detected in RV-infected cells (Sedwick and Sokol, 1970; Oker-Blom *et al.*, 1984; Vidgren *et al.*, 1987; Hemphill *et al.*, 1988; Nakhasi *et al.*, 1988). It was found to be polyadenylated by using oligo(dT) cellulose chromatography, and was translated *in vitro* into a p110 precursor of the structural proteins (see Section 3.2); Northern blot analysis of intracellular RNA with an appropriate cDNA probe showed that it represented the 3′ end of the genome (Oker-Blom *et al.*, 1984). This was later confirmed by nucleotide sequencing of complementary DNA (cDNA) of the genome (Clarke *et al.*, 1987; Vidgren *et al.*, 1987). The subgenomic RNA became prominent about 19 hr postinfection, reached a peak at 26 hr and declined thereafter; by 24 hr it was being synthesized in greater molar amounts than the genomic RNA (Hemphill *et al.*, 1988). The complete nucleotide sequence of the subgenomic RNA has been published (Clarke *et al.*, 1987; Vidgren *et al.*, 1987; Takkinen *et al.*, 1988), and the RNA was found to be over 3300 nucleotides in length with one long open reading frame, coding for 1063 amino acids.

In addition to identification of virion-sized RNA and the 24–26S messenger RNA (mRNA) in RV-infected cells, a putative replicative form of 19–20S double-stranded RNA (RNase resistant and soluble in 1 M NaCl) was identified (Wong *et al.*, 1969; Sedwick and Sokol, 1970). Furthermore, a putative 21S replicative intermediate, 61% RNase resistant, was isolated by precipitation with 1 M NaCl followed by cellulose column chromatography (Sedwick and Sokol, 1970). The relationships of these forms of RV RNA remain undefined, e.g., their kinetics of synthesis and whether the replicative form is recycled as template, as appears to occur during flavivirus RNA synthesis (Chu and Westaway, 1984). A minor amount of cytoplasmic RNA similar in length to the prominent subgenomic RNA was shown very late in infection to hybridize in Northern blots with a positive polarity probe, but it was concluded that both the rubella genomic and subgenomic RNAs are synthesized from a genome length negative-polarity RNA template (Hemphill *et al.*, 1988), as occurs in replicative intermediates of alphaviruses (Strauss and Strauss, 1983).

No evidence was obtained for synthesis of DI RNA in Vero cells infected at a multiplicity of infection of 10 plaque-forming units (PFU)/cell for 96 hr, or after further undiluted passage for 30 hr, using a sensitive Northern blot assay (Hemphill *et al.*, 1988). Similar Northern blot results were obtained by Oker-Blom *et al.* (1984). Thus, the only subgenomic RNA detectable in RV-infected cells is the 24–26S (about 3.3 Kb) mRNA. An additional RNA species of 7.5 Kb was observed, but only after the fifth serial undiluted passage of RV in Vero cells (Frey and Hemphill, 1988), and prior to the subsequent appearance of DI particles.

### 3.2. Intracellular Virus-Specified Proteins

The lack of selective inhibition of host protein synthesis in cell lines following infection with RV posed problems in the identification of virus-specified polypeptides (Payment *et al.*, 1975b; Chantler, 1979; Bowden and Westaway, 1984). Reports describing RV-specific intracellular polypeptides have come from several groups (Chantler, 1979; Pope and Van Alstyne, 1981; Oker-Blom *et al.*, 1983; Bowden and Westaway, 1984; Waxham and Wolinsky, 1985b; Hemphill *et al.*, 1988). Of these reports, one group did not relate their findings directly to RV structural proteins, thus, full characterization was not possible (Chantler, 1979; Chantler and Tingle, 1980; Chantler *et al.*, 1987). Another group obtained results with $^{125}$I-labeled immunoprecipitates from RV-infected cells and with $^{125}$I-labeled virion proteins that were similar but inconsistent with all other reports in regard to size and number of polypeptides (Pope and Van Alstyne, 1981; 1982).

Intracellular counterparts to structural proteins were positively identified in polyacrylamide gels after immunoprecipitation of labeled cytoplasm with hyperimmune rabbit antiserum to disrupted virus (Oker-Blom *et al.*, 1983), human convalescent serum (Bowden and Westaway, 1984; Hemphill *et al.*, 1988), and RV monoclonal antibodies (Waxham and Wolinsky, 1985b; Hemphill *et al.*, 1988). By contrast, the RV envelope glycoproteins were detected directly by comparative gel analysis of [$^3$H]mannose labeled cytoplasm and purified labeled virions (Fig. 3). The electrophoretic mobilities of the intracellular counterparts to E1 and C were similar to those of the corresponding virion proteins. However, instead of the diffuse band corresponding to virion E2, a sharp band of approximate ~43,000 $M_r$ was resolved from infected cytoplasm (Fig. 4, lane 11) (Oker-Blom *et al.*, 1983; Waxham and Wolinsky, 1985b; Hemphill *et al.*, 1988). This variation in migration of E2 appears to be related to possible glycosylation differences between the virion glycoproteins and their intracellular counterparts. Bowden and Westaway (1984) therefore proposed that the latter be designated GP59(E1) and GP43(E2).

The glycosylation of E1 and E2, and of GP59(E1) and GP43(E2), has been further examined by glycopeptide analysis after pronase digestion (Bowden and Westaway, 1985). Thus, GP59(E1) produced glycopeptides in two size classes with 1500 and 2100 $M_r$. Digests of E1 also yielded these two size classes and an additional glycopeptide of 2700 $M_r$. Similar digestion of GP43(E2) yielded three glycopeptides of 1500, 2100, and 2700 $M_r$; E2 was found to contain these and an

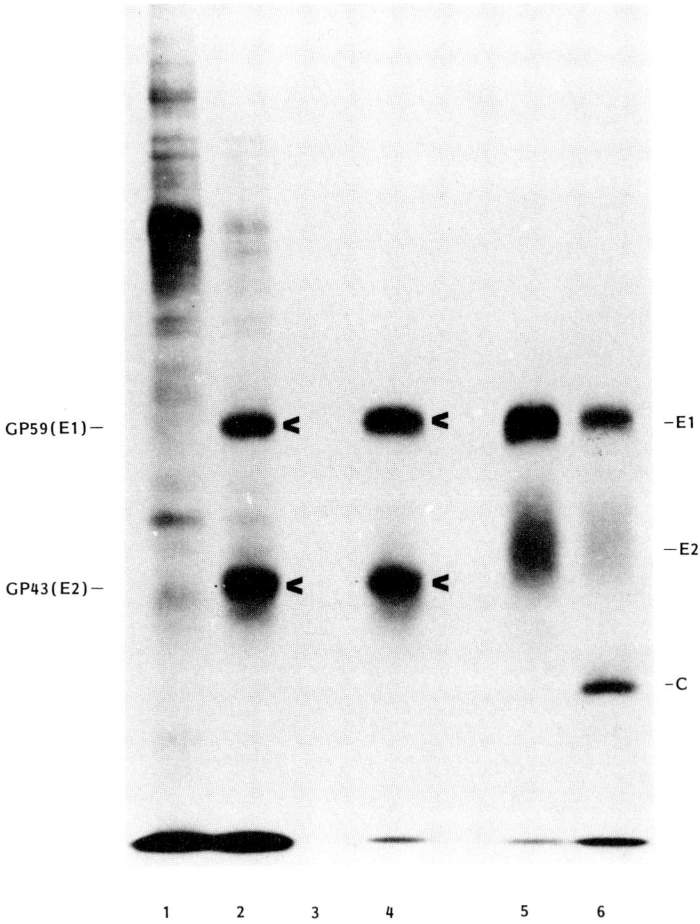


**FIGURE 3.** Fluorogram of an 8% slab gel showing intracellular counterparts to E1 and E2 of RV labeled with [$^3$H]mannose. Lane 1, mock-infected cells; lane 2, RV-infected cells; lane 3, mock-infected cells after immunoprecipitation; lane 4, RV-infected cells after immunoprecipitation; lanes 5 and 6, proteins of purified RV labeled with [$^3$H]mannose and [$^{35}$S]methionine, respectively. GP59(E1) and GP43(E2), indicated by arrowheads, are the intracellular counterparts to E1 and E2. (From Bowden and Westaway, 1984.)

additional glycopeptide of 3300 $M_r$. The observed number of glycopeptide size classes, i.e., three in E1 and four in E2, corresponds exactly with the number of asparagine-linked glycosylation sites demonstrated by nucleotide sequence data (Frey *et al.*, 1986; Nakhasi *et al.*, 1986; Clarke *et al.*, 1987; Vidgren *et al.*, 1987). Although Clarke *et al.* (1987) found only three potential glycosylation sites in E2, a single base substitution would produce a fourth site in an identical

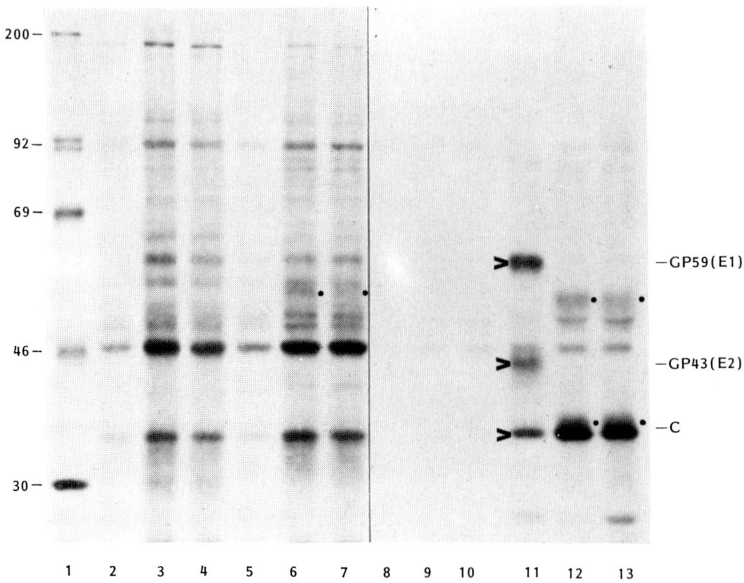


**FIGURE 4.** Fluorogram of an 8% gel showing RV-infected cell extracts labeled with [$^{35}$S]methionine in the presence of tunicamycin before and after immunoprecipitation. Lane 1, $^{14}$C-labeled molecular-weight markers; lanes 2–4 and 5–7 are mock-infected and infected cell extracts, respectively, treated with 0 (lanes 2 and 5), 1 (lanes 3 and 6), and 2 (lanes 4 and 7) μg/ml tunicamycin; lanes 8–13, the same cell extracts as in lanes 2–7, after immunoprecipitation. The intracellular counterparts to the structural proteins are indicated by arrowheads. Only the polypeptide backbone of GP59 (E1) can be detected in infected cells without immunoprecipitation. (●) Unglycosylated forms of GP59(E1) and GP43(E2).

position to that shown by Vidgren *et al.* (1987). The glycopeptides derived from the intracellular glycoproteins were all sensitive to endoglycosidase H treatment, showing that they contained only high-mannose glycans (Kornfeld and Kornfeld, 1980), whereas the glycopeptides of virion E1 and E2 contained both sensitive and resistant components, as expected if they contained both high-mannose and complex glycans. This finding, together with the detection of [$^{3}$H]galactose and of additional size classes only in E1 and E2, indicate that some changes in the glycan moieties take place during maturation of the envelope glycoproteins (Bowden and Westaway, 1985).

When infected cells labeled with [$^{3}$H]mannose and treated with increasing amounts of tunicamycin were analyzed on gels, two possible intermediates in the glycosylation of GP43(E2) were found. These partially glycosylated species migrated with 41,300 and 38,800 $M_r$ and probably lack some of the glycans of E2 (Bowden and Westaway, 1985). Analysis on gels of [$^{35}$S]methionine-labeled infected cells treated with high concentrations of tunicamycin (1 or 2 μg/ml) show the unglycosylated polypeptides of GP59(E1) and GP43(E2) (Fig. 4), which migrate with molecular weights of 53,000 and 30,000–34,000 (Oker-Blom *et al.*, 1983; Bowden and Westaway, 1985; Hemphill *et al.*, 1988). These results and the deduced translation sequence are thus consistent with N-linked attachment to E1 of a set of three glycans of total molecular weights of 6300 and to E2 of four glycans of total molecular weight 9600 (see previous paragraph).

Translation of the structural proteins was shown by Oker-Blom *et al.* (1984) to be similar to that of the alphaviruses, i.e., from a 24S subgenomic messenger with a molecular weight of $\sim 1.2 \times 10^6$, representing the 3′-end of the genome. The 24S RNA was polyadenylated and, when translated *in vitro,* yielded a protein of molecular weight 110,000, p110, (and some smaller products), precipitated by antisera to each of the three structural proteins (Oker-Blom, 1984; Oker-Blom *et al.*, 1984). As the total of the polypeptides of E1, E2, and C is close to 110,000 daltons, p110 is obviously their precursor, and no autoproteolytic cleavages occur. The gene order $NH_2$-C-E2-E1-COOH was established by pulse-labeling of infected cells, synchronized in initiation of translation after reversal of hypertonic salt inhibition (Oker-Blom, 1984). This gene order has subsequently been confirmed from the deduced translation sequence of the nucleotide sequence, obtained from cDNA cloned after copying purified RV RNA (Clarke *et al.*, 1987; Vidgren *et al.*, 1987; Takkinen *et al.*, 1988). The coding sequence is in one long open reading frame of at least 2979 nucleotides. The start points for E2 and E1 were deduced from N-terminal amino acid sequence data obtained previously by Kalkkinen *et al.* (1984). There is lack of agreement regarding the C-terminal of E1; the most likely result seems to be that of Vidgren *et al.* (1987) because the total coding content of the subgenomic mRNA would then be ~1060 amino acids, representing approximately 114,000 daltons of polypeptide, very close to the estimated sizes of the polyprotein precurser p110 and of the total polypeptides C + E2 + E1, viz. ~120,000 daltons. Thus, there appears to be no intergenic region corresponding to E3 of the alphaviruses (Strauss and Strauss, 1983).

Processing of the polyprotein representing the structural proteins has been studied by *in vitro* translation of RNA transcribed by SP6 RNA polymerase from a cDNA clone representing the 24S RNA coding region of the genome by Clarke *et al.* (1987). Immunoprecipitation of the products by RV-specific antisera yielded p110 and, in the presence of microsomes, small amounts of labeled protein equivalent in size to virion E2, E1, and C were also produced. These results suggest strongly that C does not act as an autoprotease (as it does for the alphaviruses) and that E2 and E1 are cleaved by a cellular signal peptidase within the lumen of the endoplasmic reticulum. In conformity with this strategy, the N-terminals of E2 and E1 are preceded by hydrophobic sequences that could function as signal sequences (Clarke *et al.*, 1987; Vidgren *et al.*, 1987). For E2, the signal sequence appears to be defined better by Vidgren *et al.* (1987), mainly because of some apparent frame shifting within the relevant sequence published by Clarke *et al.* (1987). E2 and E1 are probably anchored in the lipid bilayer by the hydrophobic sequences at their C-terminals serving as transmembrane domains. It is of interest that the sequence preceding the N-terminal of E1, Pro-Ala-Tyr-Gly, agrees almost exactly with that preceding E of several flaviviruses, Pro-Ala-Tyr-Ser (see Coia *et al.*, 1988).

Several large putative nonstructural proteins were detected in gels after immunoprecipitation of RV-infected cell extracts with human convalescent serum. The most prominent of these had molecular weights of 200,000 (ns200), 150,000 (ns150), 87,000 (ns87), and 75,000 (ns75); they were also detected during hypertonic salt treatment of infected cells. Pulse-chase studies indicated that the largest, ns200, was cleaved to ns150 (Bowden and Westaway, 1984). Similarly, in cells infected with the alphavirus Sindbis, p150 is an intermediate product during cleavages of p270, the product of read-through of an opal stop codon that otherwise terminates the translation of p230 (Strauss and Strauss, 1983).

Analysis of the kinetics of protein synthesis in RV-infected cells is compounded by the small percentage of infected cells present during the latent period, as well as by the lack of cell switch-off alluded to earlier. Less than 50% of cells were shown to be infected by infectious center assays during the latent period using multiplicities of infection greater than one, rapidly rising to much greater than 50% by 24–36 hr (Wong *et al.*, 1969; Sedwick and Sokol, 1970; Hemphill *et al.*, 1988). Immunofluorescence also indicates that detectable antigen production is confined to a minority of cells early in infection (see Section 4.1). However, by 16 hr, the intracellular forms of E1, E2, and C could be labeled with [$^{35}$S]methionine; the rate of synthesis appeared to be similar up to 48 hr postinfection (Hemphill *et al.*, 1988). Synthesis of ns200 and smaller ns proteins was also detectable by radiolabeling at 24 hr (Bowden and Westaway, 1984).

## 4. LOCALIZATION OF VIRAL PRODUCTS

Most of the information concerning the intracellular distribution of virus in products in RV-infected cells has been obtained using the immunofluorescence technique and specific antibodies as probes. Ultrastructural studies of infected cells have contributed little on localization, and the advent of immunogold labeling is still awaited.

### 4.1. Virus Antigens

Indirect immunofluorescence assays have been applied generally for the detection of RV antibodies in human serum. For analysis of the distribution of

RV proteins in infected cells, both human convalescent sera (Pope and Van Alstyne, 1982; Bowden *et al.*, 1987) and individual monoclonal antibodies to E1, E2, and C (Waxham and Wolinsky, 1985b) have been used as probes. RV antibodies in human convalescent sera were shown by radioimmunoprecipitation to be directed against both structural and nonstructural proteins, the former being more strongly immunoprecipitated (Bowden and Westaway, 1984).

With convalescent rubella sera, virus antigens were detected in a minority of fixed infected Vero cells as early as 12 hr postinfection, and the number of cells exhibiting fluorescence reached a peak at 48 hr (Fig. 5). Hemphill *et al.* (1988) obtained similar results in Vero cells using human anti-RV serum (positive only by 24 hr) and RV monoclonal antibodies. Infected monolayers examined at 12 hr showed stained cells in isolated clusters, suggesting that the mode of spread of virus may be directly from one cell to adjacent or surrounding cells (Bowden *et al.*, 1987). The characteristic cytopathology of RV-infected RK-13 cells, in the form of microfoci, also appeared to form by a progressive involvement of cells adjacent to those originally infected, perhaps by direct cell-to-cell spreading via cellular bridges (McCarthy, 1969; Kouri *et al.*, 1974). Another mechanism by which RV infection could spread to contiguous cells includes the passage of infectious particles during mitosis; alternatively, surrounding cells may simply have a higher probability of being infected.

The pattern of fluorescence in RV-infected cells, using convalescent sera and some monoclonal antibodies to E1 and monoclonal antibodies specific for E2 and C, shows mainly perinuclear and reticular staining similar to the expected distribution of the rough endoplasmic reticulum (Fig. 5) (Pope and Van Alstyne, 1982; Waxham and Wolinsky, 1985b). Occasional intense staining of a region adjacent to the nucleus, possibly corresponding to the Golgi complex (Louvard *et al.*, 1982), was also observed (Bowden *et al.*, 1987). Similar patterns of fluorescence have been described with antibodies to alphavirus E1 and E2 in Semliki forest virus (SFV)-infected BHK-21 cells (Green *et al.*, 1981). The reticular staining probably corresponds to the location of the polypeptides translated in the rough endoplasmic reticulum and the staining adjacent to the nucleus may correspond to an accumulation of virus glycoproteins in the Golgi complex or in pre- or post-Golgi complex vacuoles, similar to those recently shown to function in the transport of SFV glycoproteins in BHK-21 cells (Saraste and Kuismanen, 1984). Intense staining of vacuoles was a feature of the immunofluorescence observed by Waxham and Wolinsky (1985b) with some monoclonal antibodies to E1. These antibodies were shown to inhibit virus infectivity, and it was proposed that they are staining virus particles accumulating after budding into the vacuoles. Thin-section electron microscopy shows that RV can bud from internal membranes into vacuoles in the region of the Golgi complex in a variety of host–cell systems (see Section 5.2).

Another pattern of immunofluorescence detected by a further class of mono-

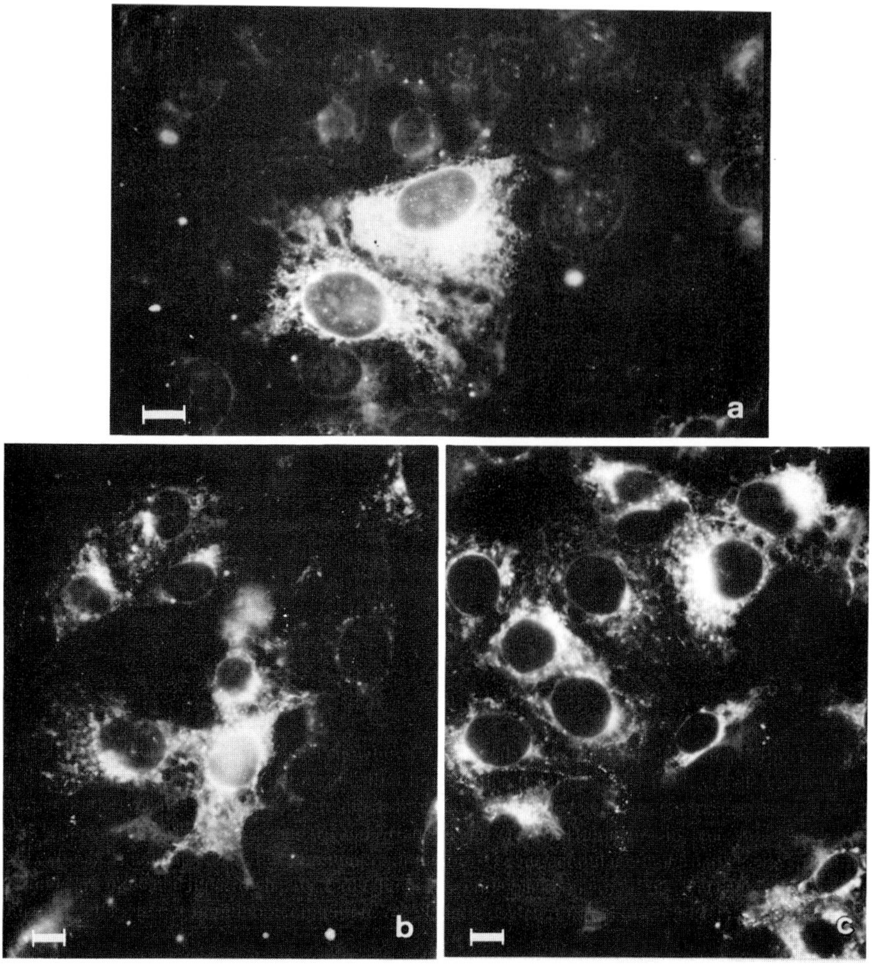

**FIGURE 5.** Immunofluorescence observed in RV-infected Vero cells with human convalescent serum. Cells infected at a multiplicity of infection of 1 were fixed in acetone and treated with antiserum at (a) 12 hr, (b) 24 hr, and (c) 48 hr, respectively. Bar = 10 μm. (From Bowden *et al.*, 1987.)

clonal antibodies identifies E1 antigen on the surface of the plasma membrane (Waxham and Wolinsky, 1985b) and is similar to the specific fluorescence produced with human convalescent serum in unfixed infected cells (Fig. 6). The latter presumably also identifies the location of envelope glycoprotein on the plasma membrane. Nucleotide sequence data show that the N-terminals of E1 and E2 are preceded by hydrophobic sequences possibly able to function as

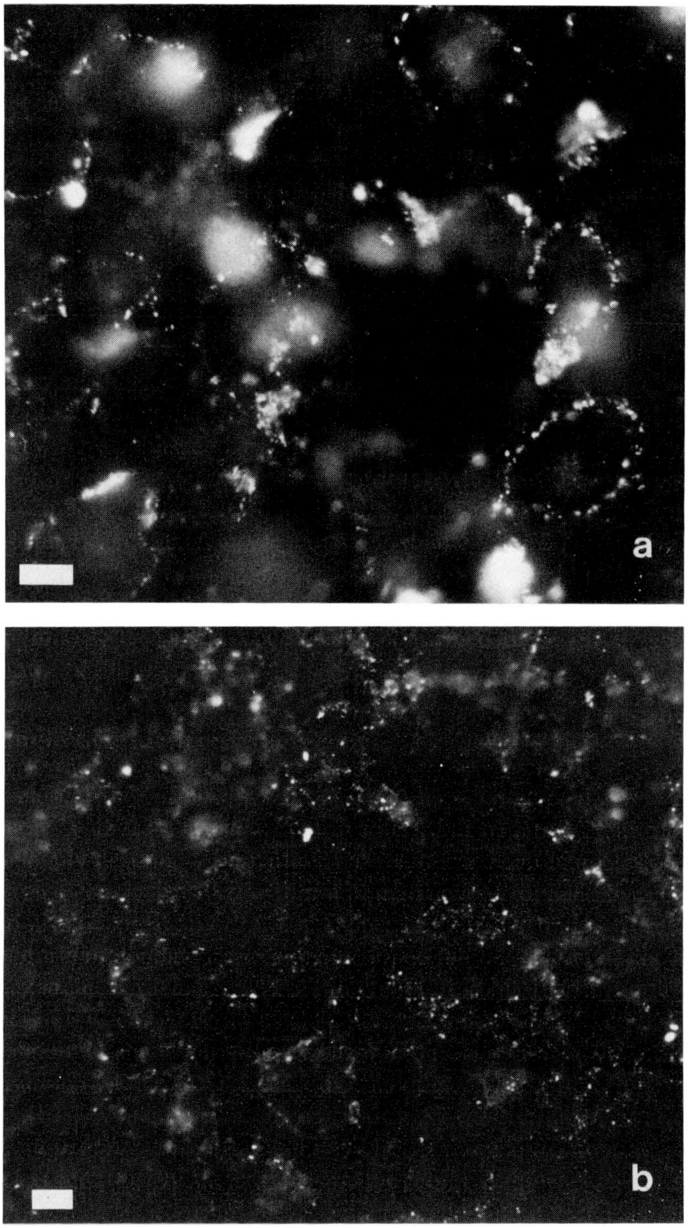


**FIGURE 6.** Surface immunofluorescence observed in unfixed RV-infected Vero cells with human convalescent serum. (a) Cells infected at a multiplicity of 1 treated with antiserum at 48 hr. (b) Cells infected at the same multiplicity and maintained in the presence of 5 μg/ml of tunicamycin until treatment with antiserum at 48 hr. Bar = 10 μm.

signal peptides for membrane translocation, and that the C-terminal of each is preceded by a hydrophobic sequence that may represent membrane-spanning domains for anchoring E1 and E2 (Clarke *et al.*, 1987; Vidgren *et al.*, 1987). In the presence of tunicamycin, there is a considerable decrease in the intensity of cell-surface fluorescence (Fig. 6b), but staining is still evident, indicating that glycosylation is probably not required for correct insertion of the viral glycoproteins in the host cell membrane.

No viral factories or accumulation of viral cores have been detected in RV-infected cells by electron microscopy, although late in infection cores have been reported near the site of budding (Holmes *et al.*, 1969; von Bonsdorff and Vaheri, 1969). Some prominent cytoplasmic foci were detected by immunofluorescence at 48 hr using a high multiplicity of infection, but these may represent an accumulation of viral proteins in cell organelles or phagocytic vacuoles during late CPE (Bowden and Westaway, unpublished observations).

### 4.2. Viral RNA

The sites of RNA synthesis in RV-infected cells have been studied by immunofluorescence using a rabbit antiserum to synthetic double-stranded RNA (Bowden *et al.*, 1987). Specific staining with this reagent was observed at 12 hr postinfection (Fig. 7a) and showed discrete immunofluorescent foci in the cytoplasm of infected cells. These probably represent the replicative complexes of the virus. This finding is consistent with the results reported by Hemphill *et al.* (1988), who detected genomic and subgenomic labeled RNA species by 12 hr. At 24 hr, more than one half the cells showed small intensely stained foci mainly in one or more large regions of cytoplasm (Fig. 7b), and at higher multiplicities of infection (Fig. 7c,d) their distribution closely resembled that reported for double-stranded RNA in Sindbis virus-infected cells, e.g., additional foci in linear arrays, some near the plasma membrane (Stollar and Stollar, 1970; Ng *et al.*, 1983; Bowden *et al.*, 1987). In Sindbis virus infection, these foci may be contained in specific membranous structures known as cytoplasmic vacuoles type 1 (Grimley *et al.*, 1972). No equivalent structure has been reported in RV-infected cells.

### 4.3. Interaction of Viral Products with Cytoskeleton

Microfilaments, together with microtubules and intermediate filaments, make up the cytoskeleton and assist in organization of the cytoplasm (Lewin, 1980; Brinkley, 1982). Drastic changes in the appearance of actin-containing microfilaments have been observed using anti-actin antibodies as probes in RV-infected cells (Bowden *et al.*, 1987). In uninfected Vero cells, microfilaments were clearly seen by fluorescence as long cables often extending across the cell

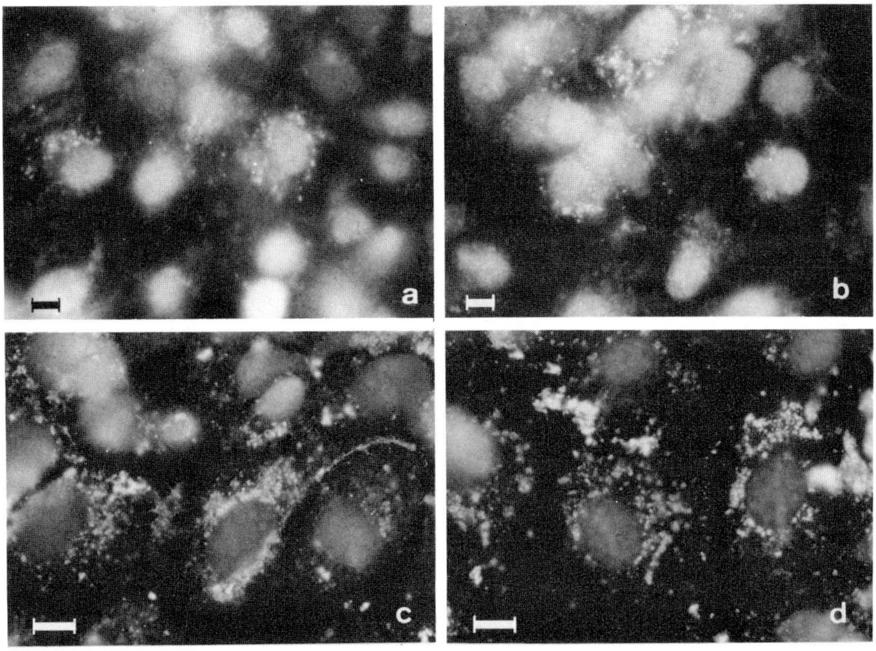


**FIGURE 7.** Intracellular location of sites of RNA synthesis in RV-infected Vero cells visualized by immunofluorescence using antibody to double-stranded RNA. Cells were infected at a multiplicity of infection of 1, fixed, and treated at (a) 12 hr, and (b) 24 hr. (c,d) Cells were infected at a multiplicity of 10, fixed, and treated at 24 hr. Bar = 10 μm. (From Bowden *et al.*, 1987.)

(Fig. 8a). Infected Vero cells examined at 16 hr showed some depolymerization or disintegration of actin filaments (Fig. 8b), which progressed until at 48 hr the filamentous structures were completely disaggregated and were localized as pockets of fluorescence near the plasma membrane of the cell (Fig. 8d). A similar result was obtained in infected BHK-21 cells, although the depolymerization was more limited (Bowden *et al.*, 1987).

Components of the cytoskeleton are associated with replication of several viruses and in some cases may be required for replication (Luftig, 1982). Alterations to the cellular microfilament distribution have been reported in a variety of host–cell systems. For example, disruption or decrease of microfilaments has been observed in cells infected with several paramyxoviruses (Fagraeus *et al.*, 1978), vesicular stomatitis virus (Rutter and Mannweiler, 1977), and cytomegalovirus (Lösse *et al.*, 1982). The specific effect on microfilaments depends on the cell type in the cases of Newcastle disease virus (Rutter and Mannweiler, 1977; Fagraeus *et al.*, 1978) and vaccinia virus (Hiller *et al.*, 1979; Meyer *et al.*,

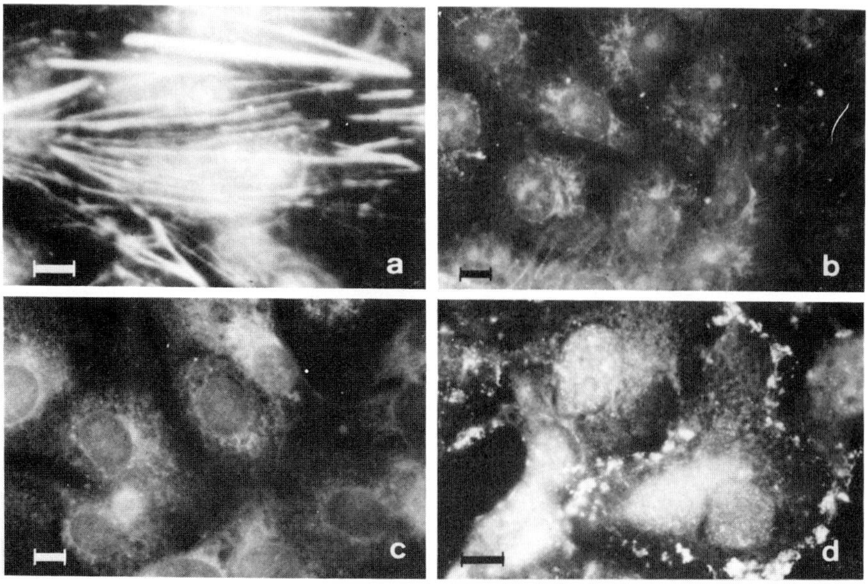


**FIGURE 8.** Intracellular arrangement of actin in RV-infected Vero cells visualized by immunofluorescence using anti-actin antibody. (a) Fixed and treated mock-infected cells, (b–d) cells infected at a multiplicity of infection of 1, fixed and treated at (b) 16 hr, (c) 24 hr, and (d) 48 hr. Bar = 10 μm. (From Bowden *et al.*, 1987.)

1981). Rearrangement of microfilaments also takes place in cells transformed by simian virus 40 SV40 (Graessmann *et al.*, 1980) and in neuronal cells infected with herpes simplex virus (HSV) (Winkler *et al.*, 1982). Because no depolymerization of microfilaments was detected until 16 hr, prior synthesis of some viral product is probably required. For example, depolymerization could be brought about by direct interaction of a viral product on the filamentous actin or by interference with one or more of the actin-associated proteins involved in actin assembly and disassembly (Schliwa, 1981; Brinkley, 1982).

It is believed that newly formed actin fibres provide some of the motive force for the mitotic chromosome movement (Lewin, 1980). If infection with RV prevents or inhibits repolymerization of actin, a corresponding inhibition of cell mitosis could result. Thus Bowden and Westaway (1987) postulated that the observed depolymerization could be related to reports of mitotic inhibition in RV-infected cells (see Dudgeon, 1969; Horzinek, 1981) and may be associated with the congenital rubella syndrome (see Section 1.2).

No dramatic reorganization of polysomes was evident by immunofluorescence in RV-infected Vero cells using antiribosomal antibodies, and similarly antimitochondrial antibodies detected no changes in the distribution of mitochondria after RV infection (Bowden *et al.*, 1987).

## 5. MORPHOLOGY AND MORPHOGENESIS

The morphology and morphogenesis of RV have been investigated in numerous reports and these studies have played a major role in the placement of RV in the *Togaviridae* family.

### 5.1. Negative-Contrast Electron Microscopy

Early descriptions of RV from negative-contrast electron microscopy were unsatisfactory mainly because of the crude preparations used for examination (Norrby *et al.*, 1963; Phillips *et al.*, 1965). The first convincing electron micrographs of RV were published by Best *et al.* (1967), closely followed by those of Holmes and Warburton (1967), who identified viral particles ranging in size from 50 to 75 nm in diameter, comprising an amorphous envelope surrounding an internal spherical component of ~40 nm in diameter. In addition, Holmes and Warburton (1967) detected spikes of 5–6 nm projecting from the envelope surface and proposed that RV appeared to have the morphology of an arbovirus (later togavirus). Subsequent reports have confirmed these findings (Murphy, 1980; Horzinek, 1981) (Fig. 9).

### 5.2. Thin-Section Electron Microscopy

Estimates of the virion dimensions have been conducted in several thin-section studies and are consistent with results from negative-contrast staining. Results generally describe viral particles as having a mean diameter of ~60 nm containing an electron-dense core 30 nm in diameter surrounded by a unit membrane. Virions are spherical or oval in shape, but elongated and distorted variations in the form of double virions and enveloped multicore containing structures are often apparent, particularly late in infection (Higashi, 1973; Murphy, 1980). This feature of RV morphogenesis is in contrast to alphaviruses and flaviviruses, in which abnormal shape variation is uncommon.

Maturation of RV can take place by budding both from the plasma membrane (Fig. 10a), which appears to be modified by insertion of viral envelope proteins (Oshiro *et al.*, 1969; Bardeletti *et al.*, 1979), and from intracytoplasmic membranes (Fig. 10b). In the cytoplasm, the Golgi complex may be a site for virus maturation because virions have been observed within vesicles in the Golgi region and infection produces a proliferation of their membrane content (Holmes *et al.*, 1969; von Bonsdorff and Vaheri, 1969; Higashi, 1973; Kouri *et al.*, 1974;

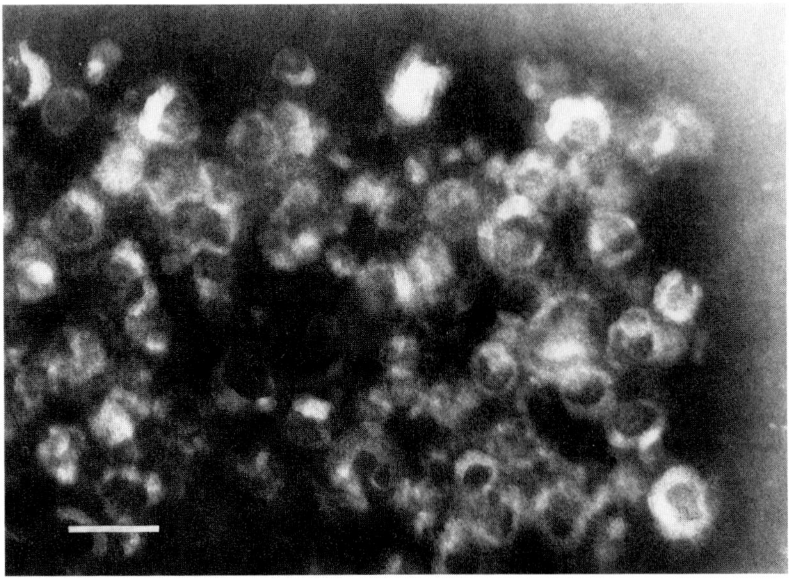

**FIGURE 9.** An aggregation of RV-like particles from infected cell culture fluids visualized using the technique of negative staining immunoelectron microscopy. Bar = 100 nm.

Bardeletti *et al.*, 1979). Involvement of mitochondria in budding and replication of the virus was also suggested based on studies of the viral lipid content and energy metabolism of infected BHK-21 cells (Bardeletti, 1977). However, this suggestion has not been supported by any thin-section electron microscopy. The diversity of maturation sites for RV clearly distinguishes it from the alphaviruses, which in vertebrate cells bud essentially from the plasma membrane (Murphy, 1980).

## 6. CONCLUSION

Rubella virus has the structure, gene order, and replication strategy of a typical togavirus. The spherical virion comprises a nucleocapsid core of RNA complexed with the core protein C, enclosed within a lipid envelope with protruding spikes of two glycoproteins E1 and E2. The approximate molecular weights of E1, E2, and C are 60,000, 44,000–55,000 (heterogeneous), and 34,000, respectively, and they appear to be in equal molar proportions. The glycans attached to E1 and E2 occur in several size classes, both simple (high mannose) and complex, and are N-linked to asparagine at three (E1) or four (E2)

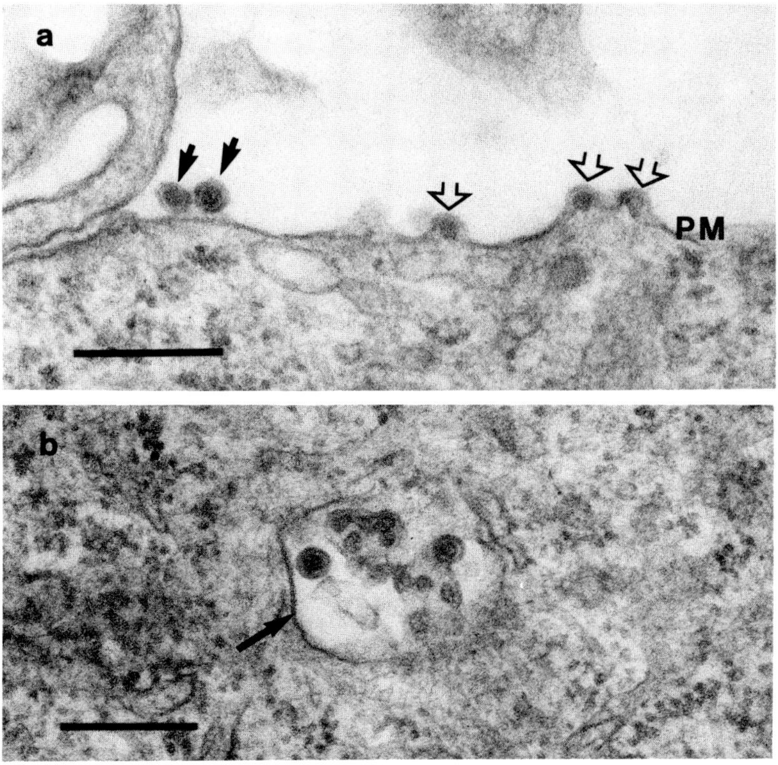


**FIGURE 10.** Thin-sectioning electron microscope observations of RV morphogenesis in Vero cells 48 hours after infection. (a) Budding of RV from the plasma membrane (PM). Note the mature virions (solid arrows) and budding forms (open arrows). (b) Cytoplasmic membrane bound vacuole (arrow) containing RV particles; this electron micrograph indicates that budding of RV also occurs in cytoplasmic vacuoles although a section through an infolding of the plasma membrane could give a similar image. Bars = 250 nm.

glycosylation sites. Both E1 and E2 have epitopes involved in viral neutralization, and E1 functions as the virus hemagglutinin. Protein C has a high proportion of arginine residues which probably provide positive charges for binding of C to virion RNA. The RNA sediments at 38–40S, has an estimated molecular weight of $3.8 \times 10^6$, and is single-stranded, polyadenylated, capped, and infectious.

Viral replication occurs slowly in the cytoplasm (latent period 8–12 hr) and involves a subgenomic 24–26S messenger, with a molecular weight of $\sim 1.3 \times 10^6$, a replicative intermediate and a replicative form. Maximum RNA synthesis and viral production are reached at ~26 hr, and then decline. The subgenomic

RNA of ~3400 nucleotides represents the 3′-end of the genome; it is polyadenylated and is translated from one long open reading frame into the three structural proteins in the order $NH_2$-E2-E1-C-COOH via a precursor p110. The molecular weights of the polypeptide backbones of E2 and E1 are 53,000 and 30,000–34,000, respectively. The N-terminals of E2 and E1 are preceded by putative signal sequences, and thcy are probably cleaved from the nascent polyprotein in the lumen of the ER by a cell-signal peptidase. Hydrophobic sequences at the C-terminals of E2 and E1 presumably serve as transmembrane domains and anchor them in the lipid bilayer. Several large putative nonstructural proteins have been observed, the most prominent of which are ns200, ns150, ns87, and ns75. The largest is cleaved to yield ns150, but their functions are unknown.

The distribution of structural proteins shown by immunofluorescence within infected cells is mainly perinuclear, in reticular membranes, in the Golgi region, and on the plasma membrane. Antibodies to double-stranded RNA produced intensely staining foci in the cytoplasm, probably representing replicative complexes. Mature virions appear to bud into cytoplasmic vesicles and from the plasma membrane. The details of morphogenesis are obscure. Cell organelles are largely unaffected during infection, but dramatic changes occur in the distribution of microfilaments. Anti-actin probes show that these elements of the cytoskeleton become progressively disintegrated. It is suggested that this depolymerization could be related to reports of mitotic inhibition in infected cells and may be associated with the congenital rubella syndrome. Further progress in analysis of RV replication awaits the introduction of immunogold labeling to thin sections of infected cells, the completion of the nucleotide sequence, and characterization of the viral RNA polymerase. Future studies of the molecular biology of RV may provide a rationale for developing chemotherapeutic agents to treat acute, congenital, and persistent postvaccination infections. A genetically engineered rubella vaccine may avoid the adverse manifestations associated with present vaccines such as polyarthritis. If the present pace of RV research is maintained, such applications may soon be attainable.

ACKNOWLEDGMENT. The authors would like to thank Dr. John Marshall for producing the electron micrographs shown in Figs. 9 and 10.

## 7. REFERENCES

Baltimore, D., Burke, D. C., Horzinek, M. C., Huang, A. S., Kääriäinen, L., Pfefferkorn, E. R., Schlesinger, M. J., Schlesinger, S., Schlesinger, W. R., and Scholtissek, C., 1976, Proposed nomenclature for alphavirus polypeptides, *J. Gen. Virol.* **30**:273.

Banatvala, J. E., 1977, Rubella vaccines, in: *Recent advances in Clinical Virology* (A. P. Watson, ed.), pp. 171–190, Churchill Livingstone, Edinburgh.

Bardeletti, G., 1977, Respiration and ATP levels in BHK21/13S cells during the earliest stages of rubella virus replication, *Intervirology* **3:**100–109.

Bardeletti, G., and Gautheron, D. C., 1976, Phospholipid and cholesterol composition of rubella virus and its host cell BHK21 grown in suspension cultures, *Arch. Virol.* **52:**19–27.

Bardeletti, G., Kessler, N., and Aymard-Henry, M., 1975, Morphology, biochemical analysis and neuraminidase activity of rubella virus, *Arch. Virol.* **49:**175–186.

Bardeletti, G., Tetkoff, J., and Gautheron, D., 1979, Rubella virus maturation and production in two host cell systems, *Intervirology* **11:**97–103.

Berger, M., and Schmidt, M. F., 1985, Protein fatty acyltransferase is located in the rough endoplasmic reticulum, *FEBS Lett.* **187:**289–294.

Best, J. M., Banatvala, J. E., Almeida, J. D., and Waterson, A. P., 1967, Morphological characteristics of rubella virus, *Lancet* **2:**237–239.

Bohn, E. M., and Van Alstyne, D., 1981, The generation of defective interfering rubella virus particles, *Virology* **111:**549–554.

Bowden, D. S., and Westaway, E. G., 1984, Rubella virus: Structural and non-structural proteins, *J. Gen. Virol.* **65:**933–943.

Bowden, D. S., and Westaway, E. G., 1985, Changes in glycosylation of rubella virus envelope proteins during maturation, *J. Gen. Virol.* **66:**201–206.

Bowden, D. S., Pedersen, J. S., Toh, B. H., and Westaway, E. G., 1987, Distribution by immunofluorescence of viral products and actin-containing cytoskeletal filaments in rubella virus infected cells, *Arch. Virol.* **92:**211–219.

Brinkley, B. R., 1982, Summary: Organization of the cytoplasm, *Cold Spring Harbor Symp. Quant. Biol.* **46:**1029–1040.

Brodersen, M., and Thomssen, R., 1969, Einbau von $^3H$=uridin in rubellaviruspartikel, *Arch. Ges. Virusforsch.* **26:**118–126.

Casals, J., 1957, The arthropod-borne group of animal viruses, *Trans. NY Acad Sci.* **19:**219–235.

Chantler, J. K., 1979, Rubella virus: Intracellular protein synthesis, *Virology* **98:**275–278.

Chantler, J. K., and Tingle, A. J., 1980, Replication and expression of rubella virus in human lymphocyte populations, *J. Gen. Virol.* **50:**317–328.

Chantler, J. K., and Davies, M. A., 1987, The effect of antibody on rubella virus infection in human lymphoid cells, *J. Gen. Virol.* **68:**1277–1288.

Chu, P. W. J., and Westaway, E. G., 1987, Characterization of Kunjin virus RNA-dependent RNA polymerase: Reinitiation of synthesis *in vitro, Virology* **151:**330–337.

Clarke, D. M., Loo, T. W., Hui, I., Chong, P., and Gillam, S., 1987, Nucleotide sequence and *in vitro* expression of rubella virus 24S subgenomic messenger RNA encoding the structural proteins E1, E2 and C, *Nucl. Acids Res.* **15:**3041–3057.

Coia, G., Parker, M. D., Speight, G., Byrne, M. E., and Westaway, E. G., 1988, Nucleotide and complete amino acid sequences of Kunjin virus: Definitive gene order and characteristics of the virus specified proteins, *J. Gen. Virol.* **69:**1–21.

Cooper, L. Z., Ziring, P. R., Weiss, H. J., Malkis, B. A., and Krugman, S., 1969, Transient arthritis after rubella vaccination, *Am. J. Dis. Child.* **118:**218–225.

Dorsett, P. H., Miller, D. C., Green, K. Y., and Byrd, F. I., 1985, Structure and function of rubella virus proteins, *Rev. Infect. Dis.* **7:**S150–S156.

Fagraeus, A., Tyrrell, D. L. J., Norberg, R., and Norrby, E., 1978, Actin filaments in paramyxovirus-infected human fibroblasts studied by indirect immunofluorescence, *Arch. Virol.* **57:** 291–296.

Fenner, F., 1975/76, The classification and nomenclature of viruses, *Intervirology* **6:**1–12.

Fenner, F., Pereira, H. G., Porterfield, J. S., Joklik, W. K., and Downie, A. W., 1974, Family and generic names for viruses approved by the International Committee on Taxonomy of Viruses, *Intervirology* **3:**193–198.

Forbes, J. A., 1969, Rubella: Historical aspects, *Am. J. Dis. Child.* **118:**5–11.

Frey, T. K., and Hemphill, M. L., 1988, Generation of defective-interfering particles by rubella virus in vero cells, *Virology* **164:**22–29.

Frey, T. K., Marr, L. D., Hemphill, M. L., and Dominquez, G., 1986, Molecular cloning and sequencing of the region of the rubella virus genome coding for glycoprotein E1, *Virology* **154:** 228–232.

Graessmann, A., Graessmann, M., Tjian, R., and Topp, W. C., 1980, Simian virus 40 small-t protein is required for loss of actin cable networks in rat cells, *J. Virol.* **33:**1182–1191.

Green, K. Y., and Dorsett, P. H., 1986, Rubella virus antigens: Localization of epitopes involved in hemagglutination and neutralization by using monoclonal antibodies, *J. Virol.* **57:**893–898.

Green, J., Griffiths, G., Louvard, D., Quinn, P., and Warren, G., 1981, Passage of viral membranes through the Golgi complex, *J. Mol. Biol.* **152:**663–698.

Gregg, N. M., 1941, Congenital cataract following german measles in the mother, *Trans. Ophthalmol. Soc. Aust.* **3:**35–46.

Grimley, P. M., Levin, J. G., Berezesky, I. K., and Friedman, R. M., 1972, Specific membranous structures associated with the replication of group A arboviruses, *J. Virol.* **10:**492–503.

Heggie, A. D., and Robbins, F. C., 1969, Natural rubella acquired after birth, *Am. J. Dis. Child.* **118:**12–17.

Hemphill, M. L., Forng, R.-Y., Abernathy, E. S., and Frey, T. K., 1988, Time course of rubella virus-specific macromolecular synthesis during rubella virus infection in Vero cells, *Virology* **162:**65–75.

Herrmann, K. L., 1979, Rubella virus, in: *Diagnostic Procedures for Viral, Rickettsial and Chlamydial Infections* (E. H. Lennette and N. J. Schmidt, eds.), pp. 725–766, American Public Health Association, Washington, D. C.

Higashi, N., 1973, Electron microscopy of viruses in thin sections of cells grown in culture, *Prog. Med. Virol.* **15:**331–379.

Hiller, G., Weber, K., Schneider, L., Parajsz, C., and Jungwirth, C., 1979, Interaction of assembled progeny pox viruses with the cellular cytoskeleton, *Virology* **98:**142–153.

Holmes, I. H., and Warburton, M. F., 1967, Is rubella an arbovirus?, *Lancet* **2:**1233–1236.

Holmes, I. H., Wark, M. C., and Warburton, M. F., 1969, Is rubella an arbovirus? II. Ultrastructural morphology and development, *Virology* **37:**15–25.

Horzinek, M. C., 1981, *Non-Arthropod-Borne Togaviruses,* Academic, London.

Ho-Terry, L., and Cohen, A., 1980, Degradation of rubella virus envelope components, *Arch. Virol.* **65:**1–13.

Ho-Terry, L., and Cohen, A., 1982, Rubella virion polypeptides: Characterisation by polyacrylamide gel electrophoresis, isoelectric focusing and peptide mapping, *Arch. Virol.* **72:**47–54.

Ho-Terry, L., and Cohen, A., 1984, The role of glycosylation on haemagglutination and immunological reactivity of rubella virus, *Arch. Virol.* **79:**139–146.

Ho-Terry, L., Terry, G. M., Cohen, A., and Londesborough, P., 1986, Immunological characterisation of the rubella E1 glycoprotein, *Arch. Virol.* **90:**145–152.

Hovi, T., and Vaheri, A., 1970a, Rubella virus-specific ribonucleic acids in infected BHK-21 cells, *J. Gen. Virol.* **6:**77–83.

Hovi, T., and Vaheri, A., 1970b, Infectivity and some physicochemical characteristics of rubella virus ribonucleic acid, *Virology* **42:**1–8.

Kalkkinen, N., Oker-Blom, C., and Pettersson, R. F., 1984, Three genes code for rubella virus structural proteins E1, E2a, E2b and C, *J. Gen. Virol.* **65:**1549–1557.

Katow, S., and Matsuno, T., 1980, Base composition of rubella virus RNA, *Arch. Virol.* **65:**67–70.

Kornfeld, R., and Kornfeld, S., 1980, Structure of glycoproteins and their oligosaccharide units, in: *The Biochemistry of Glycoproteins and Proteoglycans* (W. J. Lennarz, ed.), pp. 1–34, Plenum, New York.

Kouri, G., Aguilera, A., Rodriguez, P., and Korolev, M., 1974, A study of microfoci and inclusion bodies produced by rubella virus in the RK-13 cell line, *J. Gen. Virol.* **22**:73–80.

Lewin, B., 1980, *Gene expression.* Vol. 2: *Eucaryotic Chromosomes,* 2nd ed., Wiley, New York.

Liebhaber, H., and Gross, P. A., 1972, The structural proteins of rubella virus, *Virology* **47**:684–693.

Lösse, D., Lauer, R., Weder, D., and Radsak, K., 1982, Actin distribution and synthesis in human fibroblasts infected by cytomegalovirus, *Arch. Virol.* **71**:353–359.

Louvard, D., Reggio, H., and Warren, G., 1982, Antibodies to the Golgi complex and rough endoplasmic reticulum, *J. Cell. Biol.* **92**:97–107.

Luftig, R. B., 1982, Does the cytoskeleton play a significant role in animal virus replication?, *J. Theor. Biol.* **99**:173–191.

Maes, R. A., Vaheri, A., Sedwick, D., and Plotkin, S., 1966, Synthesis of virus and macromolecules by rubella-infected cells, *Nature (Lond.)* **210**:384–385.

McCarthy, K., 1969, Cell cultures useful for the study of rubella, *Am. J. Dis. Child.* **118**:78–82.

Meyer, R. K., Burger, M. M., Tschannen, R., and Schäfer, R., Actin filament bundles in vaccinia virus infected fibroblasts, *Arch. Virol.* **67**:11–18.

Murphy, F. A., 1980, Togavirus morphology and morphogenesis, in: *The Togaviruses* (W. Schlesinger, ed.) pp. 241–326, Academic Press, New York.

Nakhasi, H. L., Meyer, B. C., and Liu, T.-Y., 1986, Rubella virus cDNA. Sequence and expression of E1 envelope protein, *J. Biol. Chem.* **261**:16616–16621.

Nakhasi, H. L., Zheng, D., Hewlett, I. K., and Liu, T.-Y., 1988, Rubella virus replication: Effect of interferons and actinomycin D, *Virus Res.* **10**:1–15.

Ng, M. L., Pedersen, J. S., Toh, B. H., and Westaway, E. G., 1983, Immunofluorescent sites in Vero cells infected with the flavivirus Kunjin, *Arch. Virol.* **78**:177–190.

Norrby, E., Magnusson, P., Friding, B., and Gard, S., 1963, A note on the morphology of rubella virus, *Arch. Ges. Virusforsch.* **13**:421–424.

Norval, M., 1979, Mechanism of persistence of rubella virus in LLC–MK2 cells, *J. Gen. Virol.* **43**:289–298.

Ogra, P. L., and Herd, J. K., 1971, Arthritis associated with induced rubella infection, *J. Immunol.* **107**:810–813.

Oker-Blom, C., 1984, The gene order for rubella virus structural proteins is $NH_2$-C-E2-E1-COOH, *J. Virol.* **51**:354–358.

Oker-Blom, C., Kalkinnen, N., Kääriäinen, L., and Pettersson, R. F., 1983, Rubella virus contains one capsid protein and three envelope glycoproteins, E1, E2a and E2b, *J. Virol.* **46**:964–973.

Oker-Blom, C., Ulmanen, I., Kääriäinen, L., and Pettersson, R. F., 1984, Rubella virus 40S genome RNA specifies a 24S subgenomic mRNA that codes for a precursor to structural proteins, *J. Virol.* **49**:403–408.

Oshiro, L. S., Schmidt, N. J., and Lennette, E. H., 1969, Electron microscopic studies of rubella virus, *J. Gen. Virol.* **5**:205–210.

Parkman, P. D., Buescher, E. L., and Artenstein, M. S., 1962, Recovery of rubella virus from army recruits, *Proc. Soc. Exp. Biol. Med.* **111**:225–230.

Parkman, P. D., Meyer, H. M., Kirschstein, R. L., and Hopps, H. E., 1966, Attenuated rubella virus. I. Development and laboratory characterization, *N. Engl. J. Med.* **275**:569–574.

Payment, P., Ajducovic, D., and Pavilanis, V., 1975a, Le virus de la rubéole. I. Morphologie et protéines structurales, *Can. J. Microbiol.* **21**:703–709.

Payment, P., Ajducovic, D., and Pavilanis, V., 1975b, Le virus de la rubéole. II Réplication dans les cellules Vero et effets de l'acnomyciene D et du cycloheximide, *Can. J. Microbiol.* **21**:701–717.

Pettersson, R. F., Oker-Blom, C., Kalkkinen, N., Kallio, A., Ulmanen, I., Kääriäinen, L., Partanen, P., and Vaheri, A., 1985, Molecular and antigenic characteristics and synthesis of rubella virus structural proteins, *Rev. Infect. Dis.* **7**:S140–S149.

Phillips, C. A., Behbehani, A. M., Johnson, L. W., and Melnick, J. L., 1965, Isolation of rubella virus. An epidemic characterised by rash and arthritis, *JAMA* **191:**615–616.

Pope, D. D., and Van Alstyne, D., 1981, Evidence for restricted replication of rubella virus in rat glial cells in culture, *Virology* **113:**776–780.

Pope, D. D., and Van Alstyne, D., 1982, Further studies on the restriction of rubella virus replication in rat glial cells in culture, *Can. J. Microbiol.* **28:**1404–1409.

Rutter, G., and Mannweiler, K., 1977, Alteration to actin-containing structures in BHK-21 cells infected with Newcastle disease virus and vesicular stomatitis virus, *J. Gen. Virol.* **37:**233–242.

Saraste, J., and Kuismanen, E., 1984, Pre- and post-Golgi vacuoles operate in the transport of Semliki Forest virus membrane glycoproteins to the cell surfaces, *Cell* **38:**535–549.

Schliwa, M., 1981, Proteins associated with cytoplasmic actin, *Cell* **25:**587–590.

Schmidt, M. F. G., and Schlesinger, M. J., 1980, Relation of fatty acid attachment to the translation and maturation of vesicular stomatitis and Sindbis virus membrane glycoproteins, *J. Biol. Chem.* **255:**3334–3339.

Sedwick, W. D., and Sokol, F., 1970, Nucleic acid of rubella virus and its replication in hamster kidney cells, *J. Virol.* **5:**478–489.

Simpson, R. E. H., 1940, Rubella and polyarthritis, *Br. Med. J.* **1:**830.

Stollar, B. D., and Stollar, V., 1970, Immunofluorescent demonstration of double-stranded RNA in the cytoplasm of Sindbis virus-infected cells, *Virology* **42:**276–280.

Strauss, E. G., and Strauss, J. H., 1983, Replication strategies of the single-stranded RNA viruses of eukaryotes, *Current Topics Microbiol. Immunol.* **105:**1–98.

Takkinen, K., Vidgren, G., Ekstrand, J., Hellman, U., Kalkkinen, N., Wernstedt, C., and Pettersson, R. F., 1988, Nucleotide sequence of the rubella virus capsid protein gene reveals an unusually high G/C content, *J. Gen. Virol.* **69:**603–612.

Terry, G. M., Ho-Terry, L., Cohen, A., and Londesborough, P., 1985, Rubella virus RNA: Effect of high multiplicity passage, *Arch. Virol.* **86:**29–36.

Terry, G. M., Ho-Terry, L., Londesborough, P., and Rees, K. R., 1988, Localization of the rubella E1 epitopes, *Arch. Virol.* **98:**189–197.

Toivonen, V., Vainionpää, R., Salmi, A., and Hyypiä, T., 1983, Glycopeptides of rubella virus, *Arch. Virol.* **77:**91–95.

Trudel, M., Nadon, F., Comtois, R., Payment, P., Bonneau, A., and Lecomte, J., 1982, Identification of rubella virus structural proteins by immunoprecipitation, *J. Virol. Methods* **5:**191–197.

Vaheri, A., and Hovi, T., 1972, Structural proteins and subunits of rubella virus, *J. Virol.* **9:**10–16.

Van Alstyne, D., Krystal, G., Kettyls, D., and Bohn, E. M., 1981, The purification of rubella virus (RV) and determination of its polypeptide composition, *Virology* **108:**491–498.

Vidgren, G., Takkinen, K., Kalkkinen, N., Kääriäinen, L., and Pettersson, R. F., 1987, Nucleotide sequence of the genes coding for the membrane glycoproteins E1 and E2 of rubella virus, *J. Gen. Virol.* **68:**2347–2357.

Voiland, A., and Bardeletti, G., 1980, Fatty acid composition of rubella virus and BHK21/13S infected cells, *Arch. Virol.* **64:**319–328.

Von Bonsdorff, C.-H., and Vaheri, A., 1969, Growth of rubella virus in BHK21 cells: Electron microscopy of morphogenesis, *J. Gen. Virol.* **5:**47–51.

Waxham, M. N., and Wolinsky, J. S., 1983, Immunochemical identification of rubella virus hemagglutinin, *Virology* **126:**194–203.

Waxham, M. N., and Wolinsky, J. S., 1985a, A model of the structural organization of rubella virions, *Rev. Infect. Dis.* **7:**S133–S139.

Waxham, M. N., and Wolinsky, J. S., 1985b, Detailed immunologic analysis of the structural polypeptides of rubella virus using monoclonal antibodies, *Virology* **143:**153–165.

Weller, T. H., and Neva, F. A., 1962, Propagation in tissue culture of cytopathogenic agents from patients with rubella-like illness, *Proc. Soc. Exp. Biol. Med.* **111:**215–225.

Wesselhoeft, C., 1947, Rubella (german measles), *N. Engl. J. Med.* **236**:943–950.

Westaway, E. G., Brinton, M. A., Gaidamovich, S. Ya., Horzinek, M. C., Igarashi, A., Kääriäinen, L., Lvov, D. K., Porterfield, J. S., Russell, P. K., and Trent, D. W., 1985a, Togaviridae, *Intervirology* **24**:125–139.

Westaway, E. G., Brinton, M. A., Gaidamovich, S. Ya, Horzinek, M. C., Igarashi, A., Kääriäinen, L., Lvov, D. K., Porterfield, J. S., Russell, P. K., Trent, D. W., 1985b, Flaviviridae, *Intervirology* **24**:183–192.

Winkler, M., Dawson, G. J., Elizan, T. S., and Berl, S., 1982, Distribution of actin and myosin in a rat neuronal cell line infected with herpes simplex virus, *Arch. Virol.* **72**:95–103.

Wong, K. T., Robinson, W. S., and Merigan, T. C., 1969, Synthesis of viral-specific ribonucleic acid in rubella virus-infected cells, *J. Virol.* **4**:901–903.

Woods, W. A., and Robbins, F. C., 1968, Effect of actinomycin D on growth of rubella virus in tissue cultures, *J. Gen. Virol.* **3**:43–49.

*Chapter 8*

# Association of Bluetongue Virus with the Cytoskeleton

Bryan T. Eaton and Alex D. Hyatt

## 1. INTRODUCTION

Eukaryotic cells contain an extensive filamentous network referred to as the cytoskeleton, and it is currently believed that all animal viruses may use the cytoplasmic or nuclear skeletal matrix of cells for at least part of their replication cycle (Luftig, 1982; Penman, 1985). Analysis of cells infected with double-stranded RNA-containing reoviruses has done much to stimulate interest in the role played by the cytoskeleton in viral replication (Dales, 1963; Spendlove *et al.*, 1963; Dales *et al.*, 1965; Sharpe *et al.*, 1982). The purpose of this chapter is to describe the results of experiments done in our laboratory to analyze cytoskeleton involvement in the replication of another double-stranded RNA virus, bluetongue virus (BTV), a member of the orbivirus genus in the Reoviridae family.

Electron microscopic examination of thin sections of BTV-infected cells has shown that the cytoplasm is characterized by the presence of three morphological structures. In addition to small numbers of virus-like particles, the cytoplasm contains virus-specific tubules, whose function is unknown, as well as virus inclusion bodies (VIB), which are large, dense, filamentous structures, believed

**Bryan T. Eaton and Alex D. Hyatt** Australian Animal Health Laboratory, CSIRO, Geelong, Victoria 3220, Australia.

to be the site of virus morphogenesis (Bowne and Jochim, 1967; Lecatsas, 1968; Cromack *et al.,* 1971). The potential involvement of the cytoskeleton in the replication and/or morphogenesis of BTV became apparent in 1987, when it was shown that at least a proportion of the VIB, tubules and virus-like particles bind to the cytoskeleton of infected cells (Eaton *et al.,* 1987; Hyatt *et al.,* 1987). The cytoskeletal location of these structures has facilitated study of their viral protein content using monoclonal antibodies in immunogold procedures (Eaton *et al.,* 1988; Hyatt and Eaton, 1988; Gould *et al.,* 1988). This chapter summarizes what is currently known about the cytoskeletal location and viral protein content of VIB, tubules and virus-like particles in BTV-infected cells. The data point to involvement of the intermediate and microfilament networks in BTV replication and provide information on some of the previously unrecognized steps involved in the final stages of viral morphogenesis.

The following sections (1) present a brief introduction to the cytoskeleton and indicate how data from many virus–cell systems show cytoskeletal involvement in viral replication, (2) outline the structure of BTV and the cytoskeleton-associated structures found in BTV-infected cells, (3) describe how virus proteins have been localized to specific cytoskeletal-associated structures in BTV infected cells, and (4) use the information derived from these experiments to draw some conclusions on the site of addition of the outer coat protein VP2 to BTV particles during morphogenesis and speculate on the origin of the tubules found in the cytoplasm of orbivirus-infected cells.

### 1.1. Cytoskeleton

Treatment of eukaryotic cells with nonionic detergents such as NonidetP-40 (NP40) or Triton X-100 results in the solubilization of the cell membrane and release of soluble cytoplasmic components (Brown *et al.,* 1976; Lenk *et al.,* 1977). The resulting insoluble matrix is known as the cytoskeleton. It is composed of anastomosing networks of protein filaments and associated proteins, some of which serve to link filaments to each other or to other cell components (Penman *et al.,* 1983). The cytoskeleton is believed to be involved in a wide variety of functions, such as maintenance of cell shape, motility, and internal cellular organization. The three major classes of protein filament are the actin-based microfilaments, the tubulin-containing microtubules and the intermediate filaments whose protein composition depends on the cell type.

#### 1.1.1. Microfilaments

Monomeric actin, a globular protein, can assemble reversibly into filaments consisting of two chains of subunits wound around each other, forming a helix, the diameter of which is ~7 nm with an axial repeat of 37 nm. The asymmetrical shape of the actin molecule and the fact that monomers in the filament are orientated in the same direction means that each end of the filament has different domains exposed.

Microfilaments are found in cultured cells in a variety of forms. Bundles of closely packed microfilaments course through cells, often oriented parallel to the long axis of the cell. Such stress fibers are often associated with contacts between the plasma membrane of the cell and the substratum. They run from these positions back into the cytoplasm, rising from the lower to the upper surface of the cell and usually end near the nucleus. Microfilaments are also found in individual bundles in filipodia. They are also present in the cortical layer of cytoplasm underlying the cell plasma membrane. Unlike the two forms mentioned above in which microfilaments run parallel to each other, the cortical layer contains a highly branched network. The form taken by microfilaments, i.e., their length and degree of crosslinking, depends on the nature of the proteins interacting with them and a very large number of proteins that bind to filamentous (and globular) actins have been isolated (Hartwig *et al.*, 1985).

### 1.1.2. Microtubules

Microtubules consist predominantly of two globular proteins, α- and β-tubulin, which alternate along the length of the protofilaments constituting the microtubule. In most microtubules from a wide variety of sources, there are usually 13 protofilaments arranged in a circular fashion. Thus, in transverse section, microtubules usually appear as electron-dense rings ~5 nm diameter surrounding an electron-translucent central core ~15 nm diameter. Microtubules are present in the cell both above and below the nucleus and often seem to radiate from one or more regions of the cell, known as microtubule-organizing centers (MTOC). The major MTOC of interphase cells is the centrosome, which encompasses the centrioles and surrounding centrosphere.

Microtubules have been shown to contain a variety of high-molecular-weight associated proteins (MAP) and a group of proteins with an approximate molecular weight of 60 kDa, called τ proteins. Both MAP and τ proteins stimulate microtubule assembly *in vitro*, and MAP form side arms that extend from the microtubule surface. It is likely that such proteins control the growth and disassembly of microtubules and play a role in the crosslinking of microtubules to other cytoskeletal matrices (Wiche, 1985).

### 1.1.3. Intermediate Filaments

The third major cytoskeletal network in eukaryotic cells consists of filaments that contain α-helical-rich proteins and have a diameter ranging from 6 to 11 nm. Although such so-called intermediate filaments look very similar in the microscope, it is clear that their chemical composition differs in epithelial, neural, glial, muscle, and mesenchymal cells. For example, the intermediate filaments in epithelial cells contain keratins, a large and heterogeneous collection of proteins whose molecular weights range from 40 to 68 kDa (Moll *et al.*, 1982). By contrast, the major protein in the intermediate filament network of fibroblasts and other cells of mesenchymal origin has a molecular weight of 52 kDa, called vimentin. A number of cell types are known to have intermediate filament subunits of more than one type. In several varieties of cultured epithelial cells, keratin and vimentin are believed to exist in separate intermediate filament arrays (Franke *et al.*, 1979).

Vimentin and keratin fibers are characteristically wavy and are especially abundant around the cell nucleus. From this location, they appear to radiate throughout the cytoplasm and, in the case of keratin-based filaments, terminate in desmosomes (Drochmans *et al.*, 1978). Intermediate filaments may also interact with microtubules because disruption of the latter with drugs such as vinblastine and colchicine induces the formation of perinuclear whorls of most intermediate filament networks (Goldman, 1971; Blose and Chacko, 1976; Yen and Fields, 1981).

## 1.2. The Cytoskeleton and Viral Replication

Early recognition of a possible association between viruses and the cytoskeleton came from the work of Dales (1963), who analyzed thin sections of virus factories in reovirus-infected cells. This showed virus particles to be associated with microtubules and enmeshed in masses of dense twisted or kinky filaments. The association of reovirus particles with microtubules was also demonstrated *in vitro* by Babiss *et al.* (1979), who showed that reovirus type 3 has a low affinity and reovirus type 1, a high affinity for microtubules. The greater *in vitro* edge binding of reovirus type 1 to microtubules was correlated with the presence of type 1 δ1 protein, the hemagglutinin and neutralization antigen. The relevance of the association of reovirus type 1 with microtubules is not clear because addition of colchicine to infected cells only results in a delayed release of maximum virus titres (Babiss *et al.*, 1979).

An association between adenovirus and microtubules was also observed shortly after infection of HeLa cells with adenovirus type 5, prior to moving to the nucleus, where virion uncoating occurs at nuclear pores (Dales and Chardonnet, 1973; Miles *et al.*, 1980). This association has also been observed *in vitro* (Luftig and Weihing, 1975). The benefit of being able to quantitate the binding of adenovirus to microtubules *in vitro* was clearly seen when Weatherbee *et al.* (1977) prepared MAP-deficient microtubules and showed them to have a reduced ability to bind adenovirus. Thus, the link between adenovirus and microtubules may be mediated by MAP.

The interaction between reovirus and adenovirus particles with microtubules was initially detected because of the ease with which the spindle apparatus and microtubules could be seen in thin sections of infected cells. The ability to examine involvement of the cytoskeleton in the replication of these and other viruses was markedly enhanced when it was realized that a very simple procedure, i.e., the addition of a nonionic detergent to cells in either suspension or monolayer culture released the soluble cytoplasmic constituents and showed the insoluble nuclear cytoskeleton framework (Brown *et al.*, 1976; Lenk *et al.*, 1977).

### 1.2.1. Binding of Virus Proteins to the Cell Matrix

Two procedures, immunofluorescence and radiolabeling, have been used to show that virus proteins bind to the cytoskeleton of infected cells. Immunofluorescent probing, initially of whole cells, showed co-localization of virus antigen and cytoskeletal components such as vimentin in reovirus- (Sharpe *et al.*, 1979) and actin in Newcastle disease virus (NDV)-infected cells (Rutter and Mannweiller, 1977). However, more definitive evidence for virus protein–cell matrix association came from the finding that polyclonal and monoclonal anti-virus antisera react in immunofluorescence tests with detergent-derived cytoskeletons of virus infected cells (Eaton *et al.*, 1987; Mora *et al.*, 1987).

The radiochemical detection of virus-specific proteins associated with the detergent-insoluble cytoskeletal fraction of infected cells has also provided evidence for the involvement of the cell matrix in the replication of viruses such as herpes (Ben-Ze'ev *et al.*, 1983; Yamauchi *et al.*, 1985; Rose *et al.*, 1986), reovirus (Mora *et al.*, 1987), vesicular stomatitis virus (VSV) (Chatterjee *et al.*, 1984), rotavirus (Musalem and Espejo, 1985), retrovirus (Edbauer and Naso, 1983), measles virus (Bohn *et al.*, 1986), and poliovirus (Weed *et al.*, 1985). Radiolabeling at different times postinfection (p.i.) has been used to show temporal control of virus protein binding to the cytoskeleton (Mora *et al.*, 1987). Pulse-chase experiments have provided information on the kinetics with which virus proteins are transported from the cytoplasmic framework to the nuclear matrix in herpes virus-infected cells (Ben-Ze'ev *et al.*, 1983; Knipe and Spang, 1982; Yamauchi *et al.*, 1985) and shown preferential localization of the noncapsid proteins of poliovirus on the cytoskeleton (Weed *et al.*, 1985). However, polyacrylamide gel electrophoresis (PAGE) of labeled cytoskeleton-associated viral proteins and immunofluorescent probing of cytoskeletons with virus-specific antisera demonstrate only the presence of virus protein on the cell matrix. They yield no information about the structures containing the viral proteins.

### 1.2.2. Visualization of Cytoskeleton-Associated Structures

To determine the nature of virus-specific structures on the cell matrix, several laboratories have used embedment-free electron microscopy. Two procedures have been developed. In whole-mount electron microscopy, cells are grown and infected on carbon-coated Parlodion (nitrocellulose) or Formvar (polyvinyl formal)-filmed gold grids. Infected whole cells may be critical point dried and viewed directly in a high voltage electron microscope. Alternatively, and more commonly, infected cells are extracted with nonionic detergents and the nucleus–cytoskeleton matrix, which remains bound to the substrate, fixed with either glutaraldehyde or paraformaldehyde, dehydrated, critical point-dried and viewed directly in a conventional electron microscope. Unstained cytoskeletons may also be shadowed prior to viewing in a conventional electron microscope. The use of some of these approaches has confirmed the association of herpes virus-like particles with the nuclear matrix (Ben-Ze'ev *et al.*, 1983) and detected vaccinia virus-like particles bound to the cytoplasmic matrix (Hiller *et al.*, 1979). Similar experiments have resulted in the visualization of putative intermediates in viral morphogenesis. For example, assembling virions and other electron-dense material have been observed in the cytoplasmic matrix-associated assembly site of the Iridovirus frog virus 3 (FV-3) (Murti *et al.*, 1985) and Bohn *et al.* (1986) provided evidence for the binding of nucleocapsids to microfilaments in measles virus-infected cells.

A second approach to resin-free electron microscopy was used by Weed *et al.* (1985) to examine the cytoskeleton of poliovirus-infected cells. Cytoskeletons of virus-infected suspension-grown cells were fixed and dehydrated in *n*-butyl alcohol (nBA). Following infiltration with diethylene glycol distearate (DGD) and solidification, sections were cut and the DGD removed by treatment with nBA. Although no virions or procapsids were observed in the sections, many clusters of spheroidal bodies (40- to 70-nm diameter) with large electron-dense cores were associated with the cytoskeleton. These may be the poliovirus replication complex.

### 1.2.3. Alterations in Cytoskeletal Organization following Viral Infection

The ready availability of monoclonal antibodies to a number of cytoskeletal proteins and their use in immunofluorescence tests have shown that viral infection often leads to either breakdown or reorganization of one or more of the three major cytoskeletal networks. A desire to understand the basis of the virus-induced cytopathic effect has been one factor motivating studies on the role of viruses in the depolymerization and reorganization of the cytoskeleton in virus-infected cells.

Actin depolymerization occurs in cells infected with cytomegalovirus (CMV) (Jones *et al.*, 1986), vesicular stomatitis virus (Rutter and Mannweiler, 1977), herpesvirus (Bedows *et al.*, 1983), and adenovirus (Jackson and Bellett, 1985), and in cells transformed by simian virus 40 (SV40) (Pollack *et al.*, 1975), Rous sarcoma virus (RSV) (Boschek *et al.*, 1981) and adenovirus (Goldman *et al.*, 1975). The role of the virus in inducing depolymerization is not understood. However, Jackson and Bellett (1975) showed that expression of an early gene (E1a) was required for the reduction in microfilament organization in adenovirus-infected cells. The phosphorylation of vinculin and talin (components of the cytoplasmic face of focal adhesion plaques that anchor microfilaments to the plasma membrane) by RSV-transforming protein $p60^{v\text{-}src}$ may be partly responsible for the disruption of microfilament organization in RSV-transformed cells (Sefton *et al.*, 1981; Pasquale *et al.*, 1986). Several DNA viruses alter the distribution of microfilaments in infected cells. For example, both vaccinia and FV-3 viruses induce the formation of numerous microfilament-containing microvilli on the cell surface (Hiller *et al.*, 1979; Murti *et al.*, 1985). These are used for the release of virus from the cell.

The microtubule network is disrupted by herpes viruses (Norrild *et al.*, 1986; Pfeiffer *et al.*, 1983) and the paramyxovirus canine distemper virus (Howard *et al.*, 1983). In herpesvirus-infected cells, tubulin (actin and myosin, but not vimentin) relocates in juxtanuclear aggregates that also contain virus glycoproteins. The distribution of herpes simplex virus (HSV) glycoproteins and formation of infectious virus is dependent on intact microtubules (Neilsen *et al.*, 1987). In FV-3-infected cells, the number of microtubules decreases progressively with the course of infection, and this was correlated with a rapid reduction in the synthesis of tubulin (Murti *et al.*, 1985).

Viral infection may also lead to changes in the organization of some of the intermediate filament networks in infected cells. There is a specific disruption of the intermediate filament network in reovirus-infected cells (Sharpe *et al.*, 1985) and late after infection of monkey kidney cells with SV40 (Ben-Ze'ev, 1984). New keratins are generated in cells infected with HSV (Neilsen *et al.*, 1987) or transformed by SV40 (Hronis, 1984). In FV-3-infected cells, vimentin is phosphorylated prior to reorganization of the intermediate filaments around the virus assembly site (Murti *et al.*, 1985).

### 1.3. Bluetongue Virus

Bluetongue virus is the type species of the orbivirus genus in the family Reoviridae. Orbiviruses are currently divided into 13 serogroups on the basis of group-reactive serological tests. The most studied orbiviruses are those of the bluetongue serogroup, which encompasses 24 serotypes distinguished on the basis of virus-neutralization tests. The orbiviruses and their classification were recently reviewed comprehensively by Gorman and Taylor (1985). Morphologically, the virus consists of a core composed of 32 capsomers arranged in icosahedral symmetry, surrounded by an outer fibrillar coat (Bowne and Ritchie, 1970; Verwoerd *et al.*, 1972). BTV serotype 10 (BTV-10) extracted by freon from infected cells, purified by zonal centrifugation in sucrose and banding to equilibrium in cesium chloride, was found to contain seven structural proteins (VP1–7) (Verwoerd *et al.*, 1972). The outer coat of the virus can be removed from the virion by treatment with mono- or divalent cations at appropriate pH and contains two proteins, VP2 and VP5 (Huismans *et al.*, 1987a). The virus core contains 10 segments of virion RNA (Verwoerd *et al.*, 1970), two major proteins VP3 and VP7, and three minor proteins VP1, VP4 and VP6. The capsomers consist, perhaps entirely, of VP7 (Huismans *et al.*, 1987c).

*In vitro* translation of individual methylmercuric hydroxide-denatured double-stranded RNA segments of BTV has demonstrated the coding capacity of each RNA from BTV-1 (from South Africa) (Mertens *et al.*, 1984), BTV-17 (from the United States) (Grubman *et al.*, 1983), and BTV-1 (Australia) (Fig. 1). The results show that in addition to the seven structural proteins, BTV codes for three additional nonstructural proteins. Two of these, NS1 and NS2, are readily detected in BTV-infected cells (Fig. 2). In contrast to NS1, which is synthesized in very large quantities, NS3 is difficult to detect in infected cells. This may be due, at least in part, to the fact that RNA 5, which codes for NS1 is transcribed at four times the rate of RNA 10, the gene for NS3 (Huismans and Verwoerd, 1973).

The procedure devised by Verwoerd *et al.* (1972) to purify BTV-10 does not appear to be applicable to all BTV. A second method, using sodium lauryl sarcosine and dithiothreitol, has been developed by Mertens *et al.* (1985) and applied to the purification of South African serotypes BTV-1 and BTV-4. In these cases, in addition to the seven structural proteins described by Verwoerd *et al.* (1972), small amounts of the nonstructural protein NS2 were found in the purified virus.

Studies on the protein content of BTV have used purified virus. The biophysical and biochemical procedures applied during virus purification may lead, however, not only to distortion or disruption of the virus particle, but also to the removal of proteins, which, although not necessary for virus infectivity, are bona fide structural proteins of released virus particles. We recently used a morphological and immunological approach, the grid cell culture technique (Hyatt *et al.*, 1987), to examine the protein content of BTV particles released from cells and not subject to purification.

The procedure makes use of the fact that cells grown and infected on electron microscopic grids release virus particles that adsorb to the grid substrate. In this location, they can be probed with monoclonal antibodies to BTV proteins using immunogold procedures (Hyatt *et al.*, 1987). The data in Fig. 3 show that released BTV-1 (Australia) virus particles react, not only with anti-VP2 monoclonal antibodies but with monoclonal antibodies to NS1 as well. It was necessary to treat the virus with methanol to enhance the labeling with anti-NS1 antibodies (Eaton *et al.*, 1988). Similar results were obtained with virus particles released from cells infected with South African serotypes BTV-1, -4, and -10. Virus and core particles purified by the method of Mertens *et al.* (1986) also

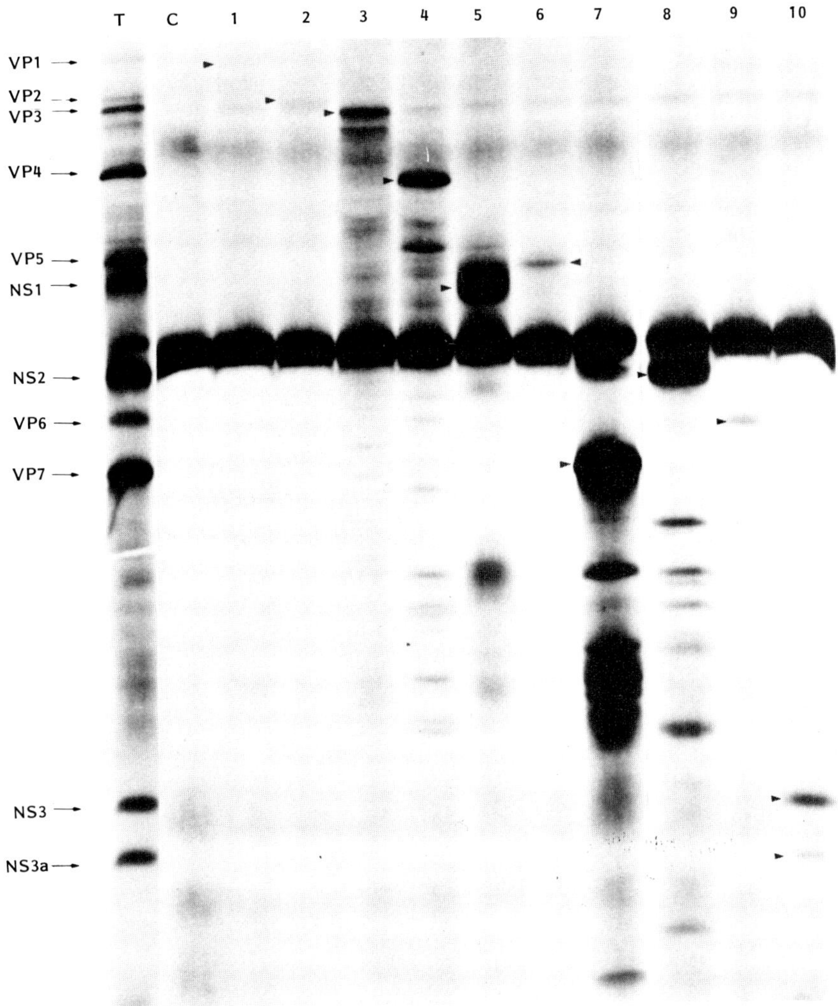


**FIGURE 1.** *In vitro* translation of individual RNA genome segments of BTV-1 (Australia). Purified virion RNA was electrophoresed in polyacrylamide gels. Individual bands were eluted from the gel, denatured in methyl mercuric hydroxide and translated in a rabbit reticulocyte lysate, C, no added RNA; T, total genomic RNA; the genome segment numbers are given on top of each lane.

react strongly with anti-NS1 antibody (Fig. 4). The protein was also detected in purified [$^{35}$S]methionine-labeled virus and core particles following gel electrophoresis (Fig. 2). These results suggest that NS1 may be associated with BTV core particles and detected in the intact virus following distortion of the outer coat of the virus by methanol. Evidence that methanol treatment alters the to-

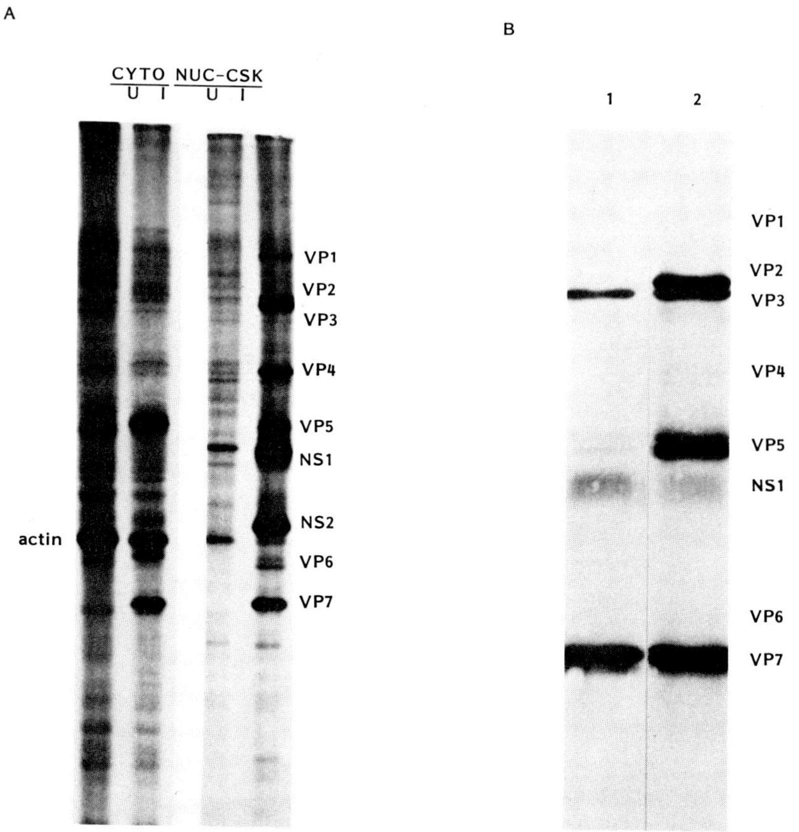


**FIGURE 2.** Polyacrylamide gel analysis (PAGE) of BTV proteins in fractions of infected (I) and uninfected (U) BHK cells and in purified virus and core particles. (A) Cells were labeled with [$^{35}$S]methionine and portions of the cytoplasmic (CYTO) and cytoskeletal (NUC-CSK) fractions analyzed in 12% polyacrylamide gels as described (Eaton *et al.*, 1987). (B) [$^{35}$S]methionine-labeled virus and core particles were purified as described by Mertens *et al.* (1987) and analyzed by PAGE. Proteins were transferred to nitrocellulose and detected by autoradiography. Lane 1, cores; lane 2, virus particles.

pography of the virus outer coat comes from the observation that while intact untreated virus particles react very weakly with gold-labeled antibodies to VP7, methanol treatment markedly enhanced the anti-VP7 labeling (Fig. 5) (Hyatt and Eaton, 1988). As NS1 and VP7 can be detected, albeit weakly, in virus particles not subjected to methanol treatment, this implies that these proteins may extend beyond the core and interact with the outer coat proteins.

Electron microscopic analysis of thin sections of BTV-infected cells has demonstrated juxtanuclear inclusions with both granular and fibrillar charac-

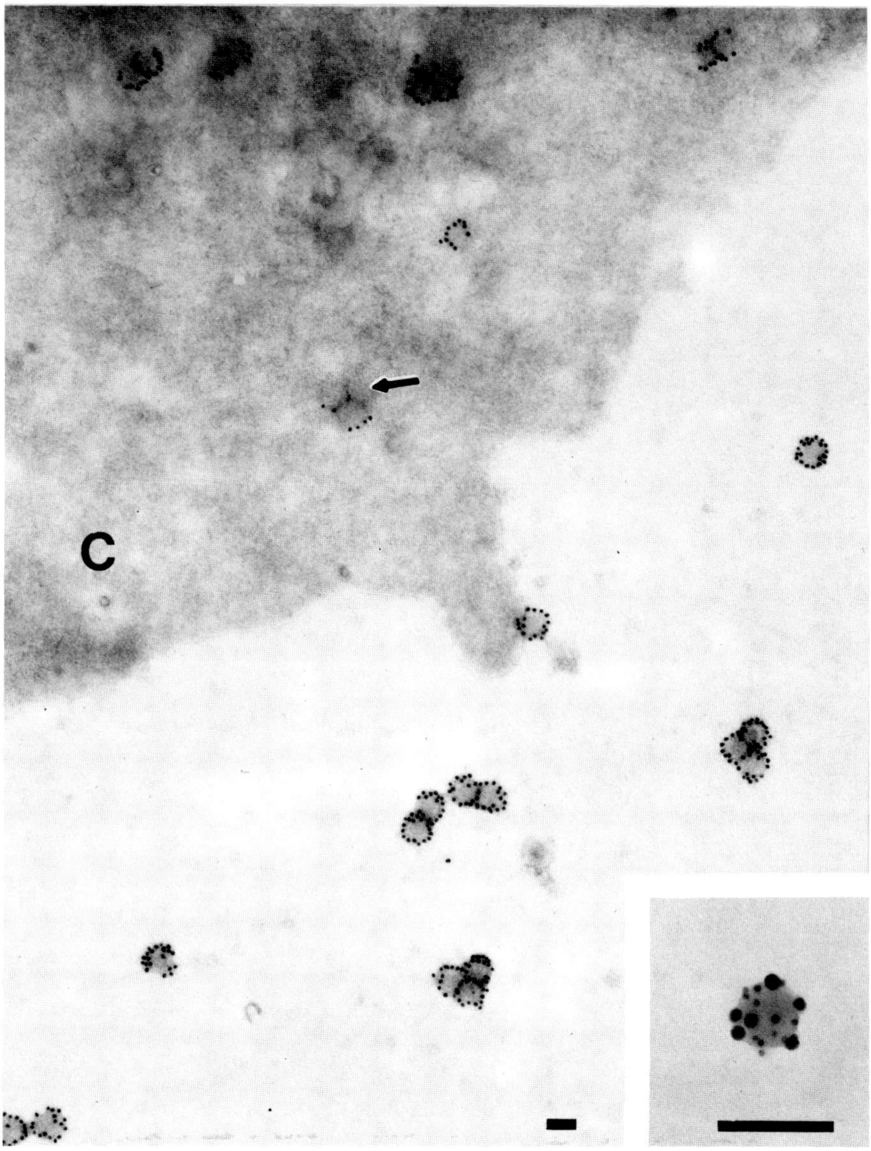


**FIGURE 3.** Immunogold-labeling of virus particles on the surface of and released from BTV-infected cells. Cells were fixed with glutaraldehyde and probed with anti-VP2 monoclonal antibody 9J3/G6 conjugated directly to gold. In the insert, virus particles were probed with gold-labeled 9J3/G6 (6-nm gold particles), treated with methanol and reacted with anti-NS1 monoclonal antibody 20 E6/A4 and protein A conjugated to 14-nm-diameter gold particles. C, cell. The arrow points to BTV particles which underlay the cell membrane and are inaccessible to the anti-VP2 antibody. Bars = 100 nm.

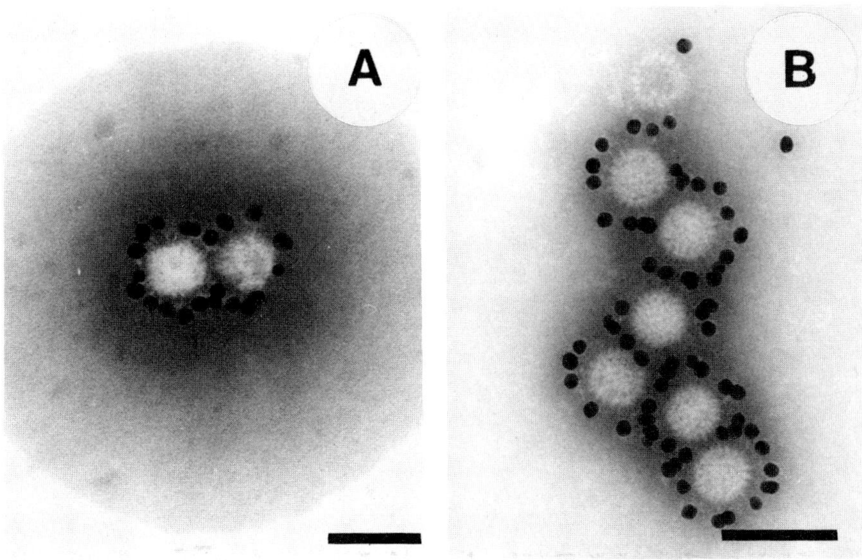


**FIGURE 4.** Presence of NS1 in purified virus and core particles. Virus (A) and core particles (B) were purified as described by Mertens *et al.* (1987), treated with methanol after absorption to the grid and probed with anti-NS1 monoclonal antibody 20E6/A4 and protein A–gold. Bars = 100 nm.

teristics called virus inclusion bodies (VIB), virus-specific tubules, and virus particles (Bowne and Jochim, 1967; Lecatsas, 1968; Cromack *et al.*, 1971) (Fig. 6). Within the VIB are small numbers of electron-dense components whose diameter (58 ± 7 nm) and associated low-density outer ring suggest that they may be virus particles. It is not possible on the basis of these criteria alone to determine whether these structures are complete virions or core particles. It seems clear, however, that VIB are the site of morphogenesis of at least virus core particles.

Tubules are found in the cytoplasm of BTV-infected cells in predominantly peri- or juxtanuclear locations. They have been isolated from infected cells by zonal sedimentation through sucrose and shown to be of varying lengths with a 68-nm diameter and a 9-nm linear periodicity. Huismans and Els (1979) detected no nucleic acid or protein other than NS1 in purified tubule preparations. The function of tubules is unknown. Tubules have also been found in close association with the centrioles of infected cells (Fig. 7). The significance of this observation and its part in generation of the cytopathic effect are unclear.

Small numbers of virus particles are found in BTV-infected cells. In spite of the proposed role of VIB in virus morphogenesis, virus particles are rarely

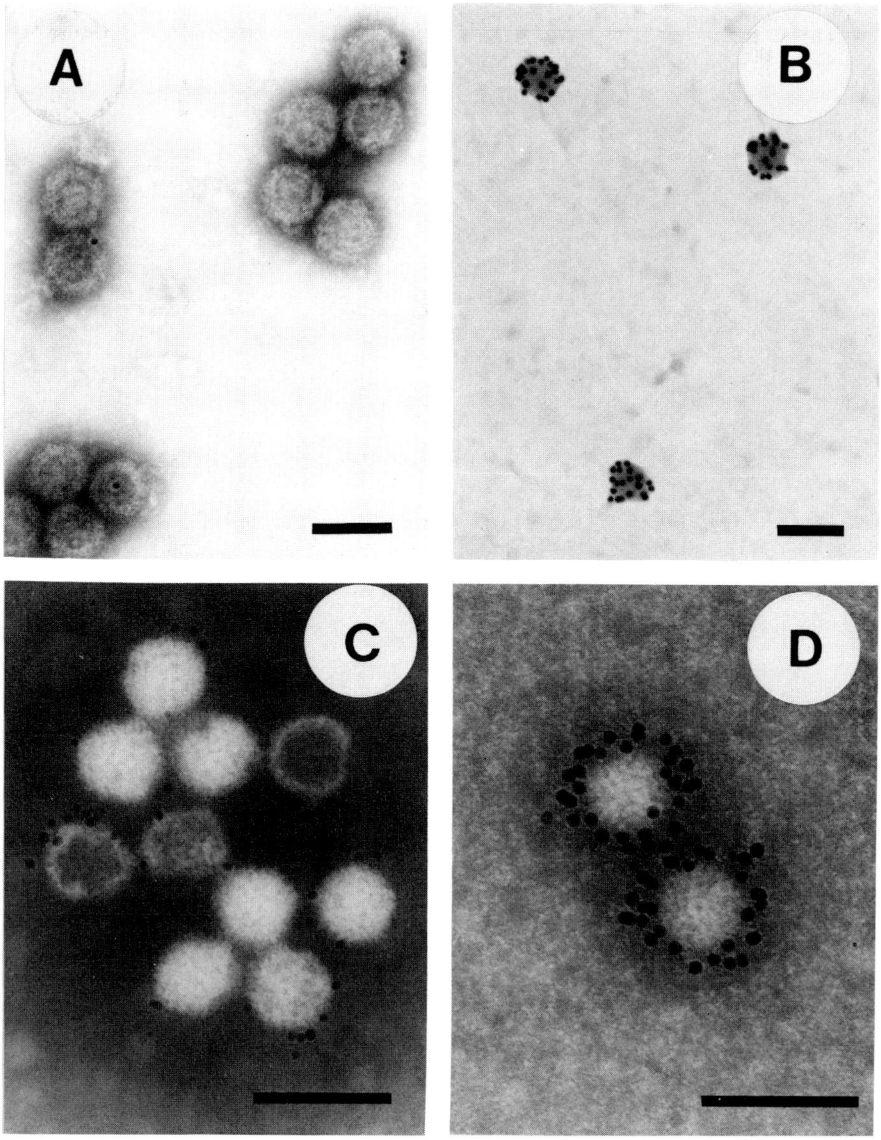


**FIGURE 5.** Presence of VP7 on virus and core particles. Anti-VP7 monoclonal antibody conjugated directly to gold was used to probe (A) virus particles released from cells in the grid cell-culture technique, (B) virus particles treated with methanol, (C) purified virus particles, and (D) purified core particles. Bars = 100 nm.

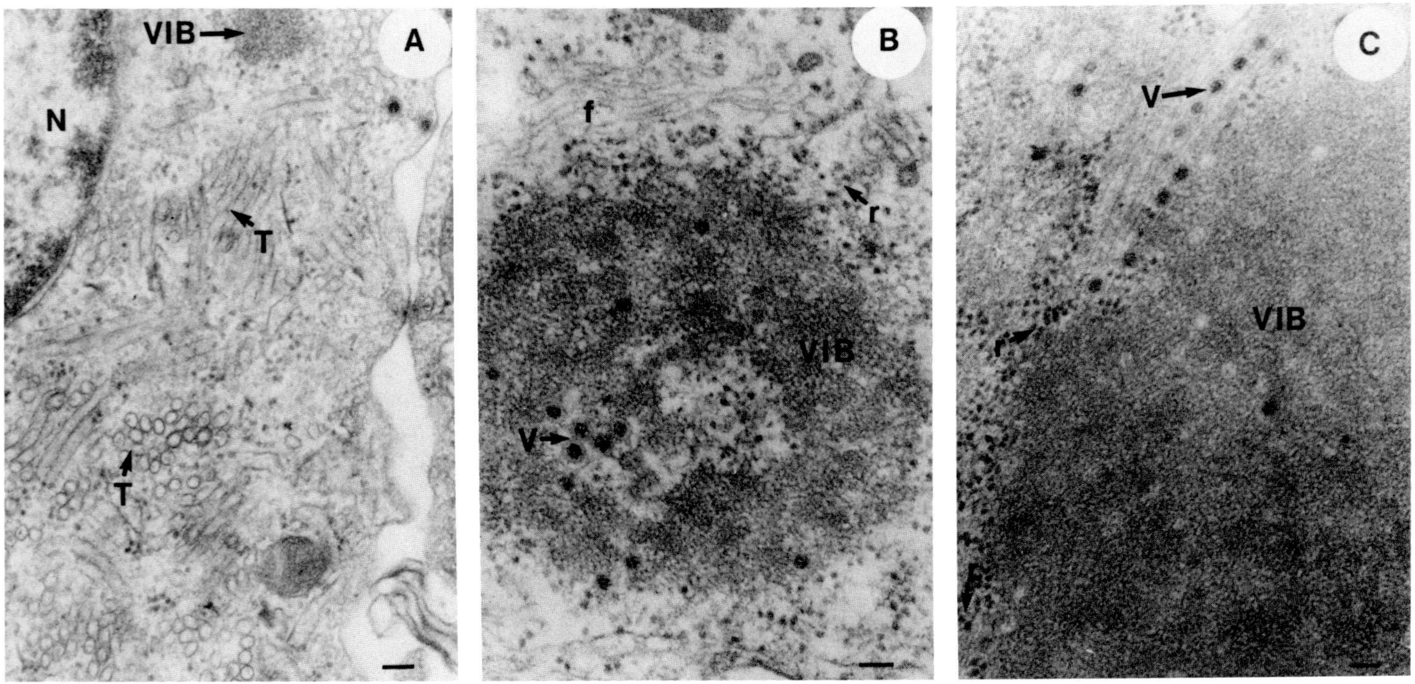


**FIGURE 6.** Tubules, VIB, and virus particles in thin sections of BTV-infected cells 20 hr p.i. N, nucleus; VIB, virus inclusion body; V, virus particles; r, ribosomes; f, filaments; T, tubules. Bars = 100 nm.

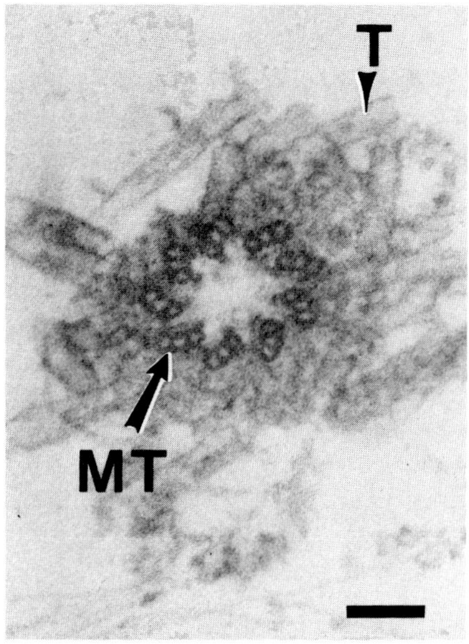


**FIGURE 7.** Association of tubules with centrioles in BTV-infected cells. T, virus tubules; MT, centriolar microtubules. Bar = 100 nm.

observed near VIB. They are most frequently detected in small aggregates distal to VIB and as single or small groups of particles near or at the edge of the cell.

## 2. CYTOSKELETON IN BTV-INFECTED CELLS

### 2.1. Cytoskeleton Organization in BTV-Infected Cells

To determine whether BTV-infection resulted in a reorganization of any of the three major cytoskeletal networks, infected cells were fixed and analyzed by immunofluorescence using monoclonal antibodies to actin, tubulin, and vimentin. The most obvious change in infected cells was the reduced staining of the microtubule-organizing center. Several minor changes were also apparent (data not shown). First, the microtubule network showed some fragmentation near the cell membrane. Second, concomitant with the spreading of infected cells and the acquisition of a flatter appearance, microfilaments were more apparent at the edges of infected cells than in control cells. Only minor changes were apparent in the intermediate filament network of infected cells compared with control cells.

### 2.2. Virus Proteins Associated with the Cell Matrix

Immunofluorescent probing of methanol-fixed BTV-infected cells with polyclonal anti-BTV antiserum indicates that viral antigens are distributed throughout the cell cytoplasm (Fig. 8A). The antigens remain associated with the cell matrix following treatment of infected cells with the nonionic detergent NP-40 (Fig. 8B). These initial studies were done using polyclonal anti-BTV antisera. The subsequent use of monoclonal antibodies to BTV proteins confirmed that at least a proportion of viral proteins VP2, VP3, VP5, VP7, NS1, and NS2 retain a cytoskeletal location following lysis of infected cells with NP-40.

The efficacy with which a monoclonal antibody can react with virus antigen in immunofluorescence tests was found to depend on the nature of the treatment used to fix virus-infected cells and cytoskeletons. The anti-VP2 antibodies, for example, react well following glutaraldehyde but not following methanol fixation. By contrast, anti-NS1 monoclonal antibodies generate detectable fluorescence only after methanol fixation or following methanol treatment of cytoskeletons fixed with glutaraldehyde. Anti-NS2 monoclonal antibodies react well on either glutaraldehyde- or methanol-fixed cells or cytoskeletons. The ability of a monoclonal antibody to react under certain fixation conditions may be due to a number of reasons. For example, glutaraldehyde, which contains two aldehyde groups that react primarily at the nucleophilic -$NH_2$ groups, may crosslink neighboring polypeptide chains (Hayat, 1986). This may alter the presentation of an epitope and render it unrecognizable by the appropriate antibody. Similarly, proteins in close proximity to an antigenic site may be altered in such a way as to mask or occlude the specific epitope. Alternatively, the change in protein conformation may serve to reveal or present an epitope and thereby enhance the antigen–antibody interaction. The mode of methanol fixation differs from that of glutaraldehyde; methanol removes water and thereby denatures protein. We believe that methanol treatment denatures VP2 epitopes in this way. By contrast,

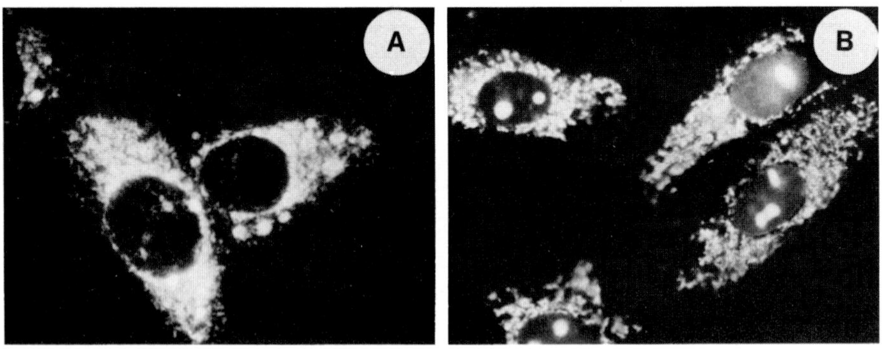


**FIGURE 8.** Immunofluorescent labeling of BTV-infected cells and cytoskeletons. (A) Fixed cells labeled with rabbit anti-BTV antiserum and fluorescein-conjugated goat antirabbit antiserum. (B) Cytoskeletons labeled with mouse polyclonal anti-BTV antiserum, biotinylated antimouse antiserum and fluorescein–streptavidin.

methanol-mediated denaturation of the viral outer coat proteins appears to provide access to proteins such as VP7 and NS1, which appear to lie predominantly underneath VP2 and VP5.

Information about the localization of viral proteins in BTV-infected cells was also obtained by labeling infected cells with [$^{35}$S]methionine for 2 hr prior to dividing the cells into the detergent-soluble cytoplasmic fraction and insoluble nuclear–cytoskeletal fractions. Analysis of the labeled proteins by PAGE (see Fig. 2) showed that most of the structural proteins VP1 and VP7 and nonstructural proteins NS1 and NS2 are found in the cytoskeletal fraction. However, a significant proportion of VP5 and VP7 and small amounts of VP1, VP4, VP6, and NS2 were detected in the cytoplasmic fraction. The data in Fig. 2 were generated using BHK cells. The distribution of virus proteins was similar in other cell types, but the amount of [$^{35}$S]methionine-labeled VP5 in the cytoplasmic fraction was decreased in BTV-infected SVP and Vero cells. Most virus proteins labeled during a 1-hr pulse with [$^{35}$S]methionine remained associated with the nuclear–cytoskeleton fraction even after chase periods of up to 10 hr.

## 2.3. Effect of Cytochalasin B and Colchicine on the Distribution of Virus Proteins in Infected Cells

The role of microfilament and microtubule networks in viral replication was monitored by determining the effect of cytochalasin B and colchicine, at concentrations which disrupt the cytoskeleton, on the titer of released virus and the distribution of [$^{35}$S]methionine-labeled viral proteins in BTV-infected cells. The results in Table I show that addition of 10 μg/ml cytochalasin B from approximately 9 hr p.i. has no discernible effect on the titer of released virus. Colchicine had no affect on the titer of released virus when added to infected cells immediately after viral infection.

**Table I**
**Effect of Cytochalasin B on the Titer of Released BTV**

| Time of cytochalasin B addition (hr p.i.) | Titer of released virus (24 hr p.i.) |
|---|---|
| 3 | $1 \times 10^5$ |
| 5 | $3 \times 10^5$ |
| 9 | $4 \times 10^6$ |
| 12 | $4.6 \times 10^6$ |
| None added | $4.4 \times 10^6$ |

The distribution of viral proteins in BTV-infected cells treated with 10 μg/ml cytochalasin B from 9 hr p.i. was determined immediately following a 2-hr pulse with [$^{35}$S]methionine at 22 hr p.i. Cytochalasin B had no effect on either the amount or the relative distribution of viral proteins between cytoplasmic and cytoskeletal fractions of infected cells (Eaton *et al.*, 1987).

The effect of colchicine on the distribution of virus proteins in BTV-infected cells was also determined by adding the drug to BHK cells immediately after viral infection and labeling control and drug-treated cells with [$^{35}$S]methionine for 2 hr at 20 hr p.i. Examination of the viral proteins in both cytoplasmic and cytoskeletal fractions indicated that colchicine did not alter the proportion or species of viral protein on the cytoskeleton of infected cells.

### 2.4. Localization of VIB, Tubules, and Virus-like Particles on the Cytoskeleton by Whole-Mount Electron Microscopy

Cells grown on electron microscope grids were infected with BTV, and at 20 hr p.i. cytoskeletons were prepared by treatment with 1% NP-40 and 0.1% glutaraldehyde. Cytoskeletons were postfixed in osmium tetroxide, dehydrated in acetone, and critical point-dried from $CO_2$. The detergent-insoluble skeletal network of uninfected cells comprised a complex anastomosing array of filaments of ~4- to 25-nm diameter. Cytoskeletons of BTV and mock-infected cells were characterized by the presence of dense areas in juxtanuclear positions (Fig. 9). Within these areas and in other perinuclear locations were VIB and tubules. Virus-like particles were often found throughout the cytoskeleton. The VIB were dense, approximately spherical structures. There did not appear to be an increase in the number or concentration of cytoskeletal filaments surrounding VIB. Virus tubules were observed in groups of approximately 10–20 and often lying parallel to neighboring filaments (Fig. 9D). Individual tubules appeared hollow, ~30 ± 2 nm wide, 300–400 nm in length. Virus-like particles were identified on the basis of their approximate hexagonal shape and size (55 nm) and found either singly, in rows, or in aggregates (Fig. 10). In addition to being observed both intracellularly and in association with the intracellular matrix, BTV-like particles can be observed at the cell surface or leaving the cell (Fig. 11).

### 2.5. Effect of Cytochalasin B and Colchicine on the Distribution of Virus-Specific Structures in Infected Cells

In an attempt to ascertain the sites of binding of virus-like particles, tubules, and VIB to the cytoskeleton, we have used monoclonal antibodies that detect these structures to probe skeletons of virus-infected cells following treatment with colchicine or cytochalasin B. In a series of preliminary experiments, we sought monoclonal antibodies that would react specifically with the three cytoskeleton-

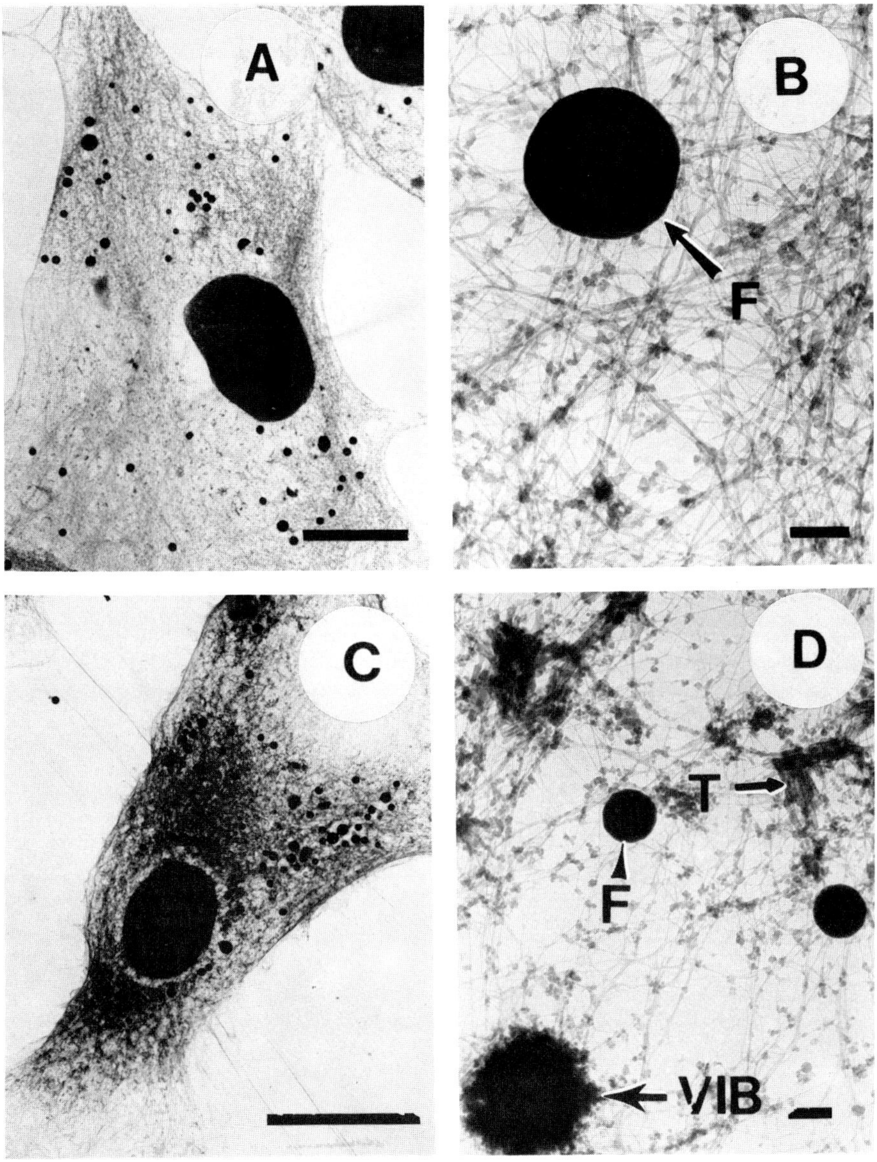


**FIGURE 9.** Cytoskeletons of uninfected and BTV-infected cells. Uninfected (A,B) and BTV-infected SVP cells 20 hr p.i. (C,D). F, fat droplet; T, tubules; VIB, virus inclusion body. Bars = 10 μm in (A,C) and 100 nm in (B,D).

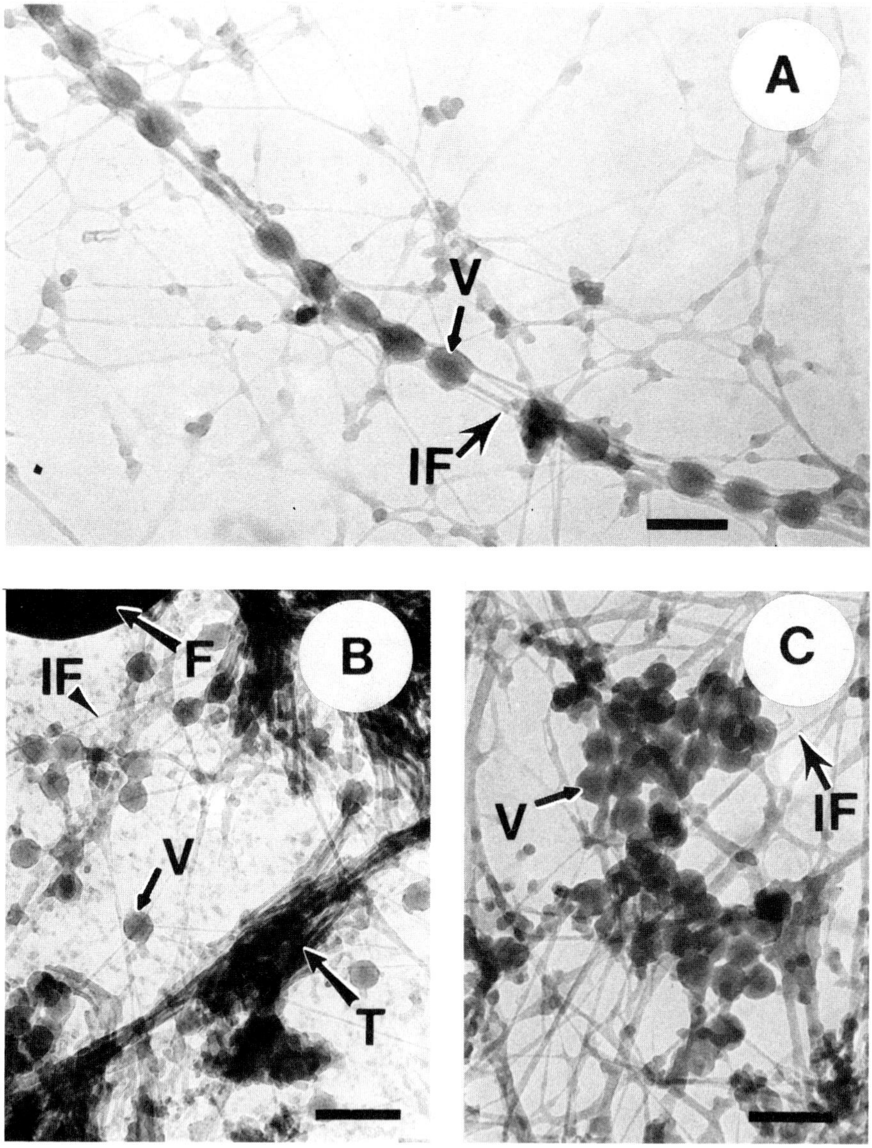


**FIGURE 10.** Cytoskeleton-associated virus-like particles in BTV-infected cells. V, virus particle; IF, intermediate filament; F, fat droplet; T, tubules. Bars = 100 nm.

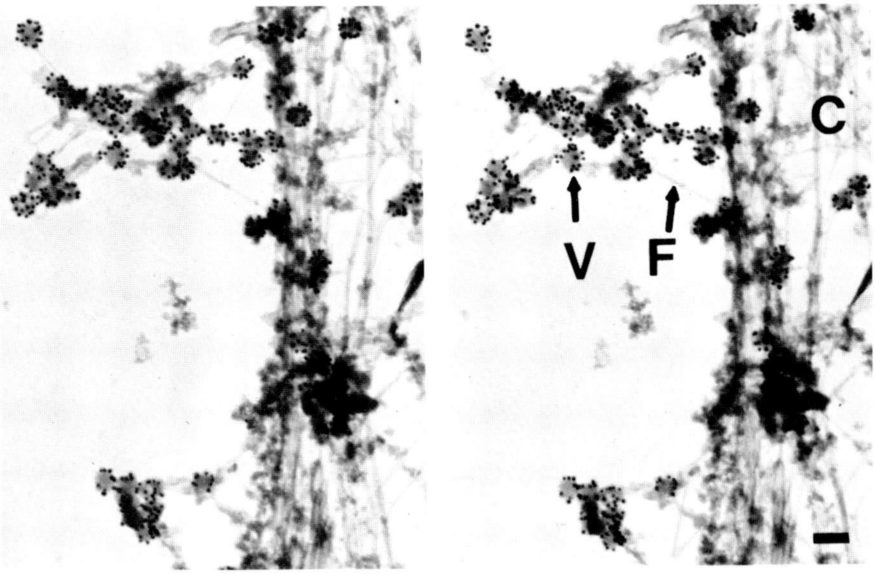


**FIGURE 11.** Release of virus particles from infected cells. Cytoskeletons of BTV-infected cells were probed with gold-labeled anti-VP2 monoclonal antibody 9J3/G6 and carbon coated. Stereomicrograph showing virus particles outside the cell but attached to the cortical layer underlying the cell membrane. V, virus particle; F, filament; C, cytoskeleton. Bar = 100 nm.

associated virus structures. The data depicted in Figs. 11 and 12 demonstrate the ability of antibodies 9J3/G6, 20H2, and 7C11/A1 to react with virus particles, VIB, and tubules, respectively. Antibody 9J3/G6 neutralizes the virus and precipitates VP2 from lysates of infected cells (White *et al.*, 1985). The results shown in Fig. 13 show the effect of cytochalasin B and colchicine on the distribution of virus particles, tubules and VIB in skeletons of infected cells. Although both drugs altered the distribution of fluorescence, they did not eliminate it from the skeletons. Thus, the binding of VIB, virus-like particles, and tubules to the skeleton is not dependent on intact microtubules or microfilaments. The distribution of fluorescence is more perinuclear in the drug-treated cells and in addition, cytochalasin B appears to remove anti-VP2-reactive material from the edge of the cell. Virus-like particles accumulate not only intracellularly on the cytoskeleton but also on the surface of virus-infected cells (see Fig. 11). The accumulation in this location but not elsewhere in the cell appears to be dependent on intact microfilaments.

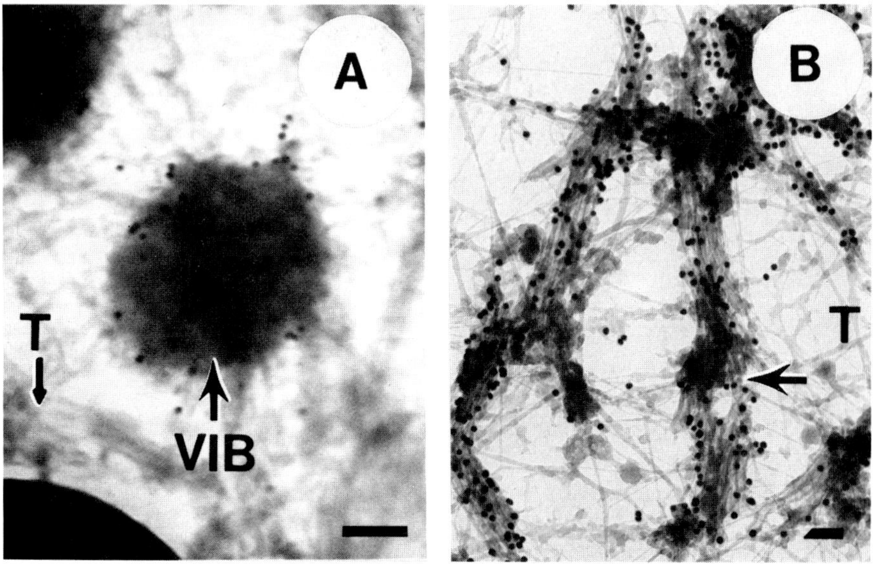


**FIGURE 12.** Immunogold labeling of VIB and tubules in cytoskeletons of BTV-infected cells. (A) Scanning transmission micrograph of VIB reacted with anti-NS2 monoclonal antibody 31G9/C2 (antibody 20H2 gives an identical result). (B) Anti-VP3 monoclonal antibody 7C11/A1. T, tubule; VIB, virus inclusion body. Bars = 100 nm.

## 2.6. Viruses, Tubules, and VIB Binding to Intermediate Filaments

An association of virus particles with filaments was observed both in thin sections and in cytoskeletal arrays of virus-infected cells (Figs. 6 and 10). The diameter of the filaments in thin section was estimated to be ~10 nm, similar to that of intermediate filaments. These data and the fact that virus, VIB, and tubules on the cytoskeleton resisted release by cytochalasin B and colchicine suggested that intermediate filaments may be the source of binding to the cytoskeletal framework. This was supported by the following observations.

Large numbers of intermediate filaments aggregate next to the nucleus after treatment of cells for prolonged periods with antimitotic drugs such as colchicine and vinblastine (Blose and Chacko, 1976; Goldman, 1971). The micrograph in Fig. 14A shows the colchicine-induced collapse of intermediate filaments around the nucleus of an uninfected SVP cell. Immunofluorescent analysis of such cells with an antivimentin monoclonal antibody confirmed the juxtanuclear location of vimentin. To determine whether virus particles, VIB, and tubules were located

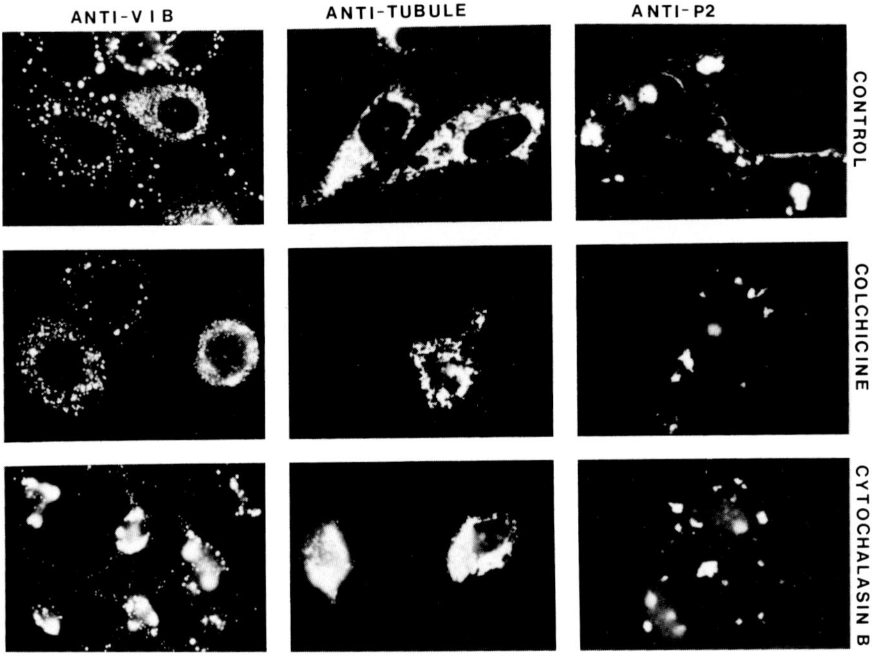


**FIGURE 13.** Fluorescent antibody analyses of virus particle, tubule, and VIB distribution in cytoskeletons of untreated control, colchicine-, and cytochalasin B-treated BTV-infected cells. Cytoskeletons were probed with monoclonal antibodies that react with VP2 in virus particles (9J3/G6), VP3 in tubules (7C11/A1), and NS2 in VIB (20H2). Antibodies were located using biotinylated antimouse antiserum and fluorescein–streptavidin.

within the juxtanuclear intermediate filament assays, SVP cells on electron microscope grids were treated with colchicine immediately after BTV infection and cytoskeletons prepared for whole-mount electron microscopy at 22 hr p.i. The results in Figs. 14B,C and 15 show that VIB, tubules, and virus particles were present both within and at the outer extremities of the intermediate filament aggregates. In particular, virus particles were associated in long trains with filaments that appeared to be part of the intermediate filament array.

Further evidence for the involvement of vimentin-based intermediate filaments in the binding of virus particles to the cytoskeleton was obtained by probing skeletons of virus-infected cells with antivimentin antibody and localizing the latter with biotinylated antimouse antibodies and avidin–gold. The data presented in Fig. 16 confirm that virus particles are associated with vimentin-rich cytoskeletal filaments.

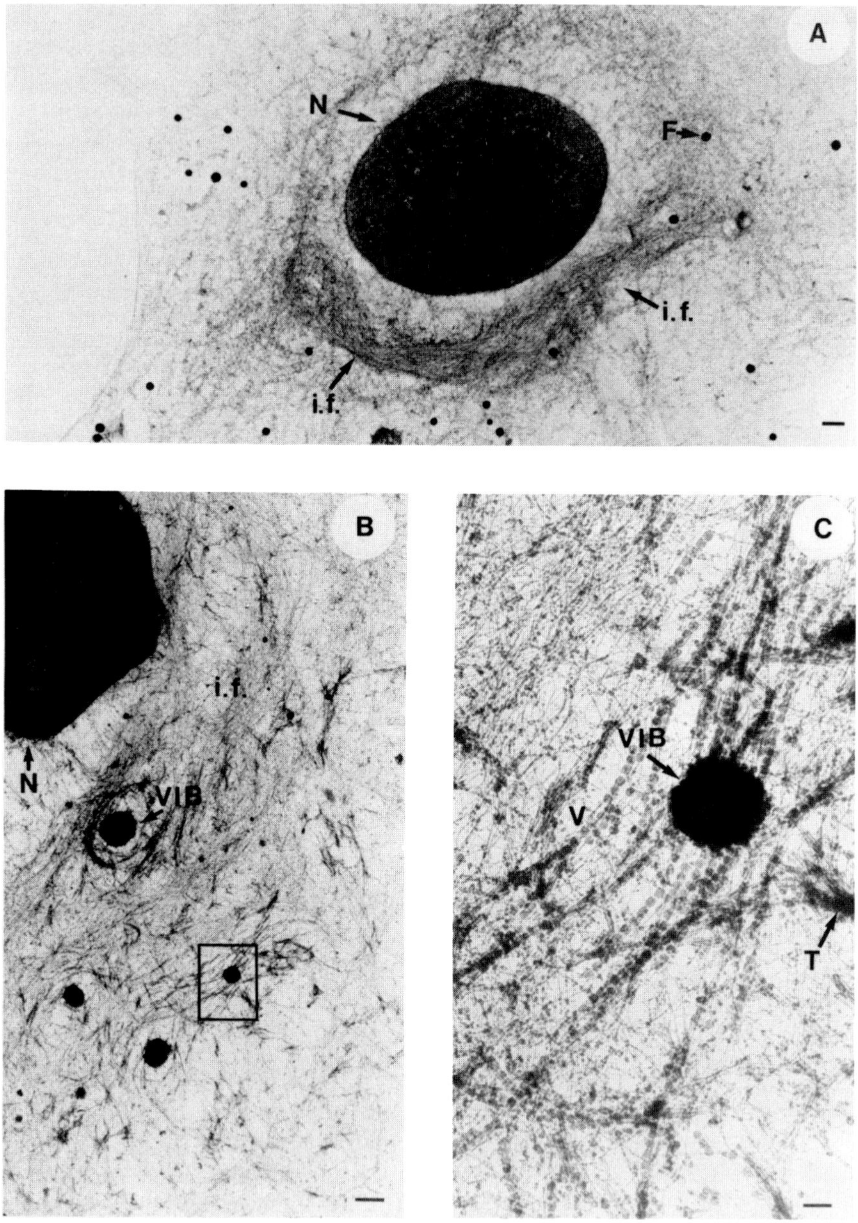


**FIGURE 14.** Effect of colchicine on the cytoskeletal network of uninfected (A) and BTV-infected (B,C) cells. Colchicine was added after mock or virus infection and cytoskeletons prepared at 22 hr p.i. N, nucleus; F, fat droplet; VIB, virus inclusion body; i.f., intermediate filament. The rectangle in (B) is enlarged in C. The VIB indicated by the arrow in (B) is enlarged in Fig. 15. Bars = 1 μm in (A) and (B) and 200 nm in (C).

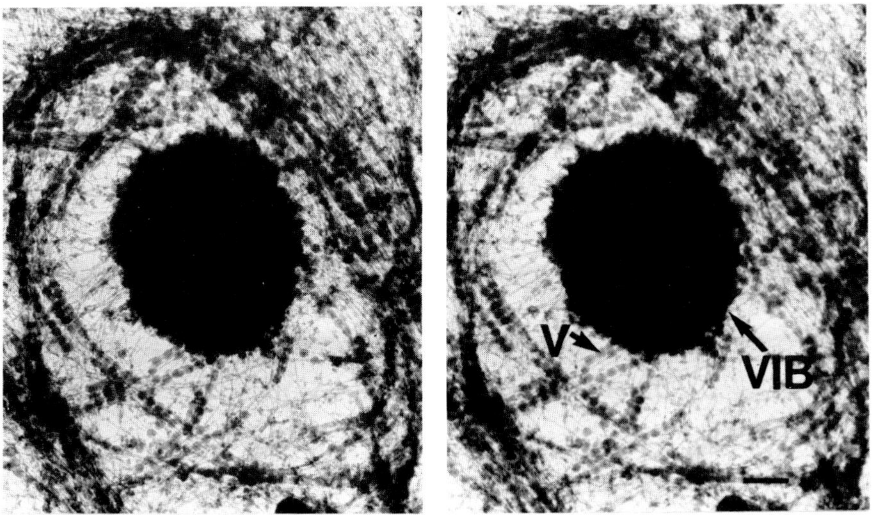


**FIGURE 15.** Stereomicrograph illustrating the distribution of virus-like particles around VIB in BTV-infected cells treated with colchicine. The position of this VIB in the infected cell is indicated by the arrow in Fig. 14. V, virus particle; VIB, virus-inclusion body. Bar = 100 nm.

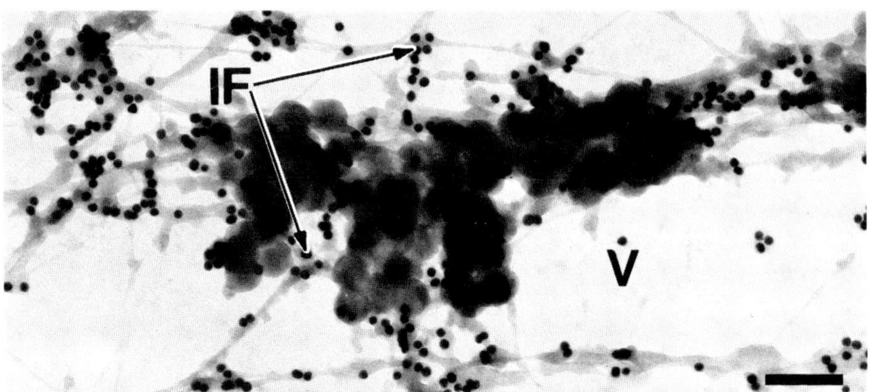


**FIGURE 16.** Localization of vimentin in cytoskeletons from BTV-infected cells. Cytoskeletons were probed with antivimentin antibody, biotinylated antimouse antiserum and avidin-gold. V, virus particles; IF, intermediate filaments. Bar = 100 nm.

## 3. ULTRASTRUCTURAL LOCALIZATION OF VIRUS PROTEINS IN CYTOSKELETON-ASSOCIATED STRUCTURES

We have used a variety of anti-BTV monoclonal antibodies to locate nonstructural and structural virus proteins in both cytoskeleton-associated virus-specific entities and released virus particles. Knowledge of the protein content of VIB, tubules, and cytoskeleton-associated virus-like particles will help clarify the role of these structures in replication and virus morphogenesis. In addition, information about the immunological reactivity of the outer coat of the virus may lead to a better understanding of the antigenic nature and variation of BTV.

Monoclonal antibodies were screened for potential use in ultrastructural studies and appropriate fixation conditions for cytoskeletons determined in preliminary immunofluorescence procedures. We have found that immunological probing of cytoskeletons from infected cells on electron microscope grids or on tissue culture flasks generally requires that the cytoskeletons be fixed with such agents as glutaraldehyde or paraformaldehyde. We routinely use 0.1% glutaraldehyde. In the absence of fixation, the skeletons fail to maintain their integrity and often come off the grid during the monoclonal antibody incubation or washing steps. In contrast to the nuclear–cytoskeleton matrix, virus particles released from cells during infection and that adsorb to the grid substrate remain there during incubation and washing steps in the absence of any fixation procedure.

Three approaches to immunogold labeling have been used. First, colloidal gold has been conjugated directly to purified monoclonal antibodies, as described by Slot and Geuze (1985), and the labeled antibody used in a one-step procedure. Second, in a two-step protocol, gold-labeled protein A has been used to locate monoclonal antibodies bound to their respective cytoskeleton-associated antigen. Third, virus antigens may also be located in a three-step procedure using monoclonal antibody, biotinylated antimouse antibody, and streptavidin or avidin–gold. Whole cells or cytoskeletons labeled by any of these procedures may be viewed in the electron microscope directly, following negative staining, critical point drying, or after embedment and thin sectioning.

### 3.1. Virus Tubules

Huismans and Els (1979) have purified tubules from BTV-10 infected cells and shown them to consist predominantly, if not entirely, of NS1. Probing of glutaraldehyde-fixed methanol-treated cytoskeletons from BTV-infected cells with six of seven anti-NS1 monoclonal antibodies confirms that tubules contain NS1. Four of the monoclonal antibodies reacted only with tubules (Fig. 17A), whereas the other two also bound to virus particles (Fig. 17B). One anti-NS1 monoclonal antibody reacted only with virus particles (Fig. 17C). The specificity of anti-NS1 monoclonal antibodies was confirmed by showing their ability to react

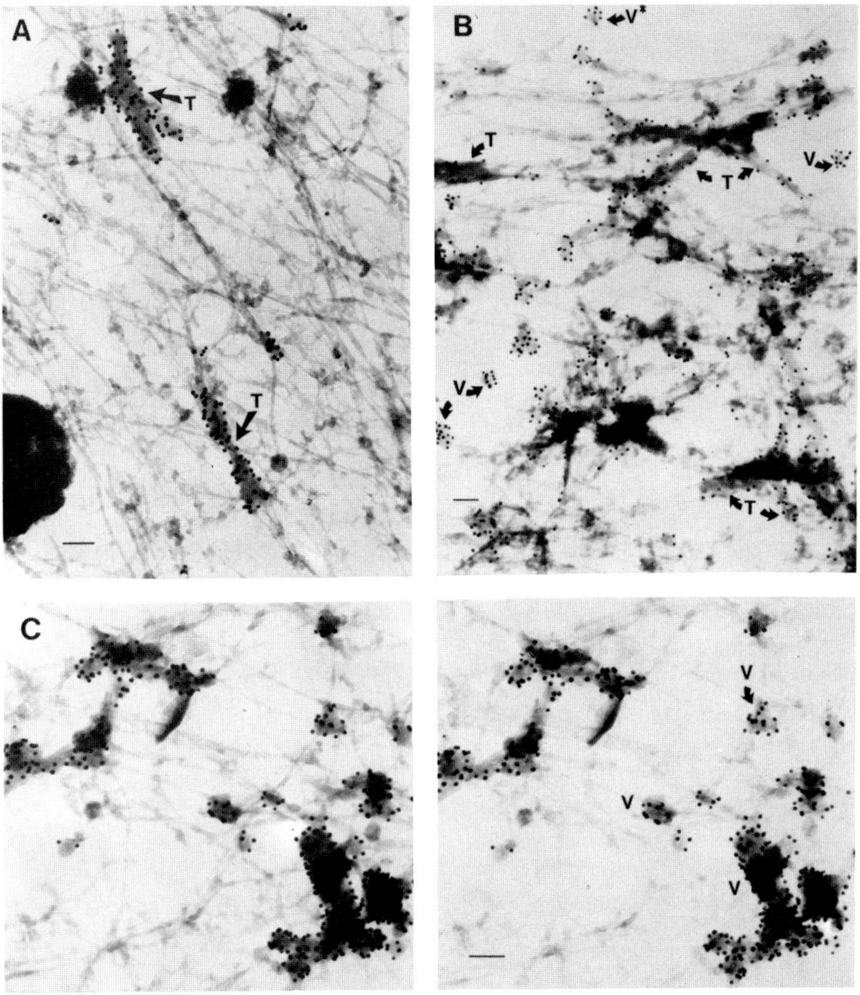


**FIGURE 17.** Immunological analysis of cytoskeletons of BTV-infected cells using monoclonal antibodies to NS1. Cytoskeletons were prepared using 1% NP-40 and 0.1% glutaraldehyde, treated with methanol, and probed with anti-NS1 monoclonal antibodies (A) 20D5, (B) 20C1/A7, and (C) 20E6/A4 (stereomicrograph). Antibodies were located with protein A–gold. T, tubule; V, virus-like particle; V*, virus particle leaving the cell. Bars = 100 nm.

in Western blots with protease-derived fragments of gel-purified NS1 (Eaton *et al.*, 1988). NS1 in tubules appears to express two epitopes unique to tubules because three of the four tubule-specific antibodies react with tubules synthesized in cells infected with nineteen different BTV serotypes (the other five serotypes are not available to us), whereas the other reacts only with tubules in BTV-1-infected cells.

Probing cytoskeletons of BTV-infected cells with monoclonal antibodies to structural proteins has shown that tubules also contain, or are associated with, VP3 (Fig. 12) and VP7 (Fig. 18). The fact that tubules isolated from infected cells and purified by sucrose-gradient centrifugation (Huismans and Els, 1979) contain no proteins, other than NS1, in detectable amounts suggests either that only small quantities of structural proteins are present or that during purification the structural proteins are removed. We have no explanation for the presence of VP3 and VP7 on tubules.

## 3.2. VIB

The presence of dense virus-like structures in VIB has led to the belief that it is within these filamentous structures that virus morphogenesis occurs. Immunofluorescent studies showed that cytoskeleton-associated VIB (identified as approximately circular dense structures under phase contrast) bound antibodies to both VP3 and VP7. Indication that these proteins were present not only at the

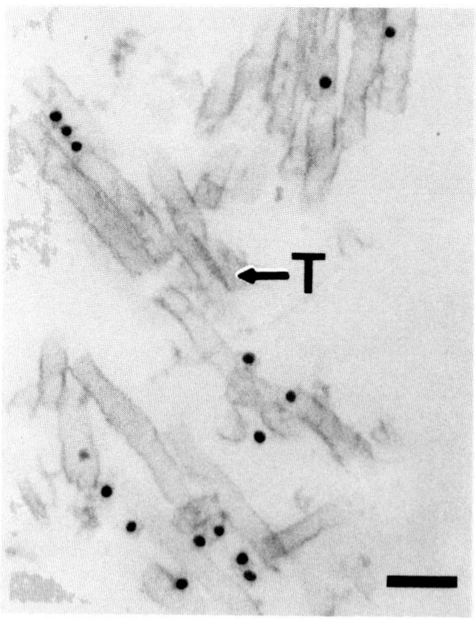


**FIGURE 18.** Presence of VP7 in tubules. Cytoskeletons of BTV-infected cells were probed with gold-labeled anti-VP7 monoclonal antibody, embedded and thin sections examined in the electron microscope. T, tubules. Bars = 100 nm.

periphery but also within the VIB matrix came from studies in which Lowicryl sections of whole cells were probed with anti-VP3 and anti-VP7 monoclonal antibodies. The gold probes associated with anti-VP7 (Fig. 19) did not appear to be associated with the dense virus-like structures in VIB. This finding suggests that VP7, in addition to its known presence within virus cores, constitutes part of the VIB matrix. Failure to label precursor virus particles in VIB with anti-VP7 antibodies may indicate that a large proportion of VP7 in VIB is present in the matrix. Lowicryl sections of VIB labeled very weakly (one or two gold particles per VIB) with anti-VP3 antibody. It seems unlikely that this protein is located solely on the periphery of VIB as indicated by immunofluorescence experiments. Thus, the low but specific level of anti-VP3 labeling points to the presence of VP3 within VIB. The occurrence of the major BTV core proteins in VIB is not surprising. What is interesting, however, is the failure of gold-labeled antibodies to VP2 to label the internal VIB matrix (data not shown). In addition, the immunofluorescent probing of cytoskeletons from infected cells shows that the phase dense VIB label very weakly with anti VP2 monoclonal antibodies. The labeling, if it is present, is limited to one or two pin-point fluorescent foci (Gould *et al.*, 1988).

The nature of interaction of VP2 with VIB was investigated by probing cytoskeletons with an anti-VP2 antibody conjugated directly to gold. Following embedment, thin sections were cut and examined under the electron microscope. The results in Fig. 20A show that anti-VP2 antibody reacted with VIB only in locations in which virus-like particles appeared to be leaving the VIB. This suggests that VP2 may be added to developing virus particles, perhaps cores or corelike particles, at the periphery of the VIB. However, virus particles outside the cell and cytoskeleton-associated virus-like particles distal to VIB (Fig. 20B) are labeled more intensely than the particles at the VIB periphery. There are two explanations for this. First, VP2 may be present in VIB in a form undetectable by the anti-VP2 antibodies used to date. The increase in the reactivity of cytoskeleton-associated virus-like particles with gold-labeled anti-VP2 antibody may reflect a reorganization in the outer coat of the virus, after release of the virus from VIB, with a subsequent increase in accessibility of the appropriate antigen on VP2. Alternatively, although some VP2 is added to corelike particles at the VIB periphery, the remainder may be added after the virus-like particle leaves the VIB either in the cytosol or following binding to the cytoskeleton.

Analysis of cells infected with a neutralization-resistant variant of BTV-1, V35B2, provides evidence in support of the suggestion that VP2 is added to corelike particles at the periphery of and outside VIB (Gould *et al.*, 1988). In cells infected with this variant, VIB appear surrounded by particles whose size and morphology are consistent with their identification as core particles. As expected, these particles label very weakly with anti-VP2 antibodies (Fig. 21a), whereas the virus-like particles elsewhere on the cytoskeleton and those released

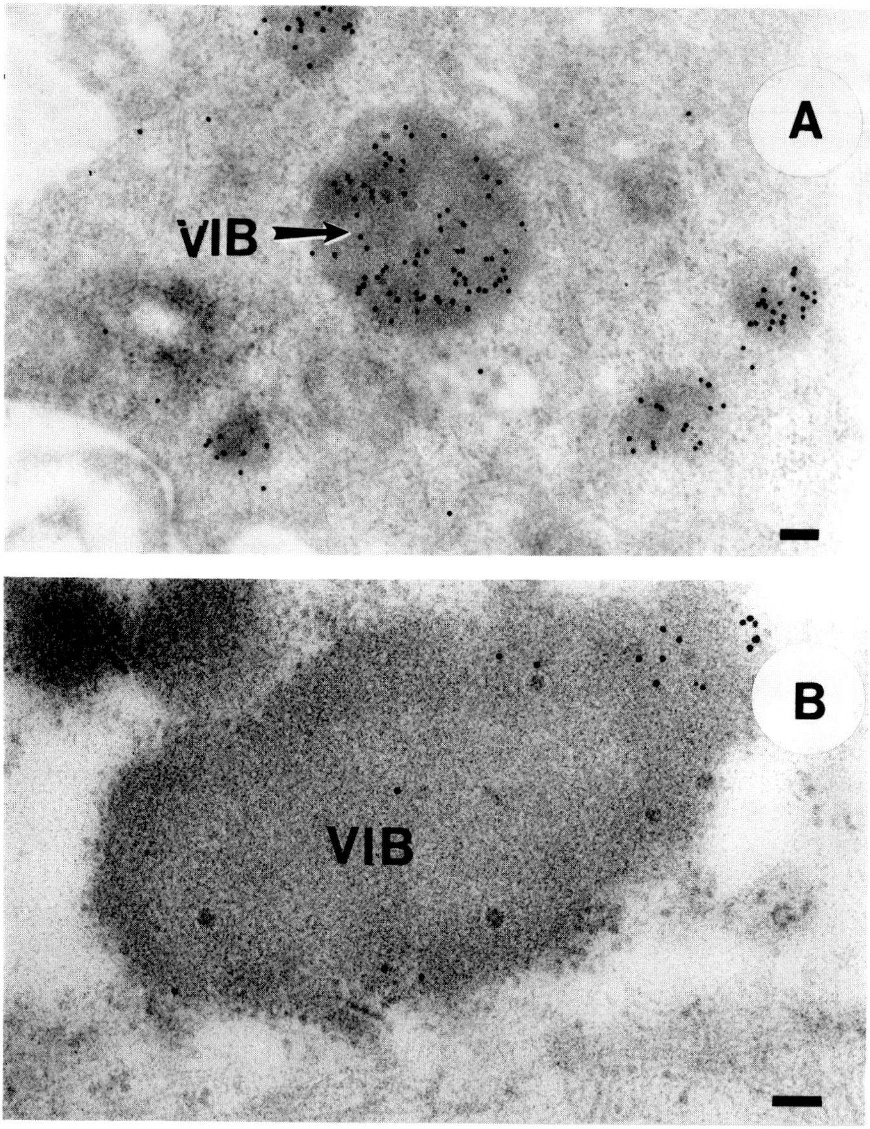


**FIGURE 19.** Presence of NS2 and VP7 within virus-inclusion bodies. Lowicryl sections of BTV-infected cells were probed with gold-labeled monoclonal antibodies to (A) NS2 and (B) VP7. Bars = 100 nm.

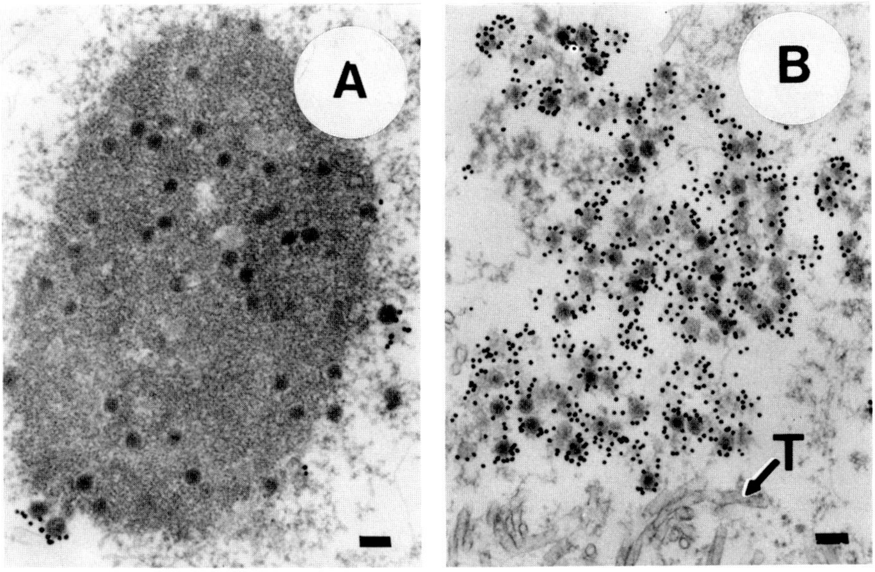


**FIGURE 20.** Localization of VP2 on VIB and cytoskeleton-associated virus-like particles in BTV-infected cells. Cytoskeletons were probed with gold labeled anti-VP2 monoclonal antibody 9J3/G6, embedded and thin sections examined in the electron microscope (A) virus-inclusion body, (B) aggregates of virus-like particles distal to virus-inclusion body. Bars = 100 nm.

from the cell label strongly with the anti-VP2 monoclonal antibody (Fig. 21B). Recent studies indicated that the difference in anti-VP2 labeling patterns is not due to an inability of the gold-labeled antibody to diffuse efficiently into the cytoskeleton and reach the VIB. Thus, the differences in number of gold probes reflect either the amount of VP2 or accessibility of the epitope on the virus-like particles.

Monoclonal antibodies to NS2 react with VIB in immunofluorescence and immunogold- (Fig. 12) labeling procedures. Labeling of the internal matrix of VIB was also achieved with anti-NS2 monoclonals (Fig. 19). NS2 is phosphorylated and has been shown to bind to single-stranded RNA (Huismans *et al.*, 1987b). Its presence within VIB is consistent with a role for this protein in RNA synthesis or sequestration of viral single-stranded RNA molecules prior to virus morphogenesis. Unlike monoclonal antibodies to NS2, anti-NS1 antibodies fail to react with the matrix of VIB in thin sections. Two anti-NS1 antibodies do react, however, with VIB in immunofluorescence tests. In an effort to delineate the nature of NS1 associated with VIB, cytoskeletons of infected cells were incubated sequentially with anti-NS1 antibody 20E6/A4 and protein A–gold. Following embedment and thin sectioning, VIB were found to be nonuniformly

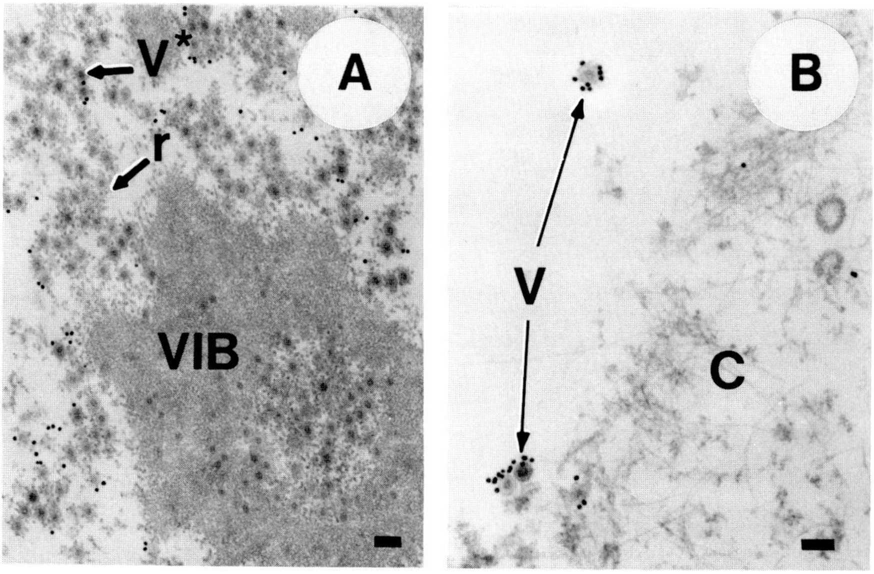


**FIGURE 21.** Aggregation of corelike particles around VIB in cells infected with BTV variant V35B2. Cytoskeletons of infected cells were probed with gold-labeled anti-VP2 monoclonal antibody 9J3/G6, embedded and thin sections examined in the electron microscope. (A) Virus-inclusion body, (B) virus particles at the edge of the cell. V, virus; r, ribosome; C, cell; V*, corelike particle. Bars = 100 nm.

labeled with gold (Fig. 22). The label appeared to be localized around and in the immediate vicinity of virus particles leaving VIB. Importantly, gold was observed in association with fibrillar arrays around virus particles observed at the edge of the VIB. There are two possible sources for the NS1-containing material at the VIB periphery. First, NS1 may be present in VIB in an immunologically undetectable form and accompany the virus core particles that mature from there. Second, NS1 may be added to the developing virus particles at the VIB periphery. This would account for the failure to detect NS1 inside VIB and its presence at the VIB periphery in association with virus particles.

Some gold-labeled anti-NS1 monoclonal antibodies react with purified BTV core particles (see Fig. 4). This is evidence for an association which may have its beginnings at the VIB periphery. The localization of anti-VP2 and anti-NS1 monoclonal antibodies at the periphery of VIB, where virus particles appear to be leaving is indirect evidence for the role of the nonstructural protein in morphogenesis.

The presence on the cytoskeleton of virus particles that lack the fibrillar material raises the possibility that following release of the virus–fibril complex

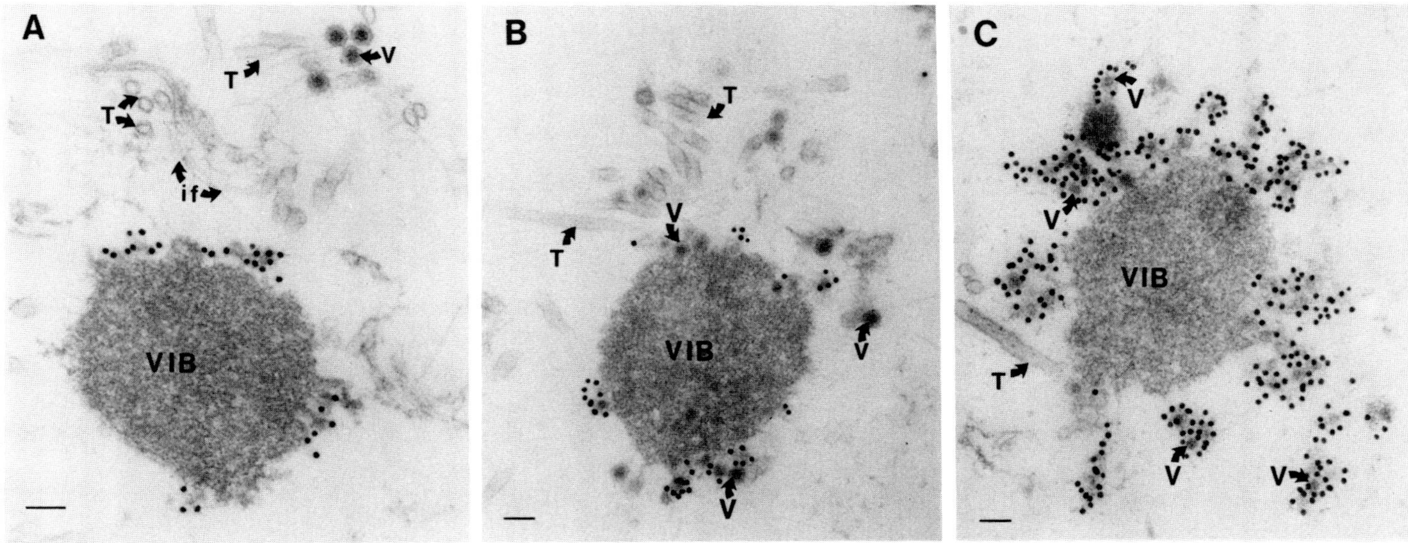


**FIGURE 22.** Thin sections of cytoskeletons probed with anti-NS1 monoclonal antibody 20E6/A4 and protein A-gold. T, tubule; V, virus particle; VIB, virus-inclusion body; IF, intermediate filaments. Bars = 100 nm.

from the VIB, a late step in morphogenesis occurs, following which the fibrillar material may be removed and the virus released from the cytoskeleton into the cytosol. This raises the question of what happens to the fibrillar material following such an event. It appears unlikely that the material is broken down into constituent soluble proteins because little, if any, soluble NS1 is detected in the cytoplasmic fraction of infected cells. An attractive possibility is that the NS1-rich fibrillar material condenses to form tubules. In this model, tubules are therefore the repository of NS1 that has been used in a prior stage of virus morphogenesis.

### 3.3. Virus-like Particles

Virus and virus-like particles in BTV-infected cells are identified in thin section on the basis of their diameter (~65 nm) and structure that encompasses a dense nucleoid (approximately 40 nm diameter) surrounded by a ring of electron translucent material. In whole mount electron microscopy, virus-like particles have a characteristic hexagonal appearance and a diameter of ~55 nm. Particles with these criteria and the ability to react with monoclonal antibodies to VP2 are observed in several locations in BTV-infected cells. In addition to those released from the cell and adsorbed to the electron microscope grid, virus-like particles are associated with the intracellular cytoskeleton, the cortical layer of the cytoplasmic matrix underlying the cell membrane; they also appear to be bound to the surface of the infected cell (see Fig. 23).

In an attempt to determine the relationship between cytoskeleton-associated virus-like particles and released viruses, we have compared and quantitated their ability to react with a monoclonal antibody to VP2 conjugated directly to gold. The number of gold particles found on extracellular and cytoskeleton-associated virus particles was determined. The results indicated that cytoskeleton-associated virus-like particles not only contained more gold probes than released virus particles but displayed a more heterogeneous distribution of probes. There are two possible explanations for this result. First, cytoskeleton-associated virus particles contain more VP2 than do released viruses; this extra VP2 is removed during morphogenesis. Alternatively, there is no difference in the amount of VP2 present on cytoskeleton-associated and released viruses, but the latter have undergone a rearrangement that obscures some of the reactive epitopes on VP2. We cannot distinguish between these two possibilities.

In attempting to understand the relationship between intracellular and released virus particles, it must be remembered that viruses and virus-like particles are also found in the cytoplasmic fraction of infected cells. The evidence for this comes from two sources. First, the cytoplasmic fraction contains high levels of infectious virus. Second, the cytoplasmic (i.e., the nonionic detergent-soluble) fraction of [$^{35}$S]methionine-labeled BTV-infected cells contains particles com-

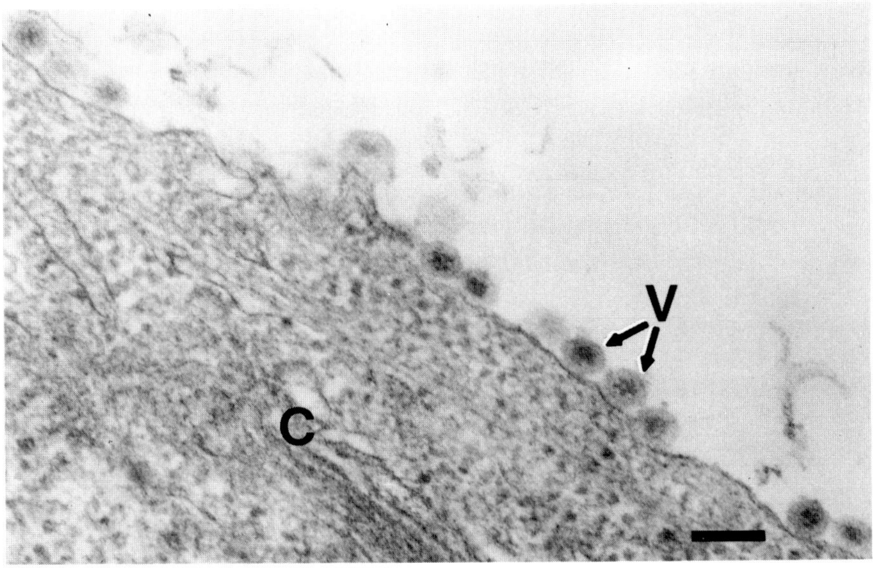


**FIGURE 23.** Virus particles on the surface of infected cells. BTV-infected cells were embedded and thin sections examined in the electron microscope. V, virus particle. Bar = 100 nm.

posed of the seven BTV structural proteins and varying amounts of NS1 and that sediment heterogeneously in sucrose gradients (data not shown). We recently devised a novel way to probe these cytoplasmic particles with anti-VP2 monoclonal antibodies under conditions identical to those used for the immunological analysis of cytoskeleton-associated and released virus particles. The procedure makes use of the observation that following lysis of infected cells with NP-40, a proportion of the released cytoplasmic particles bind to the grid substrate in close proximity to the cell. The previously released virus particles can be differentiated from the recently arrived cytoplasmic ones using two anti-VP2 monoclonal antibody preparations each conjugated with different-sized gold probes. The steps used are as follows: (1) intact but unfixed BTV-infected cells and extracellular viruses are probed with an anti-VP2 antibody conjugated to a 6-nm gold probe, (2) cells are then lysed with NP-40 and the cytoskeletons fixed with glutaraldehyde, and (3) the newly released cytoplasmic particles that have adsorbed to the grid and are unlabeled are probed with the same monoclonal antibody conjugated to a larger gold probe. Preliminary results indicate that cytoplasmic virus-like particles resemble their cytoskeleton-associated counterparts and are heterogeneous in their capacity to bind anti-VP2 monoclonal antibody. The relationship of the cytoplasmic and cytoskeleton-associated viruses to each other and their role in BTV morphogenesis is currently being investigated.

### 3.3.1. Virus Particles on the Cell Surface

Probing of glutaraldehyde-fixed BTV-infected cells in immunofluorescent procedures reveals the presence of VP2 on the cell surface (data not shown). Several findings indicate that the protein is probably present there as part of virus particles. First, immunogold probing shows the gold-labeled anti-VP2 antibody on the cell surface to be orientated in a circular fashion (see Fig. 3) with a diameter very similar to that observed on extracellular virus particles. This is consistent with antibody reacting with virus particles either ''pushing through'' the membrane or adsorbed to the cell surface. The circular orientation of the probes may reflect the collapse of the antibody around the virus particle during drying. Virus particles are also seen underlying the membrane and therefore inaccessible to gold-labeled anti-VP2 antibody added to intact cells (Fig. 3). The presence of single virus particles and aggregates on the cell surface is confirmed in thin sections (Fig. 23). Several lines of evidence indicate that in spite of their cell-surface location, virus particles retain an attachment to the cytoskeleton. First, stereoscopic analysis of cytoskeletons from infected cells reveals virus-like particles at the surface of the skeletal matrix (Figs. 11 and 24). In addition, anti-VP2 monoclonal antibody bound to virus particles on the surface of living cells

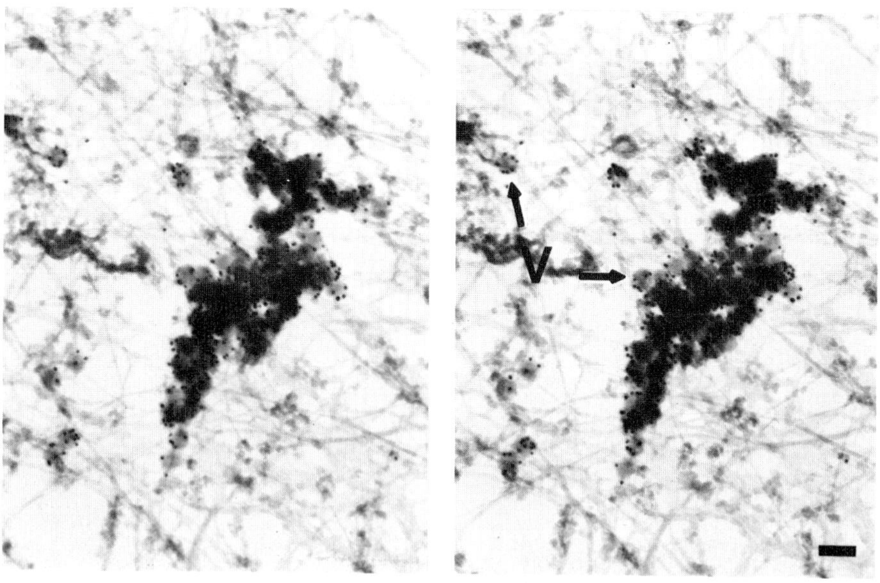


**FIGURE 24.** Stereomicrograph showing an aggregate of virus particles on the surface of BTV-infected cells. Cytoskeletons were probed with gold-labeled anti-VP2 monoclonal antibody 9J3/G6. V, virus particle. Bar = 100 nm.

remains there during preparation of the cytoskeleton by treatment with NP-40 (data not shown). A proportion of virus particles on the cell surface are associated with actin (data not shown). We are currently examining the nature of this association and its role in the release of virus particles from infected cells.

## 4. CONCLUSIONS

The data presented here indicate that VIB and a proportion of the virus tubules and virus-like particles in BTV-infected cells bind to intermediate filaments. Determination of the virus structural and nonstructural proteins present in both these cytoskeleton-associated structures and cytoplasmic and released virus particles has provided a new perspective on the morphogenesis of BTV and presented, for the first time, a rationale for the presence of tubules in BTV-infected cells. Information about the viral proteins detected by gold-labeled monoclonal antibodies in VIB, tubules, cytoskeleton-associated virus-like particles, and cytoplasmic and released virus particles can be used as a basis for speculation on the mechanism, whereby the outer coat proteins are added to core particles of BTV during morphogenesis.

The initial event in this process appears to occur at the periphery of VIB, where both fibrillar NS1 and some VP2 become associated with corelike particles. The presence of NS1 in purified core particles suggests that the addition of NS1 may precede, and therefore aid, the addition of VP2. The absence of virus particles in close proximity to VIB indicates that these incomplete particles are released into the cytosol, and it is there or after binding to the intermediate filament network that more VP2 is added.

NS1 may be removed from the virus particles either concomitantly with the addition of VP2 or as a separate and unrelated event. Some NS1 remains associated predominantly, however, with the virus core particle. The data presented raise the possibility that tubules are formed by condensation of NS1-containing fibrils released from virus-like particles at this stage of morphogenesis. The presence of VP3 and VP7 in tubules may be due to NS1-mediated removal of these proteins from the corelike particles.

Virus-like particles on the cytoskeleton or in the cytosol either lose excess VP2 or undergo rearrangement of the outer coat, which results in a decreased accessibility of VP2 and gold-labeled anti-VP2 antibody. Cytoskeleton-associated virus particles may leave the intermediate filaments following this process.

Before release from the cell, infectious virus in the cytosol binds to the actin-rich cortical layer underlying the cell membrane. Following emergence through the cell membrane, by an as yet unknown mechanism, the virus may remain attached to the cell surface through a cytochalasin B-sensitive (i.e., microfilament dependent) mechanism. Although this sequence of events explains the observations outlined herein, it is still speculative. Further work will be necessary to confirm or clarify some of these steps in BTV morphogenesis.

ACKNOWLEDGMENTS. The authors would like to thank Terry Wise, Gary Crameri, and Andrea Conolan for expert technical assistance; Vincent Haveaux for help with the photography; Anne McGlade and Debbie Monea for typing the manuscript; and Dr. Allan Gould for helpful discussions. In particular, we wish to thank John White, who supplied the monoclonal antibodies used in these studies.

## 5. REFERENCES

Babiss, L. E., Luftig, R. B., Weatherbee, J. A., Weihing, R. R., Ray, U. R., and Fields, B. N., 1979, Reovirus serotypes 1 and 3 differ in their in vitro association with microtubules, *J. Virol.* **30**:863–874.

Bedows, E., Rao, K. M. K., and Welsh, M. J., 1983, Fate of microfilaments in Vero cells infected with measles virus and herpes simplex virus type 1, *Mol. Cell. Biol.* **3**:712–719.

Ben-Ze'ev, A., 1984, Inhibition of vimentin synthesis and disruption of intermediate filaments in simian virus 40-infected monkey kidney cells, *Mol. Cell Biol.* **4**:1880–1889.

Ben-Ze'ev, A., Abulafia, R., and Bratosin, S., 1983, Herpes simplex virus and protein transport are associated with the cytoskeletal framework and the nuclear matrix in infected BSC-1 cells, *Virology* **129**:501–507.

Blose, S. H., and Chacko, S., 1976, Rings of intermediate (100A) filemant bundles in the perinuclear region of vascular endothelial cells. Their mobilization by colcemid and mitosis, *J. Cell Biol.* **70**:459–466.

Bohn, W., Rutter, G., Hohenberg, H., Mannweiler, K., and Nobis, P., 1986, Involvement of actin filaments in budding of measles virus: Studies on cytoskeletons of infected cells, *Virology* **149**: 91–106.

Boschek, C. B., Jockusch, B. M., Friis, R. R., Back, R., Grundmann, E., and Bauer, H., 1981, Early changes in the distribution and organization of microfilament proteins during cell transformation, *Cell* **24**:175–184.

Bowne, J. G., and Jochim, M. M., 1967, Cytopathologic changes and development of inclusion bodies in cultured cells infected with bluetongue virus, *Am. J. Vet. Res.* **28**:1091–1105.

Bowne, J. G., and Ritchie, A. E., 1970, Some morphological features of bluetongue virus, *Virology* **40**:903–911.

Brown, S., Levinson, W., and Spudich, J. A., 1976, Cytoskeletal elements of chick embryo fibroblasts revealed by detergent extraction, *J. Supramol. Struct.* **5**:119–130.

Chatterjee, P. K., Cervera, M. M., and Penman, S., 1984, Formation of vesicular stomatitis virus nucleocapsid from cytoskeletal framework-bound N protein: Possible model for structure assembly, *Mol. Cell. Biol.* **4**:2231–2234.

Cromack, A. S., Blue, J. L., and Gratzek, J. B., 1971, A quantitative ultrastructural study of the development of bluetongue virus in Madin-Darby bovine kidney cells, *J. Gen. Virol.* **13**:229–244.

Dales, S., 1963, Association between the spindle apparatus and reovirus, *Proc. Natl. Acad. Sci. USA* **50**:268–275.

Dales, S., and Chardonnet, Y., 1973, Early events in the interaction of adenoviruses with HeLa cells. IV. Association with microtubules and the nuclear pore complex during vectorial movement of the inoculum, *Virology* **56**:465–483.

Dales, S., Gomatos, P. J., and Hsu, K. C., 1965, The uptake and development of reovirus in strain L cells followed with labeled viral ribonucleic acid and ferritin–antibody conjugates, *Virology* **25**: 193–211.

Drochmans, P., Freudenstein, C., Wanson, J. C., Laurent, L., Keenan, T. W., Stadler, J., LeLoup, R., and Franke, W. W., 1978, Structure and biochemical composition of desmosomes and tonofilaments isolated from calf muzzle epidermis, *J. Cell Biol.* **79**:427–443.

Eaton, B. T., Hyatt, A. D., and White, J. R., 1987, Association of bluetongue virus with the cytoskeleton, *Virology* **157**:107–116.

Eaton, B. T., Hyatt, A. D., and White, J. R., 1988, Localization of the nonstructural protein NS1 in bluetongue virus infected cells and its presence in virus particles, *Virology* **163**:527–537.

Edbauer, C. A., and Naso, R. B., 1983, Cytoskeleton-associated Pr65$^{gag}$ and retrovirus assembly, *Virology* **130**:415–426.

Franke, W. W., Schmid, E., Weber, K., and Osborn, M., 1979, HeLa cells contain intermediate-sized filaments of the prekeratin type, *Exp. Cell. Res.* **118**:95–109.

Goldman, R. D., 1971, The role of three cytoplasmic fibres in BHK-21 cell motility. 1. Microtubules and the effects of colchicine, *J. Cell Biol.* **51**:752–762.

Goldman, R. D., Chang, C., and Williams, J. F., 1975, Properties and behaviour of hamster embryo cells transformed by human adenovirus type 5, *Cold Spring Harbor Symp. Quant. Biol.* **39**: 601–614.

Gorman, B. M., and Taylor, J., 1985, Orbiviruses, in: *Virology* (B. N. Fields, ed.) pp. 907–925, Raven, New York.

Gould, A. R., Hyatt, A. D., and Eaton, B. T., 1988, Morphogenesis of a bluetongue virus variant with an amino acid alteration at a neutralization site in the outer coat protein, VP2, *Virology* **165**:23–32.

Grubman, M. J., Appleton, J. A., and Letchworth, G. L., 1983, Identification of bluetongue virus type 17 genome segments coding for polypeptides associated with virus neutralization and intergroup reactivity, *Virology* **131**:355–366.

Hartwig, J. H., Niederman, R., and Lind, S. E., 1985, Cortical actin structures and their relationship to mammalian cell movements, *Subcell. Biochem.* **11**:1–49.

Hayat, M. A., 1986, *Micron Microsc. Acta* **17**:115–135.

Hiller, G., Weber, K., Schneider, L., Parajsz, C., and Jungwirth, C., 1979, Interaction of assembled progeny pox viruses with the cellular cytoskeleton, *Virology* **98**:142–153.

Howard, J. M., Eckert, B. S., and Bourguignon, L. Y. W., 1983, Comparison of cytoskeletal organization in canine distemper virus-infected and uninfected cells, *J. Gen. Virol.* **64**:2379–2385.

Hronis, T. S., Steinberg, M. L., Defendi, V., and Sun, T. T., 1984, Simple epithelial nature of some simian virus-40-transformed human epidermal keratinocytes, *Cancer Res.* **44**:5797–5804.

Huismans, H., and Cloete, M., 1987, A comparison of different cloned bluetongue virus genome segments as probes for the detection of virus-specified RNA, *Virology* **158**:373–380.

Huismans, H., and Els, H. J., 1979, Characterization of the tubules associated with the replication of three different orbiviruses, *Virology* **92**:397–406.

Huismans, H., and Verwoerd, D. W., 1973, Control of transcription during the expression of the bluetongue virus genome, *Virology* **52**:81–88.

Huismans, H., Van Der Walt, N. T., Cloete, M., and Erasmus, B. J., 1987a, Isolation of a capsid protein of bluetongue virus that induces a protective immune reponse in sheep, *Virology* **157**: 172–179.

Huismans, H., Van Dijk, A. A., and Bauskin, A. R., 1987b, *In vitro* phosphorylation and purification of a non structural protein of bluetongue virus with affinity for single-stranded RNA, *J. Virol.* **61**:3589–3595.

Huismans, H., Van Dijk, A. A., and Els, H. J., 1987c, Uncoating of parental bluetongue virus to core and subcore particles in infected L cells, *Virology* **157**:180–188.

Hyatt, A. D., and Eaton, B. T., 1988, Ultrastructural distribution of the major capsid proteins within bluetongue virus and infected cells, *J. Gen. Virol.* **69**:805–815.

Hyatt, A. D., Eaton, B. T., and Lunt, R., 1987, The grid-cell-culture technique: The direct examination of virus-infected cells and progeny viruses, *J. Microsc.* **145**:97–106.

Jackson, P., and Bellett, A. J. D., 1985, Reduced microfilament organization in adenovirus type 5-infected rat embryo cells: A function of early region 1a, *J. Virol.* **55**:644–650.

Jones, N. L., Lewis, J. C., and Kilpatrick, B. A., 1986, Cytoskeletal disruption during human cytomegalovirus infection of human lung fibroblasts, *Eur. J. Cell. Biol.* **41**:304–312.

Knipe, D. M., and Spang, A. E., 1982, Definition of a series of stages in the association of two herpesviral proteins with the cell nucleus, *J. Virol.* **43**:314–324.

Lecatsas, G., 1968, Electron microscopic study of the formation of bluetongue virus, *Onderstepoort J. Vet. Res.* **35**:139–150.

Lenk, R., Ransom, L., Kaufmann, V., and Penman, S., 1977, A cytoskeletal structure with associated polyribosomes obtained from HeLa cells, *Cell* **10**:67–78.

Luftig, R. B., 1982, Does the cytoskeleton play a significant role in animal virus replication, *J. Theor. Biol.* **99**:173–191.

Luftig, R. B., and Weihing, R. R., 1975, Adenovirus binds to rat brain microtubules *in vitro*, *J. Virol.* **16**:696–706.

Mertens, P. P. C., Brown, F., and Sangar, D. V., 1984, Assignment of the genome segments of bluetongue virus type 1 to the proteins which they encode, *Virology* **135**:207–217.

Miles, B. D., Luftig, R. B., Weatherbee, J. A., Weihing, R. R., and Weber, J., 1980, Quantitation of the interaction between adenovirus types 2 and 5 and microtubules inside infected cells, *Virology* **105**:265–269.

Moll, R., Franke, W. W., Schiller, D. L., Greiger, B., and Krepler, R., 1982, The catalog of human cytokerations: Patterns of expression in normal epithelial, tumors and cultured cells, *Cell* **31**: 11–24.

Mora, M., Partin, K., Bhatia, M., Partin, J., and Carter, C., 1987, Association of reovirus proteins with the structural matrix of infected cells, *Virology* **159**:265–277.

Murti, K. G., Goorha, R., and Chen, M., 1985, Interaction of frog virus 3 with the cytoskeleton, *Curr. Top. Microbiol. Immunol.* **116**:107–131.

Murti, K. G., Chen, M., and Goorha, R., 1985, Interaction of frog virus 3 with the cytomatrix. III. Role of microfilaments in virus release, *Virology* **142**:317–325.

Musalem, C., and Espejo, R. T., 1985, Release of progeny virus from cells infected with simian rotavirus SA11, *J. Gen. Virol.* **66**:2715–2724.

Nielsen, L. N., Forchhammer, J., Dabelsteen, E., Jepsen, A., Teglbjaerg, C. S., and Norrild, B., 1987, Herpes simplex virus-induced changes of the keratin type intermediate filament in rat epithelial cells, *J. Gen. Virol.* **68**:737–748.

Norrild, B., Lehto, V.-P., and Virtanen, I., 1986, Organization of cytoskeleton elements during herpes simplex virus type 1 infection of human fibroblasts: An immunofluorescence study, *J. Gen. Virol.* **67**:97–105.

Pasquale, E. B., Maher, P. A., and Singer, S. J., 1986, Talin is phosphorylated on tyrosine in chicken embryo fibroblasts transformed by Rous Sarcoma Virus, *Proc. Natl. Acad. Sci. USA* **83**:5507–5511.

Penman, S., 1985, Virus metabolism and cellular architecture, in: *Virology* (B. N. Fields, ed.), pp. 169–182, Raven, New York.

Penman, S., Capco, D., Fey, E., Chatterjee, P., Reiter, T., Ermish, S., and Wan, K., 1983, The three dimensional structural networks of cytoplasm and nucleus: Function in cells and tissue, in J. R. McIntosh and B. H. Satir (eds.), *Modern Cell Biology,* Vol. 2, pp. 385–415, Liss, New York.

Pfeiffer, G., Willutzki, D., Weder, D., Becker, B., and Radsak, K., 1983, Microtubular reaction in human fibroblasts infected with cytomegalovirus, *Arch. Virol.* **76:**153–159.

Pollack, R., Osborn, M., and Weber, K., 1975, Patterns of organization of actin and myosin in normal and transformed cultured cells, *Proc. Natl. Acad. Sci. USA* **72:**994–998.

Rose, D. S. C., Shriver, K., Latchman, D. S., and LaThangue, N. B., 1986, A filamentous distribution for the herpes simplex virus type 2-encoded major DNA-binding protein, *J. Gen. Virol.* **67:**1315–1325.

Rutter, G., and Mannweiler, K., 1977, Alterations of actin-containing structures in BHK21 cells infected with Newcastle Disease virus and vesicular stomatitis virus, *J. Gen. Virol.* **37:**233–242.

Sefton, B. M., Hunter, T., Ball, E. H., and Singer, S. J., 1981, Vinculin: A cytoskeletal target of the transforming protein of Rous Sarcoma virus, *Cell* **24:**165–174.

Sharpe, A. H., Chen, L. B., and Fields, B. N., 1982, The interaction of mammalian reoviruses with the cytoskeleton of monkey kidney CV-1 cells, *Virology* **120:**399–411.

Spendlove, R. S., Lennette, E. H., Knight, C. O., and Chin, J. H., 1963, Development of viral antigen and infectious virus in HeLa cells infected with reovirus, *J. Immunol.* **90:**548–553.

Verwoerd, D. W., Louw, H., and Oellermann, R. A., 1970, Characterization of bluetongue virus ribonucleic acid, *J. Virol.* **5:**1–7.

Verwoerd, D. W., Els, H. J., De Villiers, E. M., and Huismans, H., 1972, Structure of the bluetongue virus capsid, *J. Virol.* **10:**783–794.

Weatherbee, J. A., Luftig, R. B., and Weihing, R. R., 1977, Binding of adenovirus to microtubules. II. Depletion of high-molecular-weight microtubule-associated protein content reduces specificity of *in vitro* binding, *J. Virol.* **21:**732–742.

Weed, H. G., Krochmalnic, G., and Penman, S., 1985, Poliovirus metabolism and the cytoskeletal framework: Detergent extraction and resinless section electron microscopy, *J. Virol.* **56:**549–557.

White, J. R., Breschkin, A. M., and Della-Porta, A. J., 1985, Immunochemical analysis of Australian bluetongue virus serotypes using monoclonal antibodies, in: *Bluetongue and Related Orbiviruses* (T. L. Barber and M. M. Jochim, eds.), pp. 397–405, Liss, New York.

Wiche, G., 1985, High-molecular-weight microtubule associated proteins (MAPS): A ubiquitous family of cytoskeletal connecting links, *Trends Biochem. Sci.* **10:**67–70.

Yamauchi, M., Nishiyama, Y., Fujioka, H., Isomura, S., and Maeno, K., 1985, On the intracellular transport and the nuclear association of human cytomegalovirus structural proteins, *J. Gen. Virol.* **66:**675–684.

Yen, S. H., and Fields, K. L., 1981, Antibodies to neurofilament, glial filament and fibroblast intermediate filament proteins bind to different cell types of the nervous system, *J. Cell Biol.* **88:**115–126.

*Chapter 9*

# Morphogenesis of Flaviviruses

Tatsuo Hase, Peter L. Summers, Kenneth H. Eckels, and Joseph R. Putnak

## 1. INTRODUCTION

The flaviviruses consist of about 70 viruses that include some important pathogens that are responsible for a number of serious diseases, such as yellow fever, dengue fever, and various encephalitides (Porterfield, 1980; Shope, 1980). They are transmitted to humans by arthropod vectors, i.e., mosquitoes and ticks (Chamberlain, 1980) and are also called arboviruses. The number of known flaviviruses will undoubtedly increase in the future as more viruses are isolated from various hosts and their vectors. Until recently, the flavivirus genus was included with three others, alphavirus, pestivirus, and rubivirus, as part of the Togaviridae family (Porterfield *et al.*, 1978). With the accumulation of experimental data, it has become increasingly clear that flaviviruses have substantially different genomic organization and mechanisms of replication and gene expression from those of alphaviruses, the other major genus of the Togaviridae family. Therefore, the possibility that the two genera have diverged from the same ancestor is remote. Consequently, flaviviruses are now classified in their own family, Flaviviridae (Westaway *et al.*, 1985).

---

**Tatsuo Hase** Department of Ultrastructural Studies, Walter Reed Army Institute of Research, Washington, D.C. 20307. **Peter L. Summers and Kenneth H. Eckels** Department of Biologics Research, Walter Reed Army Institute of Research, Washington, D.C. 20307. **Joseph R. Putnak** Department of Viral Diseases, Walter Reed Army Institute of Research, Washington, D.C. 20307.

Using the cross-neutralization test, DeMadrid and Porterfield (1974) and Porterfield (1980) divided flaviviruses into seven subgroups.* In the division, most viruses of medical importance are contained in three groups: (1) West Nile (or Japanese encephalitis) subgroup: West Nile (WN) virus, St. Louis encephalitis (SLE) virus, Japanese encephalitis (JE) virus, Murray Valley encephalitis (MVE) virus, Kunjin (KUN) virus, and others; (2) tick-borne subgroup: tick-borne encephalitis (TBE) virus, Kyasanur Forest disease virus, Omsk hemorrhagic fever virus, louping ill virus, Powassan virus, and others; and (3) dengue subgroup: dengue types 1,2,3,4 (DEN-1,2,3,4,). Yellow fever virus is monospecific in the cross-neutralization test. The observed antigenic relationships are of more than academic interest and are relevent both from a medical and an evolutionary point of view, since within the flaviviruses there is a wide range of pathogenesis for man.

Morphologically, virions of the Flaviviridae consist of an isometric ribonucleoprotein core of 30 to 35 nm in diameter surrounded by a 7- to 10-nm-thick lipid envelope that exhibits short spikes (Murphy, 1980). The viral genome is an infectious, single-stranded RNA of (+) polarity, that immediately serves as messenger RNA (mRNA) upon uncoating from the nucleocapsid in the host-cell cytoplasm. The genomic RNA can be found in infected cells as 38-45S RNA, collectively called 44S RNA (Westway, 1980), which has a type 1 cap at the 5′-end and no poly (A) at the 3′-end and a molecular weight of approximately $4 \times 10^6$ when measured by gel electrophoresis (Boulton and Westway, 1972; Wengler *et al.*, 1979; Cleaves and Dubin, 1979; Wengler and Wengler, 1981). The genomic RNA is also shown to function as mRNA in cell-free translation systems (Wengler *et al.*, 1979; Svitkin *et al.*, 1981; Monckton and Westaway, 1982), and genome-sized RNA is the only virus-specified RNA isolated from polysomes (Naeve and Trent, 1978).

Currently, complete sequencing of genomic RNA has been accomplished for several flaviviruses (Rice *et al.*, 1985; Castle *et al.*, 1986; Mackow *et al.*, 1987; Sumiyoshi *et al.*, 1987; Hahn *et al.*, 1988). Genomic RNAs of other flaviviruses are also partially sequenced (Dalgarno *et al.*, 1986; Deubel *et al.*, 1986; Biedrzycka *et al.*, 1987; McAda *et al.*, 1987; Mason *et al.*, 1987; Trent *et al.*, 1987). Therefore, we face the increasing prospect of being able to explain virus-host cell interactions of flaviviruses from the genetic and molecular aspects. As part of this exciting development, we review herein the available data concerning flavivirus morphogenesis.

*New and old terminology of flavivirus polypeptides (Rice *et al.*, 1986b). C = V2, prM = NV2, M = V1, E = V3, NS1 = NV3, ns2a = NV2½, ns2b = NV1½, NS3 = NV4, ns4a = NVX, ns4b = NV1, NS5 = NV5.

## 2. VIRAL ENTRY

Generally speaking, viruses enter host cells by one of the following mechanisms: (1) translocation of the entire virus across the plasma membrane; (2) pinocytosis of the virus by the host cells; and (3) fusion of the plasma membrane with the virus envelope (Roizman, 1985). The entry process of flaviviruses into their target cells is largely unknown; however, as enveloped viruses, flaviviruses are thought to enter host cells by either of the last two mechanisms, in which the viral envelope fuses to pinocytic vesicular membrane or the plasma membrane just before the release of the viral genome into the cytoplasm. In addition, the entry mechanism used by a given virus may depend on (1) the passage history and resultant selection of a particular viral subpopulation, (2) the lipid and carbohydrate compositions of the viral envelope contributed by the host cell, and (3) the target cell.

### 2.1. Attachment

To initiate the entry process, a flavivirus must first attach to the host–cell surface. The specificity for attachment to certain cells is considered to be a major determinant of flavivirus tropism and pathogenesis. In other words, each pathogenic flavivirus most likely has evolved specific ligands for its mammalian target cells in a natural ecological system: for example, YF virus for liver cells, DEN virus for monocyte-macrophage cells (Theofilopoulos *et al.*, 1976; Halstead *et al.*, 1977; Peiris and Porterfield, 1979), and a neurotropic flavivirus for brain cells. In addition, arboviruses apparently recognize receptors on vector insect cells for replication. Flaviviruses display on their surfaces short spikes that are made up of the envelope glycoprotein (E) encoded by the viral genome, and this protein is believed to play the principal role in the viral attachment to the host cell by binding to cell-membrane receptor macromolecules (Gollins and Porterfield, 1985). However, little is known at present about the chemical properties of flavivirus ligands, host-cell receptors, and ligand–receptor complexes. In addition to specific binding between virus ligands and host-cell receptors, attachment of flaviviruses to those cells that display Fc receptors on their surfaces may be facilitated by host antibodies that simultaneously bind to virions and to Fc receptors on the cell surface (Daughaday *et al.*, 1981; Brandt *et al.*, 1982; Gollins and Porterfield, 1985). The phenomenon of antibody-dependent enhancement of DEN virus infection has been confirmed both in vitro and in vivo (Halstead *et al.*, 1973; Halstead, 1979); this seems to play a significant role in secondary DEN virus infections. In the infection of human monocytes with DEN virus, Daughaday *et al.* (1981) have reported evidence that viral entry is facilitated by trypsin-sensitive virus receptors and, when an enhancing antibody is present, by trypsin-resistant immune-complex receptors. Currently, it is generally believed that immunologic enhancement plays a significant role in the pathogenesis of severe dengue infections (Halstead, 1988); however, its role, if any, in neurotropic flavivirus infections is not clear.

## 2.2. Penetration

Little is known about the entry mode of flaviviruses. The only available data are those of Gollins and Porterfield (1985) and Kimura *et al.* (1986). Studying the entry process of Vero cell-generated WN virus into macrophage-like P388D1 cells, these investigators have found that under physiologic conditions the virus enters the host cells exclusively by receptor-mediated endocytosis, but at an acidic pH the virus enters the cells both at the cell surface and within coated vesicles through a membrane fusion process. The above findings, therefore, indicate that WN virus depends on an acidic endosomal environment to facilitate membrane fusion for viral entry.

The entry processes of other flavivirus-host cell combinations are not known. We have shown that DEN-2 virus, PR-159 strain, and JE virus, $SA_{14}$ strain, both of which have been generated from C6/36 mosquito cells, enter the same cells by penetrating through the plasma membrane at the cell surface; also, the DEN-2 virus enters human peripheral blood monocytes essentially in the same manner as it does the mosquito cells (Hase *et al.*, 1989). At present, therefore, it is certainly premature to assume that all the flaviviruses display a single entry mode. In view of the diverse tropism and pathogenicity exhibited among flaviviruses, a possibility exists that individual flaviviruses have evolved diverse entry modes, depending on their characteristic life cycles under variable ecological conditions. In this respect, further experimental data in the area of viral entry are sorely needed.

## 2.3. Uncoating

For alphaviruses, Wengler (1987) has noted that the large ribosomal subunit has a binding site for the viral nucleocapsid or core protein, and suggested that it may play a role in the viral uncoating process within the cytoplasm. However, existence of affinity between flaviviral nucleocapsid protein and host ribosomal subunits has not been verified, and nothing is known about molecular interactions that may take place in the uncoating of flavivirus genomes within the host-cell cytoplasm. In the receptor-mediated endocytosis of WN virus by P388D1 mouse macrophage-like cells, Gollins and Porterfield (1985) suggest that endocytosed virions lose their envelopes through fusion to the endosomal membrane at an acidic pH and release their genomic RNA through participation of host proteolytic activity in the phagosome. However, detailed information about chemical and molecular interactions that may take place in the putative viral uncoating process within the phagosome is lacking at present.

## 3. VIRAL REPLICATION

### 3.1. RNA Species

Despite accumulation of a large amount of experimental data, the events of flavivirus RNA replication are still poorly understood. The subject was extensively reviewed by Westaway (1980) and Brinton (1986). Since the viral RNA is of (+) polarity and does not carry its own polymerase in the viral core, the first step that uncoated RNA of a flavivirus must take in the host-cell cytoplasm is translation of proteins that are needed for its own replication. RNA-dependent RNA polymerase activity has been detected in extracts from cells infected with JE virus (Takehara, 1971; Zebovitz *et al.*, 1972), SLE virus (Qureshi and Trent, 1972), DEN-2 virus (Cardiff *et al.*, 1973a), and KUN virus (Chu and Westaway, 1985, 1987). Therefore, it is reasonable to assume that, after uncoating, viral RNA undergoes translation to produce the viral RNA polymerase, which then catalyzes RNA replication. In the flavivirus replication process, two major species of RNA are synthesized, one with a sedimentation coefficient of 44S, and the other with a sedimentation coefficient of 20S (Igarashi *et al.*, 1964; Stoller *et al.*, 1966, 1967; Trent *et al.*, 1969). It is now clear that 44S RNA is RNase-sensitive and made up of single-stranded RNA molecules, and 20S RNA is RNase resistant and represents a double-stranded replicative form (RF). In addition, heterodisperse and partially RNase-resistant 26S RNA that is believed to be a replicative intermediate (RI) between 44S RNA and 20S RNA was found (Stoller *et al.*, 1967; Trent *et al.*, 1969; Zebovitz *et al.*, 1972; Wengler *et al.*, 1978; Cleaves *et al.*, 1981).

### 3.2. Replication Scheme

On the basis of accumulated data, Westaway (1980) proposed a replication scheme for flaviviruses which is illustrated in Fig. 1. According to his scheme, the intracellular replication of flaviviruses occurs in two stages: replication stage 1 and replication stage 2. Replication stage 1 takes place as a slow process in the latent period, starting around 6 hr postinfection (p.i.) and continuing up to 16 to 18 hr p.i. During this stage, imput virion RNA acts as mRNA to synthesize RNA-dependent RNA polymerase and other viral proteins necessary for viral replication. Virion RNA also serves as the template for the newly synthesized polymerase to produce 20S RF RNA. The RF supposedly dissociates to release its components, 44S (+) RNA, 44S (−) RNA, and polymerase, to repeat this

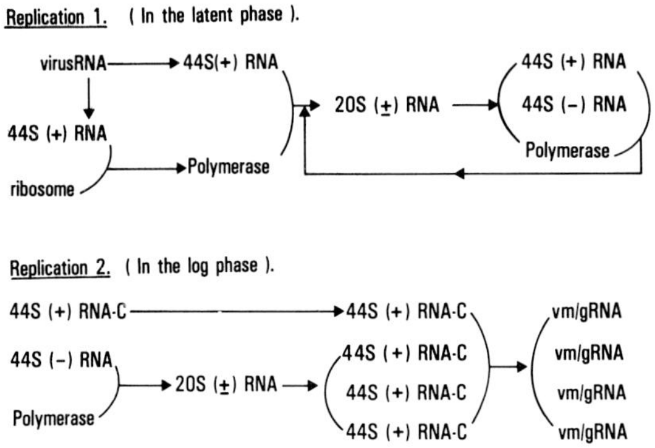


**FIGURE 1.** Replication scheme of the flavivirus (modified from Westaway, 1980). 44S(+)RNA-C, 44S(+)RNA-capsid protein complex; vm/g RNA, viral messenger or genomic RNA. See text for details.

cycle of replication. At this stage, polymerase molecules supposedly attach to either 44S (+) RNA or 44S (−) RNA nonselectively, and a slow release of the polymerase from the accumulating 20S (±) RNA provides an opportunity for control of transcription and translation.

Replication stage 2 occurs as a rapid process beginning immediately before the log phase. Westaway (1980) postulates that release or displacement of the polymerase attached to 20S (±) RNA is facilitated by the attachment of the viral capsid (C) protein, that increasingly accumulates in the cytoplasm, near the 3′-end of the 44S (+) RNA strand, forming a 44S (+)-C complex. It is assumed that the 44S (+) RNA-C complex renders its 44S (+) RNA component incapable of acting as a replication template but capable of functioning as mRNA in viral protein synthesis or in maturation to a complete, infectious virus particle. In replication stage 2, therefore, an increasing number of 44S (+) RNA strands are diverted to either viral protein synthesis or viral assembly, with the polymerase binding favorably to 44S (−) RNA strands to produce additional 44S (+) RNA. Recently, Chu and Westaway (1985) have proposed that the replication stage 2 occurs on RF semiconservatively, with the plus strand being replicated preferentially on the minus strand of the 20S (±) RNA. Westaway's replication scheme takes into account the following features of intracellular replication of flaviviruses: (a) the long latent period of 12 hr or more, (b) the slow production of detectable amounts of viral RNA and viral proteins during the latent period, and (c) a rather sudden switch to productive infection. However, more evidence needs to be accumulated before the scheme is solidly established. A particularly intriguing feature of the scheme is the assumption that selective control is exerted by the C protein over the functional directions of 44S (+) and 44S (−) RNA. The C protein contains one or more regions that complex with the genomic RNA (Rice *et al.*, 1985); accordingly, it is reasonable to believe that the C protein has an affinity for the viral RNA. However, some questions remain unanswered. For example, what is the mechanism that causes the C protein to bind selectively to 44S (+) RNA and not to 44S (−) RNA? Might a packaging sequence be present in 44S (+) RNA? Moreover, the idea that the selective attachment of the C protein to 44S (+) RNA causes diversion of the RNA from replication to viral assembly is novel and needs further supportive data.

### 3.3. Replication Site

Morphologically, there seems to be general agreement among investigators that the perinuclear region is the site where early virus-induced protein and RNA syntheses take place (Cardiff *et al.*, 1973b; Zebovitz *et al.*, 1974; Kos *et al.*, 1975; Boulton and Westaway, 1976; Takeda *et al.*, 1978; Ng *et al.*, 1983; Westaway and Goodman, 1987). The close topographic relationship between the replication site and the nucleus suggests contribution of some nuclear factors in the early stages of replication (Zebovitz *et al.*, 1974; Takeda *et al.*, 1978), although the nucleus itself does not seem to be involved in replication (Cardiff *et al.*, 1973b; Ng *et al.*, 1983). Viral protein and RNA syntheses appear to spread from the perinuclear region to the rest of the cytoplasm as replication proceeds. In an immunocytochemical study of LLC-$MK_2$ cells infected with DEN-2 virus, New Guinea C (NGC) strain, Cardiff *et al.* (1973b) found that DEN-specific fluorescence was most intense in the perinuclear area and radiated in a granular pattern of decreasing intensity into the cytoplasm over time. Perinuclear fluorescence was more strongly associated with virus-specific nonstructural antigens, and the cytoplasmic fluorescence was associated with virion-structural antigens. Most of these antigens were localized primarily in the RER cisternae and intracytoplasmic vacuoles. In analyzing the membrane fractions of BHK-21 cells infected with DEN-2 virus, NGC strain, Stohlman *et al.* (1975) found that virus-induced protein synthesis, as indicated by the incorporation of radioactive precursors, was localized on the rough endoplasmic reticulum (RER) and that glycosylation, measured by the incorporation of radiolabeled glucosamine into membrane-associated protein, was localized on the smooth endoplasmic reticulum (SER). In the infection of Vero cells with KUN virus, MR61C strain, Ng *et al.* (1983) found synthesis of envelope protein and its accumulation in a narrow rim around the nucleus at 4 hr postinfection. Afterward, accumulation of the E protein occurred in the RER. They also found immunofluorescent sites for viral RNA synthesis that extended from the perinuclear region to another fine network that included many small foci and vesicles. When microtubules were disrupted by vinblastin treatment of the cells, the distribution of the immunofluorescent sites for viral RNA synthesis was reduced by at least 10-fold. They concluded that viral RNA replication and protein synthesis occurred in different membrane systems and that microtubules provided a scaffolding for SER-associated viral RNA polymerase.

Available evidence, therefore, suggests that uncoated genomic RNA of flaviviruses moves to the perinuclear region of the host cell for initiation of the replication process. Early viral protein synthesis occurs on the perinuclear RER, whereas viral RNA replication takes place on the perinuclear SER. Thus, the host cytoplasmic membrane and cytoskeletal systems seem to play significant roles in viral replication and viral protein synthesis, possibly by providing scaffolding on which virus-related macromolecular interactions take place (Ng *et al.*, 1983; Hong and Ng, 1987). They may also serve to localize viral protein and RNA to certain areas within the cytoplasm. The reason for the selection of the perinuclear region as the early viral replication site is not clear. Chu and Westaway (1985) have recently presented evidence that, under favorable experimental conditions, flavivirus RNA replication occurs on the 44S (−) strand of a double-stranded 20S (±) RNA template, with preferential synthesis of 44S (+) RNA molecules. This is probably the viral replication mode in the log phase, which takes place extensively in the host-cell cytoplasm. By contrast, since the single-stranded viral RNA template is necessarily used in the initial phase of flavivirus infection following viral entry (see Fig. 1), it is possible that the localization of early flavivirus RNA replication to the perinuclear region may reflect this form of replication mode, with the host nucleus possibly providing some accessory factors (Zebovitz *et al.*, 1974; Takeda *et al.*, 1978).

### 3.4. Virus–Host Interactions

Flaviviruses found in nature exhibit phenotypic heterogeneity. Mixed plaque populations are readily observed for these viruses in field isolates and cell culture or animal-passaged virus seeds (Mayer, 1963; Paul, 1966; Eckels *et al.*, 1976). It is thought that these diverse virus populations are necessary to maintain the virus in nature between vertebrate and invertebrate hosts (Schlesinger, 1980). RNA viruses such as dengue (DEN) exhibit high rates of mutation in nature and in the laboratory, which account for the generation of variant viruses (Holland *et al.*, 1982). Enrichment for a selected virus phenotype can be accomplished through laboratory manipulation. Pure clones of virus can be isolated, amplified, and maintained in cell culture in the laboratory. A purified population of DEN-2 virus, the S-1 clone, has been used for preparing a live-attenuated vaccine (Eckels *et al.*, 1980). In this case small plaque size and temperature sensitivity (ts) of replication were markers that were associated with animal attenuation (Harrison *et al.*, 1977). Undiluted passage of the S-1 clone resulted in reversion to virulence in animals with the concurrent emergence of large plaque populations (Eckels *et al.*, 1976). Diluted passage of the S-1 clone was required for maintenance of the small plaque/ts phenotype.

Selection for reduced virulence may also result from serial cell culture passage of flaviviruses. A DEN-2 attenuated vaccine was recently shown to be immunogenic and not cause vaccine-related illness in humans (Bhamarapravati *et al.*, 1987). The vaccine was prepared by serial passage of a human DEN-2 isolate in primary canine kidney cell cultures. The uncloned attenuated vaccine was small plaque and *ts*. Igarashi (1979) persistently infected *A. albopictus* mosquito cells with DEN-2 virus and after multiple passages of the cells could demonstrate a predominant small plaque, ts population of virus that had a reduced virulence. The studies just mentioned used cell culture passage to enrich for attenuated virus populations for practical purposes, i.e., vaccine preparation.

Host factors play a major role in selection of dominant populations of flaviviruses. In a study using West Nile (WN) virus, it was shown that nonpermissive and permissive cells significantly influenced the composition of viral populations (Brinton, 1982). The nonpermissive cells caused depletion of wild-type (standard) virus and an enrichment of defective interfering virus particles. It is thought that the defective particles are already present in standard virus populations but are amplified in the nonpermissive cells. It was proposed by Brinton that a host protein specified by the nonpermissive allele may interact at the level of the WN virus replication complex affecting template–polymerase relationships (Brinton, 1986).

## 4. VIRAL ASSEMBLY

### 4.1. Viral Proteins

Structurally, all flaviviruses are similar; they contain single-stranded genomic RNA of ~10.5 kB, encapsidated to form a nucleocapsid believed to have icosahedral symmetry, composed of a single species of capsid protein of $M_r$ about 14,000 (Boulton and Westaway, 1977; Naeve and Trent, 1978; Wengler *et al.*, 1978; Cleaves and Dubin, 1979). The nucleocapsid is enveloped by a lipid bilayer in which are embedded two virus-specific proteins; (1) a membrane protein, M, which is nonglycosylated and has a molecular weight of ~ 8000, and (2) an envelope protein, E, with a molecular weight of ~60,000 that may or may not be glycosylated, depending on the virus. The M protein is generated by proteolytic cleavage from the glycosylated prM protein of ~ 22,000 $M_r$ during or after release of virus particles from cells (Shapiro *et al.*, 1972). The M protein molecules are removed from virions together with E by treatment with nonionic detergents (Stollar, 1969; Shapiro *et al.*, 1971; Trent and Qureshi, 1971); however, treatment with deoxycholate leaves M associated with the sedimenting particles (Westaway and Reedman, 1969; Trent and Qureshi, 1971). Thus, the M protein appears to be associated with both the envelope and the nucleocapsid. By contrast, the E-protein molecules project from the viral envelope and form spikes thought to have a role in attaching to host cell surface receptors for viral entry

(Gollins and Porterfield, 1985). Since the E protein plays a dominant role in the biological functions of the flavivirion membrane, its molecular configuration has attracted the increasing attention of investigators. The entire amino acid sequences of some flavivirus E proteins are known. For WN virus, all six disulfide bridges present in the viral E protein have been identified, and it has been reported that the E protein contains regions termed R1, L1, R2, L2, and R3, which together form the E protein ectodomain followed by a C-terminal membrane anchor region (Nowak and Wengler, 1987; Wengler *et al.*, 1987). The WN virus possesses E-protein oligomers (trimers) that are sensitive to proteolytic attack at two sites located in L1 and R3, suggesting that these segments are exposed at the viral surface. However, proteolytic cleavage in these regions neither destroys nor releases the oligomers from the viral surface, while prM and M proteins are degraded by proteases (Wengler *et al.*, 1987).

Oligosaccharides, which are integral components of many viral envelope and eukaryotic cell surface proteins, are thought to play significant roles in protein folding and specific interprotein recognition (Berger *et al.*, 1982). The significance of glycosylation of flavivirus E proteins, however, has not yet been clearly defined. The isolated E protein contains viral hemagglutinin and antigenic components and, upon immunization, induces hemagglutinin-inhibiting, neutralizing, and protective antibodies (Kitano *et al.*, 1974; Della Porta and Westaway, 1977; Trent, 1977; Russell *et al.*, 1980; Heinz *et al.*, 1981; Takegami *et al.*, 1982; Lobigs *et al.*, 1987). Evidence for the glycoprotein nature of the E proteins of several flaviviruses has been obtained by the incorporation of radiolabeled sugars (Stollar, 1969; Shapiro *et al.*, 1971; Westaway, 1975) and by its reactivity with concanavalin A (Con A) (Stohlman *et al.*, 1976; Clegg, 1982). Flavivirus E proteins are apparently glycosylated nonspecifically. Recent cloning and sequencing data indicate that the E proteins of YF virus (Rice *et al.*, 1985) and MVE virus (Dalgarno *et al.*, 1986) contain only one potential asparagine-linked glycosylation site, and that of WN virus (Wengler *et al.*, 1985) contains none in the corresponding region. In infected cells, flavivirus E proteins have been found in both glycosylated and nonglycosylated forms (Wright *et al.*, 1981; Wright, 1982; Schlesinger *et al.*, 1983), and there is evidence that the E proteins of some flaviviruses are not glycosylated at all (Wright, 1982; Winkler *et al.*, 1987). The significance of E-protein glycosylation and its role in host immune response, viral entry, and morphogenesis needs to be further clarified in the future.

The genomes of YF virus (Rice *et al.*, 1985), WN virus (Castle *et al.*, 1986), DEN-4 virus, Caribbean 814669 strain (Zhao *et al.*, 1986; Mackow *et al.*, 1987), JE virus, JaOArS982 strain (Sumiyoshi *et al.*, 1987), and DEN-2 virus, PR-159 strain (Hahn *et al.*, 1988) have been completely sequenced. They all contain a single, long, open reading frame of ~10,500 nucleotides. One open reading frame is also a feature of the sequenced portions of other flaviviruses

(Dalgarno *et al.*, 1986; Biedrzycka *et al.*, 1987; McAda *et al.*, 1987). These findings suggest that a single, long, open reading frame is a property of all flavivirus genomes and indicate that viral polypeptides are probably formed by post-translational cleavage of precursor proteins. It is proposed that both cellular and virus-encoded proteases are involved in the cleavage events (Castle *et al.*, 1985; Rice *et al.*, 1985, 1986a; Wengler *et al.*, 1985; Dalgarno *et al.*, 1986). The structure and translation strategy of the flavivirus genome have recently been reviewed by Rice *et al.* (1986b).

The genomic organization of YF virus, 17D vaccine strain, shown diagrammatically in Fig. 2, has been reported by Rice *et al.* (1985). The three structural proteins, C, prM, and E, are encoded in the 5′-quarter of the genomic RNA. It is assumed that the initiating methionine in flavivirus genome translation is removed by cellular methionine aminopeptidase (Boege *et al.*, 1983; Bell *et al.*, 1985; Castle *et al.*, 1985). C is a basic capsid protein encoded at the beginning of the long, open reading frame. C is followed by prM, which is a glycosylated precursor that is cleaved to generate M, a nonglycosylated membrane protein, during viral maturation. The coding region for prM, in turn, is followed by that of the envelope protein, E, whose amino acid sequence in the case of YF virus is typical of a glycoprotein having a C-terminal transmembrane region.

NS1, the protein that follows the E protein, is the only glycosylated nonstructural protein (Shapiro *et al.*, 1973; Westaway, 1975; Schlesinger *et al.*, 1983). At least three forms of this protein, differing in molecular weight, appear to be associated with DEN virus, New Guinea C (NGC) strain, infected cells (Putnak, unpublished data). Two of these forms are intracellular and membrane-associated, and one is secreted. All three forms are glycosylated, with sugar residues contributing ~5000 to their respective molecular weights, consistent with the use of both of two conserved signals for N-linked glycosylation. Glycosylation appears to occur in the RER before NS1 enters the Golgi apparatus. Pulse-chase experiments suggest that the high-molecular-weight form of NS1 serves as precursor to both the small intracellular and secreted forms, perhaps via proteolytic processing. The role of NS1 in viral replication has yet to be elucidated but, given the fact that it is a nonstructural glycoprotein present in large quantity in the infected cell, we may speculate that it is involved in viral maturation, perhaps organizing the membrane for viral assembly. NS1, previously designated NV3 or gp48, has been shown to be the soluble complement-fixing

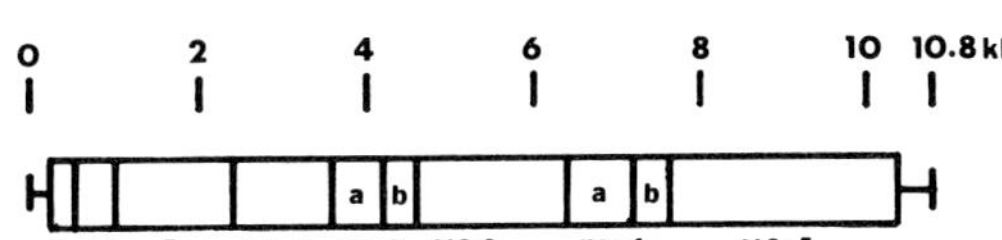


**FIGURE 2.** Genomic organization of yellow fever virus RNA (as proposed by Rice *et al.*, 1985). The open reading frame is shown by an open box. Nonstructural proteins that are proposed but not identified are labeled ns.

(SCF) antigen that is expressed on the infected cell surface (Brandt *et al.*, 1970: Russell *et al.*, 1970; Smith *et al.*, 1985). The NS1 protein elicits a protective immunity in animals (Schlesinger *et al.*, 1985, 1986, 1987). Located 3′ to the NS1 protein-coding region are nonstructural proteins designated ns2a and ns2b with a combined predicted molecular weight of ~30 kD. They are highly hydrophobic, having no fewer than 10 potential transmembrane domains. Interestingly, in the case of JE virus but not DEN virus, the ns2a protein was demonstrated to be a fusion protein with NS1 (Mason *et al.*, 1987). The role of ns2a and 2b in viral replication is not known: they may serve as membrane anchors for other, more hydrophilic, nonstructural proteins such as those involved in RNA replication, a process which appears to be tightly associated with the SER (Mason *et al.*, 1987). Following ns2a and 2b are relatively large and hydrophilic proteins designated NS3 and NS5. They are separated by putative ns4a and 4b proteins which appear to be lipophilic on the basis of their corresponding nucleotide sequences. NS5 appears to have some amino acid sequence homology with the RNA polymerases of other positive strand RNA viruses; this, and its large size of about 100 kD, suggests that it is the viral RNA-dependent RNA polymerase. Also by analogy with other RNA viruses, it might be speculated that NS3, ns4a and ns4b are polymerase-associated proteins which serve to modulate the specificity or activity of the polymerase (Rice *et al.*, 1985).

### 4.2. Viral Assembly Site

By electron microscopy, flaviviruses are usually found within the cytoplasmic vacuoles of infected cells (Ota, 1965, Filshie and Rehacek, 1968; Murphy *et al.*, 1968; Matsumura *et al.*, 1971; Cardiff *et al.*, 1973b; Murphy, 1980; Ng, 1987). Accordingly, it was generally assumed that flavivirus nucleocapsids bud at the cytoplasmic membrane into cytoplasmic vesicles. This assumption, however, had long been unproved because of the lack of convincing experimental evidence that flavivirus nucleocapsids form in the cytosol and bud at the cytoplasmic membrane. It has recently been found in the infection of C6/36 mosquito cells with DEN-2 virus, PR-159 strain, that distinct DEN nucleocapsids form in the cytosol and move to the plasma and cytoplasmic membranes for budding (Hase *et al.*, 1987a). By contrast, the infection of the same cells with JE virus, $SA_{14}$ strain, under similar experimental conditions, showed virions in the RER cisternae and within vesicles, with no nucleocapsids forming in the cytosol and no budding occurring at the host-cell membranes (Hase *et al.*, 1987b). The above findings indicate that the nucleocapsids of DEN-2 and JE viruses form in different compartments; for the DEN-2 virus they form in the cytosol and move to the cell membrane for viral envelope acquisition, and for the JE virus they form in the RER cisternae as whole virions.

### 4.3. *cis*- and *trans*-Type Maturation

The two types of maturation represented by DEN-2 virus, PR-159, and JE virus, $SA_{14}$, in C6/36 cells are named respectively *cis*- and *trans*-type maturation and are illustrated in Fig. 3 (Hase *et al.*, 1987b). Nucleocapsids of the flavivirus that undergo *cis*-type maturation form in the cytosol; the nucleocapsids then move to the host plasma or cytoplasmic membrane for budding and acquire viral envelopes in the budding process. For *trans*-type maturation, virions appear in the RER cisternae, without forming nucleocapsids in the cytosol; from there, the virions are carried within transfer vesicles to the Golgi apparatus and then appear within secretory vesicles to be released extracellularly (Leary and Blair, 1980; Hase *et al.*, 1987b).

Morphologically, cells infected with flaviviruses undergoing *cis*- and *trans*-type maturation show certain characteristic changes listed in Table I. Cells infected with a *cis*-type flavivirus usually exhibit fragmentation of RER, with dissociation of RER ribosomes (Fig. 4a). In the process, RER tends to transform into vesicles and vacuoles; at the same time, nucleocapsids appear in the cytosol (Fig. 4a). By contrast, cells infected with a *trans*-type flavivirus exhibit markedly hypertrophic RER, with prominently dilated cisternae that contain virions (Fig. 4b). The dilated RER cisternae also contain characteristic vesicles that have been variously called round membranous vesicles (Murphy *et al.*, 1968), cytopathic vesicles (Cardiff *et al.*, 1973b; Sriurairatna *et al.*, 1973), and smooth membranous structures (Leary and Blair, 1980). In this chapter, we refer to them

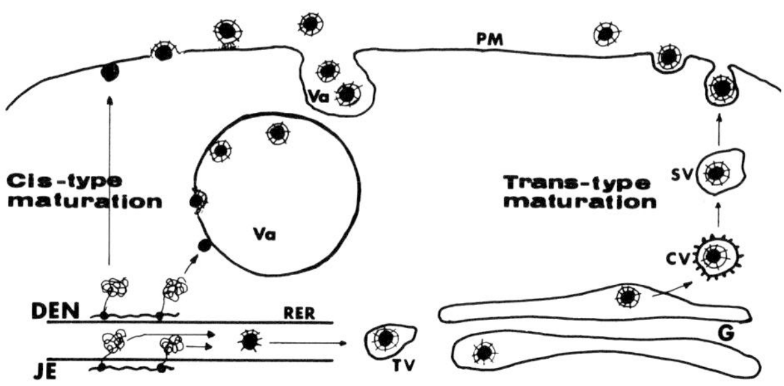


**FIGURE 3.** Schematic illustration of the *cis*-type and *trans*-type maturation of flaviviruses (as proposed by Hase *et al.*, 1987b). CV, coated vesicles; DEN, DEN-2 virus, PR-159, genome on host ribosomes, synthesizing the viral structural proteins and releasing them into the cytosol; G, Golgi apparatus, JE, JE virus, $SA_{14}$, genome on host ribosomes, synthesizing the viral structural proteins and releasing them into the cisternae; PM, plasma membrane; RER, rough endoplasmic reticulum, SV, secretory vesicle; TV, transfer vesicle; Va, cytoplasmic vacuole. See the text for detail. Printed by permission of *Arch. Virol.*

**Table I**
***cis*-Type and *trans*-Type Maturation of Flaviviruses**

| | *cis*-Type maturation | *trans*-Type maturation |
|---|---|---|
| Host-cell changes | Dissociation of RER ribosomes; vesiculation of RER membrane | Hypertrophy of RER; dilation of RER cisternae |
| Nucleocapsids | Form in the cytosol | Form in the RER cisternae |
| Viral envelope | Acquired at the plasma and cytoplasmic membranes | Acquired in RER cisternae |
| Virus release | By budding at the cell surface and by exocytosis of virus-containing vacuoles | By secretory-type exocytosis |

as endoplasmic reticulum (ER) vesicles, because we believe that they form through invagination and pinching off of RER membrane and appear in the RER cisternae in connection with the virus undergoing *trans*-type maturation. Cells infected with a *cis*-type flavivirus and those infected with a *trans*-type flavivirus are also clearly differentiable at the cell surface. The former show multiple budding virions at the plasma membrane, and the peripheral cytoplasm beneath the plasma membrane usually contains multiple nucleocapsids that are presumably moving to the membrane (Fig. 5a). In addition, within the cytoplasm nucleocapsids line the cytosolic side of cytoplasmic vacuolar membranes, and budding and mature virions are found at the membranes and within lumens, respectively. The cells infected with a *trans*-type virus show no budding virions at the plasma or cytoplasmic vacuolar membranes; the peripheral cytoplasm has prominently extended RER, the dilated cisternae of which contain virions and ER vesicles (Fig. 5b). Therefore, flaviviruses that undergo *cis*- or *trans*-type maturation can be clearly differentiated morphologically.

Flaviviruses tested in C6/36 cells are categorized in Table 2 as having either *cis*- or *trans*-type maturation. DEN-2 virus, PR-159 strain, is, so far, the only virus that has been found to exhibit distinct *cis*-type maturation, whereas many flaviviruses tested underwent *trans*-type maturation. In addition to flaviviruses that clearly exhibited either *cis*- or *trans*-type maturation, we encountered a group of flaviviruses, e.g., YF virus, 17D vaccine strain, and DEN-4 virus, Caribbean 814669 strain, which we could not classify definitively into either *cis* or *trans* type (T. Hase, unpublished data). C6/36 cell infection with a flavivirus of this group produced recognizable nucleocapsids in the cytosol but exhibited no viral budding at the plasma and cytoplasmic membranes; at the same time, it resulted in severe damage to cytoplasmic organelles, including RER and the Golgi apparatus, with subsequent rupture of the cells. It appeared that virions of this group were released mainly by cytolysis. In the infection of J-111 cells infected with DEN-1 virus, Mochizuki strain and Hawaii 32748 strain, Mat-

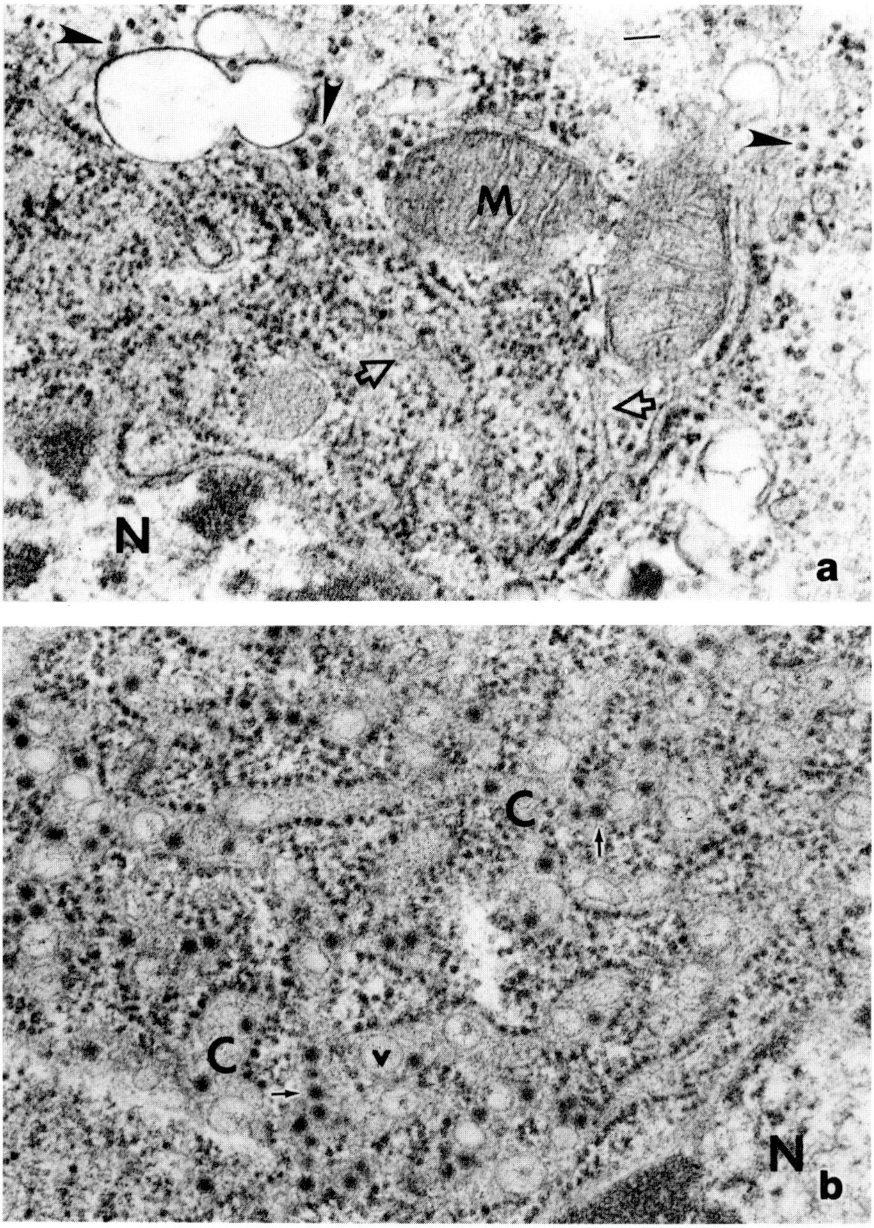


**FIGURE 4.** Comparison between *cis*-type (DEN-2 virus, PR-159 strain) and *trans*-type (JE virus, $SA_{14}$ strain) maturation in C6/36 cells. Bar = 100 nm. a shows DEN-2 virus maturation displaying fragmentation of RER (empty arrow) and dissociation of RER ribosomes. Nucleocapsids that are larger than ribosomes are seen in the cytosol (arrowheads). M, mitochondrion; N, nucleus. b shows JE virus maturation displaying hypertrophy of RER and dilation of the RER cisternae (C). N, nucleus; v, EM vesicles.

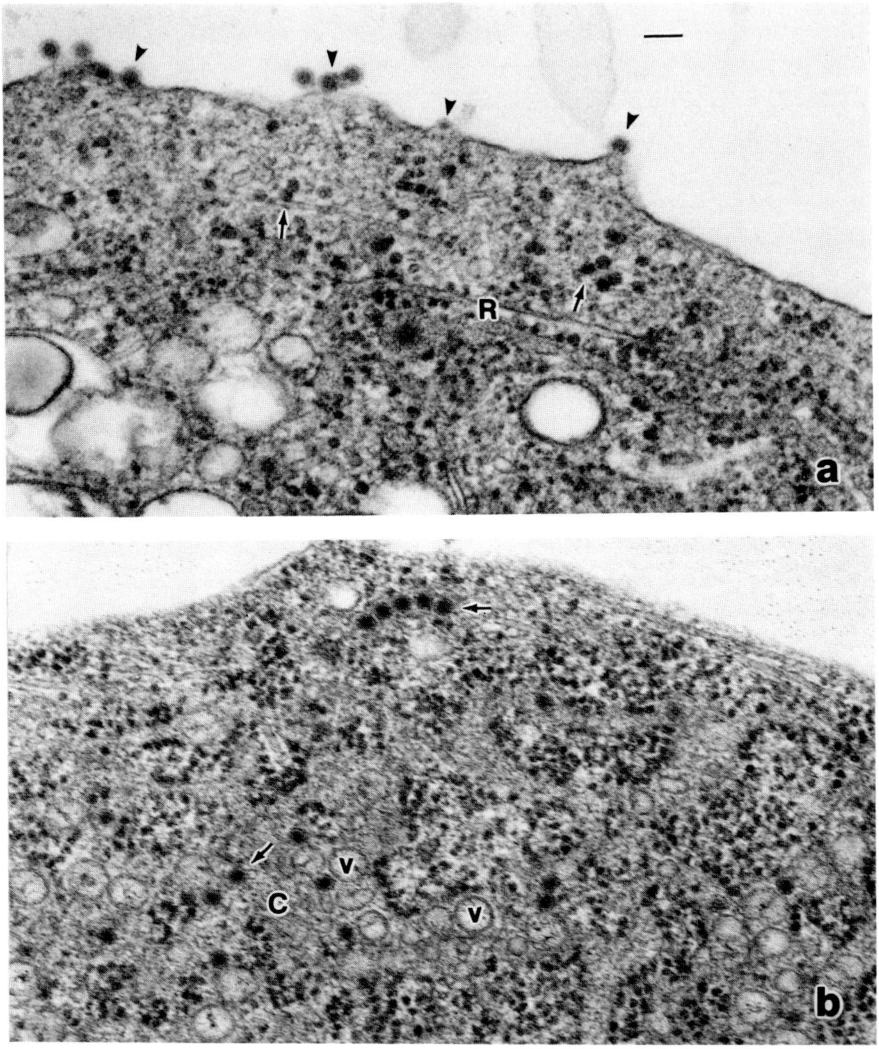


**FIGURE 5.** Comparison between *cis*-type (DEN-2 virus, PR-159 strain) and *trans*-type (JE virus, $SA_{14}$ strain) at the surface of C6/36 cells. Bar = 100 nm. a shows DEN-2 virus maturation displaying budding virions (arrowheads) at the plasma membrane and nucleocapsids (arrows) in the underlying cytoplasm. R, RER. b shows JE virus maturation with no budding virions at the plasma membrane but displaying many virions (arrows) and EM vesicles (v) within hypertrophic RER cisternae in the cytoplasm.

**Table II**
**Flavivirus Strains that Undergo *cis*- or *trans*-Type Maturation in C6/36 Cells**

| *cis*-Type maturation | *trans*-Type maturation |
|---|---|
| DEN-2 virus, PR-159 strain | DEN-2 virus, NGC strain<br>JE virus, $SA_{14}$, $SA_{14}$-14-2, Nakayama strains<br>WN virus, E101 strain<br>MVE virus<br>KUN virus |

sumura *et al.* (1982) reported observing budding virions at the cell surface and assembling virions in the RER cisternae, raising the possibility that some flaviviruses undergo both *cis*- and *trans*-type maturation simultaneously. These observations may suggest that flaviviruses have evolved diverse maturation processes, involving complex series of virus–host cell molecular interactions in their life cycles in vertebrate and invertebrate hosts. Among different strains of DEN-2 virus, the PR-159 strain underwent clear-cut *cis*-type maturation, while the NGC strain underwent clear-cut *trans*-type maturation. This raises an interesting question: Can a flavivirus switch from *cis*- to *trans*-type maturation, or vice versa? In an attempt to answer this question, the NGC strain was passaged 20 times in C6/36 cells; however, examination of the 20th-passaged virus demonstrated no change from a *trans*-type maturation (P. L. Summers and T. Hase, unpublished data). The JE viruses, $SA_{14}$ and $SA_{14}$-14-2, although one is the nonattenuated parent virus ($SA_{14}$) and the other is attenuated ($SA_{14}$-14-2), both exhibited *trans*-type maturation. Evidence obtained, so far, suggests that *cis*- or *trans*-type maturation is a fairly stable trait of an individual virus strain.

### 4.4. Assembly via *cis*- and *trans*-Type Maturation

The occurrence of *cis*- or *trans*-type maturation for a flavivirus seems to be decided at an early stage of nucleocapsid formation, possibly depending on the movement of newly synthesized viral C protein either into the cytosol or the RER cisternae, respectively. As a result, the nucleocapsid forms in the cytosol for the *cis*-type maturation, and the whole virion assembles in the RER cisternae for the *trans*-type maturation. The observation that, during *in vitro* translation of TBE virus RNA, structural proteins are produced more efficiently when membrane fractions are included (Svitkin, *et al.*, 1984) suggests that membrane-associated polysomes (RER) are involved in viral protein synthesis. It is most likely that the C protein is translated on the RER. At the same time, viral RNA synthesis proceeds on the SER membrane. Synthesis of both of these nucleocapsid components on the cytoplasmic membranes in a proximate area would promote RNA-C complexing for the formation of nucleocapsids.

For *cis*-type maturation, viral RNA and C combine, and the nucleocapsid forms in the cytosol; in *trans*-type maturation, the nucleocapsid formation per se is not observed; instead, complete virions assemble within the RER cisternae. As an early event, complexing of C to viral RNA is a necessary step. For *cis*-type maturation, it apparently occurs in the cytosol. In the infection of chick embryo fibroblasts with Sindbis virus, a typical budding alphavirus, Wirth *et al.* (1977) observed that the core protein newly translated from subgenomic 26S RNA is localized on the cytoplasmic side of the RER membrane and released into the cytosol, while newly synthesized envelope proteins are sequestered by the lipid bilayer. The same kind of post-translational cleavage of the C protein may occur in the *cis*-type maturation of a flavivirus for subsequent RNA-C complexing in the cytosol.

For *trans*-type maturation, newly synthesized C protein apparently crosses the RER membrane to form virions within the cisternae. The concept that membrane translocation of proteins from the cytoplasm to the RER cisternae for secretion is guided by the information system that each protein carries within its amino acid sequence is well established (Blobel and Dobberstein, 1975; Wickner and Lodish, 1985; Briggs and Gierash, 1986). Comparative analysis of available genomic RNA sequences between *cis*- and *trans*-type flaviviruses, however, fail to implicate a signal sequence for the membrane translocation of the C protein in *trans*-type maturation. It has become increasingly clear that the mechanism involved in the RER membrane crossing by a nascent protein is quite complex; for the actual translocation of a secretory protein the molecular structure of the protein, its relationship to the RER membrane and membrane-associated enzymes, and other factors may collectively serve as the information complex (Ferenci and Silhavy, 1987; Kaiser *et al.*, 1987). From this point of view, one possibility for the appearance of the C protein within the RER cisternae in *trans*-type maturation is that the cleavage of C from the polyprotein chain or at least from its 3′ neighbor, prM, does not occur. If this is the case, since the prM protein is a *trans*-membrane protein that is inserted into the RER membrane, the C protein may be sequestered into the membrane together with the prM protein, instead of being released into the cytosol. Leary and Blair (1980) suggested that the C protein and viral RNA of JE virus is carried into the cisternae by ER vesicles; they also proposed that the ER vesicles are the precursors of JE virions. Existence of direct morphogenetic relationship between virions and ER vesicles has been disputed (Hase *et al.*, 1987b). However, since the invagination of RER membrane and the resultant formation of ER vesicles within the RER cisternae are the characteristic features of *trans*-type flavivirus replication, it is possible that these membrane changes in connection with the RER may somehow be related to the translocation of viral RNA, C, and the RNA-C complex for subsequent virion assembly within the cisternae.

The genetic changes involved in the divergence of *cis*- and *trans*-type flaviviruses is not clear at present. In this respect, accumulated genomic RNA sequence data indicate that there is considerable diversity in amino acid sequences of the structural and nonstructural proteins among flaviviruses. In some comparative analyses of the structural proteins, C shows the lowest sequence homology, prM an intermediate, and E the highest (Bell *et al.*, 1985; Mason *et al.*, 1987; Sumiyoshi *et al.*, 1987; Hahn *et al.*, 1988).

### 4.5. Intracellular Movement of Virus

The intracellular movement of flaviviruses that undergo *cis*- and *trans*-type maturation differs markedly. The nucleocapsid of *cis*-type virus forms in the cytosol and moves to the plasma or cytoplasmic vacuolar membrane for budding and envelope acquisition. However, the presence of flaviviruses whose nucleocapsids form in the cytosol but fail to move to the host cell membrane for budding, such as YF virus, 17D vaccine strain, and DEN-4 virus, Caribbean 814669 strain, suggests that the formation of nucleocapsids in the cytosol does not automatically ensure that they will bud and that additional mechanisms are involved in nucleocapsid movement and budding. These mechanisms need to be elucidated in the future. In the consideration of molecular composition and structure of flavivirus nucleocapsids, the C protein is quite basic, containing a high proportion of lysine and arginine (Rice *et al.*, 1985), and these basic residues of the C protein are expected to complex with the phosphate backbone of the genomic RNA. There is also a hydrophobic stretch of uncharged amino acids in a middle region of the C protein, which is conserved among flaviviruses (Rice *et al.*, 1985; Dalgarno *et al.*, 1986). It is likely that, as the charged residues of the C-protein complex with the genomic RNA, a macromolecular nucleoprotein is formed with the external surface consisting of the hydrophobic portion of the C protein. By virtue of its surface hydrophobicity, then, the nucleocapsid may act like a membrane vesicle and move along a lipid gradient in the cytoplasm toward the membrane. The precise time when prM and E are incorporated into the nucleocapsid also remains vague, since the insertion of these membrane proteins into the host plasma membrane has not so far been confirmed (Hase *et al.*, 1987a).

During *trans*-type maturation, nucleocapsid formation is apparently coupled to envelope assembly, with a resultant appearance of the virion in the RER cisternae. The virion then uses the cell secretory channel to find its way to the extracellular space. Modification of the viral envelope at the Golgi apparatus may occur during *trans*-type maturation. In the infection of mouse brain cells with JE virus, it has been shown that virions leave the Golgi apparatus within coated vesicles, suggesting that some kind of ligand–receptor interaction takes place between the viral envelope and the *trans*-face membrane of the Golgi apparatus (Hase *et al.*, 1987b).

## 5. VIRAL EXIT

Before the discovery of *cis*- and *trans*-type maturation, viral release was not clearly defined. Since flavivirus virions were usually found in dilated RER cisternae and within cytoplasmic vacuoles of infected cells, it was generally assumed that virions in these locations were eventually released extracellularly by exocytosis. With the proposed model of maturation, it has become clear that flavivirus virions appear extracellularly by budding in the case of *cis*-type maturation and by secretory-type exocytosis in the case of *trans*-type maturation.

### 5.1. Budding of *cis*-Type Virus

The budding for flavivirus *cis*-type maturation resembles that of alphaviruses in that distinct nucleocapsids form in the cytoplasm and move to the host cell membranes. In the infection of C6/36 cells with DEN-2 virus, PR-159 strain, however, we could not confirm a progressive nucleocapsid enwrapment process by host cell membrane, because the envelopes of budding virions showed somewhat different structure from the adjacent host cell membrane, and the host cell membrane at the budding sites was usually damaged and did not show unequivocal continuity with the viral envelope (Hase *et al.*, 1987a) (Fig. 6). In the cytoplasmic vacuoles in which vigorous maturation of virions were taking place, we found filamentous forms of virions that probably occurred through linear, helical arrangement of DEN virus genomes and their coenvelopment by common membranes. It is difficult to explain envelopment of a filamentous form on the basis of enwrapment of budding viral nucleocapsids by existing

**FIGURE 6.** Budding virions of DEN-2 virus, PR-159, in C6/36 cells. Bar = 100 nm. (a–c) Different stages of budding virions (small arrowheads), which do not show continuity of their membrane envelopes with the plasma membrane. A large arrowhead points to an area of plasma membrane damage adjacent to the budding virions. Arrows point to the stalks of fuzzy material that still connect the budding virions to the cell surface. (d) Budding virion (small arrowhead) whose envelope shows a structural difference from the contiguous plasma membrane. Large arrowhead points to an area of plasma membrane damage. Arrows point to nucleocapsids in the underlying cytoplasm. M, mitochondrion. (e) Budding virions that accumulate on the cell surface. The virions in the right of the picture are sandwiched between two sheets of membrane as if they lifted up a part of the plasma membrane (arrowheads). Arrows point to nucleocapsids in the underlying cytoplasm. rer, rough endoplasmic reticulum.

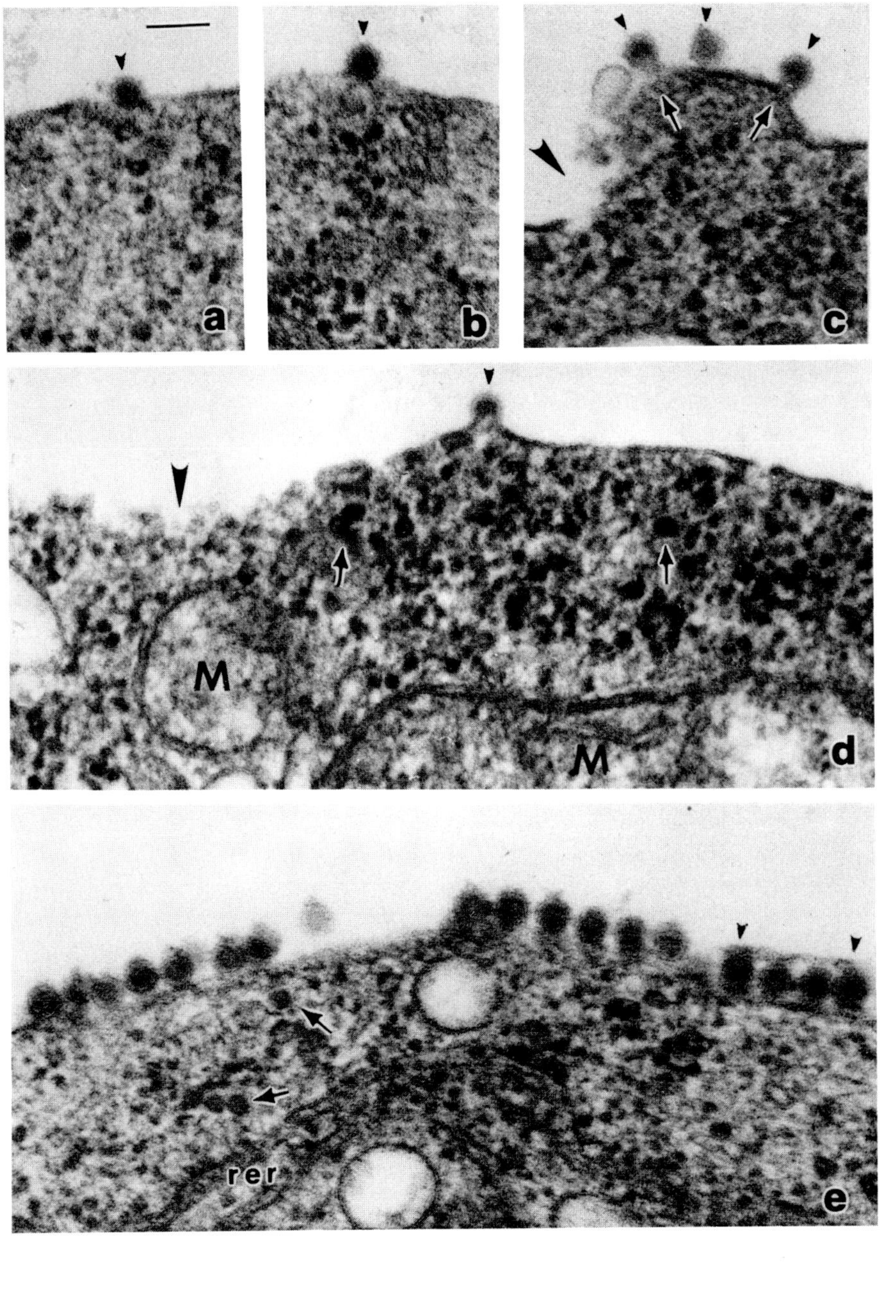
a
b
c
M
M
d
rer
e

host cell membrane. From the available evidence, we cannot rule out the possibility that flavivirus nucleocapsids assemble their envelopes *de novo* at the budding sites.

In immunoelectron microscopy of C6/36 cells infected with DEN-2 virus, PR-159, DEN-2 specific antibodies localized in areas of virus budding, with host plasma membrane and cytoplasmic components adjacent to the budding virions showing less intense staining when the cells were reacted with DEN-2 polyclonal antibody and no staining at all when the cells were reacted with DEN-2 monoclonal antibody specific for the viral envelope glycoprotein (Hase *et al.*, 1987a). The diffuse staining of cellular components by the polyclonal antibody, therefore, appears to represent reaction of antibodies with nonvirion polypeptides, particularly NS1, that is known to be expressed on the infected-cell surface (Brandt *et al.*, 1970; Russell *et al.*, 1970; Smith *et al.*, 1985). The negative reaction of the plasma membrane with an E specific monoclonal antibody may be due to a conformational change of E after its insertion in the membrane. Nonetheless, at present the absence of experimental evidence for the incorporation of viral membrane proteins into the host plasma membrane presents uncertainty about the time and location at which prM and E are docked on the nucleocapsid. Although the orientation of prM has not been proved, the conserved glycosylation sites near the N terminus and the C-terminal hydrophobic domain suggest a transmembrane protein with an external glycosylated portion and a C-terminus associated with membrane (Rice *et al.*, 1986b). No M has been detected in the infected cell; therefore, it appears that during maturation of virus particles, a glycopeptide is removed from the prM, leaving the nonglycosylated M protein embedded in the virion envelope (Shapiro *et al.*, 1972; Russell *et al.*, 1980). The role of the M or prM protein in flavivirus morphogenesis remains unclear. The orientation of E has also not been established, but again appears to be N-terminus outside and C-terminus spanning the lipid bilayer (Rice *et al.*, 1986b). The E protein forms spikes which project from the viral envelope and contain many biologically active sites.

It is not clear what kind of kinetic movement is involved at the host plasma membrane to push budding flavivirus virions outward since there is no confirmatory evidence for insertion of the viral membrane proteins into the host plasma membrane in *cis*-type maturation. This and the budding features of a *cis*-type flavivirus make it questionable that the zipper-like linking between the viral glycoproteins embedded in the host cell membrane and the nucleocapsid protein proposed for alphavirus budding (Brown, 1980; Simons *et al.*, 1980) is involved as an expulsion mechanism for flavivirus budding. Budding flavivirions tend to accumulate at the cell surface, indicating that the host cell membrane offers a substantial barrier against virus passage and that budding virions had a long transit time at the budding sites (Hase *et al.*, 1987a). It is possible that the biological significance of the nucleocapsid movement to the host cell membrane for budding is to place the nucleocapsid in the microenvironment of a high concentration of membrane material or to use existing membrane as a primer for the initiation of viral envelope assembly. Further studies are needed to explain the molecular interactions that are involved in guiding nucleocapsids to the cell surface and into the extracellular space.

### 5.2. Secretory-Type Exocytosis of *trans*-Type Virus

The initial stages of viral nucleocapsid formation for *trans*-type viruses are unclear because of the lack of identification of morphologically distinct nucleocapsids; however, it appears that whole virions assemble within the RER cisternae in the manner of, what Westaway (1980) termed, condensation (Fig. 7). Like the secretory material of host cells, assembled virions in the RER cisternae are then carried through the host secretory channel to be released extracellularly by secretory-type exocytosis (Leary and Blair, 1980; Hase *et al.*, 1987b). *trans*-Type maturation of a flavivirus therefore seems to represent a very efficient form of viral parasitism that takes full advantages of both the host biosynthetic and transport systems for viral replication and growth. Virions that have undergone *trans*-type maturation seem to be released rapidly and cleanly from host cells in the same manner as secretory material is ejected, with the host plasma membrane offering no barrier for their release; as a result, it is usually difficult to find released virions attached onto the cell surface. Viral antigens have been demonstrated on the surface of cells infected with DEN-2 virus, NGC strain, by electron microscopic immunostaining with polyclonal antibodies (Catanzaro *et al.*, 1974; Cardiff and Lund, 1976); however, demonstration of virus antigens, whether structural or nonstructural, on the surface of cells infected with a trans-type flavivirus does not seem to carry any significant implication for the viral exit mechanism.

## 6. CONCLUDING REMARKS

A comprehensive picture of flavivirus morphogenesis is emerging through studies of ultrastructure, molecular biology, and molecular genetics of these viruses. Upon entering host cells, flaviviruses replicate on the scaffolding of the cytoskeletal and SER systems and translate their proteins principally on ribosomes associated with RER. Ultrastructural studies of the sites of nucleocapsid formation and virion assembly in host cells have revealed two types of maturation for flaviviruses, termed *cis*- and *trans*-type. During *cis*-type maturation, viral nucleocapsids form in the cytosol and move to the plasma and cytoplasmic vacuolar membranes for budding, acquiring viral envelopes in the process. For *trans*-type maturation, virion assembly occurs in the RER cisternae, and as-

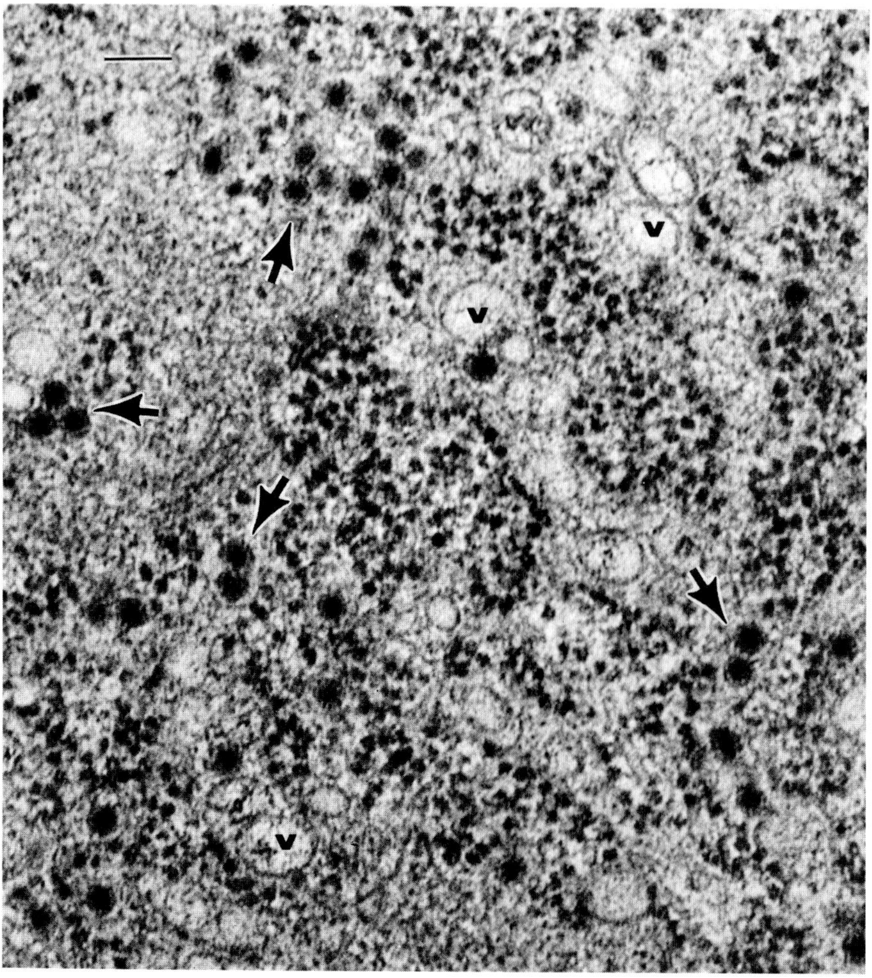

**FIGURE 7.** Assembling virions of JE virus, $SA_{14}$, in the dilated RER cisternae. All the virions possess envelopes of varying clarities and projecting spikes on their surface. v, EM vesicle. Bar = 100 nm.

sembled virions move along the host cell secretory channel to appear extracellularly by secretory-type exocytosis. The complexing of viral RNA and C protein and the subsequent intracellular movement of the complex seem to play a central role for the development of both types of maturation; however, precise molecular mechanisms involved in the process need to be elucidated in the future. The molecular interactions occurring between viral RNA and viral structural and possibly nonstructural proteins during viral assembly are also poorly understood at present. It is hoped that comparative analyses of nucleotide and amino acid sequences among flaviviruses will eventually help us to understand entire flavivirus assembly process. The mechanisms by which flaviviruses pass through the host cell in entry and exit also need to be further clarified, since an understanding of these mechanisms is crucial to the formulation of successful preventive and therapeutic measures against flavivirus infections. The two different intracellular routes taken by maturing *cis*- and *trans*-type flaviviruses make these viruses very useful morphologic markers not only for the study of intracellular viral movement but also for the study of intracellular macromolecular movement in general.

## 7. REFERENCES

Bell, J. R., Kenney, R. M., Trent, D. W., Lenches, E. M., Dalgarno, L., and Strauss, J. H., 1985, Amino-terminal amino acid sequences of structural proteins of three flaviviruses, *Virology* **143:** 224–229.

Berger, E. G., Buddecke, E., Kamerling, J. P., Kobata, A., Paulson, J. C., and Vliegenthart, J. F. G., 1982, Structure, biosynthesis, and functions of glycoprotein glycans, *Experientia* **38:**1129–1258.

Bhamarapravati, N., Yoksan, S., Chayaniyayothin, S., Angsubphakorn, S., and Bunyaratvej, A., 1987, Immunization with a live attenuated dengue-2 virus candidate vaccine (16681-PDK 53): clinical, immunological, and biological responses in adult volunteers, *Bull. WHO* **65:**189–195.

Biedrzycka, A., Cauchi, M. R., Bartholomeuz, A., Gorman, J. J., and Wright, P. J., 1987, Characterization of protease cleavage sites involved in the formation of the envelope glycoprotein and three non-structural proteins of dengue virus type-2, New Guinea C strain, *J. Gen. Virol.* **68:**1317–1326.

Blobel, G., and Dobberstein, B., 1975, Transfer of proteins across membranes. I. Presence of proteolytically processed and unprocessed nascent immunoglobulin light chains on membrane-bound ribosomes of murine myeloma, *J. Cell Biol.* **67:**835–851.

Boege, U., Heinz, F. X., Wengler, G., and Kunz, C., 1983, Amino acid compositions and amino-terminal sequences of the structural proteins of a flavivirus, European tick-borne encephalitis virus, *Virology* **126:**651–657.

Boulton, R. W., and Westaway, E. G., 1972, Comparisons of togaviruses: Sindbis virus (group A) and Kunjin virus (group B), *Virology* **49:**283–289.

Boulton, R. W., and Westaway, E. G., 1976, Replication of the flavivirus: Proteins, glycoproteins, and maturation associated with cell membranes, *Virology* **69:**416–430.

Boulton, R. W., and Westaway, E. G., 1977, Togavirus RNA: Reversible effect of urea on genomes and absence of subgenomic viral RNA in Kunjin virus-infected cells, *Arch. Virol.* **55:**201–208.

Brandt, W. E., Cardiff, R. D., and Russell, P. K., 1970, Dengue virions and antigens in brain and serum of infected mice, *J. Virol.* **6:**500–506.

Brandt, W. E., McCown, J. M., Gentry, M. K., and Russell, P. K., 1982, Infection enhancement of dengue type 2 virus in the U-937 human monocyte cell line by antibodies to flavivirus cross-reactive determinants, *Infect. Immun.* **36:**1036–1041.

Briggs, M. S., and Gierasch, L. M., 1986, Molecular mechanisms of protein secretion: The role of the signal sequence, *Adv. Protein Chem.* **38:**110–174.

Brinton, M. A., 1982, Characterization of West Nile virus persistent infections in genetically resistant and susceptible mouse cells. I. Generation of defective nonplaquing virus particles, *Virology* **116:**84–98.

Brinton, M. A., 1986, Replication of flaviviruses, in: *The Togaviridae and Flaviviridae* (S. Schlesinger and M. J. Schlesinger, eds.), pp. 327–374, Plenum, New York.

Brown, D. T., 1980, The assembly of alphaviruses, in: *The Togaviruses. Biology, Structure, Replication* (R. W. Schlesinger, ed.), pp. 437–501, Academic, New York.

Cardiff, R. D., and Lund, J. K., 1976, Distribution of dengue-2 antigens by electron immunocytochemistry, *Infect. Immun.* **13:**1699–1709.

Cardiff, R. D., Dalrymple, J. M., and Russell, P. K., 1973a, RNA polymerase in group B arbovirus (dengue-2) infected cells, *Arch. Ges. Virusforsch.* **40:**392–396.

Cardiff, R. D., Russ, S. B., Brandt, W. E., and Russell, P. K., 1973b, Cytological localization of dengue-2 antigens: An immunological study with ultrastructural correlation, *Infect. Immun.* **7:** 809–816.

Castle, E., Nowak, T., Leidner, U., Wengler, G., and Wengler, G., 1985,Sequence analysis of the viral core protein and the membrane-associated proteins V1 and NV2 of the flavivirus West Nile virus and of the genome sequence for these proteins, *Virology* **145:**227–236.

Castle, E., Leidner, U., Nowak, T., Wengler, G., and Wengler, G., 1986, Primary structure of the West Nile flavivirus genome region coding for all nonstructural proteins, *Virology* **149:**10–26.

Catanzaro, P. J., Brandt, W. E., Hogrefe, W. R., and Russell, P. K., 1974, Detection of dengue cell-surface antigens by peroxidase-labeled antibodies and immune cytolysis, *Infect. Immun.* **10:**381–388.

Chamberlain, R. W., 1980, Epidemiology of arthropod-borne togaviruses: the role of arthropods as hosts and vectors and of vertebrate hosts in normal transmission cycles, in: *The Togaviruses. Biology, Structure, Replication* (R. W. Schlesinger, ed.), pp. 175–227, Academic, New York.

Chu, P. W. G., and Westaway, E. G., 1985, Replication strategy of Kunjin virus: Evidence for recycling role of replicative form RNA as a template in semiconservative and asymmetric replication, *Virology* **140:**68–79.

Chu, P. W. G., and Westaway, E. G., 1987, Characterization of Kunjin virus RNA-dependent RNA polymerase: Reinitiation of synthesis in vitro, *Virology* **157:**330–337.

Cleaves, G. R., and Dubin, D. T., 1979, Methylation status of intracellular dengue type 2 40S RNA, *Virology* **96:**159–165.

Cleaves, G. R., Ryan, T. E., and Schlesinger, R. W., 1981, Identification and characterization of type 2 dengue virus replication intermediate and replicative form RNAs, *Virology* **111:**73–85.

Clegg, J. C. S., 1982, Glycoprotein detection in nitrocellulose transfers of electrophoretically separated protein mixtures using concanavalin A and peroxidase: Application to arenavirus and flavivirus proteins, *Anal. Biochem.* **127:**389–394.

Dalgarno, L., Trent, D. W., Strauss, J. H., and Rice, C. M., 1985, Partial nucleotide sequence of the Murray Valley encephalitis virus genome. Comparison of the encoded polypeptides with yellow fever virus structural and nonstructural proteins, *J. Mol. Biol.* **187:**309–323.

Daughaday, C. C., Brandt, W. E., McCown, J. M., and Russell, P. K., 1981, Evidence for two mechanisms of dengue virus infection of adherent human monocytes: Trypsin-sensitive virus receptors and trypsin-resistant immune complex receptors, *Infect. Immun.* **32:**469–473.

Della Porta, A. J., and Westaway, E. G., 1977, Immune response in rabbits to virion and nonvirion antigens of the Flavivirus Kunjin, *Infect. Immun.* **15:**874–882.

DeMadrid, A. T., and Porterfield, J. S., 1974, The flaviviruses (group B arboviruses): A cross-neutralization study, *J. Gen. Virol.* **23:**91–96.

Deubel, V., Kinney, R. M., and Trent, D. W., 1986, Nucleotide sequence and deduced amino acid sequence of the structural proteins of dengue type 2 virus, Jamaica genotype, *Virology* **155:**365–377.

Eckels, K. H., Brandt, W. E., Harrison, V. R., McCown, J. M., and Russell, P. K., 1976, Isolation of a temperature-sensitive dengue-2 virus under conditions suitable for vaccine development, *Infect. Immun.* **14:**1221–1227.

Eckels, K. H., Harrison, V. R., Summers, P. L., and Russell, P. K., 1980, Dengue-2 vaccine: Preparation from a small plaque virus clone, *Infect. Immun.* **27:**175–180.

Filshie, B. K., and Rehacek, J., 1968, Studies of the morphology of Murray Valley encephalitis and Japanese encephalitis viruses growing in cultured mosquito cells, *Virology* **34:**435–443.

Ferenci, T., and Silhavy, T. J., 1987, Minireview. Sequence information required for protein translocation from the cytoplasm, *J. Bacteriol.* **169:**5339–5342.

Gollins, S. W., and Porterfield, J. S., 1985, Flavivirus infection enhancement in macrophages: An electron microscopic study of viral cellular entry, *J. Gen. Virol.* **66:**1969–1982.

Hahn, Y. S., Galler, R., Hunkapiller, T., Dalrymple, J. M., Strauss, J. H., and Strauss, E. G., 1988, Nucleotide sequence of dengue 2 RNA and comparison of the encoded proteins with those of other flaviviruses, *Virology* **162:**167–180.

Halstead, S. B., 1979, In vitro enhancement of dengue infection with passively transferred antibody, *J. Infect. Dis.* **140:**527–533.

Halstead, S. B., 1988, Pathogenesis of dengue: Challenges to molecular biology, *Science* **239:**476–481.

Halstead, S. B., Chow, J. S., and Marchette, N. J., 1973, Immunological enhancement of dengue virus replication, *Nature (Lond.)* **243:**24–26.

Halstead, S. B., O'Rourke, E. J., and Allison, A. C., 1977, Dengue viruses and mononuclear phagocytes. II. Identity of blood and tissue leukocytes supporting in vitro infection, *J. Exp. Med.* **146:**218–229.

Harrison, V. R., Eckels, K. H., Sagartz, J. W., and Russell, P. K., 1977, Virulence and immunogenicity of a temperature-sensitive dengue-2 virus in lower primates. *Infect. Immun.* **18:**151–156.

Hase, T., Summers, P. L., Eckels, K. H., and Baze, W. B., 1987a, An electron and immunoelectron microscopic study of dengue-2 virus infection of cultured mosquito cells: Maturation events, *Arch. Virol.* **92:**273–291.

Hase, T., Summers, P. L., Eckels, K. H., and Baze, W. B., 1987b, Maturation process of Japanese encephalitis virus in cultured mosquito cells in vitro and mouse brain cells in vivo, *Arch. Virol.* **96:**135–151.

Hase, T., Summers, P. L., and Eckels, K. H., 1989, Flavivirus entry into cultured mosquito cells and human peripheral blood monocytes, *Arch. Virol.*, in press.

Heinz, F. X., Tuma, W., and Kunz, C., 1981, Antigenic and immunogenic properties of defined physical forms of tickborne encephalitis virus structural proteins, *Infect. Immun.* **33:**250–257.

Holland, J., Spindler, K., Horoddyski, F., Grabau, E., Nichol, S., and Van de Pol, S., 1982, Rapid evolution of RNA genomes, *Science* **215:**1577–1585.

Hong, S. S., and Ng, M. L., 1987, Involvement of microtubules in Kunjin virus replication, *Arch. Virol.* **97:**115–121.

Igarashi, A., 1979, Characteristics of *A. albopictus* cells persistently infected with dengue viruses, *Nature (Lond.)* **280:**690–691.

Igarashi, A., Fukunaga, T., and Fukai, I., 1964, Sedimentation characteristics of Japanese encephalitis virus ribonucleic acid, *Biken J.* **7:**111–119.

Kaiser, C. A., Preuss, D., Grisafi, P., and Botstein, D., 1987, Many random sequences functionally replace the secretion signal sequence of yeast invertase, *Science* **235:**312–317.

Kimura, T., Gollins, S. W., and Porterfield, J. S., 1986, The effect of pH on the early infection of West Nile virus with P388D1 cells, *J. Gen. Virol.* **67:**2423–2433.

Kitano, T., Suzuki, K., and Yamaguchi, T., 1974, Morphological, chemical, and biological characterization of Japanese encephalitis virus virion and its hemagglutinin, *J. Virol.* **14:**631–639.

Kos, K. A., Osborne, B. A., and Goldsby, R. A., 1975, Inhibition of group B arbovirus antigen production and replication in cells enucleated with cytochalasin B, *J. Virol.* **15:**913–917.

Leary, K., and Blair, C. D., 1980, Sequential events in the morphogenesis of Japanese encephalitis virus, *J. Ultrastruct. Res.* **72:**123–129.

Lobigs, M., Dalgarno, L., Schlesinger, J. J., and Weir, R. C., 1987, Location of a neutralization determinant in the E protein of yellow fever virus (17D vaccine strain), *Virology* **161:**474–478.

Matsumura, T., Stoller, V., and Schlesinger, R. W., 1971, Studies on the nature of dengue viruses. V. Structure and development of dengue virus in Vero cells, *Virology* **46:**344–355.

Mackow, E., Makino, Y., Zhao, B., Zhang, Y. M., Markoff, L., Buckler-White, A., Guiler, M., Chanock, R., and Lai, C. J., 1987, The nucleotide sequence of dengue type 4 virus: Analysis of gene coding for nonstructural proteins, *Virology* **159:**217–228.

Mason, P. W., McAda, P. C., Dalrymple, J. M., Fournier, M. J., and Mason, T. R., 1987, Expression of Japanese encephalitis virus antigens in *Escherichia coli, Virology* **158:**361–372.

Mason, P. W., McAda, P. C., Mason, T. L., and Fournier, M. J., 1987, Sequence of the dengue-1 virus genome in the region encoding the three structural proteins and the major nonstructural protein NS1, *Virology* **161:**262–267.

Mayer, V., 1963, Two variants of tick-borne encephalitis showing different plaque morphology, *Virology* **20:**372–390.

McAda, P. C., Mason, P. W., Schmaljohn, C. S., Dalrymple, J. M., Mason, T. L., and Fournier, M. J., 1987, Partial nucleotide sequence of the Japanese encephalitis virus genome, *Virology* **158:**348–360.

Monckton, R. P., and Westaway, E. G., 1982, Restricted translation of the genome of the Flavivirus Kunjin in vitro, *J. Gen. Virol.* **63:**227–232.

Murphy, F. A., 1980, Togavirus morphology and morphogenesis, in: *The Togaviruses. Biology, Structure, Replication* (R. W. Schlesinger, ed.), pp. 241–315, Academic, New York.

Murphy, F. A., Harrison, A. K., Gary, G. W., Jr., Whitfield, S. G., and Forrester, F. G., 1968, St. Louis encephalitis virus infection of mice. Electron microscopic studies of central nervous system, *Lab. Invest.* **19:**652–662.

Naeve, C. W., and Trent, D. W., 1978, Identification of St. Louis encephalitis virus mRNA, *J. Virol.* **25:**535–545.

Ng, M. L., 1987, Ultrastructural studies of Kunjin virus-infected *Aedes albopictus* cells, *J. Gen. Virol.* **68:**577–582.

Ng, M. L., Pedersen, J. S., Toh, B. H., and Westaway, E. G., 1983, Immunofluorescent sites in Vero cells infected with the flavivirus Kunjin, *Arch. Virol.* **78:**177–190.

Nowak, T., and Wengler, G., 1987, Analysis of disulfides present in the membrane proteins of the West Nile flavivirus, *Virology* **156:**127–137.

Ota, E., 1965, Electron microscopic study of the development of Japanese encephalitis virus in porcine kidney stable (PS) cells, *Virology* **25:**372–378.

Paul, S. D., 1966, Some biological properties of two variants of Kyasanur Forest disease virus, *Indian J. Med. Res.* **54:**419–424.

Peiris, J. S. M., and Porterfield, J. S., 1979, Antibody-mediated enhancement of flavivirus replication in macrophage cell lines, *Nature (Lond.)* **282:**509–511.

Porterfield, J. S., 1980, Antigenic characteristics and classification of Togaviridae, in: *The Togaviruses. Biology, Structure, Replication* (R. W. Schlesinger, ed.), pp. 13–46, Academic, New York.

Porterfield, J. S., Casals, J., Chumakov, M. P., Gaidamovich, S. Y., Hannoun, C., Holmes, I. H., Horzinek, M. C., Mussgay, M., Oken-Blom, N., Russell, P. K., and Trent, D. W., 1978, Togaviridae, *Intervirology* **9:**129–148.

Qureshi, A. A., and Trent, D. W., 1972, Saint Louis encephalitis viral ribonucleic acid replication complex, *J. Virol.* **9:**565–573.

Rice, C. M., Lenches, E. M., Eddy, S. R., Shin, S. J., Sheets, R. L., and Strauss, J. H., 1985, Nucleotide sequence of yellow fever virus: Implications for flavivirus gene expression and evolution, *Science* **229:**726–733.

Rice, C. M., Albersold, R., Teplow, D. B., Pata, J., Bell, J. R., Vorndam, V., Trent, D. W., Brandriss, M. W., Schlesinger, J. J., and Strauss, J. H., 1986a, Partial N-terminal amino acid sequences of three non-structural proteins of two flaviviruses, *Virology* **151:**1–9.

Rice, C. M., Strauss, E. C., and Strauss, J. H., 1986b, Structure of the flavivirus genome, in: *The Togaviridae and Flaviviridae* (S. Schlesinger and M. J. Schlesinger, ed.), pp. 279–326, Plenum, New York.

Roizman, B., 1985, Multiplication of viruses: An overview, in: *Virology* (B. N. Fields, D. M. Knipe, R. M. Chanock, J. L. Melnick, B. Roizman, and R. E. Shope, ed.), pp. 69–75, Raven, New York.

Russell, P. K., Chiewsilp, D., and Brandt, W. E., 1970, Immunoprecipitation analysis of soluble complement-fixing antigens of dengue viruses, *J. Immunol.* **105:**838–845.

Russell, P. K., Brandt, W. E., and Dalrymple, J. M., 1980, Chemical and antigenic structure of flaviviruses, in: *The Togaviruses. Biology, Structure, Replication* (R. W. Schlesinger, ed.), Academic, New York.

Schlesinger, J. J., Brandriss, M. W., and Monath, T. P., 1983, Monoclonal antibodies distinguish between wild and vaccine strains of yellow fever virus by neutralization, hemagglutination inhibition, and immune precipitation of the virus envelope protein, *Virology* **125:**8–17.

Schlesinger, J. J., Brandriss, M. W., and Walsh, E. E., 1985, Protection against 17D yellow fever encephalitis in mice by passive transfer of monoclonal antibodies to the nonstructural glycoprotein gp48 and by active immunization with gp48, *J. Immunol.* **135:**2805–2809.

Schlesinger, J. J., Brandriss, M. W., Cropp, C. B., and Monath, T. P., 1986, Protection against yellow fever in monkeys by immunization with yellow fever virus nonstructural protein NS1, *J. Virol.* **60:**1153–1155.

Schlesinger, J. J., Brandriss, M. W., and Walsh, E. E., 1987, Protection of mice against dengue 2 virus encephalitis by immunization with the dengue 2 virus non-structural glycoprotein NS1, *J. Gen. Virol.* **68:**853–857.

Schlesinger, R. W., 1980, Virus-host interactions in natural and experimental infections with alphaviruses and flaviviruses, in: *The Togaviruses. Biology, Structure, Replication* (R. W. Schlesinger, ed.), pp. 83–106, Academic, New York.

Shapiro, D., Brandt, W. E., Cardiff, R. D., and Russell, P. K., 1971, The proteins of Japanese encephalitis virus, *Virology* **44:**108–124.

Shapiro, D., Brandt, W. E., and Russell, P. K., 1972, Change involving a viral membrane glycoprotein during morphogenesis of group B arboviruses, *Virology* **50:**906–911.

Shapiro, D., Kos, K. A., and Russell, P. K., 1973, Japanese encephalitis virus glycoproteins, *Virology* **56:**88–94.

Shope, R. E., 1980, Medical significance of togaviruses: An overview of diseases caused by togaviruses in man and in domestic and wild vertebrate animals, in: *The Togaviruses. Biology, Structure, Replication* (R. W. Schlesinger, ed.), pp. 47–82, Academic, New York.

Simons, K., Garoff, H., and Helenius, H., 1980, Alphavirus proteins, in: *The Togaviruses. Biology, Structure, Replication* (R. W. Schlesinger, ed.), pp. 317–333, Academic Press, New York.

Sriurairatna, S., Bhamarapravati, N., and Phalavadhtana, O., 1973, Dengue virus infection of mice: Morphology and morphogenesis of dengue type-2 virus in suckling mouse neurons, *Infect. Immun.* **8:**1017–1028.

Smith, G. W., and Wright, P. J., 1985, Synthesis of proteins and glycoproteins in dengue type 2 virus-infected Vero and *Aedes albopictus* cells, *J. Gen. Virol.* **66:**559–571.

Stohlman, S. A., Wisseman, C. L. Jr., Eyler, O. R., and Silverman, D. J., 1975, Dengue virus-induced modification of host cell membranes, *J. Virol.* **16:**1017–1026.

Stohlman, S. A., Eyler, O. R., and Wissemann, C. L., Jr., 1976, Isolation of the dengue virus envelope glycoprotein from membranes of infected cells by concanavalin A affinity chromatography, *J. Virol.* **18:**132–140.

Stollar, V., 1969, Studies on the nature of dengue viruses. IV. The structural proteins of type 2 dengue virus, *Virology* **39:**426–438.

Stollar, V., Stevens, T. M., and Schlesinger, R. W., 1966, Studies on the nature of dengue viruses. II. Characterization of viral RNA and effects of inhibitors of RNA synthesis, *Virology* **30:**303–312.

Stollar, V., Schlesinger, R. W., and Stevens, T. M., 1967, Studies on the nature of dengue viruses. III. RNA synthesis in cells infected with type-2 dengue virus, *Virology* **33:**650–658.

Sumiyoshi, H., Mori, C., Fuke, I., Morita, K., Kuhara, S., Kondou, J., Kikuchi, Y., Nagamatu, H., and Igarashi, A., 1987, Complete nucleotide sequence of the Japanese encephalitis virus genome RNA, *Virology* **161:**497–510.

Svitkin, Y. V., Ugarova, T. Y., Chernovskaya, T. V., Lyapustin, V. N., Lashkevich, V. A., and Agol, V. I., 1981, Translation of tick-borne encephalitis virus (flavivirus) genome in vitro: Synthesis of two structural polypeptides, *Virology* **110:**26–34.

Svitkin, Y. V., Lyapustin, V. N., Lashkevich, V. A., and Agol, V. I., 1984, Differences between translation products of tick-borne encephalitis virus RNA in cell-free system from Krebs-2 cells and rabbit reticulocytes: Involvement of membranes in the processing of nascent precursors of flavivirus structural proteins, *Virology* **135:**536–541.

Takegami, T., Miyamoto, H., Nakamoto, H., and Yasui, K., 1982, Biological activities of the structural proteins of Japanese encephalitis virus, *Acta Virol.* **26:**312–320.

Takehara, M., 1971, Comparative studies on nucleic acid synthesis and virus-induced RNA polymerase activity in mammalian cells infected with certain arboviruses, *Arch. Ges. Virusforsch.* **34:**266–277.

Takeda, H., Oya, A., Hashimoto, K., Yasuda, T., and Yamada, M., 1978, Association of virus specific replicative ribonucleic acid with nuclear membrane in chick embryo cells infected with Japanese encephalitis virus, *J. Gen. Virol.* **38:**281–291.

Theofilopoulos, A. N., Brandt, W. E., Russell, P. K., and Dixon, F. T., 1976, Replication of dengue-2 virus in cultured human lymphoblastoid cells and subpopulations of human peripheral leukocytes, *J. Immunol.* **117:**953–961.

Trent, D. W., 1977, Antigenic characterization of flavivirus structural proteins separated by isoelectric focusing, *J. Virol.* **22:**608–618.

Trent, D. W., Swensen, C. C., and Qureshi, A. A., 1969, Synthesis of Saint Louis encephalitis virus ribonucleic acid in BHK-21/13 cells, *J. Virol.* **3:**385–394.

Trent, D. W., Kinney, R. M., Johnson, B. J., Vorndam, A. V., Grant, J. A., Deubel, V., Rice, C. M., and Hahn, C., 1987, Partial nucleotide sequence of St. Louis encephalitis virus RNA: Structural proteins, NS1, ns2a, and ns2b, *Virology* **156:**293–304.

Wengler, G., 1987, The mode of assembly of alphavirus cores implies a mechanism for the disassembly of the cores in the early stages of infection. Brief review, *Arch. Virol.* **94:**1–14.

Wengler, G., and Wengler, G., 1981, Terminal sequences of the genome and replicative form RNA of the flavivirus West Nile virus: Absence of poly (A) and possible role in RNA replication, *Virology* **113:**544–555.

Wengler, G., Wengler, G., and Gross, H. J., 1978, Studies on virus-specific nucleic acids synthesized in vertebrate and mosquito cells infected with flaviviruses, *Virology* **89:**423–437.

Wengler, G., Beato, M., and Wengler, G., 1979, In vitro translation of 42S virus-specific RNA from cells infected with the flavivirus West Nile virus, *Virology* **96:**516–529.

Wengler, G., Castle, E., Leidner, U., Nowak, T., and Wengler, G., 1985, Sequence analysis of the membrane protein V3 of the flavivirus West Nile virus and of its gene, *Virology* **147:**264–274.

Wengler, G., Wengler, G., Nowak, T., and Wahn, K., 1987, Analysis of the influence of proteolytic cleavage on the structural organization of the surface of the West Nile flavivirus leads to the isolation of a protease-resistant E protein oligomer from the viral surface, *Virology* **160:**210–219.

Westaway, E. G., 1975, The proteins of Murray Valley encephalitis virus, *J. Gen. Virol.* **27:**283–293.

Westaway, E. G., 1980, Replication of flaviviruses, in: *The Togaviruses. Biology, Structure, Replication* (R. W. Schlesinger, ed.), pp. 531–581, Academic, New York.

Westaway, E. G., and Goodman, M. R., 1987, Variation in distribution of the three flavivirus-specified glycoproteins detected by immunofluorescence in infected Vero cells, *Arch. Virol.* **94:** 215–228.

Westaway, E. G., and Reedman, B. M., 1969, Proteins of the group B arbovirus Kunjin, *J. Virol.* **4:** 688–693.

Westaway, E. G., Brinton, M. A., Gaidamovich, S. Y., Horznek, M. C., Igarashi, A., Kaariainen, L., Lvov, D. K., Porterfield, J. S., Russell, P. K., and Trent, D. W., 1985, Flaviviridae, *Intervirology* **24:**183–192.

Wickner, W. T., and Lodish, L. M., 1985, Multiple mechanisms of protein insertion into and across membranes, *Science* **230:**400–407.

Winkler, G., Heinz, F. X., and Kunz, C., 1987, Studies on the glycosylation of flavivirus E proteins and the role of carbohydrate in antigenic structure, *Virology* **159:**237–243.

Wirth, D. F., Katz, F., Small, B., and Lodish, H. F., 1977, How a single Sindbis virus mRNA directs the synthesis of one soluble protein and two integral membrane glycoproteins, *Cell* **10:** 253–263.

Wright, P. J., 1982, Envelope protein of the flavivirus Kunjin is apparently not glycosylated, *J. Gen. Virol.* **59:**29–38.

Wright, P. J., Warr, H. M., and Westaway, E. G., 1981, Synthesis of glycoproteins in cells infected by the Flavivirus Kunjin, *Virology* **109:**418–427.

Zebovitz, E., Leong, J. K. L., and Doughty, S. C., 1972, Japanese encephalitis virus replication. A procedure for the selective isolation and characterization of viral RNA species, *Arch. Ges. Virusforsch.* **38:**319–327.

Zebovitz, E., Leong, J. K. L., and Doughty, S. C., 1974, Involvement of the host cell nuclear envelope membranes in the replication of Japanese encephalitis virus, *Infect. Immun.* **10:**204–211.

Zhao, B., Mackow, E., Buckler-White, A., Markoff, L., Chanock, R. M., Lai, C. J., and Makino, Y., 1986, Cloning full-length dengue type 4 viral DNA sequences: Analysis of genes coding for structural proteins, *Virology* **155:**77–88.

*Chapter 10*

# Signals for Membrane-Associated Transport in Eukaryotic Cells

Heiner Niemann, Thomas Mayer, and Teruko Tamura

## 1. INTRODUCTION

The signals that target proteins of eukaryotic cells from the site of synthesis to their final destination have attracted much attention since Blobel and Dobberstein (1975a,b) proposed the classic signal hypothesis. Today, however, the picture is more complex than originally anticipated, and the sorting machinery of eukaryotic cells is far from being understood at the molecular level.

Each of the roughly $10^5$ proteins synthesized in the cytosol of a mammalian cell is synthesized on the same class of ribosomes. The messenger RNA (mRNA) binds via its 5′-cap-structure to the ribosome and translation is generally initiated at the first AUG codon. The efficiency of translation of a given mRNA can be predicted to some extent by applying Kozak's rule, which stresses the role of the 6-nucleotides upstream from the AUG codon (emphasizing an A or G residue in position-3, C in positions-2 and -1, and a G immediately following the AUG codon) (Kozak, 1987). About 40 amino acids of a nascent polypeptide chain have to be synthesized before its N-terminal end leaves the ribosomes and can begin to interact with other proteins or itself to fold into a characteristic structure.

---

**Heiner Niemann, Thomas Mayer, and Teruko Tamura** Institute for Medical Virology, Justus-Liebig University, D-6300 Giessen, Federal Republic of Germany.

Such immediate folding of the polypeptide chain may cause difficulties in the translocation of this protein through membrane barriers that have to be passed by the peptide to reach its final destination in the cell. Therefore, folding may counteract specific targeting signals on the protein making specific factors (e.g., heat-shock proteins) and energy in form of adenosine triphosphate (ATP) necessary to unfold the protein again into a translocation- and/or transport-competent structure.

This chapter describes primarily the signals involved in the secretory and membrane-associated protein transport from the rough endoplasmic reticulum (RER) through the Golgi complex to further organelles. In general, we focus on the viral proteins that served as model compounds to study intracellular biosynthetic pathways. Recent studies show that the mechanisms behind protein transport into various organelles in mammalian and yeast cells are more closely related than had been originally expected. Consequently, this chapter also encompasses data from yeast systems, where appropriate. In addition, the current view of transport into other cell organelles is briefly described to demonstrate similarities and differences in the mechanisms leading to recognition of the target organelle.

## 2. MEMBRANE TRANSLOCATION MECHANISMS

### 2.1. Import into the Nucleus

The nuclear pore [90–110 Å in diameter) (Peters, 1986) poses no barrier to small molecules and proteins. The passage of large molecules, however, depends on the presence of a signal sequence within the polypeptide chain. The signal sequences are not proteolytically removed upon translocation into the nucleus, and the position of the signal is variable. Precise signal sequences of several nuclear proteins are known and are listed in Table I. The signal of the T antigen of simian virus 40 (SV40) (PKK$_{128}$KRKV) is probably the best defined.

The nuclear import of a polypeptide can be separated into two distinct steps. After a fast binding step to the cytoplasmic face of the nuclear pore complex, an event that does not require energy and occurs also at low temperature, the polypeptide is translocated in an ATP-dependent process (Newmeyer and Forbes, 1988; Richardson *et al.*, 1988). *In vitro,* this second step can be inhibited by wheat germ agglutinin, which binds to glycosylated constituents of the pore complex but does not physically block the pore. Presumably this lectin, carrying two binding sites for terminal *N*-acetylglucosamine, crosslinks the pore proteins that carry *O*-glycosidically linked *N*-acetylglucosamine residues (Davis and Blobel, 1987; Park *et al.*, 1987), preventing expansion of the pore which is capable of extending its diameter to ~200 Å (Feldherr *et al.*, 1984). It is unclear whether the energy is needed for the release of the protein from the signal

**Table I**
**Comparison of the Import of Proteins into the Nucleus, Mitochondria, Peroxisomes, and Rough Endoplasmic Reticulum**

| Membrane system | ATP | Membrane potential | Transported protein | Signal | References |
|---|---|---|---|---|---|
| Nucleus | Yes | No | SV40 T-antigen | TPPK$K_{128}$KRKV | Lanford and Butel (1984); Kalderon *et al.* (1984a,b); Lanford *et al.* (1986) |
| | | | SV40 VP1 | APT$K_4$RKGSCP | Wychowski *et al.* (1986) |
| | | | Polyomavirus T | TPPK$K_{282}$ARED | Richardson *et al.* (1986) |
| | | | | PVSR$K_{192}$RPRP | |
| | | | Adenovirus E1a | | Krippl *et al.* (1985); Richter *et al.* (1985) |
| | | | | KRPRP-COOH | Lyons *et al.* (1987) |
| | | | Influenza virus nucleoprotein | AAFEDLRVLS | Davey *et al.* (1985) |
| | | | Influenza virus NS1-protein | $D_{34}$RLRR | |
| | | | | $P_{216}$KQKRK | Greenspan *et al.* (1988) |
| | | | Nucleoplasmin | GQAK$K_{164}$KKLDK | Bürglin and DeRobertis (1987) |
| | | | v-sis mutant | $R_{237}$VTIRTVRVRRPPKGKHRK | Lee *et al.* (1987) |
| | | | Yeast ribosomal protein L3 | $P_{18}$RKR | Moreland *et al.* (1985) |
| | | | Yeast MATα2-protein | MN$K_3$IPIKDLLNPQ | Hall *et al.* (1984) |
| | | | Yeast GAL4-protein | | Silver *et al.* (1984) |
| Mitochondria | Yes | Yes | Mitochondrial proteins | Amphipatic helices | von Heijne (1986a); Allison and Schatz (1986) |
| Peroxisomes | Yes | No | Firefly luciferase, *P. pyralis* | LIKAKKGGKSK$L_{550}$-COOH | Gould *et al.* (1987); DeWet *et al.* (1987); Imanaka *et al.* (1987) |
| Endoplasmic reticulum | (Only post-transl.) | No | Proteins with signal sequences | "n-b-c" | von Heijne (1986b); von Heijne and Gavel (1988) |
| | | | RER-retention | | |
| | | | Mammalian | KDEL | Munro end Pelham (1987) |
| | | | Yeast | HDEL | Pelham *et al.* (1988) |

receptor, for the expansion of the pore, or for driving the protein through the expanded pore.

## 2.2. Mitochondrial Import

The great majority of the polypeptides that are active in mitochondria are encoded by nuclear genes, and their synthesis is completed on free ribosomes in the cytoplasm before interaction with the mitochondrial membranes (for review, see Eiler and Schatz 1988; Roise and Schatz, 1988). As described for the translocation into the nucleus, mitochondrial import also requires the presence of signal sequences that are generally present at the N-terminus of larger-sized precursor polypeptides. Precursor targeting sequences differ considerably in their structures. However, these sequences generally lack acidic amino acids; they have a high content of positively charged and of hydroxylated amino acids that could fold into amphiphilic structures in form of either α-helices or β-sheets (von Heijne, 1986a; Vassarotti *et al.*, 1987b; Roise and Schatz, 1988). Artificial presequences containing the initiator methionine and the amino acids leucine, serine, and arginine were shown to target subunit IV of cytochrome oxidase into the correct intramitochondrial location *in vivo,* provided that the arginine residues were placed at the same side of the helix winding (Allison and Schatz, 1986). However, a residual mitochondrial import (about 10% of the control) occurs apparently in a receptor-independent manner as well, as shown with protease-pretreated mitochondria (Zwizinski *et al.*, 1984; Ohba and Schatz, 1987b). The targeting sequences must provide three essential features: (1) binding to the receptor, (2) insertion into proteinaceous sites within the outer mitochondrial membrane, and (3) response to the electrical potential of the inner mitochondrial membrane. CD (circular dichroism) measurements have indicated that the amphiphilic nature of the precursor becomes unmasked only when the protein unfolds (Endo and Schatz, 1988). Mitochondrial import takes place at the contact sites between the outer and the inner mitochondrial membrane (Schleyer and Neupert, 1985). The translocation process has been dissected into a series of distinct steps that differ slightly for polypeptides carrying an N-terminal cleavable signal sequence, such as $F_1$-ATPase β and proteins that contain an internal noncleavable signal sequence such as the ADP/ATP carrier. Figure 1 describes the sequence of events in the import of a typical mitochondrial matrix protein.

1. Proteins with an N-terminal cleavable signal are synthesized as higher-molecular-weight precursors. They bind via the signal sequence to receptor proteins present at the cytoplasmic face of the outer mitochondrial membrane. The insertion of the presequence requires a potential across the inner mitochondrial membrane and results in the exposure of the N-terminus into the matrix (Schleyer and Neupert, 1985; Vassarotti *et al.*, 1987a).

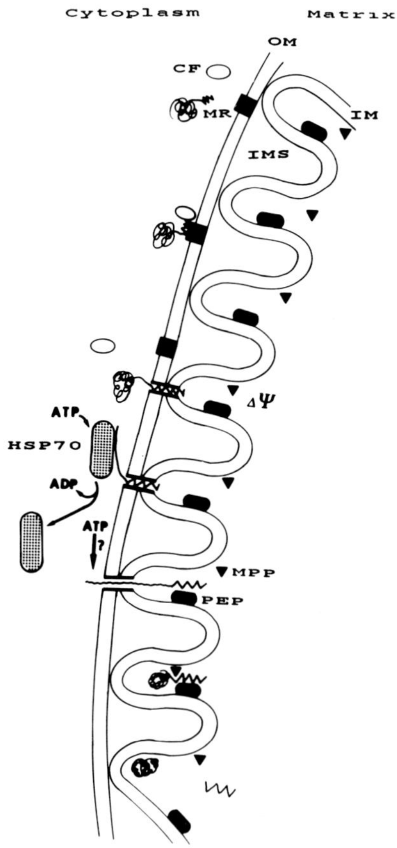


**FIGURE 1.** Steps of post-translational protein transport into the mitochondrial matrix. CF, cellular factor; HSP70, heat-shock protein 70; IM, inner mitochondrial membrane; IMS, intermembrane space; MR, membrane receptor(s); MPP, mitochondrial processing peptidase; OM, outer mitochondrial membrane; PEP, processing enhancing protein. See text for further details. (From Hawlitschek *et al.,* 1988; Pelham, 1988a.)

2. At this stage, the protein is unfolded into a protease-sensitive conformation that makes the protein translocation competent. This process is the rate-limiting step for the import. It can be blocked by folate analogues or low temperature *in vivo,* while denaturing agents such as urea enhance the process *in vitro* (Eilers *et al.,* 1988). Nucleotide triphosphates (NTP), such as hydrolyzable ATP or GTP, are required and presumably used for this unfolding step (Pfanner *et al.,* 1987; Chen and Douglas, 1987), since incompletely synthesized (and thus more translocation-competent) precursors show a decreased requirement for NTP (Verner and Schatz, 1987). The requirement for NTP can also be bypassed when the conformation of a precursor protein is altered by acid–base pretreatment (Pfanner *et al.,* 1988). Furthermore, a deletion of sequences that catalyze tetramer formation of the $F_1$-ATPase $\beta$-subunit leads to an ATP-independent import (Chen and Douglas, 1988).

Deshaies *et al.* (1988) showed that a depletion of heat-shock proteins (hsp70) in yeast cells induced cytoplasmic accumulations of the precursor of the $F_1$-ATPase β-subunit. Hsp70 is believed to bind malfolded polypeptides and by consuming ATP to convert them into a translocation competent configuration (Pelham, 1988b).

3. It is still unclear whether translocation of the unfolded polypeptide also requires ATP. Recent studies have indicated that the product of one of the yeast hsp70 genes is also transported into mitochondria. Therefore, it is tempting to speculate that hsp70 proteins might be involved in controlling refolding in the mitochondrial matrix. At any rate, translocation occurs independent of the presence of a membrane potential. During translocation, the signal sequence is recognized by a processing enhancing protein (PEP) (52,000 $M_r$) that is associated partly with the inner mitochondrial membrane. This compound is about 15-fold more abundant in mitochondria than the processing peptidase (57,000 $M_r$) purified from *Neurospora crassa* as a soluble compound present in the mitochondrial matrix (Hawlitschek *et al.*, 1988). The specificity of the protease for the vast number of signal sequences is best explained by the protein–protein interaction between the peptidase and the PEP–signal–complex, since the protease alone had low processing activity (Hawlitschek *et al.*, 1988).

## 2.3. Signals for Peroxisomes

The biogenesis of peroxisomes has unique features that distinguish it from the assembly of other organelles (Lazarow and Fujiki, 1985). Several core proteins and membrane proteins of peroxisomes are synthesized on free ribosomes in the cytoplasm, and their import into peroxisomes occurs post-translationally (Goldman and Blobel, 1978; Fujiki *et al.*, 1984; Miura *et al.*, 1984) without any detectable modification of the protein (Robbi and Lazarow, 1982; Fujiki *et al.*, 1984). A major breakthrough in the establishment of an *in vitro* import assay was achieved (Imanaka *et al.*, 1987) by applying clofibrate, a hypolipidemic drug that induces peroxisome proliferation (Reddy *et al.*, 1982) as well as peroxisome-specific mRNA (Reddy *et al.*, 1986). The sorting signal of peroxisomal luciferase from firefly *Photinus pyralis* has been identified at the molecular level (Table I). Fusion proteins containing the 12 carboxy-terminal amino acids of luciferease (De Wet *et al.*, 1987) fused in frame to the C-terminus of the cytoplasmic CAT protein were targeted into peroxisomes in mammalian cells (Gould *et al.*, 1987), indicating a high degree of evolutionary conservation of the peroxisomal sorting mechanism. Again, the import requires ATP hydrolysis, as shown for the translocation of acetyl-CoA oxidase in an *in vitro* system (Imanaka *et al.*, 1987). This latter enzyme, which is amplified in peroxisomes of *Candida tropicalis* grown on fatty acid, contains two signal sequences (present within the sequence from amino acids 1–118 and 309–427), each of which could target the cytoplasmic dihydrofolate reductase into peroxisomes (Small *et al.*, 1988). These findings suggest that the targeting signal for peroxisomes does not necessarily have to be located at the C-terminus of a protein. The C-terminal domain (the last 38-amino acid residues) has also been proposed to contain the topogenic signal of the glycosomal phosphoglycerate kinase (Swinkels *et al.*, 1988).

## 3. TRANSLOCATION ACROSS THE ENDOPLASMIC RETICULUM MEMBRANE

The cellular translocation machinery controlling the insertion of nascent proteins into or across the membranes of the endoplasmic reticulum (ER) is shared by proteins with still different final destinations. The fate of a polypeptide that has penetrated the rough endoplasmic reticulum (RER) membrane depends on the presence or absence of further sorting signals. The absence of such signals leads to secretion of the protein from the cell or to the exposure of either end of the molecule at the plasma membrane in a specific orientation, depending on the nature of the stop transfer sequence(s) that anchor the protein in the membrane. Specific signals that have not yet been identified at the molecular level may result in a post-translational replacement of a hydrophobic anchor sequence by a glycophospholipid anchor in the RER. In addition, sorting signals determine whether a protein is transported into lysosomes or secretory storage vesicles or whether transport will be arrested in one of the compartments anywhere along the transport route from the RER through the Golgi to the plasma membrane.

### 3.1. Features of Signals for ER Insertion

Originally, signal sequences for the RER were detected as transient predominantly hydrophobic sequences of 20–40 amino acids located at the N-terminus of secreted proteins (Blobel and Dobberstein 1975a). Later, membrane proteins were detected which contained either N-terminal, noncleaved signal sequences functioning simultaneously as membrane anchors, or that contained multiple permanent internal signal sequences. All these signal sequences varied considerably in primary sequence and length (Perlman and Halvorson, 1983; Watson, 1984; von Heijne, 1985a,b). Computer-assisted analyses of published signal sequences showed some common structural features with respect to hydrophobicity, second structure, the position of charges, and to the cleavage site recognized by the signal peptidase (von Heijne, 1985, 1986). Three structurally distinct domains can be distinguished in nearly all signal sequences:

1. The N-terminal part of a signal sequence (the n region) is generally positively charged carrying a mean net charge of +1.7 (von Heijne, 1985).
2. The central core sequence (h region) involves 8–20 highly hydrophobic residues. It is folded into an energetically favored α-helix or, if the length does not permit complete assembly as an α-helix, into partial α-helical and partial β-sheet structures. Natural mutations of signal sequences that abolish the signal function commonly involve the replacement of a hydrophobic amino acid by a charged residue (Rapoport, 1986).
3. Sequences flanking the core region at the carboxy-terminal end (c region) have mostly polar properties and β-sheet structure.

The fundamental structural similarities between signal sequences from organisms widely separated on the evolutionary scale have inspired experiments demonstrating that sequences from one organism are functional when inserted into proteins from another. Thus, bacterial signal sequences were shown to be functional in vertebrate systems *in vivo* and *in vitro* (Garcia *et al.*, 1987; Lingappa *et al.*, 1984; Ibrahimi and Gentz, 1987; Müller *et al.*, 1982; Wiedman *et al.*, 1984). Signals of influenza virus hemagglutinin (Jabbar and Nayak, 1987) and of human interferon (Hitzeman *et al.*, 1983) are functional in yeast; the rat preproinsulin signal functions in bacteria (Talmadge *et al.*, 1980).

Several sets of experiments show that signal sequences derived from yeast mediate translocation into mammalian membranes. These studies involve signals from yeast invertase (Bergh *et al.*, 1987), α-factor (Julius *et al.*, 1984), and killer toxin (Bostian *et al.*, 1983). Recently, however, Bird *et al.* (1987) reported that such functional equivalence of signal sequences does not apply to the signal of yeast carboxypeptidase Y. The yeast translocation machinery tolerates an h region of only five hydrophobic amino acids interrupted by two glycine residues. This N-terminal signal directs carboxypeptidase Y co-translationally, and to a minor extent also post-translationally, into the lumen of the yeast ER, where it is cleaved by the signal peptidase (Blachly-Dyson and Stevens, 1987; Johnson *et al.*, 1987). However, this sequence did not permit translocation of the enzyme in a mammalian cell-free system or *in vivo* unless the yeast signal was replaced by the signal from influenza virus hemagglutinin or either of the two glycine residues was mutated into leucine to increase the hydrophobicity of the h region (Bird *et al.*, 1987). Similar observations have been made with synthetic signal peptides that displayed different activities in mammalian and plant *in vitro* secretion systems (Duong *et al.*, 1987).

Experiments with the aim of further dissecting the functional significance of the n, h, and c regions of signal sequences are hampered by the important findings that 20% of random surrogate signal sequences replacing the normal signal sequence of yeast invertase could function as export signals (Kaiser *et al.*, 1987). In addition, Zerial *et al.* (1987) showed that foreign transmembrane anchors or completely artificial uncharged peptide sequences placed into the position of the internal hydrophobic stretch of the transferrin receptor molecule (this sequence serves the dual function of a signal and anchor sequence) permitted efficient translocation and membrane binding.

### 3.1.1. Classification of Integral Membrane Proteins

Membrane proteins have been divided into four classes based on the nature of the signal sequence and additional topogenic sequences (Wickner and Lodish, 1985; Garoff, 1985; von Heijne and Gavel, 1988):

*Class I proteins* contain a cleavable signal sequence, according to this classification (Fig. 2). Proteins are believed to penetrate the membrane in form of hair pin loops. Proteolytic removal of the signal peptide causes the newly created N-termini to become oriented toward the lumen of the ER, while their C-termini are exposed on the cytoplasmic side. Most glycoproteins of enveloped viruses such as the G protein of vesicular stomatitis virus (VSV), the hemagglutinin of influenza virus, the E1 and E2 proteins of Semliki Forest virus, the E2 protein of coronaviruses, as well as growth factor receptors, histocompatibility antigens, and the membrane-bound immunoglobulins (Ig) belong to this class. Anchorage of these proteins is mediated by a C-terminally located stop transfer sequence. A deletion of the membrane anchor will result in secretion of the corresponding anchor-minus mutant into the lumen of the RER.

*Class II proteins* acquire the opposite orientation in the membrane and carry an uncleavable signal peptide that functions simultaneously as a membrane anchor, since it cannot be cleaved off by the signal peptidase. In general proteins of this class, like those of class I, carry positively charged amino acids in the short N-terminal segments immediately preceding the actual signal sequence. The

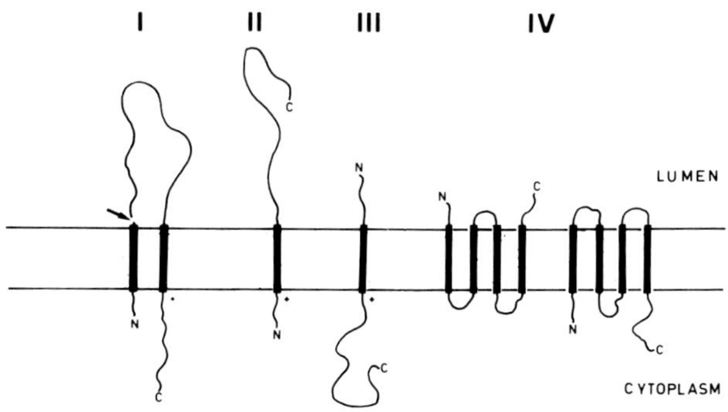


**FIGURE 2.** Membrane topology of integral membrane proteins.

influenza virus neuraminidase (Blok *et al.*, 1982), the hemagglutinin-neuraminidase of paramyxovirus (Wilson *et al.*, 1987; Hiebert *et al.*, 1985), the asialoglycoprotein receptor (Spiess and Lodish, 1986; Holland and Drickamer, 1986), and the invariant chain (Iγ-chain) (Lipp and Dobberstein, 1986), are members of this group of membrane proteins.

*Class III membrane proteins* are distinguished from class I and II proteins by the distribution of charged residues in the border regions flanking the hydrophobic segment of the signal sequence. Thus, in class III proteins, the N-terminal translated domains generally lack positively charged residues. Adversely, such basic amino acids are found often in clusters, within the C-terminal sequence flanking the hydrophobic segment. The influenza virus M2 (Hull *et al.*, 1988) and the human cytochrome P450 (Beaune *et al.*, 1986) are prominent members of this class of proteins.

*Class IV membrane proteins* contains multiple internal noncleaved signal sequences. The topology of these proteins is presumably determined by the nature of the first membrane-spanning domain carrying a signal character (Mayer *et al.*, 1988). Prominent members of this group are bovine opsin (Friedlander and Blobel, 1985), the anion transporter of erythrocytes (Lopito and Lodish, 1985), human B-cell CD20 receptor (Einfeld *et al.*, 1988), and the E1-glycoprotein of coronaviruses (Niemann *et al.*, 1984a; Armstrong *et al.*, 1984). Depending on the nature of the first membrane-spanning segment and the distribution of charged residues in its border regions, the N-terminus of a class IV protein can reside on either the cytoplasmic or the lumenal side of the ER membrane.

## 3.2. Translocation Machinery of the Rough Endoplasmic Reticulum

### 3.2.1. Signal-Recognition Particle

The availability of *in vitro* systems based on dog pancreas microsomal membranes (Blobel and Dobberstein, 1975a,b) and reticulocyte or wheat germ lysates as cell-free protein synthesizing systems have provided excellent tools for dissection of the translocation process. Such studies have shown that membrane insertion of a protein requires the presence of a signal sequence that is recognized by a ribonucleoprotein complex (Walter *et al.*, 1981; Walter and Blobel, 1981) termed the signal-recognition particle (SRP). The SRP has been purified to homogeneity from a salt extract of dog pancreas membranes (Walter and Blobel, 1980). A similar SRP-like compound has been isolated and functionally characterized from wheat germ extracts (Prehn *et al.*, 1987).

The SRP is composed of six proteins, of 9000, 14,000, 19,000, 54,000, 68,000, and 72,000 $M_r$, and a single RNA molecule of 300 nucleotides (7 SL RNA or SRP-RNA) (Fig. 3). While the 19- and 54-kDa proteins are present as

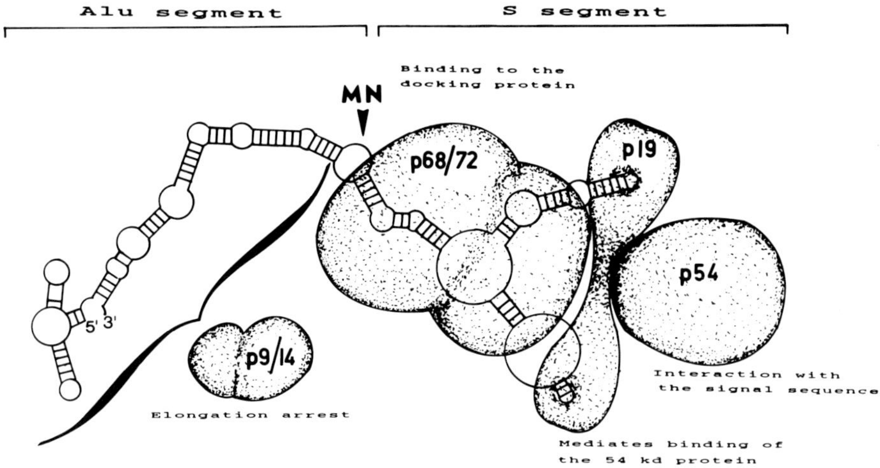


**FIGURE 3.** Structural arrangement and function of individual subunits of the signal-recognition particle. Cleavage site by micrococcal nuclease (MN) is indicated by the arrow head. See text for details. (Redrawn according to Siegel and Walter, 1988a,b.)

monomers, the other four polypeptides form heterodimers involving the 19- and 54-kDa species, and the 68- and 72-kDa polypeptides respectively (Scoulica *et al.*, 1987). The 19-kDa subunit of SRP was cloned and sequenced (Lingelbach *et al.*, 1988).

About 100 nucleotides at the 5′-end and 45 nucleotides at the 3′-end of the SRP-RNA are homologous to human β-Alu DNA, highly repetitive elements of the human genome (Ullu *et al.*, 1982; Ullu and Tschudi, 1984). The secondary structure of the 7SL RNA has been established (Gundelfinger *et al.*, 1984; Zwieb, 1985) revealing a rodlike structure flanked on each side by two stem-loop structures. The rod is formed by base pairing of nucleotides 48–118 and 233–299. The Alu sequences form the two small stem-loop structures and are involved in the rod formation (Fig. 3).

Alu-RNA synthesized *in vitro* was shown to bind to the 68-kDa peptide (Andrews and Kole, 1987). By contrast, this component is clearly associated with the core part of the SRP-RNA after partial digestion with micrococcal nuclease, which removes most of the Alu-specific sequences (Gundelfinger *et al.*, 1983; Siegel and Walter, 1986). The binding sites of the 19- and the 68/72-kDa proteins on the SRP-RNA were recently determined by protein-RNA footprinting (Siegel and Walter, 1988b), revealing the structure shown in Fig. 3.

Three different activities involving binding to the signal sequence, arrest of protein synthesis, and binding to the SRP receptor (docking protein) can be assigned to SRP in the wheat germ cell-free system:

1. Recognition of signal sequences and binding to ribosomes synthesizing secretory proteins: SRP binds via its 54-kDa subunit to signal sequences as shown by crosslinking of this protein to signal sequences (Krieg *et al.*, 1986; Kurzchalia *et al.*, 1986). Reconstituted SRP containing a 54-kDa moiety in which the free sulfhydryl groups were alkylated with *N*-ethylmaleimide failed to mediate all three features normally provided by a functional SRP (Siegel and Walter, 1988).
2. Arrest of protein synthesis, a useful artifact observed *in vitro:* Block of polypeptide chain elongation was originally proved for preprolactin (Walter *et al.*, 1981) and the light chain of IgG (Meyer *et al.*, 1982). In both instances, peptides of about 70 amino acids in length could be isolated and chased into full-sized proteins upon addition of translocation-competent membranes. However, no arrested peptides could be demonstrated from most other proteins, even though the addition of SRP to translation assays lacking microsomal membranes induced a reduction in the formation of full-length precursors. Furthermore, under similar experimental conditions some proteins were shown not to be SRP arrested (Mueckler and Lodish, 1986), and the SRP-mediated arrest could not be demonstrated with reticulocyte and HeLa cell lysates (Meyer, 1985) or with SRP isolated from wheat germ (Prehn *et al.*, 1987). Lipp *et al.* (1987) reported recently that SRP induced a block in chain elongation in a transient manner at multiple sites of the protein chain, leading to a set of lower-molecular-weight precursors, all of which could be chased into the full-length precursor by the addition of microsomal membranes. These findings are consistent with functional studies on reconstituted SRP lacking the 9- and 14-kDa heterodimer. Such particles were still capable of promoting signal binding and translocation of the peptide across the RER membrane (Siegel and Walter, 1985). Similar properties were observed with SRP pretreated with micrococcal nuclease and thus lacked the Alu-like sequences in addition to the 9/14-kDa subunit (Siegel and Walter, 1986). From these data, it may be concluded that the 9/14-kDa heterodimer causes the transient arrest of chain elongation, and that elongation arrest is not a prerequisite for membrane translocation.
3. Promotion of membrane translocation by binding to the SRP receptor (docking protein). The SRP-controlled targeting of polypeptides into microsomal membranes has been exclusively observed for chains that are still associated with the ribosome (Mueckler and Lodish, 1986; Perara *et al.*, 1986). Arrested preprolactin fragments or fragments obtained by short-term translation and subsequent inhibition of protein synthesis are still covalently linked to transfer RNA (tRNA) and can thus be precipitated by cetyltrimethylammoniumbromide (Cetavlon) (Hobden and Cundliffe, 1978). Siegel and Walter (1988) demonstrated that such peptides could subsequently be targeted into microsomal membranes by the post-translational addition of SRP. By contrast, reconstituted SRP, carrying an alkylated 68/72-kDa heterodimer,

failed to mediate membrane association of the truncated peptides. These data strongly support the hypothesis that the 68/72-kDa subunit of SRP interacts with the SRP receptor at the cytoplasmic face of the RER.

### 3.2.2. SRP Receptor (Docking Protein) and Signal-Sequence Receptor

The SRP receptor (69,684 $M_r$) comprises an integral membrane protein of the ER that, in conjunction with SRP, ensures correct and efficient insertion of polypeptides carrying appropriate signal sequences into the membranes. The receptor was purified from pancreatic membranes (Gilmore *et al.*, 1982; Siegel and Walter, 1985) and its primary sequence has been established by sequencing cDNA clones from a pancreatic library (Lauffer *et al.*, 1985). The polypeptide is anchored with its hydrophobic N-terminus in the ER membrane, while most of the protein is exposed at the cytoplasmic face of the ER (Fig. 4).

What is the fate of the polypeptide after binding of the peptide–SRP–ribosome complex to the SRP receptor? The translocation process itself is still a black box. Wiedman and co-workers (1987) have used a crosslinking approach to follow the fate of the preprolactin signal sequence. The label was introduced into the nascent peptide chain via Lys-tRNA carrying the photoactive crosslinker at the ε-amino group of the amino acid (Kurzchalia *et al.*, 1986). With this approach, a signal sequence receptor (SSR) constituting an integral membrane glycoprotein of the RER (≈35,000 $M_r$) was identified (Wiedmann *et al.*, 1987). Crosslinking of the signal sequence to SSR was strictly dependent on a previous interaction of the peptide–SRP–ribosome–complex with the SRP receptor and occurred only after the release of SRP from the signal sequence and from the ribosome. Therefore, the SRP receptor and SSR act sequentially in the SRP-dependent translocation process of nascent proteins (Walter, 1987). The role of SSR in the translocation process is not understood at the molecular level. SSR could mediate the interaction of the signal sequence with the lipids, it could mediate the transfer or open a channel through which the protein traverses the membrane, or it could be part of a channel itself. There is increasing evidence that translocation does not proceed directly through the hydrophobic interior of

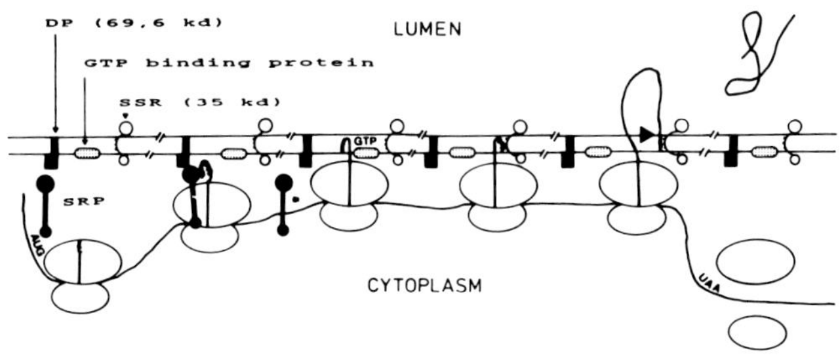


**FIGURE 4.** SRP-dependent translocation of protein across the membrane of the RER.

the lipid bilayer but involves a proteinaceous pore. Gilmore and Blobel (1985) showed that translocation of secretory proteins across the microsomal membrane occurs through an environment that is accessible to aqueous perturbants, such as 4 M urea. Partially translocated peptides created *in vitro* by oligonucleotide-mediated hybrid arrest could be extracted from the membranes either with 4 M urea or at alkaline pH. In addition, proteins containing up to 8-arginine residues within the minimum length of 12 amino acids required to span a membrane, have been found to become translocated across the microsomal membranes (Garcia *et al.*, 1988). Translocation of such domains through a merely lipid-containing environment could only be achieved at a significant energy cost.

The formation of a functional ribosome membrane junction requires the participation of a GTP-binding protein, as shown by binding studies of *in vitro*-assembled polysomes to canine microsomal membranes that had been depleted of ATG and GTP (Connolly and Gilmore, 1986). Ribonucleotides were not required for an initial low-salt-insensitive binding of the ribosome to the membrane. However, without GTP, the nascent secretory chains were readily extractable with either EDTA or 0.5 M potassium acetate and were sensitive to protease digestion. By contrast, the nascent peptides became resistant to extraction or protease when GTP or hydrolysis-resistant GTP analogues were added. GDP, a competitive inhibitory ligand of GTP-binding proteins (Yamanaka *et al.*, 1986), could inhibit the GTP-mediated conversion into the tightly bound complex and subsequent translocation (Connolly and Gilmore, 1986).

### 3.3. Co-Translational versus Post-Translational Translocation

The experimental tools used to uncouple translocation from translation involve the *in vitro* synthesis of capped RNA from suitable transcription vectors using bacteriophage SP6 or T7 polymerases (Krieg and Melton, 1984) or *Escherichia coli* polymerase (Stueber *et al.*, 1984) and subsequent translation of the RNA in one of the *in vitro* translation systems. RNA encoding the full-sized protein of interest are obtained by linearizing the templates downstream from the coding region. The ribosomes will dissociate from the RNA at the termination codon. Alternatively, the DNA template can be linearized with restriction nucleases that cleave anywhere within the coding region. In this instance, translation in one of the *in vitro* translation systems yields an arrested fragment of the desired length in form of a peptidyl-RNA that remains attached to the ribosome (Perara *et al.*, 1986). For establishing SRP requirement, translation reactions are performed in wheat germ extract, and SRP (isolated from canine pancreatic membranes by salt extraction) may be added exogeneously (Walter and Blobel, 1980).

The lack of further protein synthesis can be ensured at any time point by the addition of aurintricarboxylic acid (ATA) (to inhibit initiation) or emetine (to block elongation) or by the addition of cycloheximide. At this stage, a fraction of the translation mixture can be treated with puromycin to release the peptide from the ribosome. Microsomal membranes can be added before and after translation. In addition, preparations of hsp70 and ATP may be included to test the needs for post-translational unfolding (Chirico *et al.*, 1988). Finally, the translocation competence of the individual peptides can be assessed by a combination of the following criteria:

1. Protection against exogeneously added proteases that do not penetrate the microsomal membrane unless detergent is added (Blobel and Dobberstein, 1975a).
2. Removal of a cleavable signal sequence indicating lumenal exposure of the cleavage site.
3. Glycosylation of potential *N*-glycosylation sites [$NH_2$–Asn–X–Ser(Thr)–COOH, in which X may be any amino acid with the exception of proline]. En bloc transfer of $Glc_3Man_9GlcNAc_2$-oligosaccharides onto asparagine residues within these tripeptides consensus sequences will increase the molecular weight of the translocated peptides by ~3000 kDa per added carbohydrate side chain. A subsequent treatment with endo-β-*N*-acetylglucosaminidase H in the presence of detergent will result in cleavage of the chitobiose linkage and thus lead to a protein retaining only one *N*-acetylglucosamine residue per side chain. Alternatively, glycosylation can be stepwise inhibited by the addition of acceptor peptides that freely pass through the membrane and compete as substrates for the oligosaccharidyltransferase. Upon glycosylation, however, they become trapped in the lumen of the RER (Bause *et al.*, 1986).
4. Co-sedimentation of the newly synthesized peptides with the microsomal membrane fraction on sucrose gradients at pH 7 or pH 11.5 (Gilmore and Blobel, 1985). Nonintegral proteins remain in the supernatant at either pH, whereas integral proteins and secretory proteins sediment together with the membranes through a sucrose cushion at neutral pH. At pH 11.5, secretory proteins are released from the interior of the vesicles, while integral membrane proteins remain associated with the membranes and are thus found in the pellet fraction (Fujiki *et al.*, 1982).

Studies based on such experimental approach have shown that with the exception of short peptides in the size range of about 70 amino acids, which apparently can be translocated through the membrane in an SRP-independent manner and after release from the ribosome (Watts *et al.*, 1983; Zimmermann and Meyer, 1986; Müller and Zimmermann, 1988; Schlenstedt and Zimmermann, 1987), most other proteins make use of the translocation machinery involving SRP and the SRP receptor.

Several reports have clearly demonstrated that translocation across the ER membrane, although SRP dependent, does not have to be co-translational, in the sense that it depends on further chain elongation (reviewed by Zimmermann and Meyer, 1986; Perara *et al.,* 1986). Thus, the driving force for translocation through the membrane cannot rely on energy from the ribosome and protein synthesis.

Even larger segments of proteins, such as the amino-terminal 340 amino acids from the human glucose transporter (Mueckler and Lodish, 1986), the immunoglobulin light chain (Ainger and Meyer, 1986), human placental lactogen (Caulfield *et al.,* 1986), or the influenza virus hemagglutinin (Chao *et al.,* 1987), have been reported to be translocated in the absence of ongoing protein synthesis. Several groups have shown that prepro-α-factor, a secretory polypeptide containing 165 amino acids, was translocated across yeast microsomal membranes post-translationally in a ribosome independent manner (Hansen *et al.,* 1986; Rothblatt and Meyer, 1986; Rothblatt *et al.,* 1987; Waters and Blobel, 1986), while translocation of the same protein was ribosome dependent with canine microsomal membranes (Hansen *et al.,* 1986). Recently, Wiedmann and co-workers (1988) reported that the transport of cytochrome P450, an integral membrane protein, occurred co- and post-translationally across *Candida maltosa* membranes, whereas it was exclusively cotranslationally with dog pancreas membranes.

Hansen and Walter (1988) showed that preprocarboxypeptidase Y and a truncated form of preinvertase, in contrast to intact invertase, were post-translationally translocated across microsomal membranes from *Saccharomyces cerevisiae* after release from ribosomes. Although continuous attachment of the peptide to the ribosome accelerated the translocation process about fivefold, these data demonstrate that the translocation machinery of the yeast ER, like those of mitochondria, chloroplasts, and the bacterial plasma membrane, is principally capable of sorting polypeptides from a ribosome-free soluble pool. The notion that translocation of other arrested fragments requires the presence of a signal sequence and the attachment to ribosomes may then indicate that SRP and the ribosome serve basic functions in maintaining a translocation-competent conformation. At physiological SRP concentrations (10 nM), the affinity of SRP for presecretory proteins appeared to depend on the length of the nascent preprolactin chain and to drop drastically when the peptide was longer than 140 amino acids (Siegel and Walter, 1988a). By contrast, high concentrations of SRP (270 nM) were able to target all sequences (Siegel and Walter 1988a).

Differences have been reported concerning the requirement for NTP in the post-translational translocation of different proteins, however. Arrested fragments, comprising the 156 N-terminal residues of opsin, were presented to microsomal membranes attached to ribosomes. Post-translational membrane integration and concomitant N-glycosylation was only obtained when GTP or (although acting less efficiently) nonhydrolyzable GTP analogues were included. By contrast, ATP did not show a promoting activity (Hoffman and Gilmore, 1988; Connolly and Gilmore, 1986). It was concluded that a G protein played an essential role in the integration process of proteins into the membranes of the ER.

By contrast, the SRP-independent post-translational translocation of prepromellitin was shown to be ATP dependent (Müller and Zimmermann, 1988). In addition, charged amino acids within the N-terminal domain had to be counterbalanced by amino acids of the opposite charge at the carboxy-terminal, and a disturbance of this ratio resulted in different levels of ATP requirement. Müller and Zimmermann (1988) suggest that this energy was required for maintaining a translocation-competent conformation mediated by a cellular factor.

It is intriguing to speculate that indeed for small peptides and perhaps for large molecules under certain conditions, functions generally provided by SRP and the ribosome might in part be taken over by other soluble cellular proteins known to have unfolding activities. Such antifolding activities as prerequisites of membrane translocation have been demonstrated for the secA protein (Baker *et al.*, 1987) and secB protein of *E. coli* (Collier *et al.*, 1988).

Recently, new direct evidence came from three laboratories demonstrating that the 70-kDa heat-shock proteins stimulate post-translational protein translocation across microsomal membranes *in vitro*. Zimmermann *et al.* (1988) showed that the procoat protein of the bacteriophage M13, as synthesized in a bacterial cell-free extract, was post-translationally translocated into canine microsomal membranes in the presence of purified compounds classified as members of the hsp70 family together with a cytosolic factor. This latter soluble component was also present in reticulocyte lysate and acted synergistically with hsp70 in the presence of ATP. These findings indicate that indeed heat shock proteins can confer translocation competence to folded proteins. However, a still undefined cytosolic cofactor and ATP (or through the action of nucleoside diphosphate kinases other NTPs) enhance this activity.

Chirico *et al.* (1988) demonstrated that yeast cytosol contains at least two distinct activities capable of mediating post-translational translocation of prepro-α-factor across yeast microsomal membranes. The two activities showed different sensitivity toward *N*-ethylmaleimide. The resistent activity co-purified with the constitutively expressed SSA1 and SSA2 gene products, both of which are closely related to one another. Again, these peptides were found to be members of the 70-kDa heat-shock proteins and to show immunological cross-reactivity with hsp 70 from *Drosophila*. The 70-kDa protein bound specifically to ATP. Post-translational translocation of yeast prepro-α factor synthesized in wheat germ extract through yeast microsomal membranes was stimulated eightfold in the presence of the 70-kDa proteins.

Deshaies *et al.* (1988) studied the *in vivo* role of the SSA1-gene product in yeast mutants carrying chromosomal defects of the SSA1, SSA2, and the SSA4 genes. The SSA1 gene was expressed in these cells by introducing a centromeric plasmid that contained a single copy of the SSA1 gene under control of the yeast GAL1 promoter. This promoter remains repressed when cells are grown in medium containing glucose, while expression can be induced by growing cells in galactose medium. A sixfold overexpression of the SSA1 gene was obtained in galactose-containing medium. Shifting the cells to glucose medium to shut off the GAL1 promoter resulted in simultaneous cytoplasmic accumulations of prepro-α factor and, interestingly, of the β-subunit of the mitochondrial $F_1$ATPase. This observation indicates that targeting of prepro-α-factor to the RER as well as targeting of the β subunit into mitochondria required the presence of functional SSA1 protein. Thus, this finding stresses that initial steps in the translocation of proteins across membranes of the ER and of mitochondria may rely on the presence of the same cytoplasmic factors and heat-shock proteins, which are required to establish a translocation-competent configuration of some proteins for translocation through RER membranes.

## 3.4. Co-Translocational Events in the ER Lumen

### 3.4.1. Removal of Cleavable Signal Sequences

The signal peptidase has been purified from canine pancreatic membranes (Evans *et al.*, 1986) and from hen oviduct (Baker *et al.*, 1986). The peptidase exists in a complex of six proteins with molecular masses of 12, 18, 21, 22, 23, and 25 kDa. The 22- and the 23-kDa species were shown to be glycoproteins carrying endoglycosidase-H-sensitive carbohydrate side chains. Phospholipids are required for biological activity (Jackson and White, 1981). So far, the catalytic subunit has not yet been defined. Prokaryotic signal sequences are processed correctly by the eukaryotic signal peptidase, which appears to have a substrate specificity similar to that of the *E. coli* signal peptidase I (Garcia *et al.*, 1987). The latter peptidase is a single-chain integral transmembrane protein (37 kDa) of the bacterial plasma membrane, exposing both the N- and C-terminus into the periplasm and a cytoplasmic loop that constitutes a translocation poison sequence when attached to normally functional N- or C-terminal signal sequences (von Heijne *et al.*, 1988).

The substrate specificity of the eukaryotic signal peptidase was recently probed using mutants of the human pre(Δpro)apolipoprotein A-II, as obtained by site-directed mutagenesis (Folz *et al.*, 1988). This study and the work of others quoted therein support the (−3,−1) -rule by von Heijne (1983, 1984), termed A–X–B rule by Perlman and Halvorson (1983). According to this rule, an ideal cleavage site following the hydrophobic signal sequence should contain a β-turn located 5–6 residues toward the N-terminus and small uncharged amino acids (Ala, Gly, Ser, Thr, Cys) in position −1 and with somewhat less stringency in position −3.

A variation of the position of a signal sequence within a protein by extending the amino-terminal sequences does not influence its function and cleavability as shown for the gag–fms fusion protein of the Susan McDonough strain of feline sarcoma virus (SM-FeSV). In this instance, the signal sequence of v-fms is preceded by the 55-kDa truncated gag sequences fused in frame to the $NH_2$-side of the fms sequence. As observed with the cellular proto-oncogene c-fms, which encodes the receptor for colony-stimulating factor 1 (CSF1) of macrophages, the fms-specific signal sequence is cleaved in the gag–fms fusion protein, thereby releasing the gag-specific sequence together with the fms-encoded signal sequence (Wheeler *et al.*, 1986). Internal signal sequences, contained within the 6 kDa polypeptides, have been reported for the E1 protein of Semliki Forest virus (Garoff *et al.*, 1980) and Sindbis virus (Rice and Strauss, 1981), respectively. Internal signal sequences have also been reported for the V3 protein of West Nile virus (Wengler *et al.*, 1985) and for the G2 protein of Uukuniemi virus (Rönnholm and Pettersson, 1987). Similar observations were made by Rottier *et al.* (1987), who analyzed the effect of internalizing the signal sequence of vesicular stomatitis virus both *in vivo* and *in vitro*. Again, artificial presequences had no effect on the cleavability *in vivo*.

### 3.4.2. *N*-Glycosylation: A Game Involving Giving, Taking, and Giving

The biosynthetic pathway describing the steps from single sugar nucleotides to a dolichol-linked precursor carrying mannose-rich side chains that are transferred en bloc onto nascent peptides in the RER and subsequently converted into complex-type carbohydrates as present on mature glycoproteins has been studied in great detail (for reviews, see Kornfeld and Kornfeld, 1985; Datema *et al.*, 1987; Lennarz, 1988). Another pathway that begins only post-translationally in the *cis*-cisternae of the Golgi has been established for the stepwise acquisition of *O*-glycosidically linked side chains. However, no universal answer can be given to the question of why the eukaryotic cell spends such enormous genetically and energetically costly efforts to maintain these complex biosynthetic pathways.

Because each of these processing steps of the two pathways (Fig. 5) takes place in defined intracellular compartments that harbor the corresponding processing enzymes (Roth, 1987), the presence of a particular sugar is direct evidence that the protein has previously passed through the corresponding compartment. However, the absence of particular sugar constituents does not allow conclusions on the intracellular transport route a protein has taken.

## 3.5. Post-Translocational Glypiation and Acylation

Inositol-containing glycophospholipids of the general structure lipid–phosphate–inositol–glycan have been detected in a wide variety of biological species,

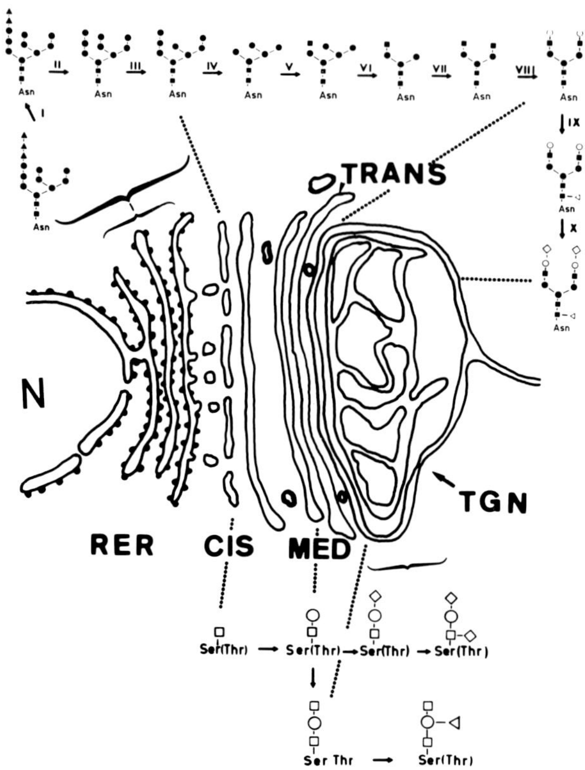


**FIGURE 5.** *N*- and *O*-glycosylation as markers of intracellular transport of integral and secretory proteins in mammalian cells. Roman numerals indicate reactions catalyzed by trimming enzymes or sugar transferases during transport from the rough endoplasmic reticulum (RER) through the Golgi into the *trans*-Golgi network (TGN). ■, *N*-Acetylglucosamine; ●, mannose; ▲, glucose; ○, galactose; ◇, sialic acid; △, fucose; □, *N*-acetylgalactosamine. Dotted lines coarsely define the subcompartments in which specific modifications take place. (For subcellular localization of the individual enzymes, see Roth, 1987.)

including bacteria, yeast, protozoa, and mammalia, but have not as yet been reported as membrane anchors for viral glycoproteins (for review, see Cross, 1987). The attachment of glycophospholipids involves a still undefined reaction mechanism (endopeptidase-transamidase), by which a hydrophobic stretch of amino acids is removed and the newly created C-terminus of the peptide becomes covalantly linked to the amino group of the terminal ethanolamine residue. So far there is no identifiable signal or consensus sequence for this glypiation reaction. The reaction occurs shortly after translation in the RER (Conzelmann *et al.*, 1988), and the complete structure of the VSG-glycophospholipid anchor has been elucidated (Schmitz *et al.*, 1987). One can only speculate about the biological function of this type of membrane anchor. The detection of phospholipases that specifically cleave the inositol phosphodiester linkage to yield diacylglycerol and the peptide–ethanolamine–glycan supported speculations that this mode of membrane anchoring provided a mechanism by which mammalian or protozoan cells could quickly and specifically change their surface features. Other hypotheses are related to the potential role of diacylglycerol as second messengers either in the activation of protein kinase C, or after conversion by diacylglycerol kinase into phosphotidic acid (Molenaar *et al.*, 1986).

Direct acylation has been reported for a large number of viral glycoproteins which, typically, exhibit fusion activity (Schmidt, 1983). Acylation of viral glycoproteins occurs within the carboxy-terminal domain reaching into the cytoplasm and can occur already in the RER (Repp *et al.*, 1985; Berger and Schmidt, 1985). In all instances in which the acylated amino acid has been identified, cysteine residues were found to be the recipient residues, and the bond formed was a thioester bond (Rose *et al.*, 1984). The physiological significance of acylation of viral glycoproteins remains to be shown. Acylation of glycoproteins does not appear to have any targeting function, glycoproteins that lack fatty acid are glycosylated and transported to the plasma membrane normally (Rose *et al.*, 1984). In addition, virus particles that contain nonacylated glycoproteins are infectious.

## 4. VIRAL PROTEINS ASSOCIATED WITH THE RER

The E19 protein of adenoviruses is a glycoprotein residing in the RER of infected cells. It never appears in the virus particle and it is not required for propagation of the virus *in vitro*. It has been shown to abrogate the transport of class I histocompatibility antigens by forming complexes with them in the ER (Severinsson and Petersen, 1985). This mechanism might be designed to deplete the surface of the infected cell of HLA antigens in order to prevent attack by cytotoxic T cells.

The E19 protein is a class I glycoprotein carrying two *N*-glycosylation sites (substituted with high-mannose-type oligosaccharides) within its aminoterminal domain. It is retained at the membrane via a membrane-anchor sequence. It has a C-terminal tail of 15 residues exposed into the cytosole (Wold *et al.*, 1985), and it shows the ER-specific transport properties also in cells that do not express class I antigens (Pääbo *et al.*, 1987). The retention signal has been assigned to a short linear sequence, presumably involving the last three amino acid residues (–KMP–COOH), which are identical in adenoviruses type 2, 3, and 5. Shortening of the tail domain of the E19 protein by eight residues allowed the protein to exit from the ER while an extension by 5 amino acids altered the intracellular distribution (Pääbo *et al.*, 1987). Alterations of the C-terminus diminished the affinity for class I antigens, whereas only shortening of the tail altered the transport properties. This observation suggests that the C-terminal domain plays a critical role both in establishing a conformation that allows for interaction with class I proteins and other unknown cellular factors involved in retaining the E19 proteins in the RER.

The glycoprotein VP7 of human rotavirus plays a role in the morphogenesis of virus particles which bud into the lumen of the RER and transiently acquire a lipid membrane. The VP7 gene is potentially bicistronic, containing a second AUG codon in frame separated from the first by 87 nucleotides. Each of the corresponding methionine residues precedes a hydrophobic sequence (H1 and H2) that could act as a signal for membrane translocation. It has been suggested (Poruchynsky *et al.*, 1985) that the hydrophobic sequence H2 contributed a positive signal for the retention of VP7 in the RER. However, recent *in vitro* transcription/translation studies show that even though both AUG codons are used, albeit the second at a much lower rate than the first, both primary translation products were processed by the signal peptidase into peptides of identical size (Stirzaker *et al.*, 1987). Determination of the N-terminal amino acid of the mature protein demonstrated that proteolytic cleavage occurred between $Ala_{50}$ and $Gln_{51}$ and thus removed H1 as well as H2. Therefore, the specific signal that makes VP7 a resident RER protein has yet to be characterized.

Human hepatitis B virus gives rise to synthesis of two core proteins, the major core protein (21 kDa), and a precore protein (25 kDa) which contains the entire sequence of the major core protein. This latter protein is translated from a separate mRNA containing additional 5′-sequences that are absent in the mRNA that encodes the major core protein (Enders *et al.*, 1985). The amino-terminal precore sequence comprises 29 amino acids that exhibit the typical features of a signal sequence for translocation through the RER membrane. In particular, it contains a positively charged amino acid followed by a cluster of 10 hydrophobic residues. Garcia *et al.* (1988) used *in vitro* transcription/translation to assess the biological significance of the precore signal sequence. Their data show that the signal sequence is indeed efficiently engaged in targeting all the nascent peptides into the microsomal membranes. However, only 30% of the newly synthesized molecules were translocated completely into the lumen of the ER resulting in a secretory peptide of 22 kDa. For the rest of the molecules, the translocation process was aborted irreversibly after proteolytic removal of the signal sequence and the protein was directed back into the cytosolic compartment.

These findings are in agreement with the observation that fragments of the core protein (HBV e antigen) containing the same N-terminus as the 22-kDa species are detected in the serum of infected patients (Ou *et al.*, 1986). A secretion of N-terminal pieces of the core protein was also detected upon expression of the entire core gene in mammalian cells (McLachlan *et al.*, 1987) or

*Xenopus* oocytes (Standring *et al.*, 1988). Although the precore protein does not belong to the resident ER proteins this aborted translocation process restricts its intracellular distribution. This in turn might provide a hitherto unnoticed mode by which processes can be controlled at the cytoplasmic face of the ER membrane during virus assembly.

## 5. FROM THE RER TO THE GOLGI

### 5.1. Role of Carbohydrates

A biantennary oligosaccharide side chain containing 11 monomeric residues interlinked to one another to form the structure shown in Fig. 5 represents only one out of many theoretically possible combinations of these monomers. Although nature only allows a small number of such structures to occur, it is clear that a tremendous quantity of information could be stored in the structure of carbohydrate side chains. However, with the exception of the lysosomal mannose 6-phosphate marker no direct access has been found to decipher the code of the carbohydrate language. Structural analyses of oligosaccharides derived from single glycosylation sites of the influenza virus hemagglutinin have demonstrated a wide structural heterogeneity (Niemann *et al.*, 1985; Keil *et al.*, 1987). Therefore, the presence of a heterologous pattern of oligosaccharides at particular attachment sites reflects the accessibility of this site for the modifying enzymes, rather than suggesting specific signals for the individual structures.

Numerous reports have documented the role of the carbohydrates in the folding process of particular proteins that occurs before and during transport of the protein from the RER to its final destination (Kornfeld and Kornfeld, 1985; Datema *et al.*, 1987). The complete absence of N-linked carbohydrate side chains as induced by mutation of the *N*-glycosylation site or by tunicamycin, but also the introduction of minor structural changes within the oligosaccharides, as mediated by the application of trimming inhibitors, may have unpredictable consequences for particular features of viral glycoproteins. Such alterations may involve the stability against proteases (Schwarz *et al.*, 1976; Bosch *et al.*, 1982), incorrect formation of intermolecular disulfide bonding (Machamer and Rose 1988b), intracellular transport (Gibson *et al., 1979; Repp et al.*, 1985; Schlesinger *et al.*, 1985; Machamer and Rose, 1988a), proteolytic processing (Datema *et al.*, (1984), masking of potential epitopes (Kaluza *et al.*, 1980; Repp *et al.*, 1986; Ploegh *et al.*, 1987), and failure to establish the functional conformation (Hadwiger *et al.*, 1986). It is clear that no generalizations can be made, since each glycoprotein has its very own demands for carbohydrates (Datema *et al.*, 1987).

### 5.2. Folding and Oligomerization, Thresholds of Transport

Most viral envelope glycoproteins exist as oligomers, and their biological functions are intimately dependent on their quarternary structure. In the last few years, the VSV G protein and the influenza virus hemagglutinin served as model compounds to estimate the contribution of initial folding and subsequent assembly of subunits into oligomers on the rate and efficiency by which the proteins are transported from the RER to the Golgi complex. Both class I glycoproteins form homotrimers that are transported to the plasma membrane, where they are sequestered into budding virus (Doms *et al.*, 1988; Copeland *et al.*, 1986). These studies have indicated that the correct folding soon after translocation is a prerequisite for oligomerization, which in turn is a requirement for export from the RER. The assembly of prefolded monomers apparently occurs with a high fidelity given the low concentration of viral glycoproteins in the RER membrane.

Doms *et al.* (1987) showed that the mutant G protein of VSV *ts*045 carrying a single-point mutation in the ectodomain (Gallione and Rose, 1985), accumulated in aggregates at the nonpermissive temperature (39°C) thus preventing exit from the RER. Upon downshift to permissive temperature (32°C) these aggregates assembled in an ATP-dependent manner. The oligomerization of wild-type G protein was determined to occur in the RER with $t_{1/2}$ of 6–8 min. No monomers were detected in the Golgi indicating that the formation of trimers is indeed a prerequisite to pass the quality control mechanism that allows exit from the RER.

However, trimerization alone is apparently not sufficient for transport of VSV G protein from the ER to the Golgi. Various mutations introduced into the C-terminal tail (29 amino acids in length) showed no effect on the rate of trimerization, but caused a significant delay in transport of the mutant G proteins. In particular, a mutant in which the C-terminal domain was replaced by 12 foreign amino acids trimerized as rapidly as wild-type G protein, but it never left the RER (Doms *et al.*, 1988). In a recent study, Guan *et al.* (1988) showed that these specific transport properties could be transplanted onto human chorionic gonadotropin-α. The membrane anchor plus the foreign C-terminal tail were fused in frame to the carboxy-terminus of the hormone to create a membrane-anchored fusion protein. Sedimentation studies indicated that the latter behaved (like the wild-type hormone) as a monomer on sucrose gradients. However, the transport of the fusion protein in transfected COS-cells was arrested in the RER. These important findings indicate the RER-specific retention signals for transmembrane proteins can principally reside on the cytoplasmic side of the ER membrane and that such retention signals can be uncoupled from signals mediated by the folding of the ectodomain. Furthermore, the knowledge about such transport retarding or accelerating signals may help identify escort proteins involved in regulating vesicular traffic between the RER and the Golgi complex.

Under normal growth conditions, trimerization of the influenza virus hemagglutinin occurs 7–10 min after synthesis, presumably close to the budding compartments of the RER. These compartments have been identified by electron microscopy as in part rough, in part smooth cisternae of the ER, from which transitory vesicles emerge (Gething *et al.*, 1986). Prior to trimerization the hemagglutinin undergoes extensive folding and the intrachain disulfide bonds are formed (Copeland *et al.*, 1988). Trimerization does not require ATP as determined in ATP-depleted cells *in vivo* or as studied by *in vitro* transcription/translation in the presence of microsomal membranes. Attempts to form heterotrimers from HA subunits of various HA subtypes failed. However, the formation of mixed trimers containing HA monomers stemming from the same HA subtype has been observed (Boulay *et al.*, 1988). These findings indicate that trimerization is highly sensitive to subtle changes of particular amino acids which are involved in the noncovalent interaction between the subunits and requires more than just the conserved location of disulfide bonds, the presence of conserved carbohydrate side chains and a cleavable fusion peptide (Boulay *et al.*, 1988).

As observed for the VSV G protein, no monomers were detected in the Golgi. Most of the known hemagglutinin mutants that exhibited retarded transport kinetics between the RER and the Golgi complex contained mutations located in the ultimate vicinity of the membrane anchor domain (Scholtissek *et al.*, 1984; Doyle *et al.*, 1985, 1986; Roth *et al.*, 1986).

Gething *et al.* (1986) showed that an HA mutant with alterations in the cytoplasmic tail failed to trimerize, while a second mutant, carrying a mutation at the base of the fibrous stalk, formed trimers that failed to fold into the native configuration. Such malfolded HA proteins were found in association with a cellular protein of 77 kDa (Bip or hsp78) (Gething *et al.*, 1986). Kozutsumi *et al.* (1988) recently showed that synthesis of this cellular protein was induced by the presence of malfolded proteins in the ER.

Studies on the role of individual domains of glycoproteins in the consecutive events of folding, oligomerization, and transport are most conveniently performed on isolated cDNA genes that can easily be manipulated *in vitro*. Unfortunately, the two model glycoproteins described here, the G protein of VSV and the influenza virus hemagglutinin, are derived from negative-stranded RNA viruses, making it difficult to reimplant the information of the manipulated DNA gene into the viral genome. In addition, studies from Scholtissek's group have indicated that temperature-sensitive transport properties of the influenza virus hemagglutinin could also be repaired by interaction of the hemagglutinin with other virus-encoded proteins (Mücke and Scholtissek, 1987). Transport of the mutant hemagglutinin was determined to be related to the loss of a glycosylation site, due to a mutation of Thr(480) to Ile. Revertants exhibiting normal transport properties were isolated. Sequencing revealed that some of these were structurally identical to the mutant HA. Mücke and Scholtissek attributed the restoration of normal transport properties to an extragenic suppressor mutation involving another, yet undefined, viral protein that normally comigrates with the HA-trimers in virus-infected cells. Clearly, these studies should be complemented by co-expression experiments involving individual viral genes in addition to the HA gene.

### 5.3. Resident Proteins of the RER

The export of different secretory and membrane-associated proteins from the RER to the Golgi occurs at quite different rates and some proteins including soluble proteins present in the lumen seem not leave the RER at all. The primary sequences of three resident proteins, the disulfide isomerase (Freedman 1984; Edman *et al.*, 1985) and the glucose-regulated proteins grp78 (Munro and Pelham, 1986) and grp94 (Kulomaa *et al.*, 1986; Sorger and Pelham, 1987) have been established. The three proteins share a common C-terminal motive, KDEL. Attachment of this sequence to the C-terminus of foreign secretory proteins conferred retention of the fusion proteins in the RER, as shown for the chicken lysozyme (Munro and Pelham, 1987). The C-terminal location of this retention signal is critical, since extension by additional random amino acids again resulted in secretion (Munro and Pelham, 1987).

A similar sequence, HDEL, has recently been identified as a functional ER-retention signal in *Saccharomyces cerevisiae,* indicating that yeast relies on a similar sorting machinery. This yeast-specific signal, however, was not an efficient retention signal in mammalian cells (Pelham *et al.*, 1988).

Grp78 has previously been recognized as the IgG heavy-chain binding protein, Bip, known to associate posttranslationally with the $C_H1$ domain until the heavy chain assembles with the light chain (Haas and Wabl, 1983; Hendershot *et al.*, 1987). Recently a more general function of grp78 and grp94 have been detected: Apparently synthesis of both proteins is induced severalfold in response stress conditions that specifically lead to the production of malfolded transport-incompetent proteins in the RER (Gething *et al.*, 1986; Lee, 1987; Pelham 1988a,b; Kassenbrock *et al.*, 1988; Kozutsumi *et al.*, 1988).

Kassenbrock *et al.*, (1988) showed that aberrantly glycosylated invertase or incorrectly disulfide-bonded prolactin formed tight complexes with grp78. However, such complexes were formed only with completed chains and not with nascent chains, indicating that grp78 does not participate in the folding process of a protein but acts as a quality-control substance that discriminates between correctly folded and malfolded proteins after the folding of peptides has taken place.

Direct evidence of a transcriptional activation of the grp78 gene was reported by Kozutsumi *et al.* (1988). The authors showed by Northern blot analyses that the synthesis of ER-specific transport mutants of the influenza virus hemagglutinin resulted in enhanced levels of grp78 RNA. Only mutants that gave rise to malfolded proteins in the RER (either because they were incapable of forming correct intrachain disulfide bonds due to a substitution of Cys(67) by Ser or because they contained a six-residue extension at the C-terminal tail of the HA molecule), induced an enhanced transcription of grp78 mRNA. By contrast, neither wild-type hemagglutinin nor a mutant accumulating in a post-Golgi compartment showed similar effects.

The detection of soluble resident RER proteins lacking a hydrophobic membrane-anchor sequence raised the question of how these proteins are retained from the pool of secretory proteins in the ER. While the required presence of a linear C-terminal sequence suggests the interaction with a KDEL receptor in the RER, several lines of evidence argue against such a direct retention mechanism. First, no such interaction has been demonstrated *in vitro* (Munro and Pelham, 1987). Second, the presence or absence of KDEL did not affect the diffusion rate of grp78 within the ER lumen of injected *Xenopus* oocytes (Ceriotti and Colman, 1988). Third, fusion of KDEL to the C-terminus of cathepsin D, a lysomal enzyme that undergoes mannose 6-phosphorylation by a sequence of reactions occurring in the Golgi complex (Kornfeld and Kornfeld, 1985; von Figura and Hasilik, 1986) led to the expected retention of cathepsin D in the ER but did not prevent phosphorylation of its carbohydrates (Pelham, 1988b). Fourth, an invertase fusion protein containing the retention signal showed processing of the oligomannosidic side chains which are typically induced by Golgi resident outer chain mannosidases (Pelham *et al.*, 1988). From these data it may be concluded that the signal-mediated ER-accumulations are achieved by a continuous and efficient retrieval of the corresponding proteins from a post ER compartment, presumably involving the *cis*-Golgi (Pelham, 1988b).

Presumably, elucidation of the retrieval or retention mechanism will be facilitated by the characterization of yeast mutants that are deficient in this sorting mechanism. Pelham *et al.* (1988) isolated 60 mutant strains of *Saccharomyces cerevisiae* that failed to retain an invertase-FEHDEL fusion protein in the ER. Eight of these mutants were determined to be alleles of the same erd1-gene (Pelham *et al.*, 1988).

An example for a receptor-mediated RER retention of a polypeptide has been reported for the compartmentalization of mouse liver β-glucuronidase (Medda *et al.*, 1987). Most of the total glucuronidase is generally found not in lysosomes but in the RER and here in complexes with one to four molecules of egasyn, an esterase that is specifically inhibited by organophosphorous compounds (Medda and Swank, 1985). Identical inhibitory compounds induce a rapid secretion of β-glucuronidase from the RER indicating that the esterase active site of egasyn is involved in binding β-glucuronidase (Medda *et al.*, 1987).

### 5.4. Vesicles as Carriers of Bulk Flow

Wieland *et al.* (1987) have demonstrated that the secretion of small peptides from the RER through the Golgi complex into the medium occurs with a $t_{1/2}$ of about 10 min. In these experiments, tripeptides containing a potential *N*-glycosylation site of the structure *N*-acyl-Asn-[$^{125}$]-Tyr–Thr were added exogenously to CHO or HepG2 cells. The amino-terminus of the peptide was blocked (a precondition required by the specificity of the oligosaccharidyl transferase) by acyl substituents of various lengths and the carboxyl group of Thr was converted into the amide. Iodinated tripeptides were readily taken up by the cells and those carrying short acyl groups ($C_1$–$C_8$) were efficiently *N*-glycosylated. Peptides subsequently recovered at various times from the supernatant carried processed oligosaccharides of the complex type, indicating that they had been secreted by the normal export route through the Golgi (Fig. 5). The half-time for secretion of these peptides was much shorter than observed for secretory proteins in the same cells, suggesting that no specific transport signals are required for a protein to migrate from the RER to the Golgi, through the individual stacks and from here to the plasma membrane.

There is increasing evidence that the transport of newly synthesized proteins from the RER via the Golgi to the cell surface involves the movement of carrier vesicles (Lodish, 1988). First reports on vesicular transport from the part rough, part smooth regions of the ER to the Golgi used electron microscopy to demonstrate the presence of 60-nm vesicles (for review, see Farquhar, 1985). Studies on the transport properties of Semliki Forest virus glycoproteins at low temperature (16°C) showed that the viral polypeptides accumulated in smooth-walled vesicles at the *cis*-side of the Golgi (Saraste and Kuismanen, 1984). Serial sections of uninfected sac(−) cells showed the truly vesicular nature of such transport organelles (Tooze *et al.*, 1988). The cytoplasmic face of these vesicles is coated by a structurally organized coat that fails to label with clathrin-specific antibodies. It is unknown whether the coat protein(s) on transient vesicles between the RER and the Golgi are identical to those present on vesicles that mediate the traffic between the individual Golgi stacks.

Transitory vesicles have recently been isolated and biochemically characterized. Lodish *et al.* (1987) studied the secretion of $\alpha_1$-antitrypsin in HepG2 cells by a combination of pulse-chase and cell-fractionation experiments. A vesicle fraction, lighter than the RER or Golgi fractions, was separated on shallow $D_2O$–sucrose gradients. The presence of $\alpha_1$-antitrypsin in such vesicles was transient and was only detected after a short chase period. Proteins known to be secreted from hepatoma cells at different rates entered the light vesicles in proportion to these specific rates. The carbohydrate structures of $\alpha_1$-antitrypsin from these vesicles (predominantly $Man_8GlcNAc_2$) reflected a pattern characteristic for processing by RER exoglycosidases and showed that the protein had not been a substrate for mannosidase I, an enzyme located in the medial Golgi.

Several groups have reported the development of *in vitro* systems that allow studies on the transfer of proteins from the RER to the Golgi. Haselbeck and Schekman (1986) employed donor microsomal membranes from a yeast strain carrying mutations in two genes (sec 18; mnnl): Owing to a temperature-sensitive transport block between the RER and the Golgi complex, this strain accumulated invertase in the RER at nonpermissive temperature. In addition, the donor cells were deficient in a Golgi specific α-(1,3)-mannosyltransferase. A Golgi-enriched fraction from an invertase-deficient (ΔSUC2; mnnl wild-type) strain was used to provide the recipient membranes. Transfer of prelabeled invertase from the donor membranes into acceptor membranes was judged by incorporation of outer chain mannose residues. This transfer was sensitive to temperature and detergent and required the presence of ATP, GDP-mannose, and magnesium, as well as divalent manganese cations. The invertase was shown to remain associated with the membrane throughout the experiment. Apparently, the transfer was specified by proteins residing at the cytoplasmic face of the donor membranes, since transfer could be abolished by pretreatment of donor, but not of recipient membranes, with *N*-ethylmaleimide or trypsin.

A similar approach, based on a mammalian cell system, was taken by Balch *et al.*, (1987). For the preparation of RER donor membranes, mutant CHO cells that lack a functional α-mannosidase I were infected with the temperature-sensitive mutant *ts*045 of vesicular stomatitis virus. Because of the mannosidase I deficiency, these donor cells are not capable of catalyzing trimming of the oligosaccharides beyond the $Man_8GlcNAc_2$ state (Fig. 5). Infected donor cells were incubated at nonpermissive temperature to allow for the temperature-sensitive accumulations of VSV G protein in the ER. The recipient Golgi fraction was isolated from a noninfected mutant CHO cell line that contained a functional α-mannosidase I but carried a defect in GlcNAc-transferase I, mitigating against further processing of the $Man_5GlcNAc_2$-oligosaccharides into complex-type carbohydrates. The transfer of the G protein was assayed by estimating the trimming of the oligosaccharides of prelabeled G protein. Again, this transfer required the presence of ATP and of one or more cytosolic factors.

The reconstitution of a cell-free transfer system involving transition vesicles was reported by Nowack *et al.* (1987). In their system, Golgi-specific acceptor membranes from rat liver cells were immobilized on nitrocellulose. Transfer of radiolabeled proteins from donor microsomal membranes was dependent on the presence of cytosolic factors and an ATP-regenerating system. GTP also stimulated the transfer. By contrast, no transfer of radiolabeled material was observed, when both RER or both Golgi membranes were used simultaneously as donor and acceptor membranes in the same experiment.

## 6. VESICULAR TRAFFIC THROUGH THE GOLGI COMPLEX INVOLVES G PROTEINS

The Golgi complex consists of a series of defined subcompartmentary cisternae coarsely divided into *cis*-, *medial*-, and *trans*-Golgi, as obtained by Golgi subfractionation (Saraste *et al.*, 1986) and by the distribution of specific marker enzymes acting on transitory proteins at early, intermediate, and late stages (Roth, 1987). However, the precise number of Golgi subcompartments is not clear; morphologic and cytochemical studies indicate that there might be many more than three subcompartments (Farquhar, 1985; Griffiths and Simons, 1986). Proteins that pass the individual Golgi stacks include (1) membrane proteins and proteins that are constitutively secreted, (2) proteins that carry signals for the transport into lysosomes, and (3) proteins that are destined for secretory granules and are subsequently released via the regulated pathway.

Vesicles that are coated with a pentagonal–hexagonal basketwork of clathrin have been identified previously as carriers in receptor-mediated endocytosis (Goldstein *et al.*, 1979) and more recently as transport vesicles budding from the *trans* side of the Golgi stack (Orci *et al.*, 1984, 1985; Griffiths *et al.*, 1985). Such vesicles are believed to be involved in the formation of secretory granules (Orci *et al.*, 1985) or in transport to lysosomes (Novikoff *et al.*, 1977). The finding that clathrin-coated vesicles were devoid of VSV G protein suggests that these vesicles are not involved in the constitutive transport to the plasma membrane. In addition, little if any clathrin has been detected in association with the RER or the internal cisternae of the Golgi (Griffiths *et al.*, 1985).

A new type of coated vesicles was recently proposed as carriers of constitutively transported material. Orci *et al.* (1986) described that isolated Golgi membranes formed numerous coated vesicles upon incubation with ATP and a cytosolic protein fraction. Under similar conditions, transport of VSV G protein was observed from donor membranes lacking *N*-acetylglucosaminyl transferase I (of the *medial*-Golgi compartment, Dunphy *et al.*, 1985) into acceptor membranes derived from cells containing the transferase (Orci *et al.*, 1986). The concentration of G protein present in the vesicles corresponded to that of the neighboring cisternae. The coat of these vesicles differed morphologically and immunocytochemically from the clathrin coat (Orci *et al.*, 1986). Taken together, these data suggest that these newly detected coated vesicles are responsible for constitutive transport of proteins through the subcompartments of the Golgi.

CHO (clone 1021) cells that lack permease (permitting entry of CMP-sialic acid into the Golgi) are incapable of synthesizing sialylated glycoproteins and glycolipids, even though they contain the normal set of sialyl transferases in the *trans*-Golgi (Briles *et al.*, 1977). Rothman (1987) used donor Golgi membranes from VSV-infected clone 1021 cells to study *in vitro* sialylation of the G protein upon incubation with acceptor membranes from normal CHO cells. Again, transport-coupled sialylation of VSV G protein was found to be dependent on the presence of ATP and a cytosolic protein fraction. In addition, a *N*-ethylmaleimide-sensitive factor, extractable from Golgi membranes, and long-chain acyl coenzyme A were identified as cofactors (Rothman, 1987). Glick and Rothman (1987) proposed the requirement for acyl-CoASH and the Golgi membrane-derived factor for efficient vesicular transport. The authors found that these two activities were able to restore transport between Golgi donor and acceptor membrane fractions that had previously been inactivated by *N*-ethylmaleimide. This finding suggests that an ATP-consuming round of acylation/deacylation which apparently, however, does not involve acylation of the transported VSV G protein itself, might participate in driving vesicular protein transport from one Golgi cisterna to another (Glick and Rothman, 1987).

GTP-binding proteins have been shown to be involved in the translocation of proteins through the RER membrane (Connolly and Gilmore, 1986). Melançon *et al.* (1987) recently provided direct evidence that GTP-binding proteins play a role in the constitutive vesicular transport of proteins through the Golgi. The authors used activators of mammalian GTP-binding proteins such as GTPγS and aluminum tetrafluoride $(ALF_4)^-$ in a cell-free system. The presence of such irreversible activators led to an enhanced formation of non-clathrin-coated vesicles that failed to bind to or fuse with recipient Golgi membranes. The inhibition of vesicular transport by GTPγS required the presence of stoichiometric amounts of an additional cytosolic inhibitory component with GTPγS-binding properties. This factor specifically and irreversibly poisoned acceptor membranes and prevented subsequent fusion of non-clathrin-coated vesicles derived from the donor–membrane fraction. It is unclear whether the putative GTP-binding protein plays a regulatory role in the shutdown of constitutive transport during mitosis. It has been proposed that breakdown of the Golgi stack during mitosis could be caused by a continuous budding of Golgi vesicles while vesicle fusion is stopped (Warren, 1985).

Yeast mutants are prepared with relative ease as compared to mammalian cell mutants. The analysis of thermosensitive alleles, carrying mutations in one or more of the roughly 25 genes whose products play essential roles in the transport from the RER to the plasma membrane, has shed further light on the involvement of GTP-binding proteins directing intracellular vesicle-mediated transport.

Yeast strains carrying a temperature-sensitive mutation in the YPT1 gene cannot undergo mitosis at the nonpermissive temperature and show phenotypic alterations involving the accumulation of membranes and vesicles that seem to be derived from the Golgi (Segev and Botstein, 1987). Under nonpermissive conditions, such cells showed a partial defect in Golgi-specific glycosylation and secreted invertase at a much reduced rate. The YPT1 gene product, a 23-kDa protein highly homologous to the *ras* protein, apparently plays a role in targeting vesicle transfer through the Golgi complex. Interestingly, antibodies against YPT1 recognized a mammalian homologue in mouse fibroblasts with a Golgi-specific intracellular localization (Haubruck *et al.*, 1987; Segev *et al.*, 1988).

While the YPT1 protein seems to be important for the intercisternal traffic to and within the Golgi complex, another GTP-binding protein, the SEC4 gene product has been reported to play a role at later stages of secretion (Goud *et al.*, 1988). The SEC4 protein is a 23.5-kDa ras-related polypeptide that contains four conserved domains thought to participate collectively in GTP binding. The nascent protein is quite hydrophilic, as demonstrated by partitioning into the aqueous phase following solubilization with Triton X-114. However, SEC4 protein binds rapidly to membranes of secretory vesicles and the plasma membrane and later behaves as an integral membrane protein that is nonextractable with 6 M urea or by treatment with alkali. This membrane binding is presumably made possible by conformational changes or by a covalent modification other than acylation. At the restrictive temperature, the mutant SEC4 protein failed to bind GTP and associated with secretory vesicles that accumulated under such conditions (Goud *et al.*, 1988). It was suggested that the SEC4 protein functions on the surface of such secretory vesicles by transducing a signal needed to target the vesicle from the Golgi to the plasma membrane.

### 6.1. *trans*-Golgi Network

The joint transport route of secretory, membrane, and lysosomal proteins leads from the RER to the terminal *trans*-most cisternae of the Golgi complex. From here, proteins are sequestered into different vesicular structures that are in general characterized by a low pH. Maintenance of the high $H^+$ concentration is achieved by a membrane-associated ATP-driven proton pump (Rudnick, 1986).

Packaging of prohormones into clathrin-coated vesicles occurs from the *trans* cisternae of the Golgi (Orci *et al.*, 1987a). Such vesicles subsequently transform into mature noncoated secretory granules. Concomitant with this maturation, the vesicular interior becomes acidified to a pH of ~5.7 (Orci *et al.*, 1987a), and proteolytic removal of the prosequence takes place. Such acidic compartments can be visualized by the application of DAMP (3-[2,4,-dinitroanilino]-3′amino-*N*-methyldipropylamine) and gold-labeled antibodies detecting the dinitrophenol groups in the base (for review, see Anderson and Orci, 1988).

By contrast, markers of the constitutive pathway, such as the influenza virus hemagglutinin, comigrate with the hormones into the *trans*-most Golgi cisternae, but at this stage become condensed into clear 100- to 300-nm vesicles. These vesicles fail to label with endocytotic markers and are believed to represent the bulk carriers of constitutively secreted proteins (Orci *et al.*, 1987b). The molecular principles of such sorting mechanisms which distinguish between regulated and constitutive proteins are not understood.

## 6.2. Sorting Signals for Lysosomes

It has been well established that lysosomal enzymes generally receive a first targeting signal for their final destination already in the Golgi complex in a two-step reaction (von Figura and Hasilik, 1986; Kornfeld, 1986). In a first step which presumably takes place in the *cis*-Golgi, *N*-acetylglucosamine 1-phosphate is attached to the 6-hydroxy position of one or more mannose residues of *N*-linked carbohydrate side chains. Subsequently *N*-acetylglucosamine is released, thereby leaving the lysosomal recognition marker mannose 6-phosphate on the lysosomal enzymes. The sequences of the 46-kDa cation-dependent mannose 6-phosphate receptor (CD-MPR) and of the 215-kDa cation-independent receptor (CI-MPR) have been determined (Dahms *et al.*, 1987; Pohlmann *et al.*, 1987; Oshima *et al.*, 1988). Both receptor molecules are class I membrane glycoproteins and require two phosphomannosyl residues in monoester linkage for high-affinity binding (Hoflack *et al.*, 1987).

The CI-MPR (Oshima *et al.*, 1988) is highly related (99.4% homology at the amino acid level) to the insulin-like growth factor II receptor from HepG2 hepatoma cells (Morgan *et al.*, 1987). However, mannose 6-phosphate failed to inhibit binding of insulin-like growth factor II, indicating that the binding sites for the hormone and for lysosomal enzymes carrying mannose 6-phosphate markers are different (Morgan *et al.*, 1987). The ectodomain of the CI-MPR is composed of 15 homologous repeats (134–167 amino acids in length), each of which is strikingly homologous to a nonrepetitive region within the ectodomain of the 46-kDa CD-MRP.

At first glance, no homology is found between the C-terminal regions of the two receptors. Both domains, however, contain clusters of negatively charged amino acids ($EESEERDDH_{275}$ in the 46-kDa CD-MPR, $DDSDEDLLHT_{2491}$ in the CI-MPR) at the C-terminus. It will be interesting to see whether these sequences provide targeting signals for the receptor molecules.

Several studies have addressed the question of where the receptors exhibit their biological function of sorting lysosomal enzymes from the bulk of secretory or plasma membrane proteins. Quite different results have been published, but all data indicated that the MPR was absent from mature lysosomes. Brown and Farquhar (1987) studied the intracellular distribution of CI-MPR in various cell types and found that it is variable, ranging from a predominantly *cis*-Golgi-specific distribution in rat clone 9-cells to a more *trans*-Golgi distribution in CHO cells. Griffiths *et al.* (1988) have analyzed the distribution of the CI-MPR in NRK cells by immunogold labeling. In agreement with previous data (Willingham *et al.*, 1981), no MPR was detected in mature lysosomes and low levels of receptor were present in the Golgi, the *trans*-Golgi network, and the plasma membrane. The bulk of CI-MPR was localized to an acidic prelysosomal compartment, that also contained lysosomal enzymes and the lysosomal membrane glycoprotein lgp120. This compartment was clearly distinct from the *trans*-Golgi network, as demonstrated by double labeling with antibodies against the G protein of vesicular stomatitis which forms specific aggregates in this vesicular network at 20°C (Griffiths *et al.*, 1985). Presumably this newly detected prelysosomal compartment serves as a place of transshipment where lysosomal enzymes are released from the CI-MPR because of the acidic pH of the milieu and because endocytotic vesicles discharge their contents.

A small percentage (less than 10% of the total number of receptors) of both types of mannose 6-phosphate receptors are found at the plasma membrane and are believed to function in retrieval of secreted lysosomal enzymes back into the lysosomes. Both receptors are recycled with similar kinetics (i.e. $t_{1/2}$ 3 hr) between the plasma membrane and the *trans*-most cisternae of the Golgi containing sialyltransferase activity (Duncan and Kornfeld, 1988). However, Stein *et al.* (1987) suggested that the 46-kDa CD-MPR does not participate in this retrieval process, as antibodies blocking the ligand binding site of the small receptor had no effect on the uptake of lysosomal enzymes. Furthermore, cells that solely expressed the 46-kDa CD-MPR failed to internalize exogenous lysosomal enzymes (Mainferme *et al.*, 1985; Stein *et al.*, 1987).

Another unsolved question concerns the signals that direct lysosomal proteins that never acquire the mannose 6-phosphate marker with high fidelity into lysosomes. No such targeting signals have been defined for mammalian enzymes. Valls *et al.* (1987) selected a large number of yeast mutants that secreted the vacuolar carboxypeptidase Y into the medium due to mutations in the corresponding carboxypeptidase Y gene. All such mutations mapped within the N-terminal propeptide sequence, suggesting that this region of the molecule provides the sorting signal into the yeast vacuole. However, no experiments have been reported, in which the putative targeting signal was shown to be functional upon transplantation onto foreign proteins.

## 7. THE GOLGI COMPLEX AS A MATURATION SITE OF ENVELOPED VIRUSES

### 7.1. Role of G1 and G2 of Uukuniemi Virus

The family Bunyaviridae comprises enveloped viruses that contain a segmented RNA genome of negative polarity. The Uukuniemi virus is probably the best characterized member of this family. Its helical nucleocapsid consists of three RNA segments (L, M, and S), the nucleoprotein (25,000 $M_r$), and a few copies of the putative RNA polymerase, designated L protein ($\approx$200,000 $M_r$) (Ranki and Pettersson, 1975). The viral envelope contains two glycoproteins, G1 ($\approx$75,000 $M_r$) and G2 ($\approx$65,000 $M_r$), generated by co-translational cleavage of a

110-kDa precursor, p110 (Ulmanen *et al.*, 1981). The p110 species can only be obtained *in vitro* by translation of mRNA corresponding to the M segment (Ulmanen *et al.*, 1981). The oligosaccharide side chains of G1 are mainly of the endo H-resistant complex type, while those of G2 are predominantly of the mannose-rich type (Pesonen *et al.*, 1982). Recently, the sequences of G1 and G2 have been established from cloned cDNA (Rönnholm and Pettersson, 1987). The presence of a single open reading frame, comprising 1008 amino acid residues, verified previous observations that G1 and G2 arise from a single polypeptide by co-translational proteolytic cleavage. The mature forms of G1 and G2 show all the features of class I integral membrane proteins. Although the precise membrane integration process has not been investigated, one may speculate that it parallels the mechanism previously reported for alphaviruses (Garoff *et al.*, 1980; Rice and Strauss, 1981) or flaviviruses (Wengler *et al.*, 1985). This assumption is based on similarities in the gene arrangement between these viruses ($NH_2$–G1–G2–COOH) and on the presence of a putative internal signal sequence between G1 and G2. Indeed, co-translocational cleavage between G1 and G2 and at the C-terminal side of the G2-specific signal sequence could release a small peptide (81 amino acids long) which is reminiscent of the 6-kDa protein of alphaviruses.

Uukuniemi virus, like other bunyaviruses, has the property of budding at smooth membranes of the Golgi complex (Bishop and Shope, 1979). This specific maturation site is probably determined by the transport properties of the two viral glycoproteins, as deduced from the following lines of evidence:

1. In Uukuniemi virus-infected cells both G1 and G2 and, at later stages also the N protein, accumulate in the Golgi region which subsequently undergoes expansion and vacuolization (Kuismanen *et al.*, 1984).
2. Similar accumulations of G1 and G2 are observed in hamster kidney and in chick embryo cells infected with a temperature-sensitive mutant, ts12, of Uukuniemi virus. At the restrictive temperature (39°C), N protein failed to accumulate in the Golgi. In contrast to the situation in wild-type infected cells, G1 and G2 were retained in the Golgi and could not be chased to the plasma membrane, even when cells were incubated for 6 hr (at nonpermissive temperature) in the presence of cycloheximide (Gahmberg *et al.*, 1986).
3. Transport of G1 and G2 from the RER to the Golgi requires the presence of carbohydrate side chains. In tunicamycin-treated infected cells, the non-glycosylated proteins accumulate in the RER. Under these conditions, the N protein also associates with the RER (Kuismanen *et al.*, 1984).

It will be interesting to learn whether G1 and G2 will accumulate independently from one another in the Golgi complex when expressed separately from the individual gene segments, and what structural entities cause transport arrest in the Golgi. With the gene at hand answers to these questions will be available in the near future.

### 7.2. Role of the E1 Glycoprotein of Coronaviruses

Coronaviruses (for reviews, see Wege *et al.*, 1982, Sturman and Holmes 1985; Holmes, 1985) can cause a wide spectrum of acute infections involving infections of the respiratory tract (in humans and chickens), hepatitis (in mice), or infections of the gastrointestinal (GI) tract (in newborn cattle, swine, and humans). In addition, coronavirus mouse hepatitis virus (MHV) gives rise to persistent infections both *in vitro* and *in vivo* (demyelinating panencephalitis), which may be facilitated to some extent by the intracellular maturation of the virus particles.

Coronaviruses use a unique replication strategy involving the formation of a nested set of 3′-co-terminal mRNA molecules (Holmes, 1985; Makino *et al.*, 1986). Replication takes place exclusively in the cytoplasm and can be demonstrated in enucleated cells (Wilhelmsen *et al.*, 1981). Virus assembly occurs at membranes of the smooth and rough ER (Becker *et al.*, 1967; Tooze *et al.*, 1984). Particles contain a single-stranded RNA genome of positive polarity ($5.5 \times 10^6$ $M_r$) and three structural proteins designated N (50,000 $M_r$), E1 (26,000 $M_r$), and E2 (180,000 $M_r$) (Sturman and Holmes, 1983).

Glycoprotein E2 forms the viral spikes required for attachment of the virus particle to host-cell receptors. Furthermore, E2 is transported to the plasma membrane and causes cell fusion by which the infection is spread into neighboring tissue. E2 is a class I glycoprotein (Schmidt *et al.*, 1987), it is acylated and carries predominantly N-linked glycans (Niemann and Klenk, 1981). Depending on the infected cell type, E2 is proteolytically processed into two 90-kDa subunits to a variable extent (Sturman *et al.*, 1985). Although, similar to myxoviruses, proteolytic cleavage of E2 activates its fusion capacity (Storz *et al.*, 1981; Sturman *et al.*, 1985), no hydrophobic region becomes exposed at the newly formed N-terminus of the E2 subunit (Schmidt *et al.*, 1987; Luytjes *et al.*, 1987).

The E1-glycoprotein of coronaviruses differs from all other viral glycoproteins in its biosynthetic and structural features. E1 is a class IV integral membrane glycoprotein spanning the viral membrane three times (Niemann *et al.*, 1984a; Armstrong *et al.*, 1984). E1 is translated on membrane-associated poly-

**FIGURE 6.** Intracellular maturation of coronavirus MHV-A59 in infected 3T3 (17Cl1)-cells. Cells were infected at 50 PFU/cell and fixed *in situ* after the time indicated. (A) At 8 hr postinfection, virus particles bud preferentially into smooth walled vesicles close to the Golgi. (B) At 16 hr postinfection, virus particles are seen in the RER, inclusions of the nuclear membrane (NM) and in coated peripheral loops (CV) of the Golgi, the latter seldom found intact at this late stage of infection. (Electron microscopy by C. B. Boschek.)

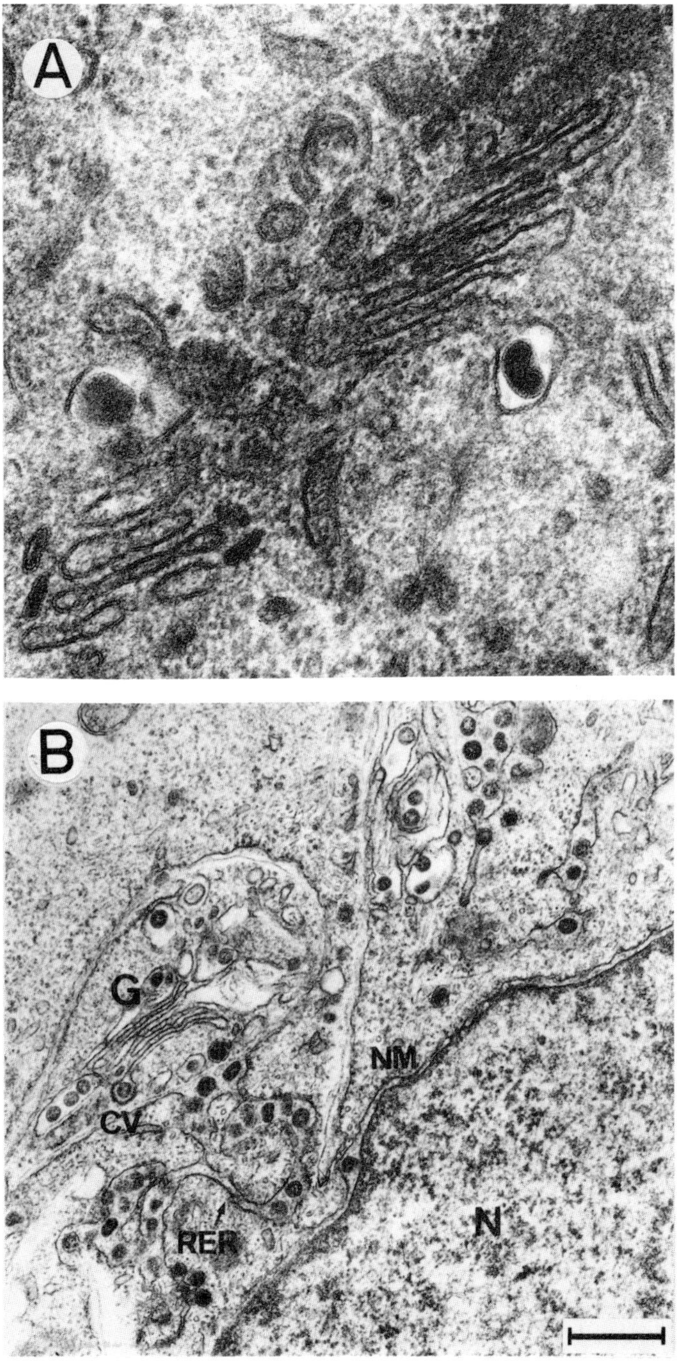
A
B
G
NM
CV
RER
N

somes (Niemann *et al.*, 1982) and assembles in microsomal membranes at the co-translational level (Rottier *et al.*, 1985). The protein is transported from the RER rapidly into budding compartments and transient vesicles, where it accumulates and where first formation of progeny virus is observed (Becker *et al.*, 1967; Tooze *et al.*, 1984) (Fig. 6). We have previously constructed a complete E1 cDNA clone (Niemann *et al.*, 1984a) and established the primary sequence (Armstrong *et al.*, 1984). Three functional domains of the E1 protein can be distinguished (Fig. 7A).

With the exception of TGEV (Laude *et al.*, 1987) all other coronaviruses contain E1 proteins which lack N-terminal cleavable signal sequences. The hydrophilic *ectodomain* of E1 from MHV (28 amino acid residues) carries a cluster of hydroxyamino acids and up to three *O*-glycosidically linked oligosaccharide side chains (Niemann and Klenk, 1981; Holmes *et al.*, 1981; Niemann *et al.*, 1984b) and exhibit blood group M activity (Niemann *et al.*, 1984b). *O*-Glycosylation occurs post-translationally, presumably when the virus particles are transported through peripheral loops of the Golgi cisternae (Niemann *et al.*, 1982; Tooze *et al.*, 1988). Recently Tooze *et al.* (1988) suggested that transfer of the internal *N*-acetylgalactosamine occurred within the transient elements, i.e., prior to the *cis*-Golgi, where the *N*-acetylgalactosaminyl transferase has been localized by means of *Helix pomatia* lectin (Roth, 1984). In addition, cell fractionation experiments and metabolic labeling of monensine-treated infected 3T3 17Cl1 cells had shown that, as with other *O*-glycosylated viral proteins (Johnson and Spear, 1983), the first step of *O*-glycosylation occurs in the Golgi (Niemann *et al.*, 1982).

The ectodomain of other coronaviruses, such as avian infectious bronchitis virus (IBV) and TGEV, contains attachment sites for asparagine-linked oligosaccharides (Stern *et al.*, 1982; Laude *et al.*, 1987).

All coronaviral E1 proteins contain *three internal membrane-spanning domains,* designated MSD-I, -II, and -III, which share a higher degree of sequence homology than the extra-membranal domains. *In vitro* transcription/translation experiments have been performed to elucidate the role of each of the individual MSDs in the integration process into microsomal membranes (Mayer *et al.*, 1988). The data indicate that MSD-I and MSD-III function independently as signal sequences and simultaneously as stop-transfer sequences in a co-translational manner (Fig. 7B). Each of the domains was capable of translocating a short N-terminal sequence across the membrane as indicated by glycosylation of a newly introduced *N*-glycosylation site (Mayer *et al.*, 1988). MSD-II of IBV alone also functions in membrane insertion *in vivo,* however, the mutant protein (lacking MSD-I and MSD-III) was not stably integrated in the membrane and could be extracted at pH 11.5 (C. Machamer, personal communication). The observation that various parts of MSD-II can be deleted without altering the overall topology of E1 further suggests that MSD-II does not constitute a topogenic signal (Mayer *et al.*, 1988).

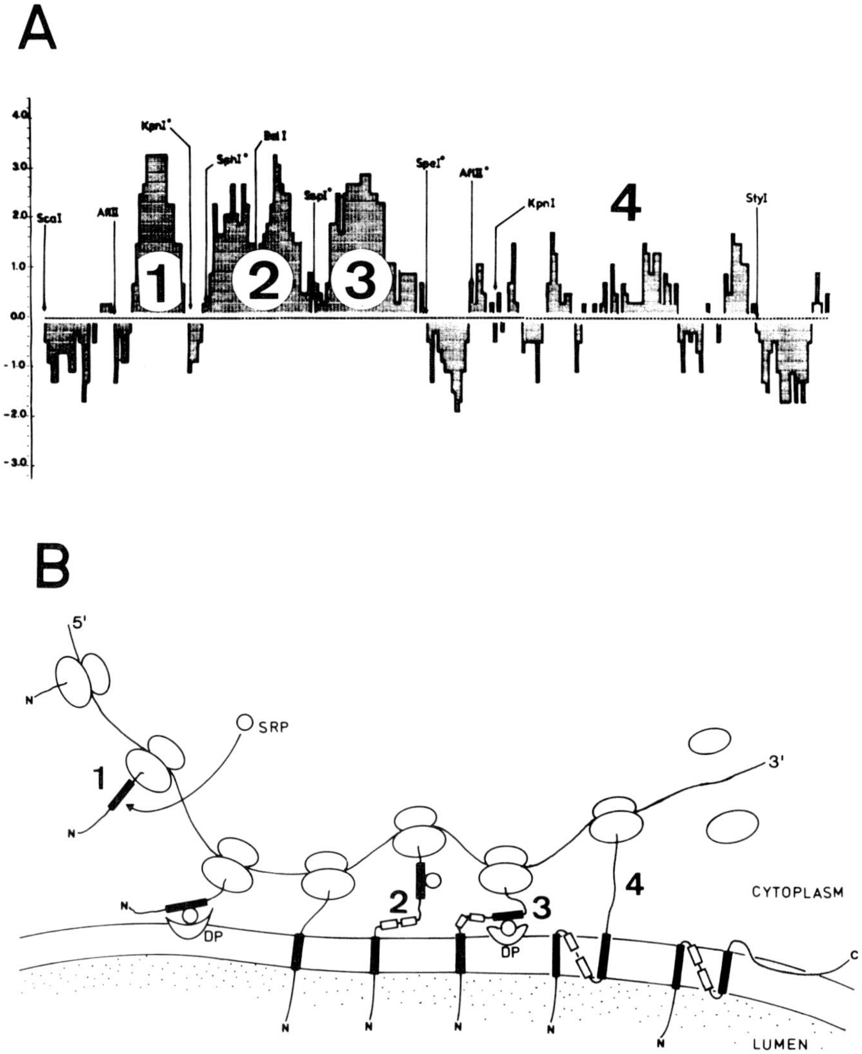


**FIGURE 7.** Hydropathy plot and membrane integration of glycoprotein E1 of coronavirus MHV-A59, a class IV integral membrane glycoprotein. (A) Hydropathy plot of the E1 polypeptide according to the program of Kyte and Doolittle (1982). The positions of restriction sites used to study *in vitro* membrane integration are indicated. Restriction sites marked with an asterisk were introduced by site directed mutagenesis. (B) Postulated membrane assembly of the E1 protein. Topogenic hydrophilic segments are indicated by solid bars. (According to Mayer *et al.*, 1988.)

The C-terminal tail of the E1 molecule interacts with the viral nucleocapsid as demonstrated by co-sedimentation of isolated E1 protein with purified nucleocapsid on sucrose gradients (Sturman *et al.*, 1980). Therefore, the E1 protein functions as a matrix protein during budding of virus particles. In MHV-infected sac(−) cells, Tooze *et al.* (1985) observed an unusual association of RER cisternae into laminated structures in which the lipid bilayers of two cisternae were closely and uniformly aligned. The formation of these structures was coincident with the accumulation of E1 in the RER. The authors suggested that this lamination was caused by a crosslinking of the two membranes competing for the same nucleocapsid structures.

Deletions within the C-terminal domain do not affect the membrane orientation of molecules retaining all three MSDs (Mayer *et al.*, 1988). Interestingly, a species lacking the three MSDs can post-translationally associate with membranes and form tight complexes that are not dissociated at pH 11.5. However, this domain did not contain a signal sequence, since it was incapable of translocating N-terminal or C-terminal sequences across the membrane. These findings may reflect the natural role of this domain of E1, i.e., to bind post-translationally to the nucleocapsid and condense it into budding virus particles (Mayer *et al.*, 1988).

Several groups have reported that the intracellular distribution of the E1 protein determines the site of virus maturation. This conclusion was based on the following lines of evidence:

1. In the infected cell, E1 accumulates in perinuclear compartments into which virus particle formation occurs. Virus budding never takes place at the plasma membrane probably because it does not contain E1 (Holmes *et al.*, 1981; Tooze *et al.*, 1984). At early stages after virus infection, E1 accumulates in and particles mature at membranes of smooth-walled transitory vesicles, assumed to mediate traffic between the RER and the Golgi. As a result of the pileup of E1 in these membranes, E1 accumulates at later stages also in the RER and virus budding occurs concomitantly also into this compartment (Tooze *et al.*, 1984).
2. The E1-gene has been expressed in the absence of other coronaviral proteins in order to determine whether the protein by itself carried a specific retention signal or whether it was merely trapped in the budding compartments due to an interaction with other viral components (Niemann *et al.*, 1987; Rottier *et al.*, 1987; Machamer and Rose, 1987). These studies clearly demonstrated that the E1 was specifically retained in perinuclear membranes that were colabeled with wheat germ agglutinin used as Golgi-specific marker (Fig. 8). The retention signal was functional also in secretory AtT20 cells which secrete adrenocorticotropic hormone (ACTH). No E1 was detected in secretory vesicles containing ACTH (Mayer *et al.*, 1988). This finding is in agreement

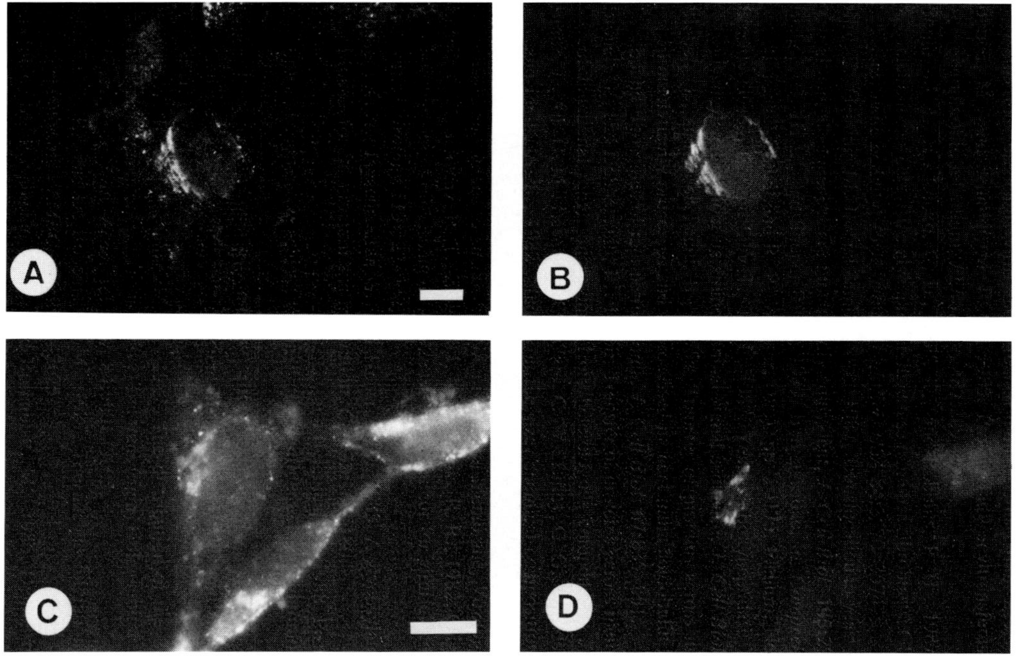

**FIGURE 8.** Glycoprotein E1 contains a retention signal for the Golgi that is functional in the absence of other viral components. Capped and polyadenylated wild-type E1-mRNA was microinjected into CV-1 cells or secretory AtT20 cells. The distribution of the E1 protein was determined by indirect immunofluorescence 6 hr after injection. (A) Cells were stained with rhodamine-conjugated Golgi-specific WGA. (B) Same cells as in (A) but stained with polyclonal E1-specific antibodies and FITC-labeled goat anti-rabbit IgG. (C) AtT20-cells stained with anti-ACTH rabbit serum and rhodamine labeled goat anti-rabbit IgG. (D) Same cells as in (C) but stained with a monoclonal E1-specific antibody and FITC-conjugated goat antimouse TgG. Bar = 20 μm. (For details see Mayer *et al.*, 1988.)

E1

Δm2,3

Δm1,2

G23

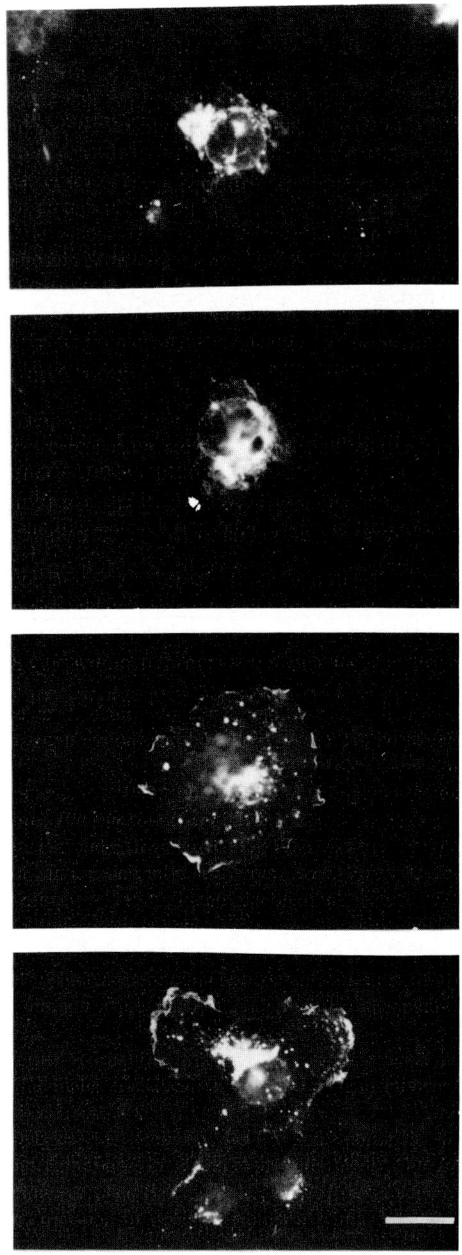

**FIGURE 9.** Transport properties of E1 mutants (from coronavirus IBV) lacking individual internal membrane spanning domains as indicated. COS-1 cells grown on coverslips were transfected with DNA encoding the wild-type E1 or mutant e1 proteins, or a hybrid protein G23 (known to be transported to the plasma membrane). Cells were fixed, permeabilized, and 44 hr after transfection, subjected to indirect immunofluorescence using affinity-purified anti-E1 antibodies as primary antibody. Bar = 10 μm. (From Machamer and Rose, 1987.)

with a previous report showing that sorting of coronaviruses from ACTH occurs at the exit of the *trans*-Golgi network (Tooze *et al.*, 1987).

The ectodomain of E1 has no influence on the Golgi-specific retention signal, as evidenced by deletions or alterations of the glycosylation sites (Mayer *et al.*, 1988). Machamer and Rose (1987) demonstrated that the complete deletion of MSD-II and MSD-III from the E1 protein of IBV did not alter the transport properties of the protein, while, conversely, a deletion of MSD-I and MSD-II resulted in transport to the plasma membrane (Fig. 9). These findings suggested that MSD-I alone contained the Golgi-specific retention signal. However, more recent unpublished data, obtained with mutant proteins containing only MSD-II, indicated that these proteins were also retained intracellularly in fairly large yet undefined vesicles close to the Golgi complex (C. Machamer, personal communication). Furthermore, several fusion proteins containing MSD-I, MSD-I together with MSD-II, or all three MSDs were not transported to the Golgi but accumulated within the RER or vesicular structures in the vicinity of the Golgi (C. Machamer, unpublished observations; Mayer *et al.*, 1988). These findings suggest that the retention signal of the E1 protein does not merely involve a linear sequence but might be contained in the quaternary structure of the E1 protein.

## 8. CONCLUDING REMARKS

The search for transport signals was initially inspired by the idea that specific linear amino acid sequences direct proteins to corresponding target organelles. Indeed, the existence of such targeting sequences which could be transplanted onto other proteins were detected for several organelles. Ironically, further elucidation of the RER translocation system that originally provided the grounds for the signal theory, led to the surprising realization that the characteristics of signal sequences are not as closely related as was originally anticipated. Random sequences could replace the signal, provided that the charge distribution in the vicinity was not altered. In addition, the process that was first considered to occur exclusively co-translationally was found under certain conditions to take place at the post-translational stage as well. This finding shed additional light on the role of a class of stress proteins that apparently constitute a quality-control device by which the cell distinguishes malfolded proteins from correctly folded proteins. Such proteins were shown to participate in the translocation process into the RER and into mitochondria, but it is feasible that they play a general role in establishing translocation and transport competence of proteins.

The association of individual protein subunits into oligomeric structures in the RER appears to be a common phenomenon for secreted and integral membrane proteins. Several studies have shown that factors or conditions that prevent an organized folding and subsequent oligomerization of the subunits cause a transport arrest of the protein in the RER and simultaneously induce synthesis of polypeptides that form complexes with the nonassembled subunits. These polypeptides are closely related to those involved in the unfolding of proteins in the cytosol. This observation peaks against the hypothesis that the proteins bear linear sequences that constitute specific transport signals. In addition, polypeptides carrying the mutations in different domains were shown to be retained in different intracellular compartments, again providing evidence against specific targeting signals. By contrast, retention signals blocking transport of proteins in the RER and in the Golgi have been identified, although retention in the Golgi compartment seemed to involve interaction of several domains of the blocked protein and not merely the presence of a linear sequence.

The recent development of *in vitro* systems that permit reconstitution of transport between the RER and the Golgi on the one hand and within the Golgi cisternae on the other hand will provide new insights into the vesicular transport mechanism as well as the cytoplasmic components that target these vesicles. The intriguing hypothesis that GTP-binding proteins control the transport of the vesicles would explain why proteins of entirely different primary sequence end up in the same intracellular compartment. As in other areas, gene technology and yeast genetics may prove superior in breaking down the barriers in elucidating the complex mechanisms controlling vesicular traffic in eukaryotic cells.

ACKNOWLEDGMENTS. We wish to thank all our colleagues who made their results available to us before publication. This work was supported by grant Nie 175/5 from the Deutsche Forschungsgemeinschaft and by Fonds der Chemischen Industrie.

## 9. REFERENCES

Ainger, K. J., and Meyer, D. I., 1986, Translocation of nascent secretory proteins across membranes can occur late in translation, *EMBO J.* **5:**951–955.

Allison, D. S., and Schatz, G., 1986, Artificial mitochondrial presequences, *Proc. Natl. Acad. Sci. USA* **83:**9011–9015.

Anderson, R. G. W., and Orci, L., 1988, A view of acidic intracellular compartments, *J. Cell Biol.* **106:**539–543.

Andrews, P. G., and Kole, R., 1987, Alu RNA transcribed *in vitro* binds to the 68-kDa subunit of the signal recognition particle, *J. Biol. Chem.* **262:**2908–2912.

Armstrong, J., Niemann, H., Smeekens, S., Rottier, P., and Warren, G., 1984, Sequence and topology of a model intracellular membrane protein, E1 glycoprotein, from a coronavirus, *Nature (Lond.)* **308:**751–752.

Baker, K., Mackman, N., Jackson, M., and Holland, I. B., 1987, Role of secA and secY in protein export as revealed by studies of TonA assembly into the outer membrane of Escherichia coli, *J. Mol. Biol.* **198:**693–703.

Baker, R. K., Bentivoglio, G. P., and Lively, M. O., 1986, Partial purification of microsomal signal peptidase from hen oviduct, *J. Cell. Biochem.* **32:**193–200.

Balch, W. E., Wagner, K. R., and Keller, D. S., 1987, Reconstitution of vesicular stomatitis virus G protein from the endoplasmic reticulum to the Golgi complex using a cell-free system, *J. Cell Biol.* **104:**749–760.

Bause, E., Günther, R., Schweden, J., and Tillmann, U., 1986, *In vitro* studies on the subcellular location of glucosidase I and glucosidase II in dog pancreas, *Biosci. Rep.* **6:**827–834.

Beaune, P. H., Umbenhauer, D. R., Bork, R. W., Lloyd, R. S., and Guengerich, F. P., 1986, Isolation and sequence determination of a cDNA clone related to human cytochrome P-450 nifedipine oxidase, *Proc. Natl. Acad. Sci. USA* **83:**8064–8068.

Becker, W. B., McIntosh, K., Drees, J. H., and Chanock, R. M., 1967, Morphogenesis of avian infectious bronchitis virus and a related human virus (strain 229E), *J. Virol.* **1:**1019–1027.

Berger, M., and Schmidt, M. F. G., 1985, Protein fatty acyltransferase is located in the rough endoplasmic reticulum, *FEBS Lett.* **187:**289–294.

Bergh, M. L. E., Cepko, C. L., Wolf, D., and Robbins, P. H., 1987, Expression of the Saccharomyces cerevisiae glycoprotein invertase in mouse fibroblasts: Glycosylation, secretion and enzymatic activity, *Proc. Natl. Acad. Sci. USA* **84:**3570–3574.

Bird, P., Gething, M.-J., and Sambrook, J., 1987, Translocation in yeast and mammalian cells: Not all signal sequences are functionally equivalent, *J. Cell Biol.* **105:**2905–2914.

Bishop, D. H. L., and Shope, R. E., 1979. Bunyaviridae, in: *Comprehensive Virology,* Vol. 14 (Fraenkel-Conrat and R. R. Wagner, eds.), pp. 1–175, Plenum, New York.

Blachly-Dyson, E., and Stevens, T. H., 1987, Yeast carboxypeptidase Y can be translocated and glycosylated without its aminoterminal signal sequence, *J. Cell Biol.* **105:**1183–1191.

Blobel, G., and Dobberstein, B., 1975a, Transfer of proteins across membranes. I. Presence of proteolytically processed and unprocessed nascent immunoglobulin light chains on membrane bound ribosomes of murine myeloma, *J. Cell Biol.* **67:**835–851.

Blobel, G., and Dobberstein, B., 1975b, Transfer of proteins across membranes. II. Reconstitution of functional rough microsomes from heterologous components, *J. Cell Biol.* **67:**852–862.

Block, J., Air, G. M., Laver, W. G., Ward, C. W., Lilley, G. G., Woods, E. F., Roxburgh, C. M., and Inglis, A. S., 1982, Studies on the size, chemical composition, and partial sequence of neuraminidase (NA) from type A influenza virus show that the N-terminal region is not processed and serves to anchor the NA in the viral membrane, *Virology* **119:**109–121.

Bosch, J. V., Schwarz, R. T., Ziemiecki, A., and Friis, R. R., 1982, Oligosaccharide modifications and the site of processing of gPr92env, the precursor for viral glycoproteins of Rous sarcoma virus, *Virology* **119:**122–132.

Bostian, K. A., Jayachandran, S., and Tipper, D. J., 1983, A glycosylated protoxin in killer yeast: Models for its structure and maturation, *Cell* **32:**169–180.

Boulay, F., Doms, R. W., Webster, R. G., and Helenius, A., 1988, Posttranslational oligomerization and cooperative acid activation of mixed influenza hemagglutinin trimers, *J. Cell Biol.* **106:** 629–639.

Briles, E. B., Li, E., and Kornfeld, S., 1977, Isolation of rat germ agglutinin resistant clones of Chinese hamster ovary cells deficient in membrane sialic acid and galactose, *J. Biol. Chem.* **252:**1107–1116.

Brown, W. J., and Farquhar, M. G., 1987, The distribution of the 215-kilodalton mannose 6-phosphate receptors within cis (heavy) and trans (light) Golgi subfractions varies in different cell types, *Proc. Natl. Acad. Sci. USA* **84:**9001–9005.

Bürglin, T. R., and De Robertis, E. M., 1987, The nuclear migration signal of *Xenopus laevis* nucleoplasmin, *EMBO J.* **6:**2617–2625.

Caulfield, M. P., Duong, L. T., and Rosenblatt, M., 1986, Demonstration of posttranslational secretion of human placental lactogen by a mammalian in vitro translation system, *J. Biol. Chem.* **261:**10953–10956.

Ceriotti, A., and Colman, A., 1988, Binding to membrane proteins within the endoplasmic reticulum cannot explain the retention of the glucose-regulated protein GRP78 in *Xenopus* oocytes, *EMBO J.* **7**:633–638.

Chao, C. C.-K., Bird, P., Gething, M.-J., and Sambrook, J., 1987, Posttranslational translocation of influenza virus hemagglutinin across microsomal membranes, *Mol. Cell. Biol.* **7**:3842–3845.

Chen, W.-J., and Douglas, M. G., 1988, An $F_1$-ATPase β-subunit precursor lacking an internal tetramer-forming domain is imported into mitochondria in the absence of ATP, *J. Biol. Chem.* **263**:4997–5000.

Chen, W.-J., and Douglas, M. G., 1987, Phosphodiester bond cleavage outside mitochondria is required for the completion of protein import into the mitochondrial matrix, *Cell* **49**:651–658.

Chirico, W. J., Waters, M. G., and Blobel, G., 1988, 70K heat shock related proteins stimulate protein translocation into microsomes, *Nature (Lond.)* **332**:805–810.

Collier, D. N., Bankaitis, V. A., Weiss, J. B., and Bassford, P. J., 1988, The antifolding activity of SecB promotes the export of the *E. coli* maltose binding protein, *Cell* **53**:273–283.

Connolly, T., and Gilmore, R., 1986, Formation of a functional ribosome–membrane junction during translocation requires the participation of a GTP-binding protein, *J. Cell Biol.* **103**:2253–2261.

Conzelmann, A., Riezman, H., Desponds, C., and Bron, C., 1988, A major 125 kd membrane glycoprotein of Saccharomyces cerevisiae is attached to the lipid bylayer through an inositol-containing phospholipid, *EMBO J.* **7**:2233–2240.

Copeland, C. S., Zimmer, K.-P., Wagner, K. R., Healey, G. A., Mellman, I., and Helenius, A., 1988, Folding trimerization and transport are sequential events in the biogenesis of influenza virus hemagglutinin, *Cell* **53**:197–209.

Copeland, C. S., Doms, R. W., Bolzau, E. M., Webster, R. G., and Helenius, A., 1986, Assembly of influenza hemagglutinin trimers and its role in intracellular transport, *J. Cell Biol.* **103**:1179–1191.

Cross, G. A. M., 1987, Eukaryotic protein modification and membrane attachment via phosphotidyinositol, *Cell* **48**:179–181.

Dahms, N. M., Lobel, P., Breitmeyer, J., Chirgwin, J. M., and Kornfeld, S., 1987, 46 kd Mannose 6-phosphate receptor: Cloning, expression, and homology to the 215 kd mannose 6-phosphate receptor, *Cell* **50**:181–192.

Datema, R., Olofsson, S., and Romero, P., 1987, Inhibitors of protein glycosylation and glycoprotein processing in viral systems, *Pharmacol. Ther.* **33**:221–286.

Datema, R., Romero, P. A., Rott, R., and Schwartz, R. T., 1984, On the role of oligosaccharide trimming in the maturation of Sindbis and influenza virus, *Arch. Virol.* **81**:25–39.

Davey, J., Dimmock, N. J., and Colman, A., 1985, Identification of the sequence responsible for the nuclear accumulation of the influenza virus nucleoprotein in *Xenopus* oocytes, *Cell* **40**:667–675.

Davis, L. I., and Blobel, G., 1987, Nuclear pore complex contains a family of glycoproteins that includes p62: Glycosylation through a previously unidentified cellular pathway, *Proc. Natl. Acad. Sci. USA* **84**:7552–7556.

Deshaies, R. J., Koch, B. D., Werner-Washburne, M., Craig, E. A., and Schekman, R., 1988, A subfamily of stress proteins facilitates translocation of secretory and mitochondrial precursor polypeptides, *Nature (Lond.)* **332**:800–805.

De Wet, J. R., Wood, K. V., DeLuca, M., Helinski, D. R., and Subramani, S., 1987, Firefly luciferase gene: Structure and expression in mammalian cells, *Mol. Cell Biol.* **7**:725–737.

Doms, R. W., Keller, D. S., Helenius, A., and Balch, W. E., 1987, Role of adenosine triphosphate in regulating the assembly and transport of vesicular stomatitis virus G protein trimers, *J. Cell Biol.* **105**:1957–1969.

Doms, R. W., Ruusala, A., Machamer, C., Helenius, J., Helenius, A., and Rose, J. K., 1988,

Differential effects of mutations in three domains on folding, quarternary structure, and intracellular transport of vesicular stomatitis virus G protein, *J. Cell Biol.* **107:**89–99.

Doyle, C., Roth, M. G., Sambrook, J., and Gething, M.-J., 1985, Mutations in the cytoplasmic domain of the influenza hemagglutinin affect different stages of intracellular transport, *J. Cell Biol.* **100:**704–714.

Doyle, C., Sambrook, J., and Gething, M.-J., 1986, Analysis of progressive deletions of the transmembrane and cytoplasmic domains of influenza hemagglutinin, *J. Cell Biol.* **103:**1193–1204.

Duncan, J. R., and Kornfeld, S., 1988, Intracellular movement of two mannose 6-phosphate receptors: return to the Golgi apparatus, *J. Cell Biol.* **106:**617–628.

Dunphy, W. G., Brands, R., and Rothman, J. E., 1985, Attachment of terminal N-acetylglucosamine to asparagine-linked oligosaccharides occurs in central cisternae of the Golgi stack, *Cell* **40:**463–472.

Duong, L. T., Caulfield, M. P., and Rosenblatt, M., 1987, Synthetic signal peptide and analogs display different activities in mammalian and plant in vitro secretion systems, *J. Biol. Chem.* **262:**6328–6333.

Edman, J. C., Ellis, L., Blacher, R. W., Roth, R. A., and Rutter, W. A., 1985, Sequence of protein disulfide isomerase and implications of its relationship to thioredoxin, *Nature (Lond.)* **317:**267–270.

Eilers, M., and Schatz,G., 1988, Protein unfolding and the translocation across biological membranes, *Cell* **52:**481–483.

Eilers, M., Hwang, S., and Schatz, G., 1988, Unfolding and refolding of a purified precursor protein during import into isolated mitochondria, *EMBO J.* **7:**1139–1145.

Einfeld, D. A., Brown, J. P., Valentine, M. A., Clark, E. A., and Ledbetter, J. A., 1988, Molecular cloning of the human B cell CD20 receptor predicts a hydrophobic protein with multiple transmembrane domains, *EMBO J.* **7:**711–717.

Enders, G. H., Ganem, D., and Varmus, H. E., 1985, Mappig the major transcripts of ground squirrel hepatitis virus: The presumptive template for reverse transcriptase is terminally redundant, *Cell* **42:**297–308.

Endo, T., and Schatz, G., 1988, Latent membrane perturbation activity of a mitochondrial precursor protein is exposed by unfolding, *EMBO J.* **7:**1153–1158.

Evans, E. A., Gilmore, R., and Blobel, G., 1986, Purification of microsomal signal peptidase as a complex, *Proc. Natl. Acad. Sci. USA* **83:**581–585.

Farquhar, G., 1985, Progress in unraveling pathways of Golgi traffic, *Annu. Rev. Cell Biol.* **1:**447–488.

Feldherr, C. M., Kallenbach, E., and Schultz, N., 1984, Movement of a karyophilic protein through the nuclear pores of oocytes, *J. Cell Biol.* **99:**2216–2222.

Folz, R. J., Nithwehr, S. F., and Gordon, J. I., 1988, Substrate specificity of eukaryotic signal peptidase, *J. Biol. Chem.* **263:**2070–2078.

Freedman, R. B., 1984, Native disulfide bond formation in protein biosynthesis: Evidence for the role of protein disulfide isomerase, *Trends Biochem. Sci.* **9:**438–441.

Friedlander, M., and Blobel, G., 1985, Bovine opsin has more than one signal sequence, *Nature (Lond.)* **318:**336–343.

Fujiki, Y., Hubbard, A. L., Fowler, S., and Lazarow, P. B., 1982, Isolation of intracellular membranes by means of sodium bicarbonate treatment: Application to endoplasmic reticulum, *J. Cell Biol.* **93:**97–102.

Fujiki, Y., Rachubinski, R. A., and Lazarow, P. B., 1984, Synthesis of a major integral membrane polypeptide of rat liver peroxisomes on free polysomes, *Proc. Natl. Acad. Sci. USA* **81:**7127–7131.

Gahmberg, N., Kuismanen, E., Keränen, S., and Pettersson, R. E., 1986, Uukuniemi virus glycoproteins accumulate in and cause morphological changes of the Golgi complex in the absence of virus maturation, *J. Virol.* **57:**899–906.

Gallione, C. J., and Rose, J. K., 1985, A single amino acid substitution in a hydrophobic domain causes temperature sensitive cell surface transport of a mutant viral glycoprotein, *J. Virol.* **54:** 374–382.

Garcia, P. D., Ghrayeb, J., Inouye, M., and Walter, P., 1987, Wild type and mutant signal peptides of *Escherichia coli* outer membrane lipoprotein interact with equal efficiency with mammalian signal recognition particle, *J. Biol. Chem.* **262:**9463–9468.

Garcia, P. D., Ou, J.-H., Rutter, W. J., and Walter, P., 1988, Targeting the hepatitis B virus precore protein to the endoplasmic reticulum membrane: After signal peptide cleavage translocation can be aborted and the product released into the cytoplasm, *J. Cell Biol.* **106:**1093–1104.

Garoff, H., 1985, Using recombinant DNA techniques to study protein targeting in the eucaryotic cell, *Annu. Rev. Cell Biol.* **1:**403–445.

Garoff, H., Frischauf, A.-M., Simons, K., Lehrach, H., and Delius, H., 1980, Nucleotide sequence of cDNA coding for Semliki Forest virus membrane glycoproteins, *Nature (Lond.)* **288:**235–241.

Gething, M.-J., McCammon, K., and Sambrook, J., 1986, Expression of wild-type and mutant forms of influenza hemagglutinin: The role of folding in intracellular transport, *Cell* **46:**939–950.

Gibson, R., Schlessinger, S., and Kornfeld, S., 1979, The nonglycosylated glycoprotein of vesicular stomatitis virus is temperature sensitive and undergoes intracellular aggregation at elevated temperatures, *J. Biol. Chem.* **254:**3600–3607.

Gilmore, R., and Blobel, G., 1985, Translocation of secretory proteins across the microsomal membrane occurs through an environment accessible to aqueous perturbants, *Cell* **42:**497–505.

Gilmore, R., Walter, P., and Blobel, G., 1982, Protein translocation across the endoplasmic reticulum. II. Isolation and characterization of the signal recognition particle receptor, *J. Cell Biol.* **95:**470–477.

Glick, B. S., and Rothman, J. E., 1987, Possible role for fatty acylcoenzyme A in intracellular protein transport, *Nature (Lond.)* **326:**309–312.

Goldman, B. M., and Blobel, G., 1978, Biogenesis of peroxisomes: Intracellular site of synthesis of catalase and uricase, *Proc. Natl. Acad. Sci. USA* **75:**5066–5070.

Goldstein, J. L., Anderson, R. G. W., and Brown, M. S., 1979, Coated pits, coated vesicles, and receptor-mediated endocytosis, *Nature (Lond.)* **279:**679–685.

Goud, B., Salminen, A., Walworth, N. C., and Novick, P. J., 1988, A GTP-binding protein required for secretion rapidly associates with secretory vesicles and the plasma membrane in yeast, *Cell* **53:**753–768.

Gould, S. J., Keller, G.-A., and Subramani, S., 1987, Identification of a peroxisomal targeting signal at the carboxy terminus of firefly luciferase, *J. Cell Biol.* **105:**2923–2931.

Greenspan, D., Palese, P., and Krystal, M., 1988, Two nuclear location signals in the influenza virus NS1 Nonstructural protein, *J. Virol.* **62:**3020–3026.

Griffiths, G., and Simons, K., 1986, The trans Golgi network: Sorting at the exit side of the Golgi Complex, *Science* **234:**438–443.

Griffiths, G., Pfeiffer, S., Simons, K., and Matlin, K., 1985, Exit of newly synthesized membrane glycoproteins from the trans cisternae of the Golgi complex to the plasma membrane, *J. Cell Biol.* **101:**949–964.

Griffiths, G., Hoflack, B., Simons, K., Mellman, I., and Kornfeld, S., 1988, The mannose 6-phosphate receptor and the biogenesis of lysosomes, *Cell* **52:**329–341.

Gruters, R. A., Neefjes, J. J., Tersmette, M., de Goede, R. E. Y., Tulp, A., Huisman, H. G., Miedema, F., and Ploegh, H. L., 1987, Interference with HIV-induced syncytium formation and viral infectivity by inhibitors of trimming glucosidase, *Nature (Lond.)* **330:**74–76.

Guan, J.-L., Ruusala, A., Cao, H., and Rose, J. K., 1988, Effects of altered cytoplasmic domains on transport of the vesicular stomatitis virus glycoprotein are transferable to other proteins, *Mol. Cell. Biol.* **8:**2869–2874.

Gundelfinger, E. D., Krause, E., Melli, M., and Dobberstein, B., 1983, The organization of the 7SL RNA in the signal recognition particle, *Nucleic Acids Res.* **11:**7363–7374.

Gundelfinger, E. D., Di Carlo, M., Zopf, D., and Melli, M., 1984, Structure and evolution of the 7SL RNA component of the signal recognition particle, *EMBO J.* **3:**2325–2332.

Haas, I. G., and Wabl, M., 1983, Immunoglobulin heavy chain binding proteins, *Nature (Lond.)* **306:**387–389.

Hadwiger, A., Niemann, H., Käbisch, A., Bauer, H., and Tamura, T., 1986, Appropriate glycosylation of the fms gene product is a prerequisite for its transforming potency, *EMBO J.* **5:** 689–694.

Hall, M. N., Hereford, L., and Herskowitz, I., 1984, Targeting of *E. coli* β-Galactosidase to the nucleus in yeast, *Cell* **36:**1057–1065.

Hansen, W., and Walter, P., 1988, Prepro-carboxypeptidase Y and a truncated form of pre-invertase, but not full-length pre-invertase, can be posttranslationally translocated across microsomal vesicle membranes from Saccharomyces cerevisiae, *J. Cell Biol.* **106:**1075–1081.

Hansen, W., Garcia, P., and Walter, P., 1986, *In vitro* protein translocation across the yeast endoplasmic reticulum: ATP-dependent post-translational translocation of the prepro-α-factor, *Cell* **45:**397–406.

Haselbeck, A., and Schekman, R., 1986, Interorganelle transfer and glycosylation of yeast invertase in vitro, *Proc. Natl. Acad. Sci. USA* **83:**2017–2021.

Haubruck, H., Disela, C., Wagner, P., and Gallwitz, D., 1987, The ras related ypt protein is an ubiquitous eukaryotic protein: isolation and sequence analysis of mouse cDNA clones highly homologous to the yeast YPT1 gene, *EMBO J.* **6:**4049–4053.

Hawlitschek, G., Schneider, H., Schmidt, B., Tropschug, M., Hartl, F.-U., and Neupert, W., 1988, Mitochondrial protein import: Identification of processing peptidase and of PEP, a processing enhancing protein, *Cell* **53:**795–806.

Hendershot, L., Bole, D., Köhler, G., and Kearney, J. F., 1987, Assembly and secretion of heavy chains that do not associate posttranslationally with immunoglobulin heavy chain-binding protein, *J. Cell Biol.* **104:**761–767.

Hiebert, S. W., Paterson, R. G., and Lamb, R. A., 1985, Hemagglutinin-neuraminidase protein of the paramyxovirus Simian Virus 5: Nucleotide sequence of the mRNA predicts an N-terminal membrane anchor, *J. Virol.* **54:**1–6.

Hitzeman, R. A., Leung, D. W., Perry, L. J., Kohr, W. J., Levine, H. L., and Goeddel, D. V., 1983, Secretion of human interferon by yeast, *Science* **219:**620–625.

Hobden, A. E., and Cundliffe, E., 1978, The mode of action of α-sarcin and a novel assay for the puromycin reaction, *Biochem. J.* **170:**57–61.

Hoffman, K. E., and Gilmore, R., 1988, Guanosine triphosphate promotes the post-translational integration of opsin into the endoplasmic reticulum membrane, *J. Biol. Chem.* **263:**4381–4385.

Hoflack, B., Fujimoto, K., and Kornfeld, S., 1987, The interaction of phosphorylated oligosaccharides and lysosomal enzymes with bovine liver cation-dependent mannose 6-phosphate receptor, *J. Biol. Chem.* **262:**123–129.

Holland, E. C., and Drickamer, K., 1986, Signal recognition particle mediates the insertion of a transmembrane protein which has a cytoplasmic $NH_2$ terminus, *J. Biol. Chem.* **261:**1286–1292.

Holmes, K. V., 1985, Replication of coronaviruses, in: *Virology* (B. N. Fields, ed.), pp. 1331–1343, Raven, New York.

Holmes, K. V., Doller, E. W., and Sturman, L. S., 1981, Tunicamycin resistant glycosylation of a coronavirus glycoprotein: Demonstration of a novel type of viral glycoprotein, *Virology* **115:** 334–344.

Hull, J. D., Gilmore, R., and Lamb, R. A., 1988, Integration of a small integral membrane protein, $M_2$, of influenza virus into the endoplasmic reticulum: Analysis of the internal signal-anchor domain of a protein with an ectoplasmic $NH_2$ terminus, *J. Biol. Chem.* **106:**1489–1498.

Ibrahimi, I., and Gentz, R., 1987, A functional interaction between the signal peptide of the translation apparatus is detected by the use of a single point mutation which blocks translocation across mammalian endoplasmic reticulum, *J. Cell Biol.* **262:**10189–10194.

Imanaka, T., Small, G. M., and Lazarow, P. B., 1987, Translocation of acyl-CoA oxidase into peroxisomes requires ATP hydrolysis but not a membrane potential, *J. Cell Biol.* **105:**2915–2922.

Jabbar, M. A., and Nayak, D. P., 1987, Signal processing, glycosylation, and secretion of mutant hemagglutinins of a human influenza virus by *Saccharomyces cerevisiae, Mol. Cell. Biol.* **7:** 1476–1485.

Jackson, R. C., and White, W. R., 1981, Phospholipid is required for presecretory proteins by detergent solubilized canine pancreatic signal peptidase, *J. Biol. Chem.* **256:**2545–2550.

Johnson, D. C., and Spear, P. G., 1983, O-linked oligosaccharides are acquired by herpes simplex virus glycoproteins in the Golgi complex, *Cell* **32:**987–997.

Johnson, L. M., Bankiatis, V. A., and Emr, S. D., 1987, Distinct sequence determinants direct intracellular sorting and modification of a yeast vacuolar protease, *Cell* **48:**875–885.

Johnson, L. M., Bankiatis, V. A., and Emr, S. D., 1987, Distinct sequence determinants direct intracellular sorting and modification of a yeast vacuolar protease, *Cell* **48:**875–885.

Julius, D., Schekman, R., and Thorner, J., 1984, Glycosylation and processing of prepro-α-factor through the yeast secretory pathway, *Cell* **36:**309–318.

Kaiser, C. A., Preuss, D., Grisafi, P., and Botstein, D., 1987, Many random sequences functionally replace the secretion signal of yeast invertase, *Science* **235:**312–317.

Kalderon, D., Roberts, B. L., Richardson, W. D., and Smith, A. E., 1984a, A short amino acid sequence able to specify nuclear location, *Cell* **39:**499–509.

Kalderon, D., Richardson, W. D., Markham, A. F., and Smith, A. E., 1984b, Sequence requirements for nuclear location of simian virus 40 large-T antigen, *Nature (Lond.)* **311:**33–38.

Kaluza, G., Rott, R., and Schwarz, R. T., 1980, Carbohydrate induced conformational changes of Semliki Forest virus glycoproteins determine antigenicity, *Virology* **102:**286–299.

Kassenbrock, C. K., Garcia, P. D., Walter, P., and Kelly, R. B., 1988, Heavy chain binding protein recognizes aberrant polypeptides translocated *in vitro, Nature (Lond.)* **333:**90–93.

Keil, W., Geyer, R., Dabrowski, J., Dabrowski, U., Niemann, H., Stirm, S., and Klenk, H.-D., 1985, Carbohydrates of influenza virus. Structural elucidation of the individual glycans of the FPV hemagglutinin by two-dimensional $^1$H n.m.r. and methylation analysis, *EMBO J.* **4:**2711–2720.

Kopito, R. R., and Lodish, H. F., 1985, Primary structure and transmembrane orientation of the murine anion exchange protein, *Nature (Lond.)* **316:**234–238.

Kornfeld, S., 1986, Trafficking of lysosomal enzymes in normal and disease states, *J. Clin. Invest.* **77:**1–6.

Kornfeld, R., and Kornfeld, S., 1985, Assembly of asparagine-linked oligosaccharides, *Annu. Rev. Biochem.* **54:**631–664.

Kozak, M., 1987, At least six nucleotides preceding the AUG initiator codon enhance translation in mammalian cells, *J. Mol. Biol.* **196:**947–950.

Kozutsumi, Y., Segal, M., Normington, K., Gething, M.-J., and Sambrook, J., 1988, The presence of malfolded proteins in the endoplasmic reticulum signals the induction of glucose-regulated proteins, *Nature (Lond.)* **332:**462–464.

Krakow, J. L., Hereld, D., Bangs, J. D., Hart, G. W., and Englund, P. T., 1986, Identification of a glycolipid precursor of the *Trypanosoma brucei, J. Biol. Chem.* **261:**12147–15153.

Krieg, P. A., and Melton, D. A., 1984, Functional messenger RNAs are produced by SP6 *in vitro* transcription of cloned cDNA, *Nucleic Acids Res.* **12:**7057–7070.

Krieg, U. C., Walter, P., and Johnson, A. E., 1986, Photocrosslinking of the signal sequence of nascent preprolactin to the 54 kd polypeptide of the signal recognition particle, *Proc. Natl. Acad. Sci. USA* **83:**8604–8608.

Krippl, B., Ferguson, B., Jones, N., Rosenberg, M., and Westphal, H., 1985, Mapping of functional domains in adenovirus E1A proteins, *Proc. Natl. Acad. Sci. USA* **82:**7480–7484.

Kuismanen, E., Bång, B., Hurme, M., and Pettersson, R. E., 1984, Uükuniemi virus maturation: Immunofluorescence microscopy with monoclonal glycoprotein-specific antibodies, *J. Virol.* **51:**137–146.

Kulomaa, M. S., Weigel, N. L., Kleinsek, D. A., Beattie, W. G., Conneely, O. M., March, C., Zarucki-Schulz, T., Schrader, W. T., and O'Malley, B. W., 1986, Amino acid sequence of a chicken heatshock protein derived from the complementary DNA nucleotide sequence, *Biochemistry* **25:**6244–6251.

Kurzchalia, T. V., Wiedmann, M., Girshovich, A. S., Bochkareva, E. S., Bielka, H., and Rapoport, T. A., 1986, A signal sequence of nascent preprolactin interacts with the 54K polypeptide of the signal recognition particle, *Nature (Lond.)* **320:**634–636.

Kyte, J., and Doolittle, R. F., 1982, A simple method for displaying the hydropathic character of a protein, *J. Mol. Biol.* **157:**105–132.

Lanford, R. E., and Butel, J. S., 1984, Construction and characterization of an SV40 mutant defective in nuclear transport of T-antigen, *Cell* **37:**801–813.

Lanford, R. E., Kanda, P., and Kennedy, R. C., 1986, Induction of nuclear transport with a synthetic peptide homologous to the SV40 T antigen transport signal, *Cell* **46:**575–582.

Laude, H., Rasschaert, D., and Huet, J.-C., 1987, Sequence and N-terminal processing of the transmembrane protein E1 of the coronavirus transmissible gastroenteritis virus, *J. Gen. Virol.* **68:**1687–1693.

Lauffer, L., Garcia, P. D., Harkins, R. N., Coussens, L., Ullrich, A., and Walter, P., 1985, Topology of signal recognition particle receptor in endoplasmic reticulum membrane, *Nature (Lond.)* **318:**334–338.

Lazarow, P. B., and Fujiki, Y., 1985, Biogenesis of peroxisomes, *Annu. Rev. Cell Biol.* **1:**489–530.

Lee, A. S., 1987, Coordinated regulation of a set of genes by glucose and calcium ionophores in mammalian cells, *Trends Biochem. Sci.* **12:**20–23.

Lee, B. A., Maher, D. W., Hannink, M., and Donoghue, D. J., 1987, Identification of a signal for nuclear targeting in platelet-derived-Growth-factor-related molecules, *Mol. Cell. Biol.* **7:**3527–3537.

Lennarz, W. J., 1987, Protein glycosylation in the endoplasmic reticulum: Current topological issues, *Biochemistry* **26:**7205–7210.

Lingappa, V. R., Chaidez, J., Yost, C. S., and Hedgpeth, J., 1984, Determinants for protein localization: β-Lactamase signal sequence directs globin across microsomal membranes, *Proc. Natl. Acad. Sci. USA* **81:**456–460.

Lingelbach, K., Zwieb, C., Hoben, P. J., Webb, J. R., Marshallsay, C., Walter, P., and Dobberstein, B., 1988, Isolation and characterization of a cDNA clone encoding the 19kDa protein of signal recognition particle (SRP): Expression and binding to 7SL RNA, *Nucl. Acids Res.* **16:** 9431–9442.

Lipp, J., and Dobberstein, B., 1986, Signal recognition particle-dependent membrane insertion of mouse invariant chain: A membrane-spanning protein with a cytoplasmically exposed amino terminus, *J. Cell Biol.* **102:**2169–2175.

Lipp, J., Dobberstein, B., and Haeuptle, M.-T., 1987, Signal recognition particle arrests elongation of nascent secretory and membrane proteins at multiple sites in a transient manner, *J. Biol. Chem.* **262:**1680–1684.

Lodish, H. F., 1988, Transport of secretory and membrane glycoproteins from the rough endoplasmic reticulum to the Golgi, *J. Biol. Chem.* **263:**2107–2110.

Lodish, H. F., Kong, N., Hirani, S., and Rasmussen, J., 1987, A vesicular intermediate in the transport of hepatoma secretory proteins from the rough endoplasmic reticulum to the Golgi complex, *J. Cell Biol.* **104:**221–230.

Luytjes, W., Sturman, L. S., Bredenbeek, P. J., Charite, J., van der Zeijst, B. A. M., Horzinek, M. C., and Spaan, W. J. M., 1987, Primary structure of the glycoprotein E2 of coronavirus MHV-A59 and identification of the trypsin cleavage site, *Virology* **161:**479–487.

Lyons, R. H., Ferguson, B. W., and Rosenberg, M., 1987, Pentapeptide nuclear localization signal in Adenovirus E1A, *Mol. Cell. Biol.* **7:**2451–2456.

Machamer, C. E., and Rose, J. K., 1987, A specific transmembrane domain of a coronavirus E1 glycoprotein is required for its retention in the Golgi region, *J. Cell Biol.* **105:**1205–1214.

Machamer, C., and Rose, J. K., 1988a, Influence of new glycosylation sites on expression of the vesicular stomatitis virus G protein at the plasma membrane, *J. Biol. Chem.* **263:**5948–5954.

Machamer, C., and Rose, J. K., 1988b, Vesicular stomatitis virus G protein with altered glycosylation sites display temperature-sensitive intracellular transport and are subject to aberrant intermolecular disulfide bonding, *J. Biol. Chem.* **263:**5955–5960.

Mainferme, F., Wattiaux, R., and von Figura, K., 1985, Synthesis, transport and processing of cathepsin D in Morris hepatoma 7777 cells and rat hepatocytes, *Eur. J. Biochem.* **153:**211–216.

Makino, S., Stohlman, S. A., and Lai, M. M. C., 1986, Leader sequences of murine coronavirus mRNAs can be freely reassorted: Evidence for the role of free leader RNA in transcription, *Proc. Natl. Acad. Sci. USA* **83:**4204–4208.

Mayer, T., Tamura, T., Falk, M., and Niemann, H., 1988, Membrane integration and intracellular transport of the coronavirus glycoprotein E1, a class III membrane glycoprotein, *J. Biol. Chem.* **263:**14956–14963.

Mazzarella, R. A., and Green, M., 1987, ERp99, an abundant, conserved glycoprotein of the endoplasmic reticulum, is homologous to the 90-kDa heat shock protein (hsp90) and the 94-kDa glucose regulated protein (GRP94), *J. Biol. Chem.* **262:**8875–8883.

McLachlan, A., Milich, D. R., Raney, A. K., Riggs, M. G., Hughes, J. L., Sorge, J., and Chisari, F. V., 1987, Expression of hepatitis virus B surface and core antigen: Influence of pre-S and precore sequences, *J. Virol.* **61:**683–692.

Medda, S., and Swank, R. T., 1985, Egasyn, a protein which determines the subcellular distribution of β-glucuronidase, has esterase activity, *J. Biol. Chem.* **260:**15802–15808.

Medda, S., Stevens, A. M., and Swank, R. T., 1987, Involvement of the esterase active site of egasyn in compartmentalization of β-glucuronidase within the endoplasmic reticulum, *Cell* **50:** 301–310.

Melançon, P., Glick, B. S., Malhotra, V., Weidman, P. J., Serafini, T., Gleason, M. L., Orci, L., and Rothman, J. E., 1987, Involvement of GTP-binding "G" proteins in transport through the Golgi stack, *Cell* **51:**1053–1062.

Meyer, D. I., 1985, Signal recognition particle (SRP) does not mediate a translational arrest of nascent secretory proteins in mammalian cell-free systems, *EMBO J.* **4:**2031–2033.

Meyer, D. I., Krause, E., and Dobberstein, B., 1982, Secretory protein translocation across membranes—The role of the docking protein, *Nature (Lond.)* **297:**647–650.

Miura, S., Mori, M., Takiguchi, M., Tatibana, M., Furuta, S., Miyazawa, S., and Hashimoto, T., 1984, Biosynthesis and intracellular transport of enzymes of peroxisomal β-oxidation, *J. Biol. Chem.* **259:**6397–6402.

Molenaar, W. H., Kruijer, W., Tilly, B. C., Verlaan, I., Bierman, A. J., and de Laat, S. W., 1986, Growth factor-like action of phosphatidic acid, *Nature (Lond.)* **323:**171–173.

Moreland, R. B., Nam, H. G., Hereford, L. M., and Fried, H. M., 1985, Identification of a nuclear localization signal of a yeast ribosomal protein, *Proc. Natl. Acad. Sci. USA* **82:**6561–6565.

Morgan, D. O., Edman, J. C., Standring, D. N., Fried, V. A., Smith, M. C., Roth, R. A., and Rutter, W. J., 1987, Insulin-like growth factor II receptor as a multifunctional binding protein, *Nature (Lond.)* **329:**301–307.

Mücke, K., and Scholtissek, C., 1987, Extragenic and intragenic suppression of a transport mutation in the hemagglutinin gene of an influenza A virus as revealed by backcross and sequence determination, *Virology* **158:**112–117.

Mueckler, M., and Lodish, H. F., 1986, The human glucose transporter can insert posttranslationally into microsomes, *Cell* **44:**629–637.

Müller, G., and Zimmermann, R., 1988, Import of honeybee prepromelittin into the endoplasmic reticulum: Structural basis for independence of SRP and docking protein, *EMBO J.* **6:**2099–2107.

Müller, M., Ibrahimi, I., Chang, C. N., Walter, P., and Blobel, G., 1982, A bacterial secretory protein requires Signal Recognition Particle for translocation across mammalian endoplasmic reticulum, *J. Biol. Chem.* **257:**11860–11863.

Munro, S., and Pelham, H. R. B., 1987, A C-terminal signal prevents secretion of luminal ER proteins, *Cell* **48:**899–907.

Munro, S., and Pelham, H. R. B., 1986, An hsp70-like protein in the ER: Identity with the 78 kd glucose-regulated protein and immunoglobulin heavy chain binding protein, *Cell* **46:**291–300.

Newmeyer, D. D., and Forbes, D. J., 1988, Nuclear import can be separated into distinct steps *in vitro:* Nuclear pore binding and translocation, *Cell* **52:**641–653.

Niemann, H., and Klenk, H.-D., 1981, Glycoprotein E1 of coronaviruses: A novel type of viral glycoprotein. *J. Mol. Biol.* **153:**993–1010.

Niemann, H., Boschek, C. B., Evans, D., Rosing, M., Tamura, T., and Klenk, H.-D., 1982, Post-translational glycosylation of coronavirus glycoprotein E1: Inhibition by monensin, *EMBO J.* **1:** 1499–1504.

Niemann, H., Heisterberg-Moutsis, G., Geyer, R., Klenk, H. D., and Wirth, M., 1984a, Glycoprotein E1 of MHV A59: Structure of the O-linked carbohydrates and construction of full length cDNA-clones, *Adv. Exp. Med. Biol.* **173:**201–213.

Niemann, H., Geyer, R., Klenk, H.-D., Linder, D., Stirm, S., and Wirth, M., 1984b, The carbohydrates of mouse hepatitis virus (MHV) A59: Structures of the O-glycosidically linked oligosaccharides of glycoprotein E1, *EMBO J.* **3:**665–670.

Niemann, H., Dabrowski, J., Dabrowski, U., Geyer, R., Keil, W., Klenk, H.-D., and Stirm, S., 1985, The major oligosaccharides in the large subunit of the hemagglutinin from fowl plague virus, strain Dutch, *Eur. J. Biochem.* **146:**523–532.

Niemann, H., Mayer, T., Wirth, M., and Tamura, T., 1987, Expression of the E1 gene of Mouse Hepatitis virus (MHV A59) *in vivo* and *in vitro, Adv. Exp. Med. Biol.* **218:**83–97.

Novikoff, A. B., Mori, M., Quintana, N., and Yam, A., 1977, Studies of the secretory processin the mammalian exocrine pancreas, *J. Cell Biol.* **75:**148–165.

Nowack, D. D., Morré, D. M., Paulik, M., Keenan, T. W., and Morré, D. J., 1987, Intracellular membrane flow, Reconstitution of transition vesicle formation and function in a cell-free system, *Proc. Natl. Acad. Sci. USA* **84:**6098–6102.

Ohba, M., and Schatz, G., 1987a, Protein import into yeast mitochondria is inhibited by antibodies raised against 45-kd proteins of the ourter membrane, *EMBO J.* **6:**2111–2116.

Ohba, M., and Schatz, G., 1987b, Disruption of the outer membrane restores protein import to trypsin treated yeast mitochondria, *EMBO J.* **6:**2117–2122.

Orci, L., Ravazzola, M., Louvard, D., and Perrelet, A., 1984, Distribution asymétrique de la clathrine au niveau de l'appareil de Golgi sécrétoire à polypeptides, *C.R. Acad. Sci.* **299:**697–702.

Orci, L., Ravazzola, M., Amherdt, M., Louvard, D., and Perrelet, A., 1985, Clathrin-immunoreactive sites in the Golgi apparatus are concentrated at the trans pole in polypeptide hormone-secreting cells, *Proc. Natl. Acad. Sci. USA* **82:**5385–5389.

Orci, L., Glick, B. S., and Rothman, J. E., 1986, A new type of coated vesicular carrier that appears not to contain clathrin: Its possible role in protein transport within the Golgi stack, *Cell* **46:**171–184.

Orci, L., Ravazzola, M., Storch, M.-J., Anderson, R. G. W., Vassalli, J.-D., and Perrelet, A., 1987a, Proteolytic maturation of insulin is a post-Golgi event which occurs in acidifying clathrin-coated secretory vesicles, *Cell* **49:**865–868.

Orci, L., Ravazzola, M., Amherdt, M., Perrelet, A., Powell, S. K., Quinn, D. L., and Moore, H.-P. H., 1987b, The trans-most cisternae of the Golgi complex: A compartment for sorting of secretory and plasma membrane proteins, *Cell* **51:**1039–1051.

Oshima, A., Nolan, C. M., Kyle, J. W., Grubb, J. H., and Sly, W. S., 1988, The human cation-independent mannose 6-phosphate receptor, *J. Biol. Chem.* **263:**2553–2562.

Ou, J. H., Laub, O., and Rutter, W. J., 1986, Hepatitis virus B gene function: The precore region targets the core antigen to cellular membranes and causes the secretion of the e antigen, *Proc. Natl. Acad. Sci. USA* **83:**1578–1582.

Pääbo, S., Bhat, B. M., Wold, W. S. M., and Peterson, P. A., 1987, A short sequence in the COOH-terminus makes an adenovirus membrane glycoprotein a resident of the endoplasmic reticulum, *Cell* **50:**311–317.

Park, M. K., D'Onofrio, M., Willingham, M. C., and Hanover, J. A., 1987, A monoclonal antibody against a family of nuclear pore proteins (nucleoporins): O-linked N-acetylglucosamine is part of the immunodeterminant, *Proc. Natl. Acad. Sci. USA* **84:**6462–6466.

Pelham, H., 1988a, Heat shock proteins. Coming in from the cold, *Nature (Lond.)* **332:**776–777.

Pelham, H. R. B., 1988b, Evidence that the luminal ER proteins are sorted from secreted proteins in a post-ER compartment, *EMBO J.* **7:**913–918.

Pelham, H. R. B., Hardwick, K. G., and Lewis, M. J., 1988, Sorting of soluble ER proteins in yeast, *EMBO J.* **7:**1757–1762.

Perara, E., Rothman, R. E., and Lingappa, V. R., 1986, Uncoupling translocation from translation: Implications for transport of proteins across membranes, *Science* **232:**348–352.

Perlman, D., and Halvorson, H. O., 1983, A putative signal peptidase recognition site and sequence in eukaryotic and prokaryotic signal peptides, *J. Mol. Biol.* **167:**621–636.

Pesonen, M., Kuismanen, E., and Pettersson, R. E., 1982, Monosaccharide sequence of protein-bound glycans of Uukuniemi virus, *J. Virol.* **41:**390–400.

Peters, R., 1986, Fluorescence microphotolysis to measure nucleocytoplasmic transport and intracellular mobility, *Biochim. Biophys. Acta* **864:**305–359.

Pfanner, N., Tropschug, M., and Neupert, W., 1987, Mitochondrial protein import: Nucleoside triphosphates are involved in conferring importcompetence to precursors, *Cell* **49:**815–823.

Pfanner, N., Pfaller, R., Kleene, R., Ito, M., Tropschug, M., and Neupert, W., 1988, Role of ATP in mitochondrial import, *J. Biol. Chem.* **263:**4049–4051.

Pohlmann, R., Nagel, G., Schmidt, B., Stein, M., Lorkowski, G., Krentler, C., Cully, J., Meyer, H. E., Grzeschik, K.-H., Mersmann, G., Hasilik, A., and von Figura, K., 1987, Cloning of a cDNA encoding the human cation-dependent mannose 6-phosphate-specific receptor, *Proc. Natl. Acad. Sci. USA* **84:**5575–5579.

Poruchynsky, M. S., Tyndall, C., Both, G. W., Sato, F., Bellamy, A. R., and Atkinson, P. H., 1985, Deletions into the $NH_2$-terminal hydrophobic domain result in secretion of rotavirus VP7, a resident endoplasmic reticulum membrane glycoprotein, *J. Cell Biol.* **101:**2199–2209.

Prehn, S., Wiedmann, M., Rapoport, T. A., and Zwieb, C., 1987, Protein translocation across wheat germ microsomal membranes requires an SRP-like component, *EMBO J.* **6:**2093–2097.

Ranki, M., and Pettersson, R. F., 1975, Uukuniemi virus contains an RNA polymerase, *J. Virol.* **16:** 1420–1425.

Rapoport, T. A., 1986, Protein translocation across and integration into membranes, *CRC Crit. Rev. Biochem.* **20:**73–137.

Reddy, J. K., Warren, J. R., Reddy, M. K., and Lalwani, N. D., 1982, Hepatic and renal effects of peroxisome proliferators: Biological implications, *Ann. NY Acad. Sci.* **386:**81–110.

Reddy, J. K., Goel, S. K., Nemali, M. R., Carrino, J. J., Laffler, M. K., Reddy, M. K., Sperbeck, S. J., Osumi, T., Hashimoto, T., Lalwani, N. D., and Sambasiva Rao, M., 1986, Transcriptional regulation of peroxisomal fatty acyl-CoA oxidase and enoyl-CoA hydratase/3-hydroxyl-CoA dehydrogenase in rat liver by peroxisome proliferators, *Proc. Natl. Acad. Sci. USA* **83:** 1747–1751.

Repp, R., Tamura, T., Boschek, C. B., Wege, H., Schwarz, R. T., and Niemann, H., 1985, The effects of processing inhibitors of N-linked oligosaccharides on the intracellular migration of glycoprotein E2 of mouse hepatitis virus and the maturation of coronavirus particles, *J. Biol. Chem.* **208:**15873–15879.

Rice, C. M., and Strauss, J. H., 1981, Nucleotide sequence of the 26 S mRNA of Sindbis virus and deduced sequence of the encoded virus structural proteins, *Proc. Natl. Acad. Sci. USA* **78:** 2062–2066.

Richardson, W. D., Roberts, B. L., and Smith, A. E., 1986, Nuclear location signals in Polyoma Virus Large-T, *Cell* **44:**77–85.

Richardson, W. D., Mills, A. D., Dilworth, S. M., Laskey, R. A., and Dingwall, C., 1988, Nuclear protein migration involves two steps: Rapid binding to the nuclear envelope followed by slower translocation through nuclear pores, *Cell* **52:**655–664.

Richter, J. D., Young, P., Jones, N. C., Krippl, B., Rosenberg, M., and Ferguson, B., 1985, A first exon encoded domain of E1A sufficient for posttranslational modification, nuclear-localization, and induction of adenovirus E3 promoter expression in *Xenopus* oocytes, *Proc. Natl. Acad. Sci. USA* **82:**8434–8438.

Robbi, M., and Lazarow, P. B., 1982, Peptide mapping of peroxisomal catalase and its precursor, *J. Biol. Chem.* **257:**964–970.

Rönnholm, R., and Pettersson, R. F., 1987, Complete nucleotide sequence of the M RNA segment of Uukuniemi virus encoding the membrane glycoproteins G1 and G2, *Virology* **160:**191–202.

Roise, D., and Schatz, G., 1988, Mitochondrial presequences, *J. Biol. Chem.* **263:**4509–4511.

Rose, J. K., Adams, G. A., and Gallione, C. J., 1984, The presence of cysteine in the cytoplasmic domain of vesicular stomatitis virus glycoprotein is required for palmitate addition, *Proc. Natl. Acad. Sci. USA* **81:**2050–2054.

Roth, J., 1984, Cytochemical localization of terminal N-acetylgalactosamine residues in cellular compartments of intestinal goblet cells: Implications for the topology of O-glycosylation *J. Cell Biol.* **98:**399–406.

Roth, J., 1987, Subcellular organization of glycosylation in mammalian cells, *Biochim. Biophys. Acta* **906:**405–436.

Roth, M. G., Gething, M.-J., and Sambrook, J., 1986, Intracellular transport and membrane insertions of the influenza viral glycoproteins, in: *The Influenza Viruses* (R. M. Klug, ed.), pp. 165–189, Plenum, New York.

Rothblatt, J. A., and Meyer, D. I., 1986, Secretion in yeast: translocation and glycosylation of prepro-α-factor *in vitro* can occur via an ATP-dependent post-translational mechanism, *EMBO J.* **5:**1031–1036.

Rothblatt, J. A., Webb, J. R., Ammerer, G., and Meyer, D. I., 1987, Secretion in yeast: Structural features influencing the posttranslational translocation of prepro-α-factor in vitro, *EMBO J.* **6:** 3455–3463.

Rothman, J. E., 1987, Transport of the vesicular stomatitis glycoprotein to trans Golgi membranes in a cell-free system, *J. Biol. Chem.* **262:**12502–12510.

Rottier, P. J. M., and Rose, J. K., 1987, Coronavirus E1 glycoprotein expressed from cloned cDNA is localized in the Golgi region, *J. Virol.* **61**:2042–2045.

Rottier, P. J. M., Armstrong, J., and Meyer, D. I., 1985, Signal recognition particle-dependent of coronavirus E1, an integral membrane glycoprotein, *J. Biol. Chem.* **260**:4648–4652.

Rudnick, G., 1986, ATP-driven $H^+$ pumping into intracellular organelles, *Annu. Rev. Physiol.* **48:** 403–415.

Saraste, J., and Kuismanen, E., 1984, Pre- and post-Golgi vacuoles operate in the transport of Semliki forest virus membrane glycoproteins to the cell surface, *Cell* **38**:535–549.

Saraste, J., Palade, G. E., and Farquhar, M. G., 1986, Temperature-sensitive steps in transport of secretory proteins through the Golgi complex in exocrine pancreatic cells, *Proc. Natl. Acad. Sci. USA* **83**:6425–6429.

Schlenstedt, G., and Zimmermann, R., 1987, Import of frog prepropeptide GLa into microsomes requires ATP but does not involve docking protein or ribosomes, *EMBO J.* **6**:699–703.

Schlesinger, S., Koyama, A. H., Malfer, C., Gee, S. L., and Schlesinger, M. J., 1985, The effects of inhibitors of glucosidase I on the formation of Sindbis virus, *Virus Res.* **2**:139–149.

Schleyer, M., and Neupert, W., 1985, Transport of proteins into mitochondria: Translocational intermediates spanning contact sides between outer and inner membranes, *Cell* **43**:339–350.

Schmidt, I., Skinner, M., and Siddell, S., 1987, Nucleotide sequence of the gene encoding the surface projection glycoprotein of coronavirus MHV-JHM, *J. Gen. Virol.* **68**:47–56.

Schmidt, M. F. G., 1983, Fatty acid binding: a new kind of posttranslational modification of membrane proteins, *Curr. Top. Microbiol. Immunol.* **102:**101–129.

Schmitz, B., Klein, R. A., Duncan, I. A., Egge, H., Gunawa, J., Peter-Katalinic, J., Dabrowski, U., and Dabrowski, J., 1987, MS and NMR analysis of the cross-reacting glycan from *Trypanosoma brucei brucei* MITat 1,6 variant specific glycoprotein, *Biochem. Biophys. Res. Commun.* **146:**1055–1063.

Scholtissek, C., Schwarz, R. T., Keil, W., and Klenk, H.-D., 1984, A mutant of fowl plague virus (influenza A) with an altered glycosylation pattern in its hemagglutinin, *Virology* **136:**1–9.

Schwarz, R. T., Rohrschneider, J. M., and Schmidt, M. F. G., 1976, Suppression of glycoprotein formation of Semliki Forest, influenza and avian sarcoma virus by tunicamycin, *J. Virol.* **19:** 782–791.

Scoulica, E., Krause, E., Meese, K., and Dobberstein, B., 1987, Disassembly and domain structure of the proteins in the signal recognition particle, *Eur. J. Biochem.* **163:**519–528.

Segev, N., and Botstein, D., 1987, The ras-like yeast YPT1 gene is itself essential for growth, sporulation and starvation response, *Mol. Cell. Biol.* **7**:2367–2377.

Segev, N., Mulholland, J., and Botstein, D., 1988, The yeast GTP-binding YPT1 protein and a mammalian counterpart are associated with the secretion machinery, *Cell* **52:**915–924.

Severinsson, L., and Petersom, P. A., 1985, Abrogation of cell surface expression of human class I transplantation antigens by adenovirus protein in *Xenopus laevis* oocytes, *J. Cell Biol.* **101:**540–547.

Siegel, V., and Walter, P., 1985, Elongation arrest is not a prerequisite for secretory protein translocation across the microsomal membrane, *J. Cell Biol.* **100:**1913–1921.

Siegel, V., and Walter, P., 1986, Removal of the Alu structural domain from signal recognition particle leaves its protein translocation activity intact, *Nature (Lond.)* **320:**81–84.

Siegel, V., and Walter, P., 1988a, Binding sites of the 19-kDa and 68/72 kDa signal recognition particle (SRP) proteins on SRP RNA as determined by protein-RNA "footprinting," *Proc. Natl. Acad. Sci. USA* **85:**1801–1805.

Siegel, V., and Walter, P., 1988b, Each of the activities of signal recognition particle (SRP) is contained within a distinct domain: Analysis of biochemical mutants of SRP, *Cell* **52:**39–49.

Silver, P., Keegan, L. P., and Ptashne, M., 1984, Amino terminus of the yeast GAL4 gene product is sufficient for nuclear localization, *Proc. Natl. Acad. Sci. USA* **81;**5951–5955.

Small, G. M., Szabo, L. J., and Lazarow, P. B., 1988, Acyl-CoA oxidase contains two targeting sequences each of which can mediate protein import into peroxisomes, *EMBO J.* **7:**1167–1173.

Sorger, P. K., and Pelham, H. R. B., 1987, The glucose-regulated protein grp94 is related to heat shock protein hsp90, *J. Mol. Biol.* **194:**341–344.

Spiess, M., and Lodish, H. F., 1986, An internal signal sequence: The asialoglycoprotein receptor membrane anchor, *Cell* **44:**177–185.

Standring, D. N., Ou, J. H., Masiarz, F., and Rutter, W. J., 1988, A single peptide encoded within the precore region of hepatitus B virus directs the secretion of a heterogeneous population of e antigens in *Xenopus* oocytes, *Proc. Natl. Acad. Sci. USA* **85:**8405–8409.

Stein, M., Zijderhand-Bleekemolen, J. E., Geuze, H., Hasilik, A., and von Figura, K., 1987, $M_r$ 46 000 mannose 6-phosphate specific receptor: Its role in targeting of lysosomal enzymes, *EMBO J.* **9:**2677–2681.

Stern, D. F., and Sefton, B. M., 1982, Coronavirus proteins: Structure and functions of the oligosaccharides of the avian infectious bronchitis virus glycoproteins, *J. Virol.* **44:**804–812.

Stirzaker, S. C., Whitfield, P. L., Christie, D. L., Bellamy, A. R., and Both, G. W., 1987, Processing of rotavirus VP7: Implications for the retention of the protein in the endoplasmic reticulum, *J. Cell Biol.* **105:**2897–2903.

Storz, J., Rott, R., and Kaluza, G., 1981, Enhancement of plaque formation and cell fusion of an enteropathogenic coronavirus by trypsin treatment, *Infect. Immunol.* **31:**1214–1222.

Stueber, D., Ibrahimi, I., Cutler, D., Dobberstein, B., and Bujard, H., 1984, A novel in vitro transcription translation system: Accurate and efficient synthesis of single proteins from cloned DNA sequences, *EMBO J.* **3:**3143–3148.

Sturman, L. S., and Holmes, K. V., 1983, The molecular biology of coronaviruses, *Adv. Virus Res.* **28:**35–112.

Sturman, L. S., and Holmes, K. V., 1985, The novel glycoproteins of coronaviruses, *Trends Biochem. Sci.* **10:**17–20.

Sturman, L. S., Ricard, C. S., and Holmes, K. V., 1985, Proteolytic cleavage of the E2 glycoprotein of murine coronavirus: Activation of cell-fusing activity of virions by trypsin and separation of the two different 90K cleavage fragments, *J. Virol.* **56:**904–911.

Sturman, L. S., Holmes, K. V., and Behnke, J., 1980, Isolation of coronavirus envelope glycoproteins and interaction with the viral nucleocapsid, *J. Virol.* **33:**449–462.

Swinkels, B. W., Evers, R., and Borst, P., 1988, The topogenic signal of the glycosomal (microbody) phosphoglycerate kinase of *Crithia fasciculata* resides in a carboxy-terminal extension, *EMBO J.* **7:**1159–1165.

Talmadge, K., Stahl, S., and Gilbert, W., 1980, Eukaryotic signal sequence transports insulin antigen in *E. coli, Proc. Natl. Acad. Sci. USA* **77:**3369–3373.

Tooze, J., Tooze, S. A., and Warren, G., 1984, Replication of coronavirus MHV A59 in sac(−) cells: Determination of the first site of budding of progeny virions, *Eur. J. Cell. Biol.* **33:**281–293.

Tooze, J., Tooze, S. A., and Warren, G., 1985, Laminated cisternae of the rough endoplasmic reticulum induced by coronavirus MHV-A59 infection, *Eur. J. Cell Biol.* **36:**108–115.

Tooze, J., Tooze, S., and Fuller, S. D., 1987, Sorting of coronavirus from condensed secretory proteins at the exit of the trans Golgi network of AtT20 cells, *J. Cell Biol.* **105:**1215–1226.

Tooze, S. A., Tooze, J., and Warren, G., 1988, Site of the addition of N-acetyl-galactosamine to the E1 glycoprotein of mouse hepatitis virus, *J. Cell Biol.* **106:**1475–1487.

Ullu, E., and Tschudi, C., 1984, Alu sequences are processed 7SL RNA genes, *Nature (Lond.)* **312:**171–172.

Ullu, E., Murphy, S., and Melli, M., 1982, Human 7SL RNA consists of a 140 nucleotide middle-repetitive sequence inserted in an Alu sequence, *Cell* **29:**195–202.

Ulmanen, I., Seppälä, P., and Pettersson, R. F., 1981, *In vitro* translation of Uukuniemi virus-specific RNAs: Identification of a nonstructural protein and a precursor to the membrane glycoproteins, *J. Virol.* **37:**72–79.

Valls, L. A., Hunter, C. P., Rothman, J. H., and Stevens, T. H., 1987, Protein sorting in yeast: the localization determinant of yeast vacuolar carboxypeptides Y resides in the propeptide, *Cell* **48:** 887–897.

Vassarotti, A., Chen, W.-J., Smagula, C., and Douglas, M. G., 1987a, Sequences distal to the mitochondrial targeting sequences are necessary for the maturation of the $F_1$-ATPase β-subunit precursor into mitochondria, *J. Biol. Chem.* **262:**411–418.

Vassarotti, A., Stroud, R., and Douglas, M. G., 1987b, Independent mutations at the amino terminus of a protein act as surrogate signals for mitochondrial import, *EMBO J.* **6:**705–711.

Verner, K., and Schatz, G., 1987, Import of an incompletely folded precursor protein into isolated mitochondria requires an energized inner membrane but no added ATP, *EMBO J.* **6:**2449–2456.

von Figura, K., and Hasilik, A., 1986, Lysosomal enzymes and their receptors, *Annu. Rev. Biochem.* **55:**167–193.

von Heijne, G., 1983, Patterns of amino acids near signal sequence cleavage sites, *Eur. J. Biochem.* **133:**17–21.

von Heijne, G., 1984, How signal sequences maintain cleavage specificity, *J. Mol. Biol.* **173:**243–251.

von Heijne, G., 1985, Signal sequences. The limits of variation, *J. Mol. Biol.* **184:**99–105.

von Heijne, G., 1986a, Mitochondrial targeting sequences may form amphiphilic helices, *EMBO J.* **5:**1335–1342.

von Heijne, G., 1986b, Towards a comparative anatomy of N-terminal topogenic signal sequences, *J. Mol. Biol.* **189:**239–242.

von Heijne, G., and Gavel, Y., 1988, Topogenic signals in integral membrane proteins, *Eur. J. Biochem.* **174:**671–678.

von Heijne, G., Wickner, W., and Dalbey, R. E., 1988, The cytoplasmic domain of *Escherichia coli* leader peptidase is a translocation poison sequence, *Proc. Natl. Acad. Sci. USA* **85:**3363–3366.

Walter, P., 1987, Signal recognition: Two receptors act sequentially, *Nature (Lond.)* **328:**763–764.

Walter, P., and Blobel, G., 1980, Purification of a membrane associated protein complex required for protein translocation through the endoplasmic reticulum, *Proc. Natl. Acad. Sci. USA* **77:** 7112–7116.

Walter, P., and Blobel, G., 1981, Translocation of proteins across the endoplasmic reticulum. III. Signal recognition protein (SRP) causes signal sequence and site specific arrest of chain elongation that is released by microsomal membranes, *J. Cell Biol.* **91:**557–561.

Walter, P., Ibrahimi, I., and Blobel, G., 1981, Translocation of proteins across the endoplasmic reticulum. I. Signal recognition protein (SRP) binds to *in vitro* assembled polysomes synthesizing secretory proteins, *J. Cell Biol.* **91:**545–550.

Warren, G., 1985, Membrane traffic and organelle division, *Trends Biochem. Sci.* **10:**439–443.

Waters, M. G., and Blobel, G., 1986, Secretory protein translocation in a yeast cell-free system can occur post-translationally and requires ATP-hydrolysis, *J. Cell Biol.* **102:**1543–1550.

Watson, M. E. E., 1984, Compilation of published signal sequences, *Nucleic Acids Res.* **12:**5145–5164.

Watts, C., Wickner, W., and Zimmermann, R., 1983, M13 procoat and a preimmunoglobulin share processing specificity, but use different membrane receptor mechanisms, *Proc. Natl. Acad. Sci. USA* **80:**2809–2813.

Wege, H., Siddell, S., and terMeulen, V., 1982, The biology and pathogenesis of coronaviruses. *Curr. Top. Microbiol. Immunol.* **99:**165–200.

Wengler, G., Castle, E., Leidner, U., Nowak, T., and Wengler, G., 1985, Sequence analysis of the membrane protein V3 of the Flavivirus West Nile virus and of its gene, *Virology* **147:**264–274.

Wheeler, E. F., Roussel, M. F., Hampe, A., Walker, M. H., Fried, V. A., Look, A. T., Rettenmier, C. W., and Sherr, C. J., 1986, The amino-terminal domain of the v-fms oncogene product includes a functional signal peptide that directs synthesis of a transforming glycoprotein in the absence of feline leukemia virus gag sequences, *J. Virol.* **59:**224–233.

Wickner, W. T., and Lodish, H. F., 1985, Multiple mechanisms of protein insertion into and across membranes, *Science* **230:**400–407.

Wiedmann, M., Kurzchalia, T., Hartmann, T., and Rapoport, T., 1987, A signal sequence receptor in the endoplasmic reticulum membrane, *Nature (Lond.)* **328:**830–833.

Wiedmann, M., Wiedmann, B., Voight, S., Wachter, E., Müller, H.-G., and Rapoport, T., 1988, Post-translational transport of proteins into microsomal membranes of *Candida maltosa, EMBO J.* **7:**1763–1768.

Wiedmann, M., Huth, A., and Rapoport, T. A., 1984, *Xenopus* oocytes can secrete bacterial beta-lactamase, *Nature (Lond.)* **309:**637–639.

Wieland, F. T., Gleason, M. L., Serafini, T. A., and Rothman, J. E., 1987, The rate of bulk flow from the endoplasmic reticulum to the cell surface, *Cell* **50:**289–300.

Wilhelmsen, K. C., Leibowitz, J. L., Bond, C. W., and Robb, J. A., 1981, The replication of murine coronaviruses in enucleated cells, *Virology* **110:**225–230.

Willingham, M. C., Pastan, I. H., Sahagian, G. G., Jourdian, G. W., and Neufeld, E. F., 1981, Morphologic study on the internalization of a lysosomal enzyme by the mannose 6-phosphate receptor in cultured Chinese hamster ovary cells, *Proc. Natl. Acad. Sci. USA* **78:**6967–6971.

Wilhelmsen, K. C., Leibowitz, J. L., Bond, C. W., and Robb, J. A., 1981, The replication of murine coronaviruses in enucleated cells, *Virology* **110:**225–230.

Wilson, C., Gilmore, R., and Morrison, T., 1987, Translation and membrane insertion of the hemagglutinin-neuraminidase glycoprotein of Newcastle disease virus, *Mol. Cell. Biol.* **7:** 1386–1392.

Wold, W. S. M., Cladaras, C., Deutscher, S. L., and Kapoor, Q. S., 1985, The 19 kDa glycoprotein coded by region E3 of adenovirus, *J. Biol. Chem.* **260:**2424–2431.

Wychowski, C., Benichou, D., and Girard, M., 1986, A domain of SV40 capsid polypeptide VP1 that specifies migration into the cell nucleus, EMBO J. **5:**2569–2576.

Yamanaka, G., Eckstein, F., and Stryer, L., 1986, Stereochemistry of the guanyl nucleotide binding site of transducin probed by phosphorothioate analogues of GTP and GDP, *Biochemistry* **24:** 8094–8101.

Zerial, M., Huylebroeck, D., and Garoff, H., 1987, Foreign transmembrane peptides replacing the internal signal sequence of transferrin receptor allow its translocation and membrane binding, *Cell* **48:**147–155.

Zimmermann, R., and Meyer, D. I., 1986, 1986: A year of new insights into how proteins cross membranes, *Trends Biochem. Sci.* **11:**512–515.

Zimmermann, R., Sagstetter, M., Lewis, M. J., and Pelham, H. R. B., 1988, Seventy kilodalton heat shock proteins and an additional component from reticulocyte lysate stimulate import of M13 procoat protein into microsomes, *EMBO J.* **7:**2875–2880.

Zwieb, C., 1985, The secondary structure of the 7SL RNA in the signal recognition particle: Functional implications, *Nucleic Acids Res.* **13:**6105–6124.

Zwizinski, C., Schleyer, M., and Neupert, W., 1984, Proteinaceous receptors for the import of mitochondrial precursor proteins, *J. Biol. Chem.* **259:**7850–7856.

*Chapter 11*

# Membrane Biology in Yeast as Probed with Enveloped Viruses

Hannele Sareneva and Marja Makarow

## 1. INTRODUCTION

The vast genetic and biochemical information available for the yeast *Saccharomyces cerevisiae* (Fig. 1) has made this organism an important eukaryotic model system for molecular as well as cell biology. As a mostly haploid organism, yeast offers the possibility to isolate mutants, which permit cloning of the gene responsible for the biochemical reaction of interest. In yeast, basic eukaryotic biological mechanisms may be studied without interference of parameters arising from differentiation. Enveloped animal viruses have turned out to be an excellent tool to study membrane biology of the mammalian cell. Many features of exocytic and endocytic transport and of post-translational modifications of membrane glycoproteins have been elucidated using viral envelope glycoproteins as probes. Viral envelope proteins are still the only known membrane fusion promoting cellular polypeptides and have thus been used as fusogens in studies on membrane fusion as well as in liposome-mediated transfer of macromolecules into cells. Since yeast does not have endogenous viruses, envelope proteins of animal viruses have lately been used in research on membrane biology of *S. cerevisiae*. Here we have reviewed these first attempts,

**Hannele Sareneva and Marja Makarow** Institute of Biotechnology, University of Helsinki, Valimotie 7, 00380 Helsinki, Finland.

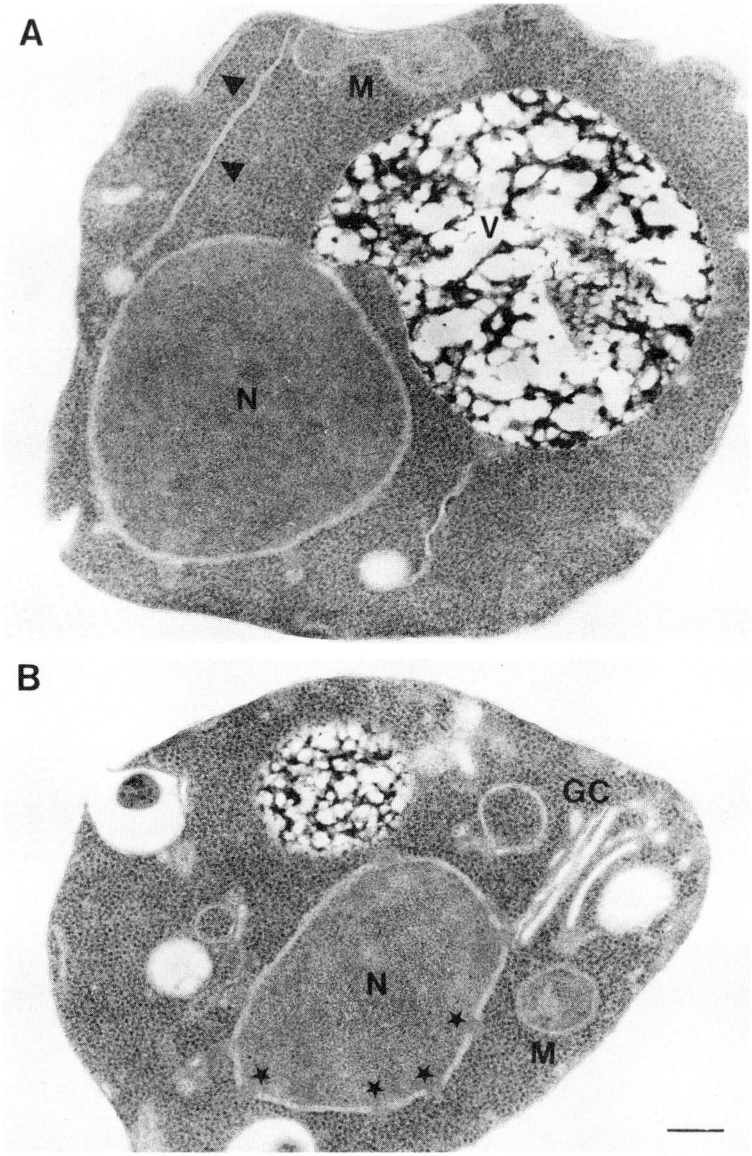


**FIGURE 1.** Thin-section electron microscopy of wild-type *Saccharomyces cerevisiae*. The cell wall has been removed to expose the plasma membrane. (A) Vacuole (V), nucleus (N), and a dividing mitochondrion (M). Arrowheads point to stretches of endoplasmic reticulum studded with ribosomes. (B) Golgi complex (GC) with three cisternae and flanking vesicular-appearing structures. The asterisks mark areas of the nuclear membrane that have bursted open in the fixation procedure, apparently at the regions of nuclear pores, since the outer and inner membranes form there a continuum. Bar = 0.2 μm.

emphasizing those aspects of the lower eukaryote, which may not be self-evident for the cell biologist studying mammalian cell systems.

## 2. FUSION OF ANIMAL VIRUSES WITH THE YEAST PLASMA MEMBRANE

### 2.1. Target Membrane

The plasma membrane of *S. cerevisiae* appears to have a normal eukaryotic phospholipid composition; however, it is covered with a prominent cell wall. The wall consists mainly of glucan and extensively glycosylated mannoproteins. Chitin is the major component of the primary septum in budding cells and its deposition at the cell surface is restricted to the area in which the bud will emerge (Cabib *et al.*, 1982). The cell wall presents a permeability barrier for virus particles. Thus, it has to be removed for studies on the interaction of viruses with the yeast plasma membrane.

#### 2.1.1. Cell Wall

**2.1.1a. Structure.** Yeast glucan consists of backbone chains where glucopyranose units are linked to each other by β(1–3) bonds. The backbone chains are interlinked by β(1–6) bonds (Manners *et al.*, 1973a,b). Glucan appears to form a network covering the plasma membrane. β-Glucanase activity has been found in the cell wall (Farkas, 1979). Since growth and bud formation requires flexibility of the cell wall, this activity may be involved in local degradation of the glucan network to allow its expansion. The biosynthesis of the glucan polymer is thought to occur *in situ* at the cell surface (Lopez-Romero and Ruiz-Herrera, 1977; Shematek *et al.*, 1980).

The mannoproteins are soluble molecules retained at the cell surface, probably by intercalation within the glucan network. The often modest and hitherto ill-defined polypeptide moieties of the mannoproteins are glycosylated with *N*-glycosidic and *O*-glycosidic glycans. The structure of the *N*-glycans has been elucidated in detail for a number of yeast strains (see Ballou, 1982). The *N*-glycosidic chains are attached to asparagine residues of the consensus sequence Asn-X-Ser(Thr) (Bause and Lehle, 1979; Lehle and Bause, 1984). The primary glycan $Glc_3Man_9GlcNac_2$ has the same structure as that of mammalian cells (Prakash and Vijay, 1982; Prakash *et al.*, 1983; see Kornfeld and Kornfeld, 1985) and is transferred at the luminal aspect of the endoplasmic reticulum (ER) membrane from the lipid carrier dolichol pyrophosphate (Dol-PP) to the polypeptide chain (Tsai *et al.*, 1984a). It appears that translation and glycosylation do not have to be tightly coupled, since prepro-α-factor, the precursor to a hormone-like secretory peptide, can be segregated through the ER membrane and glycosylated post-translationally at least *in vitro* (Waters and Blobel, 1986; Hansen *et al.*, 1986; Rothblatt and Meyer, 1986). All glucose residues and one of the nine mannose residues are then removed from the primary glycan (Byrd *et al.*, 1982; Jelinek-Kelly *et al.*, 1985) (Fig. 2A, outer box), and the glycoprotein is transported to the Golgi complex (Esmon *et al.*, 1984). In Mammalian cells, up to six mannose units may be removed from the primary glycan in further trimming events (see Fig. 2A, inner box). The oligosaccharide core structure is extended in the yeast Golgi complex (Esmon *et al.*, 1981) by a linear polymer of D-mannose units linked to each other with α(1–6) bonds. More mannose units are attached to the backbone with α(1–3) and α(1–2) linkages, forming side chains of one to three mannose residues (Jones and Ballou, 1969; Raschke *et al.*, 1973; Nakajima and Ballou, 1974a,b). The side chains may contain diester-linked mannosyl phosphate or mannobiosyl phosphate units (Karson and Ballou, 1978). The total number of mannose residues of one *N*-glycan may amount from 8 to 100–200 units. Interestingly, the filamentous fungus *Trichoderma reesei* glycosylates its major secretory proteins with mammalian cell-type *N*-glycosidic chains with the structure $Man_{5-9}GlcNAc_2$ (Salovuori *et al.*, 1987).

The *O*-glycosidic glycans of *S. cerevisiae* are different from the sialylated mucin-type *O*-glycans of mammalian cells. The yeast *O*-glycans are attached to serine or threonine like the mammalian *O*-glycans, but they consist of only mannose residues, which occur in a linear array containing one to four or five

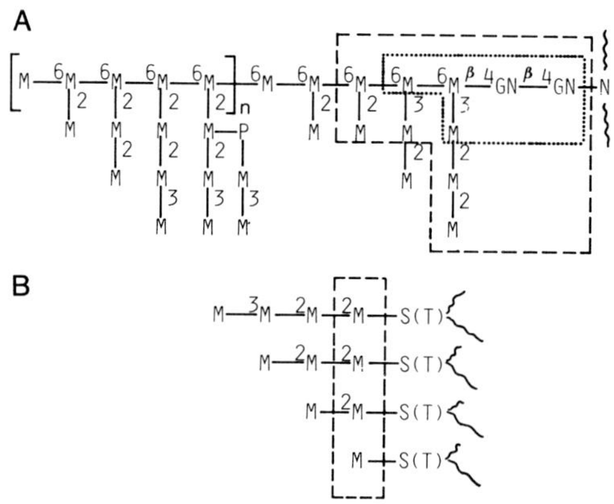

**FIGURE 2.** Structure of the protein-bound glycans of *Saccharomyces cerevisiae* (Ballou, 1982; Jelinek-Kelly *et al.*, 1985). (A) *N*-Glycosidic glycans. (B) *O*-Glycosidic glycans. GN, *N*-acetylglucosamine; M, mannose; P, phosphate. All linkages are in α-anomeric configuration except where β-configuration is indicated. (A) The structure in the inner box ( . . . . . ) is common to *N*-glycans of yeast and mammalian origin. (A,B) The residues in the outer box (——) are acquired in the rough endoplasmic reticulum and the others apparently in the Golgi complex. N, S, T, amino acids.

units (Nakajima and Ballou, 1974a; Lehle and Bause, 1984) (Fig. 2B). The mannose residue involved in the *O*-glycosidic glycopeptide linkage is transfered from Dol-P to the polypeptide in the ER, whereas the rest of the mannose units are acquired from GDP-Man, probably in the Golgi complex (Baczinski and Tanner, 1973; Sharma *et al.*, 1974; Haselbeck and Tanner, 1983; see Tanner and Lehle, 1987). The putative mating-type specific *a* cell agglutinin is the only glycoprotein reported until now to carry exclusively *O*-glycans (Orlean *et al.*, 1986).

**2.1.1b. Glycosylation-Defective Mutants.** A number of mutants defective in biosynthesis of the N-linked oligosaccharide chains are available. They provide a unique tool to elucidate structural details, biosynthetic pathways and functions of protein-bound glycans. The *alg* mutants are defective in different glycosyltransferases that assemble the primary glycan on the Dol-PP carrier in the ER (Kukuruzinska *et al.*, 1987). The *alg*7, *alg*1, and *alg*2 mutants accumulate Dol-PP intermediates with oligosaccharides smaller than $Man_5GlcNAc_2$, and the *alg*3, *alg*6, and *alg*8 mutants accumulate Dol-PP with $Man_5GlcNAc_2$ or larger intermediates of the primary glycan. The mutants of the first class are temperature sensitive, whereas those of the second class are fully viable. In the *alg*3, *alg*6, and *alg*8 mutants, the truncated glycans are not only transferred to polypeptides but are also elongated by the Golgi-specific mannosyl transferases, indicating that transfer and elongation are not very substrate specific. None of these mutations inhibits secretion of invertase (Huffaker and Robbins, 1982, 1983; Runge *et al.*, 1984; Runge and Robbins, 1986).

The *mnn* mutants are defective in outer-chain elongation of the protein-bound *N*-glycosidic oligosaccharides that occurs beyond the ER (Ballou, 1982). The *mnn*1 through *mnn*6 mutants build up *N*-glycans with aberrant side chains attached to the outer chain backbone (Raschke *et al.*, 1973; Ballou *et al.*, 1973) and *mnn*9 lacks the outer chain completely (Ballou *et al.*, 1980; Tsai *et al.*, 1984b). The mutations do not effect the cell wall mannoproteins alone but all glycoproteins, although the mutant cells have been selected according to cell-wall anomalies. The deficiencies in the side chains of the outer chains have been found not to be detrimental for growth, but the outer chain backbone appears to be indispensible. The mutants lacking the outer chains have a distorted cell wall and clumpy morphology, are sensitive to osmotic lysis, and have reduced viability (Tsai *et al.*, 1984b).

**2.1.1c. Cell Wall as a Permeability Barrier.** It has been suggested that the cell wall presents a permeability barrier for very small molecules (Scherrer *et al.*, 1974). However, many large glycoproteins, endogenous ones, as well as heterologous ones expressed from recombinant plasmids, are secreted through the cell wall to the growth medium. Invertase is a glycosylated secretory enzyme of *S. cerevisiae,* the synthesis of which is under catabolite repression control. It is assembled in the ER and secreted to the exterior of the cell in homo-octameric form (Chu *et al.*, 1983; Tammi *et al.*, 1987; Esmon *et al.*, 1987). The monomer has a molecular weight of 120 kDa, half of which is due to nine heterogeneous *N*-glycosidic oligosaccharide chains (Frevert and Ballou, 1982; Lehle *et al.*, 1979). The octamer remains entrapped within the cell wall due to its very large size, whereas dimeric invertase, which is formed under glycosylation-inhibiting conditions, is able to diffuse through the cell wall to the growth medium (Esmon *et al.*, 1987; Tammi *et al.*, 1987). Also, other high-molecular-weight secretory glycoproteins, like the "barrier" activity or α-factor antagonist (> 200 kDa; MacKay *et al.*, 1988) and oligomeric α-galactosidase (300 kDa) (Lazo *et al.*, 1978), reach the extracellular fluid (ECF). Periplasmic α-galactosidase and acid phosphatase have been found to be oligomeric and have a molecular mass of 550–600 kDa (Esmon *et al.*, 1987; Mildner *et al.*, 1976).

Heterologous proteins in the molecular-weight range of 50–100 kDa, like the human granulocyte/macrophage colony-stimulating factor (Moonen *et al.*, 1987), antibodies (Wood *et al.*, 1985), glucoamylase of *Aspergillus awamori* (Innis *et al.*, 1985), α-amylase of *Bacillus amyloliquefaciens* (Ruohonen *et al.*, 1987), and salivary α-amylase (Nakamura *et al.*, 1986; Tokunaga *et al.*, 1987), are reported to be secreted through the cell wall to the growth medium. Even the envelope glycoproteins gp350 of Epstein–Barr virus (EBV) and HA1 of influenza virus, which have apparent molecular weights of >350 kDa and >250 kDa, respectively, penetrate the cell wall and reach the medium (Schultz *et al.*, 1987; Jabbar and Nayak, 1987) (see Section 3.2). No information about oligomerization of heterologous proteins is available; thus, their cell wall penetration or retention cannot as yet be correlated with the size of the native form of the protein. The glycosylation pattern seems not to be important for diffusion through the cell wall, since core-glycosylated and hypermannosylated proteins are found in the secreted proteins as well as in the entrapped ones.

**2.1.1d. Removal of the Cell Wall.** The cell wall of *S. cerevisiae* can be released by digestion with enzymes attacking the glucan layer. Concomitantly, the soluble mannoproteins are also released, exposing the plasma membrane. A high turgor prevailing within the yeast cell presses the plasma membrane against the cell wall. When the cell wall is removed, the resulting spheroplasts have to be protected from osmotic lysis by counteracting the intracellular pressure by providing an osmotic stabilizer in the medium (Fig. 3A). Removal of the cell wall changes the morphology of the cells. The spheroplasts have a round shape instead of an ellipsoid one, and no bud profiles can be recognized [Fig. 3B(b)]. The spheroplasts synthesize a new cell wall, but regeneration occurs only in solid medium, as the mannoproteins are lost into liquid medium (Necas, 1971; Pastor *et al.*, 1982; Valentin *et al.*, 1984; see Sentandreu *et al.*, 1985). This may be attributable to physical factors, the immature glucan layer not being able to retain the mannoproteins, although affinity factors are not excluded (Valentin *et al.*, 1986). After more than 10 hr in solid medium, the cell wall has regenerated to ensure osmotic resistance to the cells, and budding is recovered.

Several commercial cell wall-degrading preparations are available. Their principal activity is β(1–3) glucanase, contaminated to different extents with protease, chitinase, mannanase, lipase, and nuclease activities. Treatment with reducing agents raises the susceptibility of the cell wall toward glucanase (Kuo and Yamamoto, 1975). Reduction of disulfide bonds in the cell wall polypeptides is believed to promote access of the glucanase to the glucan layer. A lyticase activity, consisting of an endo-β(1–3) glucanase and an alkaline protease, has been isolated from *Oerskovia xantholytica* (Scott and Schekman, 1980). The glucanase is active in cell wall removal only in the presence of reducing agent or the alkaline protease.

Zymolyase is a widely used β(1–3) glucanase preparation from *Arthrobacter luteus* (Shibata *et al.*, 1983, 1984; Valentin *et al.*, 1984), which also displays proteolytic activity (Zlotnik *et al.*, 1984). We have optimized the release of the cell wall for our purposes of virus–spheroplast interaction studies by following the release of radioactivity from 2-[$^3$H]mannose-labeled cells by zymolyase in the absence of reducing agent (Makarow, 1985; Makarow *et al.*, 1987). Under

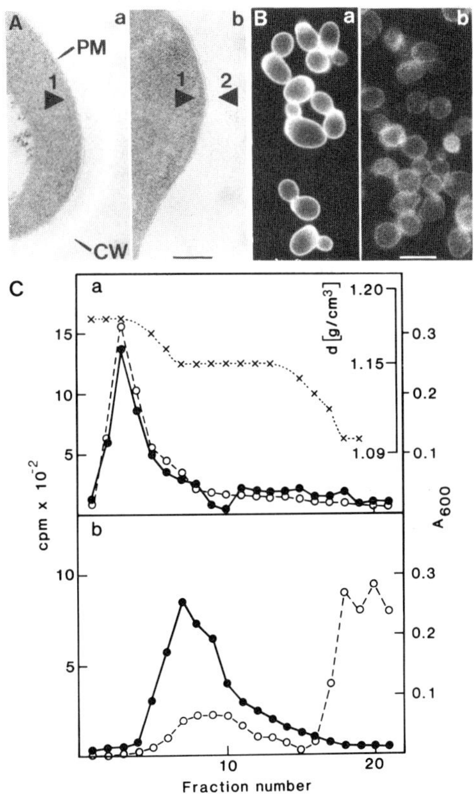


**FIGURE 3.** Release of the cell wall. (A) Cells were examined by thin-section electron microscopy before (a) and after (b) zymolyase digestion. PM, plasma membrane; CW, cell wall. The turgor (arrowhead 1) prevailing in the cell has to be counteracted after removal of the cell wall by osmotic stabilizer in the medium (arrowhead 2). Bar = 0.2 μm. (B) Intact cells (a) and spheroplasts (b) were surface-stained with rhodamine-conjugated concanavalin A after formaldehyde-fixation. Bar = 5 μm. (C) Cells were labeled with [$^3$H]mannose, which is largely incorporated into the cell wall mannoproteins, and subjected before (a) and after (b) zymolyase digestion to centrifugation in 60% Percoll density gradients containing 1.2 M sorbitol as an osmotic stabilizer. (●——●) Optical density, (○——○) $^3$H radioactivity, (× . . . ×) density. The spheroplasts sediment at lower density than the intact cells and the released cell wall components flotate at the top of the gradient.

nonreducing conditions, the degradation of the cell wall is slowed down and is thus easier to control. Our protocol yields a reasonably homogenous and intact spheroplast population as judged from electron microscopy, surface staining by fluorescent concanavalin A (Con A), and density-gradient centrifugations (Fig. 3).

### 2.1.2. Plasma Membrane

**2.1.2a. Lipids.** The lipid metabolism of *S. cerevisiae* has been widely studied, and a great number of mutants defective in various reactions of the metabolic pathways have been isolated (Henry, 1982; Henry *et al.*, 1984). The yeast plasma membrane has been found to contain the same phospholipids as the mammalian plasma membrane, in roughly similar proportions. The plasma membrane of wild-type log-phase *S. cerevisiae* cells grown in glucose contains about 26–34% phosphatidylcholine (PC), 22–24% phosphatidylethanolamine (PE), 8% phosphatidylserine (PS), 28–35% phosphatidylinositol (PI), and 2% phosphatidic acid (Rank *et al.*, 1978; Henry, 1982). In these studies, the phospholipid composition of the plasma membrane was not found to deviate much from that of the total yeast cell homogenate; however, these analyses may suffer from problems of purification of the subcellular compartments.

One remarkable example of the versatility of yeast is that an important membrane phospholipid such as PS is not essential for viability. The *cho*1 mutant is defective in phosphatidylserine synthase and thus lacks PS. This also leads to distortions of the metabolism of other phospholipids (Atkinson *et al.*, 1980a,b; Letts and Henry, 1985). The mutant grows quite normally, provided that the growth medium is supplemented with choline or ethanolamine. These compounds do not, however, correct the PS deficiency but are used for PC and PE synthesis. Here alternate pathways (Kennedy and Weiss, 1956; Waechter and Lester, 1971, 1973) are used, since PS serves as precursor for both PC and PE in wild-type cells (Atkinson *et al.*, 1980b). The membrane-associated phosphatidylserine synthase has been purified (Bae-Lee and Carman, 1984) and the gene encoding it has been cloned (Letts *et al.*, 1983). Expression of the *CHO*1 gene on a multicopy plasmid resulted in a seven- to eightfold overproduction of the synthase, which in turn led to increased proportions of PS, PC, and PI and to a decreased proportion of PE. This again appeared not to be detrimental for cell growth (Letts *et al.*, 1983). Membranes of different intracellular compartments of the mammalian cell, and even of the two domains of the plasma membrane in polarized epithelial cells, have at least to some extent differing lipid compositions (Simons and Fuller, 1985). The tolerance of *S. cerevisiae* of major changes in its lipid composition challenges suggestions that specific phospholipid compositions may have universal roles, e.g., in intracellular transport or organelle biogenesis. The availability of yeast mutants defective in lipid synthesis will permit the study of the relationships of phospholipid transfer between cellular compartments with intracellular vesicular transport (Letts and Dawes, 1983; Ramirez *et al.*, 1983; Atkinson and Ramirez, 1984; Daum *et al.*, 1986).

There are at least two major differences in the lipids of the plasma membrane of wild-type *S. cerevisiae* as compared with that of mammalian cells. First, the yeast cells do not contain cholesterol at all. Rather, they synthesize another sterol, ergosterol, which differs from cholesterol in three structural features: (1)the side chain has one additional methyl group at C-24, (2) there is a double bond between C-22 and C-23 in the side chain, and (3) there is a double bond between C-7 and C-8 in the B ring (Parks, 1978). Second, the structural membrane glycolipids of yeast are different from the mammalian ones. Yeast lacks gangliosides, no transferases for sialic acid, galactose, or galactosamine have been found. The gangliosides have been proposed to have a role in intracellular signalling in the mammalian cell system, which appears unnecessary for the unicellular yeast cells. Nevertheless, *S. cerevisiae* membranes have inositol-containing sphingolipids, which include inositol phosphorylceramide, mannosyl inositol phosphorylceramide, and mannosyldi(inositolphosphoryl)ceramide (Smith and Lester, 1974; Becker and Lester, 1980). The sphingolipids have been localized to fractions enriched in plasma membrane, ER and unidentified vesicles (Hanson, 1984).

Another particular feature of the yeast cell is its adaptability to different environmental conditions. This is reflected in the lipid composition of the cell, which can vary according to changes in such factors as temperature (Hunter and Rose, 1972), availability of molecular oxygen (Andreasen and Stier, 1953, and 1954), the physiological condition of the cell (Getz *et al.*, 1970; Homann *et al.*, 1987), and lipid precursor supplement of the growth medium (Waechter and Lester, 1971). Yeast cells tolerate a broad temperature range and grow even below 10°C, albeit slowly. Wild-type *S. cerevisiae* cells grown aerobically on glucose contain mainly saturated and monounsaturated C-14, C-16, and C-18 fatty acids (Jollow *et al.*, 1968). When the temperature is lowered, the cells ensure the fluidity of their membranes by changing rapidly the chain length and degree of saturation of the acyl chains. At 30°C or above, the ratio of palmitate (C-16) to stearate (C-18) is about 1.5. This ratio increases with decreasing temperature, reaching a value of 5.5 at 5°C. Part of the newly synthesized palmitate is converted to the monounsaturated palmitoleate. The activation energies of the palmityl and stearyl transfer activities are different at different temperatures. The short term regulation of fatty acid synthesis is suggested to be based on this and appears to be independent of *de novo* protein synthesis (Okuyama *et al.*, 1979; Hori *et al.*, 1987).

**2.1.2b. Proteins.** The glycosylation pattern of plasma membrane proteins of *S. cerevisiae* is poorly known. The amino acid sequences of four plasma membrane proteins, the arginine permease (Hoffmann, 1985, 1987), the receptor of the hormone-like α-factor (Burkholder and Hartwell, 1985), the proton-translocating ATPase (Serrano *et al.*, 1986) and chitin synthase 1 (Bulawa *et al.*, 1986), have been deduced from the nucleotide sequences of the respective cloned genes. Each has potential acceptor sites for *O*- and *N*-glycans, but no information about their eventual glycosylation is available. They all span the plasma membrane probably several times, but the N-terminal amino acid sequences do not resemble known signal sequences and are not cleaved off. The mode of transport is known for arginine permease and ATPase, which use the secretory pathway for plasma membrane delivery (Tschopp *et al.*, 1984; Holcomb *et al.*, 1988). Nevertheless, the plasma membrane contains glycosylated polypeptides, since it binds concanavalin A (Novick and Schekman, 1983) (see Fig. 3B). It is reasonable to assume that at least some of them are hypermannosylated and that they may give special features to the yeast plasma membrane surface as compared with the mammalian cell. The mammalian cell surface has a net negative charge due to protein- and lipid-linked sialic acid. There are no sialylglycoconjugates in the yeast cell, but the phosphate residues in *N*-glycosidic glycans provide negative charge to the plasma membrane surface.

Thus, in terms of lipid composition, the plasma membrane of *S. cerevisiae* appears compatible as a fusion target for animal viruses, although the lack of cholesterol is interesting, as it is required for fusion with Semliki Forest virus (see Section 2.5). There are no specific receptors for viruses to be attached to before fusion; also, sialyloligosaccharides of plasma membrane gangliosides, which influenza virus uses as nonspecific receptors (Suzuki *et al.*, 1985), are missing. Moreover, the glycocalyx may provide special features for the surface in case plasma membrane proteins are hypermannosylated or there are remnants of the cell wall left. Finally, the fusion reactions and subsequent manipulations have to be carried out in media containing an osmotic stabilizer to protect the spheroplasts from osmotic lysis.

## 2.2. Low pH-Dependent Fusogenicity of Enveloped Animal Viruses

Many enveloped animal viruses, like Semliki Forest virus (SFV), influenza virus and vesicular stomatitis virus (VSV), are fusogenic at low pH: the interaction of the virus with a target membrane in acidic milieu leads to fusion of the membranes and to mixing of the contents of the virus and the target (White and Helenius, 1980; White *et al.*, 1980, 1981). A conformational change in the viral spike proteins activates the fusion reaction, a prerequisite for fusogenicity being that the proteins are implanted in a virosomal, cellular, or liposomal membrane (White *et al.*, 1982; Kondor-Koch *et al.*, 1983; Florkiewicz and Rose, 1984; Riedel *et al.*, 1984; Huang and Uslu, 1986). The viruses use their fusogenic activity to infect host cells. After uptake into the cell by endocytosis, the acidic pH of the endosome triggers the fusion of the viral envelope with the endosomal membrane, releasing the viral genome into the cytoplasm to start infection (Kielian and Helenius, 1986).

The mechanism of the fusion reaction is still unknown. The most fruitful model at present to study fusogenicity is the hemagglutinin (HA) of influenza virus, because the three-dimensional structure of its ectodomain has been determined by X-ray crystallography (Wilson *et al.*, 1981). The HA molecule is synthesized as a single polyprotein, cleaved to HA1 and HA2, which remain interlinked with disulfide bonds. The HA molecules are assembled in trimers (Wiley and Skehel, 1987). HA2 has a conserved hydrophobic N-terminal sequence, which is exposed at acidic pH and is responsible for the fusogenic activity of HA. This has been shown by demonstrating binding under acidic conditions of the HA2 region and of synthetic N-terminal HA2 peptides to liposomes, by proteolytic digestion experiments, and by showing that site-specific mutagenization of HA2 affects fusogenicity (Skehel *et al.*, 1982; Doms *et al.*, 1985; Gething *et al.*, 1986a; Lear and DeGrado, 1987). Low pH causes an irreversible conformational change not only in HA2 but in the whole HA molecule as well. This has been detected by studying the amphiphilicity and protease sensitivity of HA (Skehel *et al.*, 1982; Doms *et al.*, 1985). Monoclonal antibodies have shown alterations in antigenicity upon acidification (Webster *et al.*, 1983; Yewdell *et al.*, 1983). Electron microscopy has detected changes in the stem tertiary structure of the trimer (Ruigrok *et al.*, 1986). Recently, anitpeptide antibodies were used to detect sequential conformational changes in HA, which are suggested to start with the exposure of the fusogenic HA2 peptide, followed by alterations in the stem region, followed in turn by changes in the globular head region of the trimer (White and Wilson, 1987).

## 2.3. Fusion of VSV with Yeast

### 2.3.1. Interactions of VSV with Yeast at Low pH

Vesicular stomatitis virus has one envelope glycoprotein (G), which is responsible for fusogenicity (White *et al.*, 1981; Florkiewicz and Rose, 1984). The viral envelope encloses the negative-polarity single-stranded RNA genome together with two structural proteins, the nucleocapsid protein (N) and the matrix protein (M), which lines the inner surface of the membrane (Strauss and Strauss, 1983). G protein is susceptible to proteinase K digestion in the cold. Thus, any G-protein-mediated superficial interactions between the virus particle and the cell surface should be relieved by proteinase K digestion. In case the virus particles are fused with the target membrane, proteinase K digestion should degrade the G protein and leave the internal proteins of the virus intact within the cell, providing an assay for fusion (Matlin *et al.*, 1982; Pesonen and Simons, 1983).

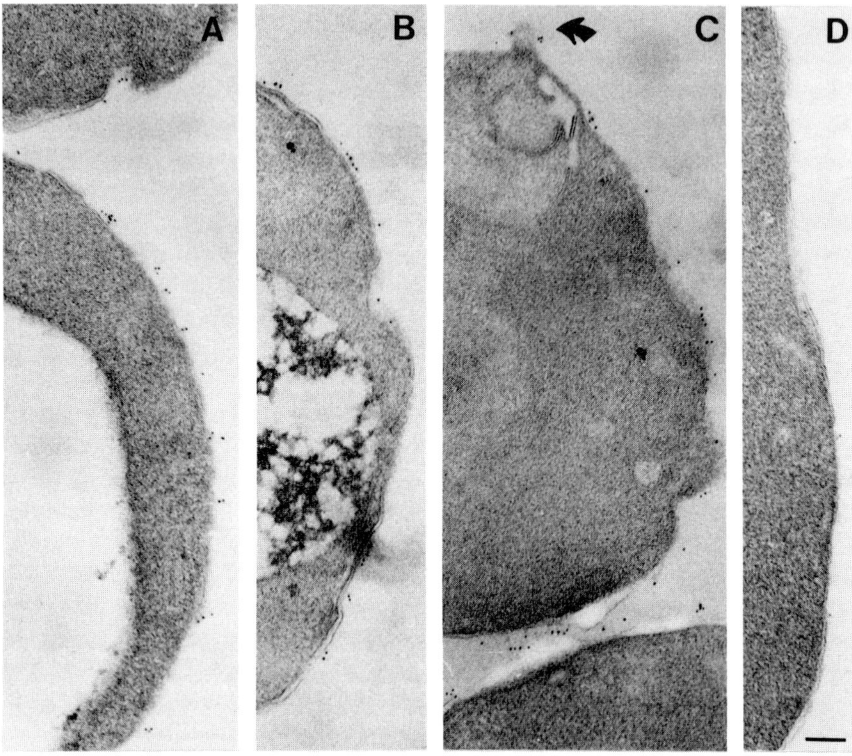


**FIGURE 4.** Visualization of the VSV G protein in the spheroplast plasma membrane. (A–C) VSV was fused with the spheroplasts at pH 4 at 37°C. The fixed spheroplasts were subjected to immunoelectron microscopy using anti-G protein antibody followed by protein A–gold. The gold particles mark the G protein molecules implanted into the spheroplast membrane. Usually the gold particles occurred in rows (A) or groups (B) of two to six. A possible fusion profile of VSV is indicated with an arrowhead in (C). (From Makarow *et al.*, 1987.) (D) VSV was omitted from the treatments. Bar = 0.1 μm.

Weak binding between VSV particles and *S. cerevisiae* spheroplasts, which was reversible with neutral pH, could be differentiated at low pH in the cold. In contrast to low temperature, at 37°C the virus particles associated irreversibly with the spheroplasts at pH 4. Proteinase K digestion of the spheroplast surface degraded the G protein completely but left the M and N proteins intact and spheroplast associated, indicating that at 37°C fusion between the VSV particles and the spheroplast plasma membrane had occurred. Thin-section electron microscopy showed that no virus particles remained attached to the spheroplast surface. After the fusion reaction, the plasma membrane could be stained by indirect immunofluorescent labeling using anti-VSV antiserum. The implanted G protein could be vizualized at the spheroplast surface by immunogold electron microscopy, using specific anti-G protein antibody (Fig. 4). For any interaction between the virus particles and the yeast cells, it was indispensable to remove the cell wall. In case an intact cell wall covered the plasma membrane, no association of the virus with the cells, binding or fusion, could be detected (Makarow *et al.*, 1987).

### 2.3.2. Characteristics of the Fusion Reaction

The pH dependence of the fusion between VSV and two different *S. cerevisiae* strains, a wild-type cell (derivative of S288C) and an auxotroph strain (OL1, Boy-Marcotte and Jacques, 1982), was similar: When the pH of the medium was decreased below 5, increasing amounts of VSV were fused with the spheroplasts. The optimum temperature for fusion was 30–42°C. The kinetics of the fusion reaction differed for the two cell types. In the case of the wild-type strain, maximal fusion was obtained in 10 sec at 37°C. In the case of the auxotroph strain, a plateau in fusion efficiency was obtained only in 5 min (Table I). There was a limit to the number of virus particles that could be fused per

**Table I**
**Parameters of the Fusion Reactions between Enveloped Viruses and Different Target Membranes**

| Virus | Target membrane | pH optimum | Kinetics (sec) | Reference |
|---|---|---|---|---|
| Influenza | BKH-21 | 5.5 | 30–60 | White *et al.* (1980) |
| | CEF | 5.0–5.8 | — | Huang *et al.* (1981) |
| | Erythrocyte | 5.0–5.8 | — | Maeda and Ohnishi (1980) |
| | Erythrocyte ghost | 5.1–5.2 | — | Stegmann *et al.* (1986) |
| | Liposome | 4.5–5.1 | 60–180 | Stegmann *et al.* (1985); Wharton *et al.* (1986) |
| SFV wild type | BHK-21 | 5.5 | 30–60 | White *et al.* (1980) |
| | Erythrocyte | 5.8 | 120 | Väänänen *et al.* (1981); Väänänen and Kääriäninen (1980) |
| | Liposome | <6.0 | 120 | White and Helenius (1980) |
| | Yeast | <4.7 | 30 | Makarow *et al.* (1987) |
| Sindbis | Erythrocyte | 5.8 | — | Väänänen and Kääriäninen (1980) |
| | Liposome | 5.0 | — | Scheule (1987) |
| VSV | BHK-21 | 6.0 | 30–60 | White *et al.* (1980) |
| | KB | 6.0–6.2 | — | Yamada and Ohnishi (1986) |
| | MDCK | 5.4 | 30–60 | Matlin *et al.* (1982) |
| | Erythrocyte ghost | 5.5–6.0 | 180 | Yamada and Ohnishi (1986) |
| | Yeast | <4.7 | 10–300 | Makarow *et al.* (1987) |
| | Liposome | 5.5 | 180 | Yamada and Ohnishi (1986) |

spheroplast; this limit also varied with the yeast strain used. On the average, maximally about 45 VSV particles could be fused into the membrane of one wild-type spheroplast, whereas the auxotroph strain could accommodate about 90 virus particles (Makarow *et al.*, 1987). The cellular diameters of the two yeast strains are similar. The reasons for the saturability of the plasma membrane, and for the variation of saturation between the two cell types are not known. Saturation might be due to residual cell wall that may partly cover the plasma membrane to different extents in the two strains.

### 2.4. Fusion of SFV with Yeast

The envelope glycoproteins of SFV form heterotrimers consisting of two transmembrane glycoproteins, E1 and E2, and one peripheral glycoprotein, E3 (Kääriäinen and Pesonen, 1982). The fusion activity resides apparently in both E1 and E2, since both undergo a conformational change in acidic milieu (Kondor-Koch *et al.*, 1983; Kielian and Helenius, 1985). SFV could be fused with wild-type *S. cerevisiae* spheroplasts at low pH (Makarow *et al.*, 1987). The characteristics of the fusion reaction between SFV and wild-type *S. cerevisiae* spheroplasts were similar to those described for VSV. Fusion required a pH value below 5 (Fig. 5A). The optimum temperature was 30–42°C (Fig. 5B), and the fusion reaction leveled off at 37°C in about 30 sec (Fig. 5C). Maximally 60 SFV particles, on the average, could be fused with one wild type spheroplast (Fig. 5D).

### 2.5. Comparison of Viral Fusion Targets

Table I summarizes characteristics of the fusion reactions between the best known low pH-dependent fusogenic viruses and different target membranes. The initial steps of low pH-dependent interactions of SFV and VSV with cultured mammalian cells include superficial binding in the cold, which is reversible with neutral pH, EDTA treatment, or proteolytic degradation. Binding occurs at slightly less acidic pH than what is required for fusion. The fusion of SFV, Sindbis virus, VSV, and influenza virus with cultured cells occurs at elevated temperatures, and the pH thresholds vary between 5 and 6.2 (White *et al.*, 1980, 1981; White and Helenius, 1980; Maeda and Ohnishi, 1980; Huang *et al.*, 1981; Matlin *et al.*, 1982; Yamada and Ohnishi, 1986; Scheule, 1987). Interaction of SFV with red blood cells (RBC) displays somewhat different characteristics, as incubation in the cold at low pH results in irreversible binding. This may be due to hydrophobic interaction of the envelope glycoprotein with the plasma membrane, rather than to superficial binding mediated by cellular surface proteins. The low pH-dependent interaction in the cold results in fusion even at neutral pH, once the temperature is elevated (Väänänen and Kääriäinen, 1980; Väänänen *et al.*, 1981). The reasons for the variation of the fusion pH with the

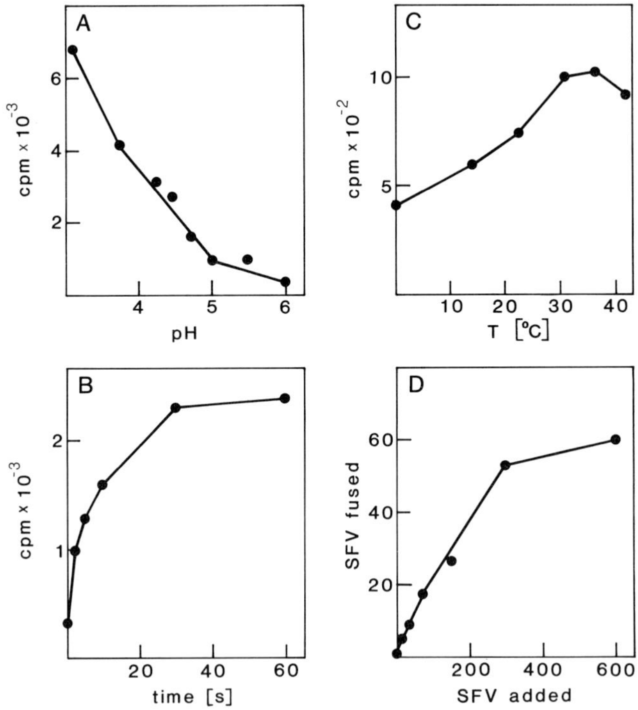


**FIGURE 5.** Characterization of the fusion reaction between SFV and wild-type spheroplasts. (A) pH dependency: [$^{35}$S]Methionine-labeled SFV ($2 \times 10^4$ cpm) was incubated with spheroplasts for 5 min at 37°C in buffer solutions with different pH values. Free virus particles were removed by washing with neutral buffer and surface attached particles by proteinase K digestion in the cold (Helenius *et al.*, 1980), and the spheroplast-associated viral radioactivity was counted. (B) Kinetics: Spheroplasts were incubated with radioactive SFV at pH 4 at 37°C for different periods of time, cleared from surface-bound particles and counted for viral radioactivity. (C) Temperature dependency: Treatments were carried out as above for 5 min in pH 4 buffer, but at different temperatures. (D) Saturability: Radioactive SFV was mixed with different amounts of nonlabeled SFV and incubated with the spheroplasts for 5 min in pH 4 buffer at 37°C. From the irreversibly spheroplast-associated viral radioactivity, the average number of virus particles fused with one spheroplast was calculated and plotted against the number of virus particles added per spheroplast.

target membrane are not known. The first step in the fusion process, the conformational change of the spike glycoprotein, occurs at a characteristic pH value independent of the target membrane, although the change in the SFV E1 protein appears to involve an interaction with cholesterol molecules in target liposomes (Kielian and Helenius, 1985). The second step, annealing of the two membranes to be fused, may be influenced by the surface properties of the target membrane, since repulsion, which does vary with lipid composition, has to be overcome.

Site-directed mutagenization of the HA2 region has yielded hemagglutinin molecules with abolished fusogenic activity or with raised threshold pH for fusion (Gething *et al.*, 1986a). A naturally occurring influenza variant strain has been isolated, which exhibits elevated pH dependency for fusion, which is due to a change in HA2 (Doms *et al.*, 1986). Mutations in HA1 or HA2, which affect the intersubunit contacts of HA, or the stability of the N-terminus of HA2, lead to elevated values of fusion-pH (Daniels *et al.*, 1985). Mutagenization of SFV resulted in mutants having a fusion threshold of pH 5.5, 0.7 pH units lower than that of the wild-type virus (Kielian *et al.*, 1984). One of the SFV fusion mutants has been used to probe the kinetics of the acidification of endosomes *in vivo* (Kielian *et al.*, 1986).

Liposomes have been used to study structural requirements of the target membrane for virus fusion (White and Helenius, 1980; White *et al.*, 1983; Yamada and Ohnishi, 1986; Nussbaum *et al.*, 1987). There are no specific phospholipid requirements, but negatively charged phospholipids appear to enhance the fusion process (Stegmann *et al.*, 1985). Haywood and Boyer (1985) reported fusion between influenza virus and negatively charged liposomes even at pH 7.5. This may have been due to nonspecific electrostatic interactions rather than to the fusogenicity of HA. A high content of cardiolipin in the liposomal target membrane promotes influenza virus fusion, which does not exhibit a strict dependence on the conformational change of HA (Nir *et al.*, 1986; Stegmann *et al.*, 1986). A high proportion of PE appears to promote the viral fusion activity (Scheule, 1987). This has been suggested to be due to the low degree of hydration of PE-containing membranes, which results in a decrease of the repulsive barrier between the viral and the target membrane. Liposome studies have revealed an absolute stereospecific requirement for cholesterol in the target membrane of SFV. One important structural detail of the sterol is the presence of a 3-β-hydroxyl group (White and Helenius, 1980; Kielian and Helenius, 1984). In the yeast plasma membrane, which is competent for fusion with SFV, cholesterol is replaced by ergosterol (see Section 2.1.2a). However, ergosterol does have the crucial 3-β-hydroxyl group. Because influenza virus does not exhibit a sterol requirement, cholesterol may be needed for specific interaction of one of the SFV spike proteins, E1, with the lipid bilayer (Kielian and Helenius, 1985, 1986).

## 3. EXPRESSION OF ENVELOPE GLYCOPROTEINS OF ANIMAL VIRUSES IN YEAST

Enveloped RNA viruses have provided an excellent tool for the study of intracellular transport and post-translational modification of mammalian membrane glycoproteins (Kääriäinen and Pesonen, 1982; Simons and Fuller, 1985; Pfeffer and Rothman, 1987). This has been mainly attributable to the current availability of antibodies, mutants, cloned genes, and sequence data and earlier to the fact that in many cases the viral infections shut off host protein synthesis, enabling facile detection of the viral gene products. In the yeast system well-characterized convenient markers to study intracellular transport and glycosylation of integral plasma membrane glycoproteins are not yet available. The envelope glycoproteins of RNA viruses, maturing at the plasma membrane of the mammalian host cell, might serve as potential markers.

## 3.1. Expression from the Viral Genome

### 3.1.1. Expression of the Structural Proteins of VSV

After fusion of VSV particles with the plasma membrane of *S. cerevisiae*, the spheroplasts remained viable at least for 4 hr. During this period, protein and RNA synthesis, as well as protein glycosylation occurred with similar efficiency as in control cells. This prompted us to ask whether the viral genome introduced into the yeast cytoplasm in the fusion process could be expressed and whether the G protein would be transported to the plasma membrane. The VSV particle contains its own transcriptase (Baltimore *et al.*, 1970), which is introduced in the fusion process into the cytoplasm together with the nucleocapsid. In case primary transcription were to take place, the yeast cell machinery would be required to carry out translation only.

Spheroplasts were fused with a maximum number of VSV particles (90/cell) and incubated for 2 hr in the presence of [$^{35}$S]methionine. Sodium dodecyl sulfate–polyacrylamide gel electrophoresis (SDS–PAGE) analysis after immunoprecipitation of the spheroplast lysate with an antiserum directed against all structural proteins of VSV revealed four *de novo* synthesized polypeptides that were not seen in control lysates, i.e., the VSV M protein and N proteins, and two forms of the G protein with apparent molecular weights of 60 and 70 kDa. Tunicamycin reduced the molecular weight of both forms of G protein to that of the polypeptide backbone of mature authentic G protein, 55 kDa, indicating that the 60-kDa and 70-kDa forms were *N*-glycosylated. The 55-kDa species partitioned into Triton X-114, showing that it was amphipathic, hence probably membrane associated (Makarow *et al.*, 1986).

### 3.1.2. Intracellular Transport and Post-Translational Modifications of the VSV G Protein

The signal sequence of the VSV G protein was recognized by *S. cerevisiae*, as G protein was guided to the secretory pathway and glycosylated. In *S. cerevisiae* the secretory pathway, presumably the Golgi complex, harbors an endopeptidase activity, which apparently is encoded by the *KEX*2 gene (Julius *et al.*,

1984b). The *KEX2* gene product processes prepro-α-factor at Lys-Arg-pairs. Thus, polypeptides traversing the yeast secretory pathway may be cleaved by the endopeptidase if susceptible sites are available (Innis *et al.*, 1985). G protein escapes this hazard, since its single pair of basic amino acids resides in the cytoplasmic carboxyterminal sequence (Rose and Gallione, 1981).

The authentic G protein has two potential glycosylation sites, both of which are occupied by complex type *N*-glycans. Assuming that both sites are occupied in the yeast-made G protein, the 60-kDa species would contain core glycans and the 70-kDa species glycans with roughly 45 mannose residues, indicating that at least the 70-kDa species was transported to the Golgi complex. However, G-protein molecules were taken even to the plasma membrane, as shown by immunofluorescent staining of spheroplast surface using anti-VSV antibody. Moreover, spheroplasts expressing the viral genome could be triggered by low pH treatment to fuse with each other to form polykaryons, confirming surface expression of biologically active G protein (Fig. 6).

Uukuniemi virus is an enveloped animal RNA virus, the envelope glycoproteins of which appear to contain a signal for retention in the Golgi complex (von Bonsdorff *et al.*, 1970; Kuismanen *et al.*, 1982; Gahmberg *et al.*, 1986). In case the Uukuniemi virus genome could be expressed in *S. cerevisiae,* either after fusion of the virion at the plasma membrane or from recombinant plasmids, it would be interesting to learn whether organelle-specific retention signals of viral glycoproteins would be recognized even in the yeast cell.

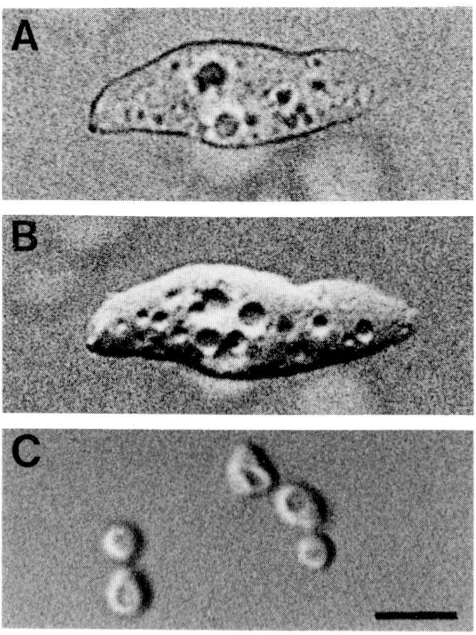


**FIGURE 6.** Polykaryon formation. The genome of VSV was introduced into the yeast cytoplasm by fusing the virion with the yeast plasma membrane at low pH. Extracellular G protein was removed by proteolytic digestion and the structural proteins of the virus were expressed for 90 min. The spheroplasts were then treated (A,B) or not treated (C) with low pH medium. (From Makarow *et al.*, 1986). (A) Phase-contrast and (B,C) Nomarski optics. Bar = 5 μm.

### 3.1.3. Exocytosis in Yeast

Secretory glycoproteins of *S. cerevisiae* traverse by vesicular membrane traffic the ER, the Golgi apparatus, and the secretory vesicles successively, to be finally externalized at the plasma membrane in a polar fashion at the growing tip of the bud (Tkacz and Lampen, 1973; Field and Schekman, 1980; Novick *et al.*, 1981; Esmon *et al.*, 1981). The secretory pathway has been elucidated by a genetic approach, isolation, and characterization of 23 complementation groups of conditional-lethal mutants where intracellular transport from the ER to the Golgi complex, from the Golgi complex to the secretory vesicles or from the secretory vesicles to the exterior of the cell is temperature sensitive (Novick *et al.*, 1980; Schekman, 1985). In addition to secretory glycoproteins, several cell wall mannoproteins and plasma membrane proteins have been shown to use the intracellular pathway defined by the *SEC* genes (Novick and Schekman, 1983; Bussey *et al.*, 1983; Tschopp *et al.*, 1984; Julius *et al.*, 1984a). All intracellular traffic appears to be constitutive in yeast, and at least the integral plasma membrane protein ATPase and the secretory enzyme acid phosphatase are co-transported in a single vesicle species (Holcomb *et al.*, 1988). The *sec* mutants enable cloning of genes responsible for intracellular transport and synthesis of the respective gene products, and thereby the study under *in vitro* conditions of the molecular mechanisms underlying intracellular transport (Bernstein *et al.*, 1985; Ruohola and Ferro-Novick, 1987; Salminen and Novick, 1987).

Several major differences in exocytosis have been encountered so far between the mammalian and the yeast cell. The glycosylating apparatus of the Golgi complexes differ profoundly (see Section 2.1.1.a). The yeast ER membrane is capable of post-translational translocation of newly synthesized secretory proteins (see Section 2.1.1a) whereas in the case of mammalian microsomes, this is possible only for polypeptides of exceptionally small size (Watts *et al.*, 1983; Schlenstedt and Zimmermann, 1987). In at least some mammalian cell types, two parallel exocytic routes co-function. The AtT-20 pituitary cells have a regulated as well as a constitutive outgoing pathway (Gumbiner and Kelly, 1982; Kelly, 1985; Pratt *et al.*, 1988), and the polar epithelial Madin–Darby canine kidney (MDCK) cells appear to direct vesicular traffic to the apical and the basolateral plasma membrane domains along routes which diverge at or after the late Golgi subcompartment (Rodriguez-Boulan and Sabatini, 1978; Rindler *et al.*, 1984; Fuller *et al.*, 1985). Sorting signals may differ in yeast and mammalian cells. The structural feature guiding the hydrolase, CPY, into the lysosome-like vacuole of yeast does not reside in carbohydrate, although it does in the mammalian lysosomal hydrolases (Schwaiger *et al.*, 1982; Sly and Fischer, 1982). The vacuolar targeting signal of CPY has been localized to the N-terminal amino acid sequence (Bankaitis *et al.*, 1986; Johnson *et al.*, 1987; Valls *et al.*, 1987). Finally, a striking difference in the regulation of intracellular transport between the yeast and the mammalian cell was recently found; i.e., in

*S. cerevisiae,* exocytosis goes on efficiently during mitosis (Makarow, 1988; Nevalainen and Makarow, 1988), whereas in mammalian cells membrane traffic comes to an abrupt halt at the onset of mitosis (Berlin *et al.*, 1978; Warren *et al.*, 1983; Warren, 1985). Interestingly, a yeast cytosolic fraction has been reported to be able to drive vesicular transport *in vitro* between subcompartments of the mammalian Golgi complex (Dunphy *et al.*, 1986).

### 3.2. Expression from Recombinant Plasmids

Envelope glycoproteins of several viruses have been expressed from recombinant plasmids in *S. cerevisiae* and shown to be segregated to the secretory pathway (Table II). Expression of viral surface components is obviously also aimed at antigen production in yeast cells for vaccination and diagnostic purposes. Synthesis of many viral membrane proteins from recombinant plasmids has proved toxic for *S. cerevisiae* cells, although the levels of expression have generally been low. Constitutive synthesis of viral protein slows down the growth of yeast cells (Schultz *et al.*, 1987). Regulated promoters appear not to solve this problem, since the burst of virus protein synthesis after induction also inhibits growth of the cells (Wen and Schlesinger, 1986; Wen *et al.*, 1986). In the system we have developed, in which synthesis of the viral proteins is brought about by introducing the viral genome into the yeast cytoplasm, the cells do not have to regenerate. Intracellular transport of the marker glycoprotein can be

**Table II**
**Intracellular Transport and Glycosylation of Virus Envelope Glycoproteins Expressed in Yeast[a]**

| Virus | Protein | Carbohydrate | Destination | Membrane anchor | MW (kDa) |
|---|---|---|---|---|---|
| VSV | G | Core[a] | Surface | + | |
| VSV[b] | G | Outer chain[b] | Surface | + | 70 |
| Influenza | HA1 | Outer chain | Medium | − | >250 |
| | HA | Core | Intracellular | − | 69 |
| | HA | Core | ? | + | 70 |
| Sindbis | E1 | Core | Surface | + | — |
| | p62 | Core | Intracellular | + | — |
| SFV | E1 | Outer chain | ? | + | 65–69 |
| EBV | gp350 | N[c] | Medium | − | >350 |
| BLV | gp51 | Core | Intracellular | + | 45 |
| HTLV-1 | env | N[c] | ? | + | 52–60 |

[a]For references, see text.
[b]All proteins were expressed from recombinant plasmids except in this case, where expression was from the viral genome introduced into the cytoplasm by fusion of the virus particle with the yeast plasma membrane.
[c]The core of the N-glycosidic oligosaccharide synthesized in the endoplasmic reticulum.
[d]The core was extended with outer chains synthesized in the Golgi complex.

followed during transient expression for some hours and thus problems arising from toxicity can be avoided.

### 3.2.1. Vesicular Stomatitis Virus

The VSV G protein has been expressed in *S. cerevisiae* from yeast shuttle vector containing the galactokinase gene (*GAL*1) promoter of *S. cerevisiae* upstream from the cDNA encoding G protein (Wen and Schlesinger, 1986). Induction in galactose-containing medium resulted in *de novo* synthesis of VSV G protein, about one half of which appeared to reach the cell surface. In the *sec* 18 mutant, where intracellular membrane traffic is blocked in the ER (Novick *et al.*, 1980), surface expression of the G protein was decreased but not inhibited. According to molecular weight reduction by endoglycosidase H digestions, G protein carried *N*-glycosidic core glycans. The G protein could not be labeled with [$^3$H]palmitic acid, although it is acylated in mammalian cells and yeast is able to acylate endogenous proteins (Wen and Schlesinger, 1984). The G protein formed (apparently already in the yeast ER) disulfide-linked trimers or tetramers, which could be dissociated with reducing agent but not with SDS. Because all of the G protein appeared to occur as oligomers, it must have been aggregates that were transported to the cell surface. The authors could not find similar oligomer formation in mammalian cells, and thus they suggest that the oligomers in yeast represent aberrant aggregates (Wen *et al.*, 1986). However, G protein has been found by others to form disulfide-linked dimers or trimers in mammalian cells (Kreis and Lodish, 1986). Oligomerization is considered a prerequisite for intracellular transport and maturation of the G protein. Similar results have been obtained for influenza HA, which assembles into trimers within the mammalian ER 5–10 min after synthesis (Gething *et al.*, 1986b; Copeland *et al.*, 1986; Boulay *et al.*, 1988). Even cellular transmembrane proteins, like the nicotinic acetylcholine receptor (AChR) (Smith *et al.*, 1987) and HLA heavy chains (Owen *et al.*, 1980; Sege *et al.*, 1981), require homo- or hetero-oligomerization, respectively, for intracellular transport. Copeland *et al.* (1986) found that 10% of the HA trimers had an aberrant conformation, leading to retention within the ER. Taking together the results on oligomerization of viral and cellular transmembrane proteins in mammalian cells and the fact that *S. cerevisiae* oligomerizes endogenous secretory glycoproteins (see Section 2.1.1c), it will be interesting to learn whether oligomerization is required for proper intracellular transport of transmembrane proteins in yeast.

### 3.2.2. Influenza Virus

Different deletion mutants of influenza HA have been expressed in *S. cerevisiae* under the control of the consitutive alcohol dehydrogenase I gene (*ADH*1) promoter. Complementary DNA (cDNA) encoding the entire HA molecule with its signal sequence produced a 70-kDa core-glycosylated protein, the intracellular location of which remained unclear (Jabbar *et al.*, 1985). When the cDNA contained the whole HA1 sequence with its signal sequence, plus the HA2 sequence with its fusion peptide but lacking the transmembrane region and the C-terminal cytoplasmic tail, a 69-kDa protein was recognized by antiviral antibodies. Even this polypeptide was core glycosylated but, in spite of translocation to the secretory pathway, it remained intracellular and was not transported to the plasma membrane. When the recombinant plasmid contained only the HA1 region with its signal sequence, thus lacking fusogenic and membrane-spanning sequences, the fate of the molecules was different: HA1 was segregated to the secretory pathway, hyperglycosylated and secreted to the growth medium apparently without retention in the cell wall. The molecular weight of secreted HA1 was >250 kDa, which is surprising, since the molecular weight of apoHa1 is 42 kDa, and it contains only four potential N-glycosylation sites. This means that one *N*-glycan contained on the average as many as 320 hexose units, a size that has not been described before. N-terminal amino acid sequencing showed that the signal sequence of HA was correctly cleaved in yeast (Jabbar and Nayak, 1987).

### 3.2.3. Sindbis Virus and Semliki Forest Virus

The genome strategies of the two closely related alphaviruses, SFV and Sindbis virus, are similar (see Kääriäinen *et al.*, 1987; Strauss and Strauss, 1983). In infected mammalian cells, a subgenomic 26S RNA is synthesized and translated into a polyprotein precursor of the structural proteins. The N-terminal capsid protein (C) is cleaved off rapidly after synthesis. The remaining intermediate yields two polypeptides, the p62 protein (pE2 in Sindbis virus) and the E1 protein. Both have their own signal sequences that guide them to the secretory pathway to be glycosylated and transported to the plasma membrane. Further proteolytic cleavages process p62 into the mature E2 and E3 proteins (pE2 yields E2) (see Kääriäinen and Pesonen, 1982).

The entire sequence of the 26S cDNA that codes for the structural polyprotein of Sindbis virus was expressed in a galactose-regulated way under the control of the yeast *GAL*1 promoter (Wen and Schlesinger, 1986). After galactose induction, the C, pE2, and E1 proteins were synthesized. This finding indicates that the proteolytic cleavages of the polyprotein occurred in yeast. No cleavage of pE2 into mature E2 could be detected. Part of E1, but not pE2, was found at the cell surface. Surface expression was partly inhibited in the *sec*18 mutant, like that of VSV G protein. E1 was core-glycosylated but not acylated. Also E1 appeared to form disulfide-linked aggregates like the VSV G protein (Wen *et al.*, 1986).

The 26S cDNA of SFV was expressed in *S. cerevisiae* under the control of the alcohol dehydrogenase promoter. However, in this case only E1 and C could be detected by immunoprecipitation and immunoblotting experiments. This indicates that p62 must have been synthesized, but was degraded after cleavage from the polyprotein (Keränen, 1986). E1 protein was hyperglycosylated and thus transported at least to the Golgi complex. The rapid degradation of p62 may have occurred by the *KEX2* gene-coded endopeptidase (see Section 3.1.2) on the secretory pathway, as it contains five pairs of basic amino acids (Garoff *et al.*, 1980). However, the closely related pE2 protein of Sindbis virus, which contains four susceptible sites in its ectodomain (Rice and Strauss, 1981), could be found in transformed cells, although partial loss caused by degradation cannot be excluded. Another possibility is that SFV p62 was not translocated across the ER membrane but was degraded in the cytoplasm.

### 3.2.4. Bovine Leukemia Virus and Human T-Cell Leukemia Virus 1

The cDNA corresponding to 210 amino acids, 70% of the C-terminus, of the mature bovine leukemia virus (BLV) surface antigen gp51 was joined to the signal sequence portion of the yeast acid phosphatase gene (Brantl *et al.*, 1988). The construction was expressed under the control of the yeast acid phosphatase gene (*PHO5*) or 3-phosphoglycerate kinase gene (*PGK*) promoters, the level of expression being higher in the first case. Two polypeptides were detected by Western blot analysis, a 28-kDa protein, that apparently was not glycosylated, and a low level of a 45-kDa species, which was suggested to be core-glycosylated and remained intracellular. The signal sequence appeared not to be cleaved off. All four sequential antigenic determinants of the yeast-made gp51 could be detected using monoclonal antibodies.

The entire gene coding for the envelope protein of human T-cell leukemia virus type 1 (HTLV-1) has also been expressed in yeast under the control of the acid phosphatase gene promoter (Kuga *et al.*, 1986). A product of 52–60 kDa was produced, which was glycosylated, since tunicamycin reduced the apparent molecular weight of the product to 45 kDa. Intracellular transport or the nature of the oligosaccharide chains was not studied.

### 3.2.5. Epstein–Barr Virus

The envelope glycoprotein gp350 of Epstein–Barr virus (EBV) has been synthesized in *S. cerevisiae* under the control of two different regulatable promoters, the *MF*$\alpha$*1* promoter the expression of which is controlled with temperature in a suitable mutant strain, and the galactose-induced *GAL10* promoter. The gp350 sequence lacked the regions coding for the N-terminal signal sequence, the C-terminal membrane spanning region and the cytoplasmic tail. The gp350 sequence was fused to the prepro-leader of $\alpha$-factor, which provides a cleavable signal sequence. The constructions encoded in yeast a glycoprotein of the molecular weight of >350 kDa. This species was secreted to the growth medium and carried N-glycosidic glycans and perhaps *O*-glycosidic ones as well. The protein backbone of gp350 has an apparent size of 145 kDa and it contains as many as 37 potential *N*-glycosylation sites, 36 of which were present in the construction expressed in yeast. The kinetics of maximal accumulation of gp350 were very slow, 2 days when expressed under the control of the thermosensitive *MFα1* promoter and 18 hr under the control of the *GAL*10 promoter (Schultz *et al.*, 1987). Note that maximal accumulation of the Influenza virus HA1 in the growth medium was achieved in 60–90 min (Jabbar and Nayak, 1987).

### 3.2.6. Cytoplasmic Expression of Viral Envelope Proteins

Several viral envelope glycoproteins have been expressed in yeast without signal sequences. As expected, these polypeptides remained in the cytoplasm in nonglycosylated form. The cytoplasmic influenza HA1 and HA polypeptides, the latter with or without the membrane anchor, were found to be unstable. The anchorless HA polypeptide had a half-life ($t_{1/2}$) of 30 min in the cytoplasm, whereas the half-life of the glycosylated form was 2.5 hr (Jabbar and Nayak, 1987).

The glycoprotein B of the DNA virus herpes simplex virus 1 (HSV-1) was expressed in yeast, under the control of the yeast acid phosphatase gene promoter, from DNA lacking the signal sequence. Immunologically active protein was found from cell lysates, but no information of the stability of the herpesvirus gene product in the yeast cytoplasm was given (Nozaki *et al.*, 1985).

Hepatitis B virus surface antigen (HBsAg) has been synthesized in *S. cerevisiae* under the control of the *ADH*I promoter (Valenzuela *et al.*, 1982) and the acid phosphatase promoter (Miyanohara *et al.*, 1983), in both cases without the signal sequence. The viral gene product synthesized by yeast had the molecular weight of unglycosylated mature HBsAg, 23 kDa, and was assembled in the cytoplasm into immunogenic 20- to 22-nm particles similar in size and shape to the Dane particles secreted by infected human cells. A human hepatitis B vaccine has been produced from the purified HBsAg particles produced in yeast (McAleer *et al.*, 1984). This technology has been developed further into antigen engineering, where genes of other immunogens are recombined with the HBsAg gene to produce hybrid HBsAg particles (Valenzuela *et al.*, 1985a,b).

## 4. FUSION OF VIROSOMES WITH YEAST SPHEROPLASTS

Liposomes are powerful carriers of macromolecules into the cytoplasm of cultured animal cells, as they have high encapsulation capacity and are nontoxic to living cells. They have been used to transfer into cells for example fluorescent dyes (Connor and Huang, 1985), nucleic acids (Soriano *et al.*, 1983; Brzobohaty and Kovac, 1986), antibodies (Philippot *et al.*, 1983), and drugs (Gregoriadis, 1985). Moreover, liposome-mediated transfer is probably the only way to deliver transmembrane proteins and defined lipids into cellular plasma membranes (Poste *et al.*, 1980; van Meer and Simons, 1983). The availability of cloned genes, antisense RNA molecules, and monoclonal antibodies, together with techniques to introduce them into cells, will open up novel methods for research in molecular and cell biology (Doxsey *et al.*, 1985, 1987).

### 4.1. Preparation of Virosomes

For efficient transfer of liposomal content to the cell interior, the liposomes have to be fused with the plasma membrane. The fusogenic activity of viral envelope glycoproteins may be used to promote this fusion reaction. The viral fusogenic proteins can be reconstituted with lipids to form virosomes, which have been shown to fuse with red blood cells, liposomes and cultured cells (Almeida *et al.*, 1975; Helenius *et al.*, 1977; Hosaka *et al.*, 1983; Eidelman *et al.*, 1984; Metsikkö *et al.*, 1986; Scheule, 1986; Lapidot *et al.*, 1987).

Detergent dialysis method is a common procedure to introduce viral proteins into the liposomal membrane. Purified envelope proteins are mixed with lipids, followed by slow dialysis of the detergent (Petri and Wagner, 1979; Huang *et al.*, 1980; Helenius *et al.*, 1981; Eidelman *et al.*, 1984; Metsikkö *et al.*, 1986; Scheule, 1986). The resulting virosomes usually have a protein–lipid ratio similar to that of the virion, which actually may be a requirement for efficient fusion. One of the problems concerning this reconstitution procedure is the orientation of the membrane proteins. In the detergent dialysis method, some of the membrane proteins acquire a wrong orientation where the ectodomain faces the interior of the liposome (Helenius *et al.*, 1981; Metsikkö *et al.*, 1986). The nature of the detergent is important for retaining the activity of the fusogenic proteins. Metsikkö *et al.* (1986) showed for VSV G protein that virosomes made by Triton X-100 dialysis were able to fuse with BHK cells with the same pH threshold as the virus itself, while octyl glucoside dialysis resulted in the formation of fusion-incompetent virosomes. Their study suggests that the reason for the loss of fusogenic activity may be due to the so-called hairpin configuration (Cardoza *et al.*, 1984) of the spike proteins, where both the cytoplasmic and the outer domain of the protein project to the same side of the liposomal membrane.

Wrong orientation of the envelope glycoproteins can be avoided by expressing the fusogen at the surface of the plasma membrane of the target cell. This may be achieved either by infecting the cells with a virus, the newly synthesized fusogenic envelope glycoproteins of which are transported to the plasma membrane (van Meer and Simons, 1983; van Meer *et al.*, 1985) or by using transfected cells expressing such a protein at the plasma membrane (Kondor-Koch *et al.*, 1983; Florkiewicz and Rose, 1984; Doxsey *et al.*, 1985). Another way to avoid wrong orientation is to implant the fusogenic protein into the liposome membrane by fusing ultraviolet (UV)-inactivated virus particles directly with the liposomes (White and Helenius, 1980; Yamada and Ohnishi, 1986). The requirements for successful fusion of the resulting virosomes with target membranes are that the fusogenic activity of the envelope glycoprotein is reversible with pH and that a high enough surface density of the fusion protein in the liposomal membrane is achieved.

### 4.2. Transfer of Macromolecules to the Yeast Cytoplasm

To start to develop a liposome-mediated transfer technique based on viral fusogens for the yeast system, we took advantage of our earlier observations on the fusion of VSV with the plasma membrane of *S. cerevisiae* spheroplasts (Makarow *et al.*, 1987). Large unilamellar liposomes were prepared by the detergent dialysis method (Philippot *et al.*, 1983) and the G protein of VSV was implanted into the liposomal membrane (Sareneva, 1989). When virosomes loaded with $^{32}P$-labeled DNA were incubated with *S. cerevisiae* spheroplasts at low pH, part of the content marker remained associated with the spheroplasts after proteinase K digestion in the cold. Since the proteinase K treatment releases surface-bound virus particles, the association of the virosomal content marker could have been a result of membrane fusion between the virosome and the spheroplast. This was supported by experiments where fluorescent PE was transferred in a low pH-dependent way from the virosomal membrane into the yeast plasma membrane. In further studies, we prepared virosomes that were loaded with Con A, a lectin that binds to mannose containing glycoconjugates. Treatment of spheroplasts with the lectin-loaded virosomes at low pH resulted in the decrease of glycosylation of newly synthesized polypeptides nearly to the level of tunicamycin-treated spheroplasts, whereas bovine serum albumin (BSA)-loaded virosomes or free Con A had little or no effect. Preliminary sizing of the glycans that were released from the Dol-PP carrier, extracted from the spheroplast treated with lectin-loaded virosomes, showed that they were considerably smaller than $Glc_3Man_9GlcNAc_2$, which was found in spheroplasts treated with BSA-loaded virosomes.

The rationale of introducing Con A into the yeast cytoplasm was based on the finding that in mammalian cells part of the biosynthetic steps of the primary glycan assembled on Dol-PP appear to occur on the cytoplasmic face of the ER membrane (Hirschberg and Snider, 1987). It has been suggested that $GlcNAc_2$ is built up on the lipid carrier in the ER lumen, as this intermediate was not accessible to galactosyltransferase added to isolated microsomes (Hanover and Lennarz, 1978, 1982), and since UDP-GlcNAc was translocated to the lumen of the endoplasmic reticulum (Perez and Hirschberg, 1985). $Man_{1-2}GlcNAc_2$-PP-Dol was also inaccessible for Con A added to microsomal preparations (Snider and Rogers, 1984), suggesting that these glycans also face the lumen or that their affinity towards the lectin ws too low for binding. It should be noted that GDP-Man has not been found to cross the ER membrane (Perez and Hirschberg, 1986). About 50–70% of $Man_{3-5}GlcNAc_2$-PP-Dol was accessible to Con A, whereas most of $Glc_{1-3}Man_{6-9}GlcNAc_2$-PP-Dol was inaccessible to the lectin (Snider and Robbins, 1982; Snider and Rogers, 1984). This has been interpreted to mean that $Man_5GlcNAc_2$-PP-Dol is synthesized at the cytoplasmic face of the ER, changes the polarity, and is translocated to the luminal face. In yeast no studies on the polarity of glycosylation have yet been published to our knowledge. However, it has to be noted that Dol-PP-linked truncated glycans as small as $GlcNAc_2$, found in the *alg* mutants (see Section 2.1.1b), are transferred to polypeptides, which may mean that the glycan intermediates do not face the cytoplasm, or that each of the species synthesized at the cytoplasmic face can change polarity. *In vivo* studies using Con A, transferred to the cytoplasm in virosomes to trap eventual cytoplasmically oriented glycosylation intermediates may shed light on the topography of the glycosylation events in yeast.

ACKNOWLEDGMENTS. The authors would like to acknowledge the financial support of the Academy of Finland, the Nordic Yeast Research Program, the University of Helsinki, the Erna and Victor Hasselblad Foundation, and the Foundation for Biotechnical and Industrial Fermentation Research.

## 5. REFERENCES

Almeida, J. D., Brand, C. M., Edwards, D. C., and Heath, T. D., 1975, Formation of virosomes from influenza subunits and liposomes, *Lancet* **2:**899–901.

Andreasen, A. A., and Stier, T. J. B., 1953, Anaerobic nutrition of *Saccharomyces cerevisiae, J. Cell. Comp. Physiol.* **41:**23–36.

Andreasen, A. A., and Stier, T. J. B., 1954, Anaerobic nutrition of *Saccharomyces cerevisiae, J. Cell. Comp. Physiol.* **43:**271–281.

Atkinson, K. D., and Ramirez, R. M., 1984, Secretion can proceed uncoupled from net plasma membrane expansion in inositol-starved *Saccharomyces cerevisiae, J. Bacteriol.* **160:**80–86.

Atkinson, K. D., Fogel, S., and Henry, S. A., 1980a, Yeast mutant defective in phosphatidylserine synthesis, *J. Biol. Chem.* **255:**6653–6661.

Atkinson, K. D., Jensen, B., Kolat, A. I., Storm, E. M., Henry, S. A., and Fogel, S., 1980b, Yeast mutants auxotrophic for choline or ethanolanime, *J. Bacteriol.* **141:**558–564.

Babczinski, P., and Tanner, W., 1973, Involvement of dolicholmonophosphate in the formation of specific mannosyl-linkages in yeast glycoproteins, *Biochem. Biophys. Res. Commun.* **54:**1119–1124.

Bae-Lee, M. S., and Carman, G. M., 1984, Phosphatidylserine synthesis in *Saccharomyces cerevisiae, J. Biol. Chem.* **259:**10857–10862.

Ballou, C. E., 1982, Yeast cell wall and cell surface, in: *The Molecular Biology of the Yeast Saccharomyces, Metabolism and Gene Expression* (J. N. Strathern, E. W. Jones, and J. R. Broach, eds.), pp. 335–359, Cold Spring Harbor Laboratory, Cold Spring Harbor, New York.

Ballou, D. L., Cohen, R. E., and Ballou, C. E., 1980, *Saccharomyces cerevisiae* mannoprotein mutants that make mannoproteins with a truncated carbohydrate outer chain, *J. Biol. Chem.* **255**:5986–5991.

Ballou, C. E., Kern, K. A., and Raschke, W. C., 1973, Genetic control of yeast mannan structure, *J. Biol. Chem.* **248**:4667–4673.

Baltimore, D., Huang, A., and Stampfer, M., 1970, Ribonucleic acid synthesis of vesicular stomatitis virus. II. An RNA polymerase in the virion, *Proc. Natl. Acad. Sci. USA* **66**:572–576.

Bankaitis, V., Johnson, L. M., and Emr, S. D., 1986, Isolation of yeast mutants defective in protein targeting to the vacuole, *Proc. Natl. Acad. Sci. USA* **83**:9075–9079.

Bause, E., and Lehle, L., 1979, Enzymatic *N*-glycosylation and *O*-glycosylation of synthetic peptide acceptors by dolichol-linked sugar derivatives in yeast, *Eur. J. Biochem.* **101**:531–540.

Becker, G. W., and Lester, R. L., 1980, Biosynthesis of phosphoinositol-containing sphingolipids from phosphatidylinositol by a membrane preparation from *Saccharomyces cerevisiae, J. Bacteriol.* **142**:747–754.

Berlin, R. D., Oliver, J. M., and Walter, R. J., 1978, Surface functions during mitosis. I. Phagocytosis, pinocytosis and mobility of surface-bound Con A, *Cell* **15**:327–341.

Bernstein, M., Hoffmann, W., Ammerer, G., and Schekman, R., 1985, Characterization of a gene product (Sec53p) required for protein assembly in the yeast endoplasmic reticulum, *J. Cell Biol.* **101**:2374–2382.

Boulay, F., Doms, R. W., Webster, R. G., and Helenius, A., 1988, Posttranslational oligomerization and cooperative acid activation of mixed influenza hemagglutinin trimers, *J. Cell Biol.* **106:** 629–639.

Boy-Marcotte, E., and Jacquet, M., 1982, A *Dictyostelium discoideum* DNA fragment complements a *Saccharomyces cerevisiae ura*3 mutant, *Gene* **20**:433–440.

Brantl, S., Eldarov, M. A., Rössler, H., Drescher, B., Lang, H., Rosenthal, S., and Skryabin, K. G., 1988, Expression of env sequences of the bovine leukemia virus (BVL) in the yeast *Saccharomyces cerevisiae, Yeast* **4**:47–59.

Brzobohaty, B., and Kovac, L., 1986, Genetic transformation of yeast protoplasts with DNA encapsulated in liposomes, *FEMS Microbiol. Lett.* **35**:29–31.

Bulawa, C. E., Slater, M., Cabib, E., Au-Young, J., Sburlati, A., Adair, W. L., and Robbins, P. W., 1986, The *S. cerevisiae* structural gene for chitin synthase is not required for chitin synthesis *in vivo, Cell* **46**:213–225.

Burkholder, A. C., and Hartwell, L. H., 1985, The yeast α-factor receptor: Structural properties deduced from the sequence of the *STE*2 gene, *Nucleic Acids Res.* **13**:8463–8475.

Bussey, H., Saville, D., Greene, D., Tipper, D. J., and Bostian, K. A., 1983, Secretion of *Saccharomyces cerevisiae* killer toxin: Processing of the glycosylated precursor, *Mol. Cell. Biol.* **3**:1362–1370.

Byrd, J. C., Tarentino, A. L., Maley, F., Atkinson, P. H., and Trimble, R. B., 1982, Glycoprotein synthesis in yeast, *J. Biol. Chem.* **257**:14657–14666.

Cabib, E., Roberts, R., and Bowers, B., 1982, Synthesis of the yeast cell wall and its regulation, *Annu. Rev. Biochem.* **51**:763–793.

Cardoza, J. D., Kleinfeld, A. M., Stallcup, K. C., and Mescher, M. F., 1984, Hairpin configuration of H-2K$^k$ in liposomes formed by detergent dialysis, *Biochemistry* **23**:4401–4409.

Chu, F. K., Watore, W., and Maley, F., 1983, Factors affecting the oligomeric structure of yeast external invertase, *Arch. Biochem. Biophys.* **223**:543–555.

Connor, J., and Huang, L., 1985, Efficient cytoplasmic delivery of a fluorescent dye by pH-sensitive immunoliposomes, *J. Cell Biol.* **101**:582–589.

Copeland, C. S., Doms, R. W., Bolzau, E. M., Webster, R. G., and Helenius, A., 1986, Assembly of influenza hemagglutinin trimers and its role in intracellular transport, *J. Cell Biol.* **103:**1179–1191.

Daniels, R. S., Downie, J. C., Hay, A. J., Knossow, M., Skehel, J. J., Wang, M. L., and Wiley, D. C., 1985, Fusion mutants of the influenza virus hemagglutinin glycoprotein, *Cell* **40:**431–439.

Daum, G., Heidorn, E., and Paltauf, F., 1986, Intracellular transfer of phospholipids in the yeast, *Saccharomyces cerevisiae, Biochim. Biophys. Acta* **878:**93–101.

Doms, R. W., Helenius, A., and White, J., 1985, Membrane fusion activity of the influenza virus hemagglutinin. The low pH-induced conformational change, *J. Biol. Chem.* **260:**2973–2981.

Doms, R. W., Gething, M.-J., Henneberry, J., White, J., and Helenius, A., 1986, Variant influenza virus hemagglutinin that induces fusion at elevated pH, *J. Virol.* **57:**603–613.

Doxsey, S. J., Brodsky, F. M., Blank, G. S., and Helenius, A., 1987, Inhibition of endocytosis by anti-clathrin antibodies, *Cell* **50:**453–463.

Doxsey, S. J., Sambrook, J., Helenius, A., and White, J., 1985, An efficient method for introducing macromolecules into living cells, *J. Cell Biol.* **101:**19–27.

Dunphy, W. G., Pfeffer, S. R., Clary, D. O., Wattenberg, B. W., Glick, B. S., and Rothman, J. E., 1986, Yeast and mammals utilize similar cytosolic components to drive protein transport through the Golgi complex, *Proc. Natl. Acad. Sci. USA* **83:**1622–1626.

Eidelman, O., Schlegel, R., Tralka, T. S., and Blumenthal, R., 1984, pH-Dependent fusion induced by vesicular stomatitis virus glycoprotein reconstituted into phospholipid vesicles, *J. Biol. Chem.* **259:**4622–4628.

Esmon, P. C., Esmon, B. E., Schauer, I. E., Taylor, A., and Schekman, R., 1987, Structure, assembly, and secretion of octameric invertase, *J. Biol. Chem.* **262:**4387–4394.

Esmon, B., Esmon, P. C., and Schekman, R., 1984, Early steps in processing of yeast glycoproteins, *J. Biol. Chem.* **259:**10322–10327.

Esmon, B., Novick, P., and Schekman, R., 1981, Compartmentalized assembly of oligosaccharides on exported glycoproteins in yeast, *Cell* **25:**451–460.

Farkas, V., 1979, Biosynthesis of cell walls of fungi, *Microb. Rev.* **43:**117–144.

Field, C., and Schekman, R., 1980, Localized secretion of acid phosphatase reflects the pattern of cell-surface growth in *Saccharomyces cerevisiae, J. Biol. Chem.* **254:**796–803.

Florkiewicz, R. Z., and Rose, J. K., 1984, A cell line expressing vesicular stomatitis virus glycoprotein fuses at low pH, *Science* **17:**721–723.

Frevert, J., and Ballou, C. E., 1982, Yeast invertase polymorphism is correlated with variable states of oligosaccharide chain phosphorylation, *Proc. Natl. Acad. Sci. USA* **79:**6147–6150.

Fuller, S. D., Bravo, R., and Simons, K., 1985, An enzymatic assay reveals that proteins destined for the apical or basolateral domains of an epithelial cell line share the same late Golgi compartments, *EMBO J.* **4:**297–307.

Gahmberg, N., Kuismanen, E., Keränen, S., and Pettersson, R. F., 1986, Uukuniemi virus glycoproteins accumulate in and cause morphological changes of the Golgi complex in the absence of virus maturation, *J. Virol.* **57:**899–906.

Garoff, H., Frischauf, A.-M., Simons, K., Lehrach, H., and Delius, H., 1980, Nucleotide sequence of cDNA coding for Semliki Forest virus membrane glycoproteins, *Nature (Lond.)* **288:**236–241.

Gething, M.-J., Doms, R. W., York, D., and White, J., 1986a, Studies on the mechanism of membrane fusion: site-specific mutagenesis of the hemagglutinin of influenza virus, *J. Cell Biol.* **102:**11–23.

Gething, M.-J., McCammon, K., and Sambrook, J., 1986b, Expression of wild-type and mutant forms of influenza hemagglutinin: The role of folding in intracellular transport, *Cell* **46:**939–950.

Getz, G. S., Jakovcic, S., Heywood, J., Frank, J., and Rabinowitz, M., 1970, A two-dimensional thin-layer chromatographic system for phospholipid separation. The analysis of yeast phospholipids, *Biochim. Biophys. Acta* **218**:441–452.

Gregoriadis, G., 1985, Liposomes for drugs and vaccines, *Trends Biotechnol.* **3**:235–241.

Gumbiner, B., and Kelly, R. B., 1982, Two distinct intracellular pathways transport secretory and membrane glycoproteins to the surface of pituitary tumor cells, *Cell* **28**:51–59.

Hanover, J. A., and Lennarz, W. J., 1978, The topological orientation of *N,N'*-diacetylchitobiosylpyrophosphoryldolichol in artificial and natural membranes, *J. Biol. Chem.* **254**:9237–9246.

Hanover, J. A., and Lennarz, W. J., 1982, Transmembrane assembly of *N*-linked glycoproteins, *J. Biol. Chem.* **257**:2787–2794.

Hansen, W., Garcia, P. D., and Walter, P., 1986, *In vitro* protein translocation across the yeast endoplasmic reticulum: ATP-dependent post-translational translocation of the prepro-α-factor, *Cell* **45**:397–406.

Hanson, B. A., 1984, Role of inositol-containing sphingolipids in *Saccharomyces cerevisiae* during inositol starvation, *J. Bacteriol.*, **159**:837–842.

Haselbeck, A., and Tanner, W., 1983, *O*-glycosylation in *Saccharomyces cerevisiae* is initiated at the endoplasmic reticulum, *FEBS Lett.* **158**:335–338.

Haywood, A. M., and Boyer, B. P., 1985, Fusion of influenza virus membranes with liposomes at pH 7.5, *Proc. Natl. Acad. Sci. USA* **82**:4611–4615.

Helenius, A., Fries, E., and Kartenbeck, J., 1977, Reconstitution of Semliki Forest virus membrane, *J. Cell Biol.* **75**:866–880.

Helenius, A., Kartenbeck, J., Simons, K., and Fries, E., 1980, On the entry of Semliki Forest virus into BHK-21 cells, *J. Cell Biol.* **84**:404–420.

Helenius, A., Sarvas, M., and Simons, K., 1981, Asymmetric and symmetric membrane reconstitution by detergent elimination. Studies with Semliki Forest virus spike glycoprotein and penicillinase from the membrane of *Bacillus licheniformis, Eur. J. Biochem.* **116**:27–35.

Henry, S. A., 1982, Membrane lipids of yeast: biochemical and genetic studies, in: *The Molecular Biology of the Yeast Saccharomyces* (J. N. Strathern, E. W. Jones, and J. R. Broach, eds.), pp. 101–158, Cold Spring Harbor Laboratory, Cold Spring Harbor, New York.

Henry, S. A., Klig, L. S., and Loewy, B. S., 1984, The genetic regulation and coordination of biosynthetic pathways in yeast: Amino acid and phospholipid synthesis, *Annu. Rev. Genet.* **18**: 207–231.

Hirschberg, C. B., and Snider, M. D., 1987, Topography of glycosylation in the rough endoplasmic reticulum and Golgi apparatus, *Annu. Rev. Biochem.* **56**:63–87.

Hoffmann, W., 1985, Molecular characterization of the *CAN1* locus in *Saccharomyces cerevisiae, J. Biol. Chem.* **260**:11831–11837.

Hoffmann, W., 1987, *CAN1-SUC2* gene fusion studies in *Saccharomyces cerevisiae, Mol. Gen. Genet.* **210**:277–281.

Holcomb, C. L., Hansen, W. J., Etcheverry, T., and Schekman, R., 1988, Secretory vesicles externalize the major plasma membrane ATPase in yeast, *J. Cell Biol.* **106**:641–648.

Homann, M. J., Poole, M. A., Gaynor, P. M., Ho, C. T., and Carman, G. M., 1987, Effect of growth phase on phospholipid biosynthesis in *Saccharomyces cerevisiae, J. Bacteriol.* **169**: 533–539.

Hori, T., Nakamura, N., and Okuyama, H., 1987, Possible involvement of acetyl coenzyme A carboxylase as well as fatty acid synthetase in the temperature-controlled synthesis of fatty acids in *Saccharomyces cerevisiae, J. Biochem.* **101**:949–956.

Hosaka, Y., Yasuda, Y., and Fukai, K., 1983, Hemolysis by liposomes containing influenza virus hemagglutinins, *J. Virol.* **46**:1014–1017.

Huang, R. T. C., and Uslu, G., 1986, Interplay between lipids and viral glycoproteins during hemolysis and fusion by influenza virus, *J. Biol. Chem.* **261**:12911–12914.

Huang, R. T. C., Wahn, K., Klenk, H.-D., and Rott, R., 1980, Fusion between cell membrane and liposomes containing the glycoproteins of influenza virus, *Virology* **104:**294–302.

Huang, R. T. C., Rott, R., and Klenk, H.-D., 1981, Influenza viruses cause hemolysis and fusion of cells, *Virology* **110:**243–247.

Huffaker, T. C., and Robbins, P. W., 1982, Temperature-sensitive yeast mutants deficient in asparagine-linked glycosylation, *J. Biol. Chem.* **257:**3203–3210.

Huffaker, T. C., and Robbins, P. W., 1983, Yeast mutants deficient in protein glycosylation, *Proc. Natl. Acad. Sci. USA* **80:**7466–7470.

Hunter, K., and Rose, A. H., 1972, Lipid composition of *Saccharomyces cerevisiae* as influenced by growth temperature, *Biochim. Biophys. Acta* **260:**639–653.

Innis, M. A., Holland, M. J., McCabe, P. C., Cole, G. E., Wittman, V. P., Tal, R., Watt, K. W. K., Gelfand, D. H., Holland, J. P., and Meade, J. H., 1985, Expression, glycosylation, and secretion of an *Aspergillus* glucoamylase by *Saccharomyces cerevisiae, Science* **228:**21–26.

Jabbar, M. A., and Nayak, D. P., 1987, Signal Processing, glycosylation, and secretion of mutant hemagglutinins of a human influenza virus by *Saccharomyces cerevisiae, Mol. Cell. Biol.* **7:** 1476–1485.

Jabbar, M. A., Sivasubramanian, N., and Nayak, D. P., 1985, Influenza viral (A/WSN/33) hemagglutinin is expressed and glycosylated in the yeast *Saccharomyces cerevisiae, Proc. Natl. Acad. Sci. USA* **82:**2019–2023.

Jelinek-Kelly, S., Akiyama, T., Saunier, B., Tkacz, J. S., and Herscovics, A., 1985, Characterization of a specific α-mannosidase involved in oligosaccharide processing in *Saccharomyces cerevisiae, J. Biol. Chem.* **260:**2253–2257.

Johnson, L. M., Bankaitis, V. A., and Emr, S. D., 1987, Distinct sequence determinants direct intracellular sorting and modification of a yeast vacuolar protease, *Cell* **48:**875–885.

Jollow, D., Kellerman, G., and Linnane, A., 1968, The biogenesis of mitochondria. III. The lipid composition of aerobically and anaerobically grown *Saccharomyces cerevisiae* as related to the membrane systems of the cell, *J. Cell Biol.* **37:**221–230.

Jones, G. H., and Ballou, C. E., 1969, Studies on the structure of yeast mannan, *J. Biol. Chem.* **244:** 1052–1059.

Julius, D., Schekman, R., and Thorner, J., 1984a, Glycosylation and processing of prepro-α-factor through the yeast secretory pathway, *Cell* **36:**309–318.

Julius, D., Brake, A., Blair, L., Kunisawa, R., and Thorner, J., 1984b, Isolation of the putative structural gene for the Lysine–Arginine-cleaving endopeptidase required for processing of yeast prepro-α-factor, *Cell* **37:**1075–1089.

Kääriäinen, L., and Pesonen, M., 1982, Virus glycoproteins and glycolipids, in: *The Glycoconjugates* (M. I. Horowitz, ed.), Vol. IV, pp. 191–242, Academic, New York.

Kääriäinen, L., Takkinen, K., Keränen, S., and Söderlund, H., 1987, Replication of the genome of alphaviruses, *J. Cell Sci.* **7**(suppl.)**:**231–250.

Karson, E. M., and Ballou, C. E., 1978, Biosynthesis of yeast mannan: Properties of a mannosylphosphate transferase in *Saccharomyces cerevisiae, J. Biol. Chem.* **253:**6484–6492.

Kelly, R., 1985, Pathways of protein secretion in eukaryotes, *Science* **230:**25–32.

Kennedy, E. P., and Weiss, S. B., 1956, The function of cytidine coenzymes in the biosynthesis of phospholipids, *J. Biol. Chem.* **222:**193–214.

Keränen, S., 1986, Synthesis and processing of Semliki Forest virus polyprotein in *Saccharomyces cerevisiae:* A yeast type glycosylation of E1 envelope protein, *Gene* **48:**267–275.

Kielian, M. C., and Helenius, A., 1984, Role of cholesterol in fusion of Semliki Forest virus with membranes, *J. Virol.* **52:**281–283.

Kielian, M. C., and Helenius, A., 1985, pH-Induced alterations in the fusogenic spike protein of Semliki Forest virus, *J. Cell Biol.* **101:**2284–2291.

Kielian, M. C., and Helenius, A., 1986, Entry of alphaviruses, in: *The Togaviridae and Flaviviridae* (S. Schlesinger and M. J. Schlesinger, eds.) pp. 91–119, Plenum, New York.
Kielian, M. C., Keränen, S., Kääriäinen, L., and Helenius, A., 1984, Membrane fusion mutants of Semliki Forest virus, *J. Cell Biol.* **98:**139–145.
Kielian, M. C., Marsh, M., and Helenius, A., 1986, Kinetics of endosome acidification detected by mutant and wild-type Semliki Forest virus, *EMBO J.* **5:**3103–3109.
Kondor-Koch, C., Burke, B., and Garoff, H., 1983, Expression of Semliki Forest virus proteins from cloned complementary DNA. The fusion activity of the spike glycoprotein, *J. Cell Biol.* **97:**644–651.
Kornfeld, R., and Kornfeld, S., 1985, Assembly of asparagine-linked oligosaccharides, *Annu. Rev. Biochem.* **54:**631–664.
Kreis, T. E., and Lodish, H. F., 1986, Oligomerization is essential for transport of vesicular stomatitis viral glycoprotein to the cell surface, *Cell* **46:**929–937.
Kuga, T., Hattori, S., Yoshida, M., and Taniguchi, T., 1986, Expression of human T-cell leukemia virus type 1 envelope protein in *Saccharomyces cerevisiae, Gene* **44:**337–340.
Kuismanen, E., Hedman, K., Saraste, J., and Pettersson, R. F., 1982, Uukuniemi virus maturation: Accumulation of virus particles and viral antigens in the Golgi complex, *Mol. Cell Biol.* **2:** 1444–1458.
Kukuruzinska, M. A., Bergh, M. L. E., and Jackson, B. J., 1987, Protein glycosylation in yeast, *Annu, Rev. Biochem.* **56:**915–944.
Kuo, S. C., and Yamamoto, S. C., 1975, Preparation and growth of yeast spheroplasts, in: *Methods in Cell Biology* Vol. XI (D. M. Prescott, ed.), pp. 169–183, Academic, London.
Lapidot, M., Nussbaum, O., and Loyter, A., 1987, Fusion of membrane vesicles bearing only the influenza hemagglutinin with erythrocytes, living cultured cells, and liposomes, *J. Biol. Chem.* **262:**13736–13741.
Lazo, P. S., Ochoa, A. G., and Cascon, S., 1978, α-Galactosidase (melibiase) from *Saccharomyces Carlsbergensis:* Structural and kinetic properties, *Arch. Biochem. Biophys.* **191:**316–324.
Lear, J. D., and DeGrado, W. F., 1987, Membrane binding and conformational properties of peptides representing the $NH_2$ terminus of influenza HA-2, *J. Biol. Chem.* **262:**6500–6505.
Lehle, L., and Bause, E., 1984, Primary structural requirements for *N*- and *O*-glycosylation of yeast mannoproteins, *Biochim. Biophys. Acta* **799:**246–251.
Lehle, L., Cohen, R. E., and Ballou, C. E., 1979, Carbohydrate structure of yeast invertase, *J. Biol. Chem.* **254:**12209–12218.
Letts, V. A., and Dawes, I. W., 1983, Temperature-sensitive *Saccharomyces cerevisiae* mutant defective in lipid biosynthesis, *J. Bacteriol.* **156:**212–221.
Letts, V. A., and Henry, S. A., 1985, Regulation of phospholipid synthesis in phosphatidylserine synthase-deficient (*chol*) mutants of *Saccharomyces cerevisiae, J. Bacteriol.* **163:**560–567.
Letts, V. A., Klig, L. S., Bae-Lee, M., Carman, G. M., and Henry, S. A., 1983, Isolation of the yeast structural gene for the membrane-associated enzyme phosphatidylserine synthase, *Proc. Natl. Acad. Sci. USA* **80:**7279–7283.
Lopez-Romero, E., and Ruiz-Herrera, J., 1977, Biosynthesis of β-glucans by cell-free extracts from *Saccharomyces cerevisiae, Biochim. Biophys. Acta* **500:**372–384.
McKay, V. L., Welch, S. K., Insley, M. Y., Nanney, T. R., Holly, J., Saari, G. C., and Parker, M. L., 1988, The *Saccharomyces* cerevisiae *BAR1* gene encodes an exported protein with homology to pepsin, *Proc. Natl. Acad. Sci. USA* **85:**55–59.
Maeda, T., and Ohnishi, S.-I., 1980, Activation of influenza virus by acidic media causes hemolysis and fusion of erythrocytes, *FEBS Lett.* **122:**283–287.
Makarow, M., 1985, Endocytosis in *Saccharomyces cerevisiae:* internalization of enveloped viruses into spheroplasts, *EMBO J.* **4:**1855–1860.
Makarow, M., 1988, Secretion of invertase in mitotic yeast cells, *EMBO J.* **7:**1475–1482.

Makarow, M., Nevalainen, L. T., and Kääriäinen, L., 1986, Expression of the RNA genome of an animal virus in *Saccharomyces cerevisiae, Proc. Natl. Acad. Sci. USA* **83:**8117–8121.

Makarow, M., Sareneva, H., and von Bonsdorff, C.-H., 1987, Characterization of the fusion of enveloped viruses with the plasma membrane of *Saccharomyces cerevisiae* spheroplasts, *J. Biol. Chem.* **262:**1836–1841.

Manners, D. J., Masson, A. J., and Patterson, J. C., 1973a, The structure of a β-(1->3)-D-glucan from yeast cell walls, *Biochem. J.* **135:**19–30.

Manners, D. J., Masson, A. J., and Patterson, J. C., 1973b, The structure of a β-(1->6)-D-glucan from yeast cell walls, *Biochem. J.* **135:**31–36.

Matlin, K. S., Reggio, H., Helenius, A., and Simons, K., 1982, Pathway of vesicular stomatitis virus entry leading to infection, *J. Mol. Biol.* **156:**609–631.

McAleer, W. J., Buynak, E. B., Maigetter, R. Z., Wampler, D. E., Miller, J. M., and Hilleman, M. R., 1984, Human hepatitis B vaccine from recombinant yeast, *Nature (Lond.)* **307:**178–180.

Metsikkö, K., van Meer, G., and Simons, K., 1986, Reconstitution of the fusogenic activity of vesicular stomatitis virus, *EMBO J.* **5:**3429–3435.

Mildner, P., Barbaric, S., Golubic, Z., and Ries, B., 1976, Purification of protoplast-secreted acid phosphatase from baker's yeast, *Biochim. Biophys. Acta* **429:**274–282.

Miyanohara, A., Toh-E, A., Nozaki, C., Hamada, F., Ohtomo, N., and Matsubara, K., 1983, Expression of hepatitis B surface antigen gene in yeast, *Proc. Natl. Acad. Sci. USA* **80:**1–5.

Moonen, P., Mermod, J.-J., Ernst, J. F., Hirschi, M., and DeLamarter, J. F., 1987, Increased biological activity of deglycosylated recombinant human granulocyte/macrophage colony-stimulating factor produced by yeast or animal cells, *Proc. Natl. Acad. Sci. USA* **84:**4428–4431.

Nakajima, T., and Ballou, C. E., 1974a, Characterization of the carbohydrate fragments obtained from *Saccharomyces cerevisiae* mannan by alkaline degradation, *J. Biol. Chem.* **249:**7679–7684.

Nakajima, T., and Ballou, C. E., 1974b, Structure of the linkage region between the polysaccharide and protein parts of *Saccharomyces cerevisiae* mannan, *J. Biol. Chem.* **249:**7685–7694.

Nakamura, Y., Sato, T., Emi, M., Miyanohara, A., Nishide, T., and Matsubara, K., 1986, Expression of human salivary α-amylase gene in *Saccharomyces cerevisiae* and its secretion using the mammalian signal sequence, *Gene* **50:**239–245.

Necas, O., 1971, Cell wall synthesis in yeast protoplasts, *Biol. Rev.* **35:**149–170.

Nevalainen, L. T., and Makarow, M., 1988, Intracellular transport in interphase and mitotic yeast cells, *Eur. J. Biochem.*, **178:**39–46.

Nir, S., Stegmann, T., and Wilschut, J., 1986, Fusion of influenza virus with cardiolipin liposomes at low pH: Mass action analysis of kinetics and extent, *Biochemistry* **25:**257–266.

Novick, P., Ferro, S., and Schekman, R., 1981, Order of events in the yeast secretory pathway, *Cell* **25:**461–469.

Novick, P., Field, C., and Schekman, R., 1980, Identification of 23 complementation groups required for post-translational events in the yeast secretory pathway, *Cell* **21:**205–215.

Novick, P., and Schekman, R., 1983, Export of major cell surface proteins is blocked in yeast secretory mutants, *J. Cell Biol.* **96:**541–547.

Nozaki, C., Makizumi, K., Kino, Y., Nakatake, H., Eto, T., Mizuno, K., Hamada, F., and Ohtomo, N., 1985, Expression of herpes simplex virus glycoprotein B gene in yeast, *Virus Res.* **4:**107–113.

Nussbaum, O., Lapidot, M., and Loyter, A., 1987, Reconstitution of functional influenza virus envelopes and fusion with membranes and liposomes lacking virus receptors, *J. Virol.* **61:**2245–2252.

Okuyama, H., Saito, M., Joshi, V. C., Gunsberg, S., and Wakil, S. J., 1979, Regulation by temperature of the chain length of fatty acids in yeast, *J. Biol. Chem.* **254:**12281–12284.

Orlean, P., Ammer, H., Watzele, M., and Tanner, W., 1986, Synthesis of an O-glycosylated cell surface protein induced in yeast by α-factor, *Proc. Natl. Acad. Sci. USA* **83:**6263–6266.

Owen, M. J., Kissonerghis, A.-M., and Lodish, H. F., 1980, Biosynthesis of HLA-A and HLA-B antigens *in vivo*, *J. Biol. Chem.* **225:**9678–9684.

Parks, L. W., 1978, Metabolism of sterols in yeast, *Crit. Rev. Microbiol.* **6:**301–341.

Pastor, F. I. J., Herrero, E., and Sentandreu, R., 1982, Metabolism of *Saccharomyces cerevisiae* envelope mannoproteins, *Arch. Microbiol.* **132:**144–148.

Perez, M., and Hirschberg, C. B., 1985, Translocation of UDP-*N*-acetylglucosamine into vesicles derived from rat liver rough endoplasmic reticulum and Golgi apparatus, *J. Biol. Chem.* **260:** 4671–4678.

Perez, M., and Hirschberg, C. B., 1986, Topography of glycosylation reactions in the rough endoplasmic reticulum membrane, *J. Biol. Chem.* **261:**6822–6830.

Pesonen (Makarow), M., and Simons, K., 1983, Transepithelial transport of a viral membrane glycoprotein implanted into the apical plasma membrane of Madin–Darby canine kidney cells. II. Immunological quantitation, *J. Cell Biol.* **97:**638–643.

Petri, W. A., Jr., and Wagner, R. R., 1979, Reconstitution into liposomes of the glycoprotein of vesicular stomatitis virus by detergent dialysis, *J. Biol. Chem.* **254:**4313–4316.

Pfeffer, S. R., and Rothman, J. E., 1987, Biosynthetic protein transport and sorting by the endoplasmic reticulum and the Golgi, *Annu. Rev. Biochem.* **56:**829–852.

Philippot, J., Mutaftschiev, S., Liautard, J. P., 1983, A very mild method allowing the encapsulation of very high amounts of macromolecules into very large (1000 nm) unilamellar liposomes, *Biochim. Biophys. Acta* **734:**137–143.

Poste, G., Lyon, N. C., Macander, P., Porter, C. W., Reeve, P., and Bachmeyer, H., 1980, Liposome-mediated transfer of integral membrane glycoproteins into the plasma membrane of cultured cells, *Exp. Cell Res.* **129:**393–408.

Prakash, C., and Vijay, I. K., 1982, Characterization of intermediates up to lipid-linked heptasaccharide implicated in the biosynthesis of *Saccharomyces cerevisiae* mannoproteins, *Biochemistry* **21:**4810–4818.

Prakash, C., Katial, A., and Vijay, I. K., 1983, Characterization of lipid-linked octa- through undecasaccharides implicated in the biosynthesis of *Saccharomyces cerevisiae* mannoproteins, *J. Bacteriol.* **153:**895–902.

Pratt, R. E., Flynn, J. A., Hobart, P. M., Paul, M., and Dzan, V. J., 1988, Different secretory pathways of renin from mouse cells transfected with the human renin gene, *J. Biol. Chem.* **263:** 3137–3141.

Ramirez, R. M., Ishida-Schick, T., Krilowicz, B. L., Leish, B. A., and Atkinson, K. D., 1983, Plasma membrane expansion terminates in *Saccharomyces cerevisiae* secretion-defective mutants while phospholipid synthesis continues, *J. Bacteriol.* **154:**1276–1283.

Rank, G., Robertson, A., and Bussey, H., 1978, The viscosity and lipid composition of the plasma membrane of multiple drug resistant and sensitive yeast strains, *Can J. Biochem.* **56:**1036–1041.

Raschke, W. C., Kern, K. A., Antalis, C., and Ballou, C. E., 1973, Genetic control of yeast mannan structure, *J. Biol. Chem.* **248:**4660–4666.

Rice, C. M., and Strauss, J. H., 1981, Nucleotide sequence of the 26S mRNA of Sindbis virus and deduced sequence of the encoded virus structural proteins, *Proc. Natl. Acad. Sci. USA* **78:** 2062–2066.

Riedel, H., Kondor-Koch, C., and Garoff, H., 1984, Cell surface expression of fusogenic vesicular stomatitis virus G protein from cloned cDNA, *EMBO J.* **3:**1477–1483.

Rindler, M. J., Ivanov, I. E., Plesken, H., Rodriguez-Boulan, E., and Sabatini, D. D., 1984, Viral glycoproteins destined for apical or basolateral plasma membrane domains traverse the same

Golgi apparatus during their intracellular transport in doubly infected Madin–Darby canine kidney cells, *J. Chem. Biol.* **98:**1304–1319.

Rodriguez-Boulan, E., and Sabatini, D. D., 1978, Asymmetric budding of viruses in epithelial monolayers: A model system for study of epithelial polarity, *Proc. Natl. Acad. Sci. USA* **75:** 5071–5075.

Rose, J. K., and Gallione, C. J., 1981, Nucleotide sequences of the mRNA's encoding the vesicular stomatitis virus G and M proteins determined from cDNA clones containing the complete coding regions, *J. Virol.* **39:**519–528.

Rothblatt, J. A., and Meyer, D. I., 1986, Secretion in yeast: translocation and glycosylation of prepro-α-factor *in vitro* can occur via an ATP-dependent post-translational mechanism, *EMBO J.* **5:**1031–1036.

Ruigrok, R. W. H., Wrigley, N. G., Calder, L. J., Cusack, S., Wharton, S. A., Brown, E. B., and Skehel, J. J., 1986, Electron microscopy of the low pH structure of influenza virus haemagglutinin, *EMBO J.* **5:**41–49.

Runge, K. W., and Robbins, P. W., 1986, A new yeast mutation in the glucosylation steps of the asparagine-linked glycosylation pathway: Formation of a novel asparagine-linked oligosaccharide containing two glucose residues, *J. Biol. Chem.* **261:**15582–15590.

Runge, K. W., Huffaker, T. C., and Robbins, P. W., 1984, Two yeast mutations in glucosylation steps of the asparagine glycosylation pathway, *J. Biol. Chem.* **259:**412–417.

Ruohola, H., and Ferro-Novick, S., 1987, Sec53, a protein required for an early step in secretory protein processing and transport in yeast, interacts with the cytoplasmic surface of the endoplasmic reticulum, *Proc. Natl. Acad. Sci. USA* **84:**8468–8472.

Ruohonen, L., Hackman, P., Lehtovaara, P., Knowles, J. K. C., and Keränen, S., 1987, Efficient secretion of *Bacillus amyloliquefaciens* α-amylase cells by its own signal peptide from *Saccharomyces cerevisiae* host, *Gene* **59:**161–170.

Salminen, A., and Novick, P. J., 1987, A *ras*-like protein is required for a post-Golgi event in yeast secretion, *Cell* **49:**527–538.

Salovuori, I., Makarow, M., Rauvala, H., Knowles, J., and Kääriäinen, L., 1987, Low molecular weight high-mannose type glycans in a secreted protein of the filamentous fungus *Trichoderma reesei, BioTechnology* **5:**152–156.

Sareneva, H., 1989, Interaction of virosomes with spheroplasts of *Saccharomyces cerevisiae, Life Science Advances,* in press.

Schekman, R., 1985, Protein localization and membrane traffic in yeast, *Annu. Rev. Cell Biol.* **1:** 115–143.

Scherrer, R., Louden, L., and Gerhardt, P., 1974, Porosity of the yeast cell wall and membrane, *J. Bacteriol.* **118:**534–540.

Scheule, R. K., 1986, Novel preparation of functional Sindbis virosomes, *Biochemistry* **25:**4223–4232.

Scheule, R. K., 1987, Fusion of Sindbis virus with model membranes containing phosphatidylethanolamine: Implications for protein-induced membrane fusion, *Biochim. Biophys. Acta* **899:** 185–195.

Schlenstedt, G., and Zimmermann, R., 1987, Import of frog prepropeptide GLa into microsomes requires ATP but does not involve docking protein or ribosomes, *EMBO J.* **6:**699–703.

Schultz, L. D., Tanner, J., Hofmann, K. J., Emini, E. A., Condra, J. H., Jones, R. E., Kieff, E., and Ellis, R. W., 1987, Expression and secretion in yeast of a 400-kDa envelope glycoprotein derived from Epstein–Barr virus, *Gene* **54:**113–123.

Schwaiger, H., Hasilik, A., von Figure, K., Wiemken, A., and Tanner, W., 1982, Carbohydrate-free carboxypeptidase Y is transferred into the lysosome-like yeast vacuole, *Biochem. Biophys. Res. Commun.* **104:**950–956.

Scott, J. H., and Schekman, R., 1980, Lyticase: Endoglucanase and protease activities that act together in yeast cell lysis, *J. Bacteriol.* **142:**414–423.

Sege, K., Rask, L., and Peterson, P. A., 1981, Role of $\beta_2$-microglobulin in the intracellular processing of HLA antigens, *Biochemistry* **20:**4523–4530.

Sentandreu, R., Herrero, E., Martinez-Carcia, J. P., and Larriba, G., 1985, Biogenesis of the yeast cell wall, in: *Subcellular Biochemistry,* Vol. 10 (D. B. Roodyn, ed.), pp. 193–235, Plenum, New York.

Serrano, R., Kielland-Brandt, M. C., and Fink, G. R., 1986, Yeast plasma membrane ATPase is essential for growth and has homology with ($Na^+ + K^+$), $K^+$-and $Ca^{2+}$0ATPases, *Nature (Lond.)* **319:**689–693.

Sharma, C. B., Babczinski, P., Lehle, L., and Tanner, W., 1974, The role of dolicholmonophosphate in glycoprotein biosynthesis in *Saccharomyces cerevisiae, Eur. J. Biochem.* **46:**35–41.

Shematek, E. M., Braatz, J. A., and Cabib, E., 1980, Biosynthesis of the yeast cell wall, *J. Biol. Chem.* **255:**888–894.

Shibata, N., Mizugami, K., and Suzuki, S., 1984, Immunochemical properties of mannan–protein complex isolated from viable cells of *Saccharomyces cerevisiae* 4484-24D-1 mutant strain by the action of zymolyase, *Microbiol. Immunol.* **28:**1283–1292.

Shibata, N., Mizugami, K., Takano, K., and Suzuki, S., 1983, Isolation of mannan–protein complexes from viable cells of *Saccharomyces cerevisiae* X2180-1A wild type and *Saccharomyces cerevisiae* X2180-1 A-5 mutant strains by the action of Zymolyase-60,000, *J. Bacteriol.* **156:** 552–558.

Simons, K., and Fuller, S. D., 1985, Cell surface polarity in epithelia, *Annu. Rev. Cell Biol.* **1:**243–288.

Sly, W. S., and Fisher, D. D., 1982, The phosphomannosyl recognition for intracellular and intercellular transport of lysosomal enzymes, *J. Cell. Biochem.* **18:**67–85.

Skehel, J. J., Bayley, P. M., Brown, E. B., Martin, S. R., Waterfield, M. D., White, J. M., Wilson, I. A., and Wiley, D. C., 1982, Changes in the conformation of influenza virus hemagglutinin at the pH optimum of virus-mediated membrane fusion, *Proc. Natl. Acad. Sci. USA* **79:**968–972.

Smith, M. M., Lindström, J., and Merlie, J. P., 1987, Formation of the $\alpha$-bungarotoxin binding site and assembly of the nicotinic acetylcholine receptor subunits occur in the endoplasmic reticulum, *J. Biol. Chem.* **262:**4367–4376.

Smith, S. W., and Lester, R. L., 1974, Inositol phosphorylceramide, a novel substance and the chief member of a major group of yeast sphingolipids containing a single inositol phosphate, *J. Biol. Chem.* **249:**3395–3405.

Snider, M. D., and Robbins, P. W., 1982, Transmembrane organization of protein glycosylation, *J. Biol. Chem.* **257:**6796–6801.

Snider, M. D., and Rogers, O. C., 1984, Transmembrane movement of oligosaccharide-lipids during glycoprotein synthesis, *Cell* **36:**753–761.

Soriano, P., Dijkstra, J., Legrand, A., Spanjer, H., Londos-Gagliardi, D., Roerdink, F., Scherphof, G., and Nicolau, C., 1983, Targeted and nontargeted liposomes for *in vivo* transfer to rat liver cells of a plasmid containing the preproinsulin I gene, *Proc. Natl. Acad. Sci. USA* **80:**7128–7131.

Stegmann, T., Hoekstra, D., Scherphof, G., and Wilschut, J., 1985, Kinetics of pH-dependent fusion between influenza virus and liposomes, *Biochemistry* **24:**3107–3113.

Stegmann, T., Hoekstra, D., Scherphof, G., and Wilschut, J., 1986, Fusion activity of influenza virus, *J. Biol. Chem.* **261:**10966–10969.

Strauss, E. G., and Strauss, J. H., 1983, Replication strategies of the single stranded RNA viruses of eukaryotes, *Curr. Top. Microbiol. Immunol.* **105:**2–98.

Suzuki, Y., Matsunaga, M., and Matsumoto, M., 1985, *N*acetylneuraminyllactosylceramide,

$G_{M3\text{-}NeuAc}$, a new influenza A virus receptor which mediates the adsorption–fusion process of viral infection, *J. Biol. Chem.* **260:**1362–1365.

Tammi, M., Ballou, L., Taylor, A., and Ballou, C. E., 1987, Effect of glycosylation on yeast invertase oligomer stability, *J. Biol. Chem.* **262:**4395–4401.

Tanner, W., and Lehle, L., 1987, Protein glycosylation in yeast, *Biochim. Biophys. Acta* **906:**81–99.

Tkacz, J. S., and Lampen, J. O., 1973, Surface distribution of invertase on growing *Saccharomyces* cells, *J. Bacteriol.* **113:**1073–1075.

Tokunaga, M., Wada, N., and Hishinuma, F., 1987, A novel yeast secretion vector utilizing secretion signal of killer toxin encoded on the yeast linear and plasmid pGKL1, *Biochem. Biophys. Res. Commun.* **144:**613–619.

Tsai, P.-K., Ballou, L., Esmon, B., Schekman, R., and Ballou, C. E., 1984a, Isolation of glucose-containing high-mannose glycoprotein core oligosaccharides, *Proc. Natl. Acad. Sci. USA* **81:** 6340–6343.

Tsai, P.-K., Frevert, J., and Ballou, C. E., 1984b, Carbohydrate structure of *Saccharomyces cerevisiae mnn9* mannoprotein, *J. Biol. Chem.* **259:**3805–3811.

Tschopp, J., Esmon, P. C., and Schekman, R., 1984, Defective plasma membrane assembly in yeast secretory mutants, *J. Bacteriol.* **160:**966–970.

Väänänen, P., and Kääriäinen, L., 1980, Fusion and haemolysis of erythrocytes caused by three togaviruses: Semliki Forest, Sindbis and Rubella, *J. Gen. Virol.* **46:**467–475.

Väänänen, P., Gahmberg, C. G., and Kääriäinen, L., 1981, Fusion of Semliki Forest virus with red cell membranes, *Virology* **110:**366–374.

Valentin, E., Herrero, E., Pastor, F. I. J., and Sentandreu, R., 1984, Solubilization and analysis of mannoprotein molecules from the cell wall of *Saccharomyces cerevisiae, J. Gen. Microbiol.* **130:**1419–1428.

Valentin, E., Herrero, E., and Sentandreu, R., 1986, Incorporation of mannoproteins into the walls of aleucin A-treated yeast cells, *Arch. Microbiol.* **146:**214–220.

Valenzuela, P., Medina, A., and Rutter, W. J., 1982, Synthesis and assembly of hepatitis B virus surface antigen particles in yeast, *Nature (Lond.)* **298:**347–350.

Valenzuela, P., Coit, D., Kuo, C. H., 1985a, Synthesis and assembly in yeast of hepatitis B surface antigen particles containing the polyalbumin receptor, *BioTechnology* **3:**317–320.

Valenzuela, P., Coit, D., Medina-Selby, M. A., Kuo, C. H., van Nest, G., Burke, L. R., Bull, P., Urdea, M. S., and Graves, P. V., 1985b, Antigen engineering in yeast: Synthesis and assembly of hybrid hepatitis B surface antigen–herpes simplex 1 gD particles, *BioTechnology* **3:**323–326.

Valls, L. A., Hunter, C. P., Rothman, J. H., and Stevens, T. H., 1987, Protein sorting in yeast: the localization determinant of yeast vacuolar carboxypeptidase Y resides in the propeptide, *Cell* **48:**887–897.

van Meer, G., and Simons, K., 1983, An efficient method for introducing defined lipids into the plasma membrane of mammalian cells, *J. Cell Biol.* **97:**1365–1374.

van Meer, G., Davoust, J., and Simons, K., 1985, Parameters affecting low-pH-mediated fusion of liposomes with the plasma membrane of cells infected with influenza virus, *Biochemistry* **24:** 3593–3602.

von Bonsdorff, C.-H., Saikku, P., and Oker-Blom, N., 1970, Electron microscope study on the development of Uukuniemi virus, *Acta Virol. (Praha)* **14:**109–114.

Waechter, C. J., and Lester, R. L., 1971, Regulation of phosphatidylcholine biosynthesis in *Saccharomyces cerevisiae, J. Bacteriol.* **105:**837–843.

Waechter, C. J., and Lester, R. L., 1973, Differential regulation of the N-methyl-transferases responsible for phosphatidylcholine synthesis in *Saccharomyces cerevisiae, Arch. Biochem. Biophys.* **158:**401–410.

Warren, G., 1985, Membrane traffic and organelle division, *TIBS* **10:**439–443.

Warren, G., Featherstone, C., Griffiths, G., and Burke, B., 1983, Newly synthesized G protein of vesicular stomatitis virus is not transported to the cell surface during mitosis, *J. Cell Biol.* **97:** 1623–1628.

Waters, M. G., and Blobel, G., 1986, Secretory protein translocation in a yeast cell-free system can occur posttranslationally and requires ATP hydrolysis, *J. Cell Biol.* **102:**1543–1550.

Watts, C., Wickner, W., and Zimmerman, R., 1983, M13 procoat and pre-immunoglobulin share processing specificity but use different membrane receptor mechanisms, *Proc. Natl. Acad. Sci. USA* **80:**2809–2813.

Webster, R. G., Brown, L. E., and Jackson, D. C., 1983, Changes in the antigenicity of the hemagglutinin molecule of H3 influenza virus at acidic pH, *Virology* **126:**587–599.

Wen, D., and Schlesinger, M. J., 1984, Fatty acid-acylated proteins in secretory mutants of *Saccharomyces cerevisiae, Mol. Cell. Biol.* **4:**688–694.

Wen, D., and Schlesinger, M. J., 1986, Regulated expression of Sindbis and vesicular stomatitis virus glycoproteins in *Saccharomyces cerevisiae, Proc. Natl. Acad. Sci. USA* **83:**3639–3643.

Wen, D., Ding, M., and Schlesinger, M. J., 1986, Expression of genes encoding vesicular stomatitis and Sindbis virus glycoproteins in yeast leads to formation of disulfide-linked oligomers, *Virology* **153:**150–154.

Wharton, S. A., Skehel, J. J., and Wiley, D. C., 1986, Studies of influenza haemagglutinin-mediated membrane fusion, *Virology* **149:**27–35.

White, J., and Helenius, A., 1980, pH-Dependent fusion between the Semliki Forest virus membranes and liposomes, *Proc. Natl. Acad. Sci. USA* **77:**3273–3277.

White, J. M., and Wilson, I. A., 1987, Anti-peptide antibodies detect steps in a protein conformational change: Low-pH activation of the influenza virus hemagglutinin, *J. Cell Biol.* **105:**2887–2896.

White, J., Kartenbeck, J., and Helenius, A., 1980, Fusion of Semliki Forest virus with the plasma membrane can be induced by low pH, *J. Cell Biol.* **87:**264–272.

White, J., Matlin, K., and Helenius, A., 1981, Cell fusion by Semliki Forest, influenza, and vesicular stomatitis viruses, *J. Cell Biol.* **89:**674–679.

White, J., Helenius, A., and Gething, M.-J., 1982, Hemagglutinin of influenza virus expressed from a clone gene promotes membrane fusion, *Nature (Lond.)* **300:**658–659.

White, J., Kielian, M., and Helenius, A., 1983, Membrane fusion proteins of enveloped animal viruses, *Q. Rev. Biophys.* **16:**151–195.

Wiley, D. C., and Skehel, J. J., 1987, The structure and function of the hemagglutinin membrane glycoprotein of influenza virus, *Annu. Rev. Biochem.* **56:**365–394.

Wilson, I. A., Skehel, J. J., and Wiley, D. C., 1981, Structure for the haemagglutinin membrane glycoprotein of influenza virus at 3 Å resolution, *Nature (Lond.)* **289:**366–373.

Wood, C. R., Boss, M. A., Kenten, J. H., Calvert, J. E., Roberts, N. A., and Emtage, J. S., 1985, The synthesis and *in vivo* assembly of functional antibodies in yeast, *Nature (Lond.)* **314:**446–449.

Yamada, S., and Ohnishi, S.-I., 1986, Vesicular stomatitis virus binds and fuses with phospholipid domain in target cell membranes, *Biochemistry* **25:**3703–3708.

Yewdell, J. W., Gerhard, W., and Bachi, T., 1983, Monoclonal anti-hemagglutinin antibodies detect irreversible antigenic alterations that coincide with the acid activation of influenza virus A/PR/834-mediated hemolysis, *J. Virol.* **48:**239–248.

Zlotnik, H., Fernandez, M. P., Bowers, B., and Cabib, E., 1984, *Saccharomyces cerevisiae* mannoproteins form an external cell wall layer that determines wall porosity, *J. Bacteriol.* **159:** 1018–1026.

*Chapter 12*

# Vaccinia Virus Hemagglutinin

Hisatoshi Shida

## 1. INTRODUCTION

Several viruses agglutinate erythrocytes of various animal species. The components responsible for this phenomenon have been termed hemagglutinins (HA), which in many cases have been identified as glycoproteins present in virus envelopes. The HA titers have often been used as convenient indicators of the amounts of viruses. Recent studies on virus HA dealt not only with their structures and functions in virus replication but with the molecular mechanisms of their biosynthesis as well, applicable in general to membrane glycoproteins.

Vaccinia virus hemagglutinin (VV HA) is placed in a separate category from other virus HA. The VV HA titer can be dissociated from the infectivity titer but is associated with VV-infected cell membrane (Stone and Burnet, 1946; Blackman and Bubel, 1972). This nature historically stimulated several investigators to study the detailed properties and functions of the VV HA. Furthermore, this characteristic that is nonessential for viral replication provided a useful way to investigate its structure, function, and biosynthesis, as it is easy to isolate $HA^-$ mutants from the colorless plaques after staining by an overlay of an erythrocyte suspension, whereas plaques of the $HA^+$ wild type are stained red after this treatment (Oda, 1965; Shida and Matsumoto, 1983; Shida, 1986b). In addition, as VV contains double-stranded DNA consisting of many genes expressed as viral proteins without a splicing event of pre-messenger RNA (mRNA) (Venkatesan *et al.*, 1981; Moss, 1985), it is relatively easy to identify the mutation points by sequencing the

**Hisatoshi Shida** Institute for Virus Research, Kyoto University, Kyoto 606, Japan.

viral genomic DNA cloned. One of the purposes of this chapter is to summarize available information about the structure, synthesis, and function of VV HA.

Recent attention on VV is its usefulness as a eukaryotic cloning and expression vector and application to a new type of vaccine (Mackett *et al.*, 1982; Panicali and Paoletti, 1982). The expansion of knowledge on properties of the HA and its gene enabled us to use the HA gene as an available site for the insertion of foreign genes (Shida, 1986a). The second part of this chapter describes the characteristics of the HA gene as an available site to construct recombinant VV expressing the foreign genes and our trial to make a vaccine against human T cell leukemia virus type I (HTLV-I) (Shida *et al.*, 1987) as an example in which the HA site was used.

## 2. BIOCHEMICAL CHARACTERIZATION OF VV HA AND ITS STRUCTURAL GENE

### 2.1. Chemical Nature and Reconstitution of HA

VV HA was first described by Nagler (1942); subsequently the dissociation of VV HA from infective VV was demonstrated by adsorption onto red cells, centrifugation, or by column chromatography (Stone and Burnet, 1946). Early studies showed VV HA to be of a lipoprotein nature composed of a virus antigen that conferred serological specificity and a lipid responsible for attachment to the red blood cell (RBC) (Chu, 1946; Stone, 1946b; Smith *et al.*, 1973). Lecithinase, snake venom phospholipase, and trypsin destroyed the activity of VV HA (Stone, 1946a).

When partially purified HA proteins, extracted by sonic oscillation, were examined by density-gradient centrifugation, the preparations varied in density according to tissue origin (Chu, 1946). Cell fractionation experiments by a discontinuous sucrose density gradient demonstrated the co-sedimentation of VV HA activity and plasma membrane, assessed by marker enzymes such as $Na^{+}$-$K^{+}$-activated ATPase (Blackman and Bubel, 1972). These results suggested that the VV HA is a complex of host cell and VV-induced components composed of at least protein and lipids.

Although most progeny viruses persist intracellularly, some viruses wrapped with cellular membrane are released into the medium. These extracellular enveloped viruses, designated EEV versus INV (intracellular naked virus), have been shown to display HA activity. The cellular membrane wrapping the virus probably contains HA molecules (Payne, 1979). However, elucidation of the detailed properties of the VV HA had to wait for the adoption of modern methods to extract the HA.

Ichihashi (1977) solubilized $HA^{+}$ VV-infected cells by detergent, deoxycholate, and separated proteins from lipids by column chromatography. Although neither fraction showed a significant titer of HA activity, Ichihashi was able to reconstitute a high titer of HA activity after mixing the both fractions followed by dialysis to remove the detergent. While the lipid fraction could be replaced by purified phosphatidylcholine and phosphatidylethanolamine, the protein fraction could not be replaced by that extracted from $HA^-$ VV-infected or uninfected cells. The protein fraction that was dialyzed without the addition of the extraneous lipids did not exhibit the HA activity. These results clearly demonstrated the lipoprotein nature of the HA described originally and enhanced the importance of the protein moiety specified by the VV for the HA activity.

## 2.2. Identification of HA Protein

An attempt was made to identify the HA protein by a comparison of the $HA^+$ IHD-J VV and $HA^-$ IHD-W strains, variants spontaneously derived from J virus. Weintraub and Dales (1974) detected a glycoprotein of molecular weight 85,000 $M_r$ as the major difference among these virus-induced proteins. Since this glycoprotein is located on the plasma membrane and synthesized according to the time schedule similar to appearance of the HA activity, they tentatively identified this glycoprotein as the HA protein. However, other investigators proposed that different glycoproteins and sugar-free protein were involved in the HA activity (Ichihashi, 1977; Ikuta *et al.*, 1979).

To resolve this inconsistency, Shida and Dales (1981) fractionated the detergent-solubilized membranes prepared from J-infected cells using several column chromatographies and examined the protein composition and the HA activity in each fraction. All the fractions showing the HA activity contained the 85-kDa glycoprotein, and finally it was shown that the 85-kDa glycoprotein purified was alone able to reconstitute the HA activity after the addition of egg lecithin, followed by dialysis. These results conclusively indicated that the 85-kDa glycoprotein was responsible for the HA activity. In addition, all the recombinant VV that harbored foreign genes in the gene encoding the 85-kDa glycoprotein were HA negative (Shida *et al.*, 1987).

## 2.3. Carbohydrate Moiety of the HA Protein

Two types of glycosylation have been reported to occur most frequently in membrane glycoproteins. One is an oligosaccharide chain linked to the Asn residue through a *N*-glycosidic linkage; the other is to the Ser or Thr residue through an *O*-glycosidic linkage (Kornfeld and Kornfeld, 1980). To characterize the carbohydrate moiety of the 85-kDa glycoprotein, tunicamycin was used, which specifically blocks production of *N*-acetylglucosaminepyrophosphoryl dolichol essential for the transfer of N-type oligosaccharide to Asn residues of polypeptides (Kuo and Lampen, 1976; Takatsuki *et al.*, 1975; Tkacz and Lampen, 1975). The molecular weight of the HA protein synthesized in the presence of tunicamycin was reduced to approximately 62 kDa, as measured by sodium dodecyl sulfate–polyacrylamide gel electrophoresis (SDS–PAGE), implying the presence of a *N*-glycosydically linked oligosaccharide (Shida and Matsumoto, 1983). This was further confirmed by endo-β-*N*-acetylglucosaminidase H (Endo H), which cleaves N-linked oligosaccharides of the high mannose type between the two proximal N-acetylglucosamine (GlcNAc) residues (Tarentino and Maley, 1974; Tarentino *et al.*, 1974). Although this enzyme did not digest the oligosaccharide of the mature HA protein, suggesting the presence of a complex type of N-type carbohydrate, it could digest that of the nascent HA with a concomitant increase in the electrophoretic mobility of the HA indicative of decreased molecular weight (Shida and Matsumoto, 1983; Shida 1986b). The number of N-linked oligosaccharide chains attached to the HA protein was estimated by the partial digestion method of the nascent HA with Endo H. Since Endo H releases an 1800-$M_r$ oligomannosyl moiety from one N-linked oligosaccharide chain, it is possible to discern the HA polypeptides possessing a different number of the oligosaccharide chains after partial digestion. The result suggested that at least three N-linked oligosaccharide chains associate with the HA polypeptide (H. Shida, unpublished data).

It was observed that the HA protein synthesized in the presence of tunicamycin was still labeled with [$^3$H]glucosamine, whereas the G protein of the vesicular stomatitis virus (VSV) synthesized under the same condition was not (Shida and Dales, 1981). The G protein has been well established as possessing only the N-linked oligosaccharides (Rothman and Lodish, 1977). The existence of the carbohydrate moiety resistant to the tunicamycin was confirmed by the ability of the underglycosylated HA form to associate with the wheat germ agglutinin-coupled sepharose (Shida and Dales, 1981), which bound to the sialic acid (SA) and GlcNAc residues (Bhavanandan and Katlic, 1979). These results suggested that the HA protein carried O-linked oligosaccharides. This notion was confirmed by the release of low-molecular-weight radioactive compounds following the mild alkaline borohydrate treatment of [$^3$H]glucosamine-labeled HA protein (Shida and Dales, 1981). This treatment released O-linked chains by β-elimination (Spiro, 1966). It was the first report to demonstrate the viral glycoprotein carrying the O-linked oligosaccharides.

### 2.4. Gene Encoding HA Protein

Purification of the HA protein made it possible to elicit monospecific anti-HA antibody (Shida and Dales, 1982). This antibody was used to map the genetic loci encoding the HA protein into the VV genome by hybrid-arrested and hybrid-selected methods. As summarized in Fig. 1 the HA gene was located within the 1.8-kilobase (kb) fragment generated by the digestion with *Hind*III and *Sal*I at the right edge of the fragment A that is derived from the *Hind*III digestion of the whole vaccinia virus genome. The position of the HA gene was confirmed by the production of the complete HA protein in the cell transfected with the plasmid containing this region (Shida, 1986a). Finally, the most precise map locations came from nucleotide sequence analysis.

## 2.5. Primary Structure of the Protein Moiety of HA

Although mRNA of VV have the general structure of eukaryotic mRNAs, such as a 5′-capped structure and 3′ poly A tail, they differ from most nuclear transcripts in that they are not spliced; the initiating nucleotide of the primary transcripts is retained at the 5′-end of mature RNAs contiguous to the body of the coding sequence (Moss, 1985; Venkatesan *et al.*, 1981). Thus, it is possible to deduce the amino acid sequence of the HA protein from the nucleotide sequence of genomic clones. The open reading frame inspected contained a 315-amino acid sequence (Shida, 1986a). The calculated $M_r$ was 35 kDa, which was different from the 58-kDa $M_r$ unglycosylated form of HA determined by SDS–PAGE. The reliability of the nucleotide sequence was ensured by the coincidence of the amino acid composition of the HA protein analyzed with the value calculated and by the fact that the RNA transcribed *in vitro* under the control of the T7 promoter from the cloned HA gene directed the 60-kDa polypeptide in the reticulocyte

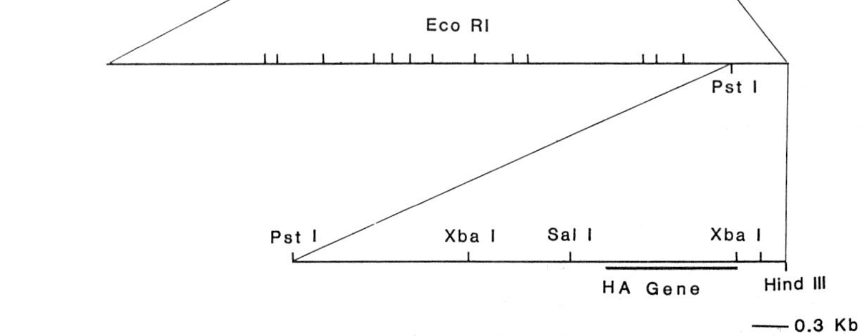


**FIGURE 1.** Position of the HA gene in vaccinia virus genome. (From Shida, 1986a.)

lysate, which possessed antigenicity reactive with the anti-HA antibody (H. Shida unpublished observation). The anomalous electrophoretic migration in SDS–polyacrylamide gel caused an apparent inconsistency. Currently, several examples have been reported that show differences between the calculated and measured molecular weight, such as scrapie PrP glycoprotein (Oesch *et al.*, 1985) and respiratory syncytial virus envelope protein (Satake *et al.*, 1985). The amino acid sequence of the HA predicted from the nucleotide sequence revealed some of the pertinent characteristics of the HA protein and provided a basis for further analysis.

### 2.5.1. Signal Sequence

Many glycoproteins are synthesized by membrane-bound ribosomes of the rough endoplasmic reticulum (RER) (Lodish and Froshauer, 1977; Morison and Lodish, 1975) and are extruded co-translationally into the RER lumen and mediated by the signal sequence at the amino-terminus of the glycoproteins (Blobel and Dobberstein, 1975; Lingappa *et al.*, 1978). The signal sequence is often cleaved off during the processing of the nascent polypeptide (reviewed by Blobel *et al.*, 1979; Kreil, 1981). The signal sequences that have considerable flexibility in length from about 15 to 60 residues, usually have one to six positively charged residues such as Arg and Lys near the amino terminus, have a core of hydrophobic residues more than eight in length, and an increase in hydrophilicity including Thr and Ser residues toward the carboxy-terminal end. The cleavage sites should reside between 4 to 11 residues downstream from the core carboxy-terminus, and cleavage would occur at the sites characterized by the existence of small, neutral residues in positions $-1$ and $-3$ (counting from the cleavage site between positions $-1$ and $+1$) (Perlman and Halvorson, 1983; Heijne, 1984).

On the basis of the analogy with well-characterized secretory and membrane proteins, it seems likely that the amino-terminal domain of the HA protein constitutes a cleavable signal sequence, because the amino-terminus of the deduced amino acid sequence has typical properties for the signal sequence described above (Fig. 2). It has one Arg residue at the third position from the amino-terminus, followed by a stretch of 8 hydrophobic amino acids (Shida, 1986a). The fact that an *in vitro* translated product of the HA mRNA in the absence of a membrane has a 2000-Da larger molecular weight than that of the unglycosylated form synthesized *in vivo,* which should have inserted into the RER, supported the presence of the cleavable signal sequence in the nascent HA polypeptide (Shida and Matsumoto, 1983). The cleavage site was assumed to be at the carboxyl side of the Ala, 16th residue of the amino-terminus, by the rule described above. The amino-terminal 14 residues were enough to function as a signal sequence because they could mediate the translocation of the foreign

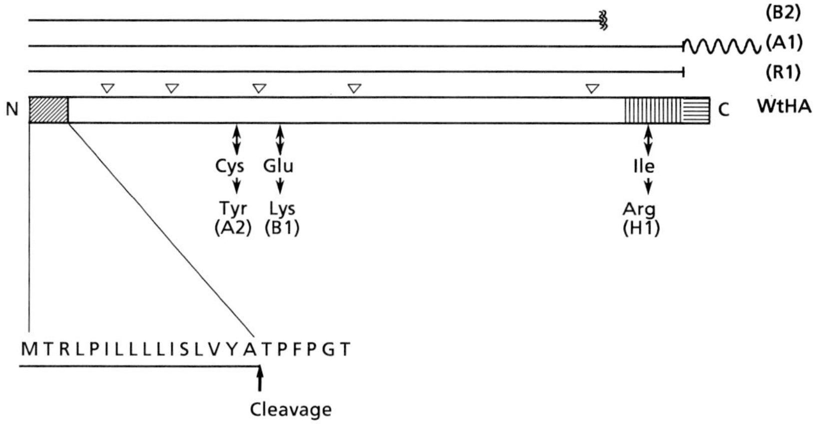


**FIGURE 2.** Diagram illustrating some of the important features of the HA protein based on the nucleotide sequence and mutated regions (Shida and Matsumoto, 1983; Shida, 1986a,b). Hatched bars represent the putative signal sequence, and the amino acid sequence of this region is shown by a one-letter symbol. Triangles indicate the positions of Asn residues that could be associated with N-linked oligosaccharides. Vertical bars represent the position of the transmembrane region. Horizontal bars represent the cytoplasmic domain.

protein linked to the carboxy-terminus of the 14th Val (M. Seki and H. Shida, unpublished observation). This finding is consistent with the facts that the hybrophobicity of both the bacterial and eukaryotic signal sequence play a crucial role in the insertion of membrane proteins (Briggs and Gierasch, 1986; Zerial *et al.*, 1987).

### 2.5.2. Potential Sites for the Addition of Oligosaccharides

N-linked oligosaccharides are added to Asn residues that are a part of the sequence Asn-X-Ser or Asn-X-Thr. X can be almost any other amino acid, except for Pro (Marshall, 1972). There are five potential glycosylation sites in the amino acid sequence deduced from the nucleotide sequence of the HA gene and at least three of them are actually glycosylated.

The HA glycoprotein also contains O-linked oligosaccharicles. Although the very high abundance of Ser and Thr residues (more than 25% of the total amino acids) in VV HA may account for the existence of the O-linked oligosaccharides, general rules have not been established to permit predictions as to which Ser or Thr residues might be used for the addition of O-linked carbohydrates (Shida, 1986a).

### 2.5.3. Transmembrane and Cytoplasmic Domains

Certain integral membrane glycoproteins have been shown to have a hydrophobic domain near the carboxy-terminus that act as an anchor in the lipid bilayer of membrane (designated stop-transfer sequence). The short hydrophilic domain containing the basic amino acid-rich sequence follows. Since this carboxy-terminal domain is oriented in the cytoplasm, it is designated the cytoplasmic domain (Sabatini *et al.*, 1982).

The amino acid sequence of HA deduced from the nucleotide sequence showed the existence of a cluster of 27 hydrophobic amino acids flanked by Lys residues, followed by the basic amino acid-rich sequence (Shida, 1986a). They are typical structures of stop transfer and cytoplasmic domain.

### 2.5.4. Orientation in the Membrane

Membrane glycoproteins are classified into three groups by their orientation with respect to the membrane (reviewed by Garoff, 1985; Wickner and Lodish, 1985). Type I membrane protein is extruded through the lipid bilayer of the hydrophobic stop-transfer sequence leaving the carboxy-terminus to be completed in the cytoplasm. Thus, type I proteins have a N-out/C-in orientation (i.e., amino-terminus exoplasmic, carboxy-terminus cytoplasmic). The G protein of VSV and hemagglutinin of the influenza virus belong to this type (Rose *et al.*, 1980; Min-Jou *et al.*, 1980). Type II membrane proteins have the opposite N-in/C-out orientation. The influenza virus neuraminidase and HLA-DR invariant chain are classified into this type (Fields *et al.*, 1981; Strubin *et al.*, 1984). Type III membrane proteins span the membrane several times. Ion transporters such as erythrocyte band 3 protein and $Na^+$ channel belong to this group (Kopito and Lodish, 1985; Noda *et al.*, 1986).

The positions of the functional domains of the HA protein determined from the amino acid sequence suggested that the HA protein belonged to the type I membrane glycoprotein, having the following sequences: amino-terminal signal sequence, N-type oligosaccharide accepter sequence, stop-transfer sequence, and cytoplasmic domain in the carboxy-terminus (Shida, 1986b). This N-out/C-in orientation was substantiated by experiments using protease. The HA protein was almost totally digested by the protease treatment of intact J virus-infected cells whereas this treatment only reduced the molecular weights of newly synthesized wild-type or mutant HA proteins, which had been inserted into RER, by 2 kDa, 5 kDa, or not at all. Sequence analysis indicated that these mutants have different sequences at their carboxy-termini corresponding to the amino acid lengths deleted by the digestion with the protease (Shida, 1986b). Since the portion of HA protein that had been inserted into the luminal side of the RER should be protected from the protease digestion by RER membrane, these results suggest that their carboxy-termini reside in the cytoplasm.

## 3. SYNTHESIS AND PROCESSING OF HA PROTEIN

### 3.1. Kinetics of Polypeptide Synthesis

VV expresses its genes in a regulated temporal fashion, controlled primarily at the transcriptional level (Moss, 1985). The early genes are transcribed from viral cores that are rapidly found in the cytoplasm soon after the adsorption and uncoating of infecting virions. Most early RNA synthesized in the presence of an inhibitor of protein synthesis are designated an immediate early class. Some residual early RNAs, a class designated delayed early, are produced before DNA replication. The late class transcripts begin to appear after replication of the parental virus DNA commences. Recent analyses showed at least two late gene temporal subsets (Pennington, 1974; Ichihashi *et al.*, 1971).

The time course of the appearance of HA was followed by titrating its biological activity and detecting the HA polypeptide by SDS–PAGE (Ichihashi *et al.*, 1971; Payne, 1979). The HA activity could be detected initially 8 hr postinfection, increasing exponentially in titer and reaching a maximum at 18 hr. The HA polypeptides were synthesized according to similar kinetics. Administration of an inhibitor of DNA replication before viral DNA synthesis completely abolished HA formation, but not after viral DNA synthesis. These results indicated that the HA is classified into the late function. Considering its very late appearance, some investigators proposed it to be classified into a late-late category. Although analyses that could directly estimate the amount of transcript have not been done, both the inhibitors of RNA and protein synthesis inhibited HA synthesis similarly depending on their administration time, suggesting that the HA synthesis was regulated mainly at the transcriptional level (Ichihashi *et al.*, 1971).

### 3.2. Promoter of HA Gene

It has been shown that the temporal regulation of gene expression is performed by the 5′ region upstream of the coding sequence (Bertholet *et al.*, 1985; Weir and Moss, 1987). When the DNA segments upstream of both early and late genes were ligated to the prokaryotic chloramphenycol acetyltransferase (CAT) gene and the chimeric genes were placed in a different portion of the VV genome, the expression mode of CAT depended solely on the DNA segments upstream of the genes (Cochran *et al.*, 1985; Weir and Moss, 1987). Comparison of the nucleotide sequences upstream of early and late genes showed that promoters of VV lacked a typical eukaryotic Hogness–Goldberg box, but instead have characteristic consensus sequences for its early or late genes, respectively (Weir and Moss, 1983, Rosel *et al.*, 1986). These characteristic signals can be recognized by the RNA polymerase of VVs but not by the cellular RNA polymerase II (Puckett and Moss, 1983).

Mars and Beaud (1987) and Vassef (1987) analyzed the structures of a number of early promoter regions with the aid of a computer; they proposed the consensus sequence of early promoters, as shown in Fig. 3. Most of the information required to locate an early RNA start site is contained within the upstream 28 nucleotide sequence. There are at least two blocks of information. The first block is found between positions −27 and −12 upstream of the RNA start site, generally composed of runs of A. Within the five nucleotides upstream of the RNA start sites is the second block, which is relatively rich in pyrimidines.

Comparison of the late promoter regions demonstrated a consensus sequence of only the TAAATG/A motif, including a codon for the initiation methionine (Rosel *et al.*, 1986; Hanggi *et al.*, 1986). Studies using site-directed mutagenesis of the putative late promoter regions indicated that clusters of A or T located 5–20 nucleotide upstream of the consensus sequence are also essential for promoter activity (Weir and Moss, 1987) (Fig. 3), although other consensus sequences have not been identified.

The region upstream of the HA coding sequence resembles over all the putative promoter regions of late genes rather than early genes, including clusters of A and T, although it contains TAATATG instead of the typical TAAATG motif (Fig. 3). The gene encoding P4b, a major late protein, has been also reported to lack the consensus sequence (Rosel and Moss, 1985). Furthermore, the foreign gene placed downstream of this sequence was expressed at the late time of the VV infection cycle (T. Tochikura and H. Shida, unpublished observation). Thus, it may be concluded that this region upstream of the HA coding sequence represents the promoter regulating the expression of the HA gene.

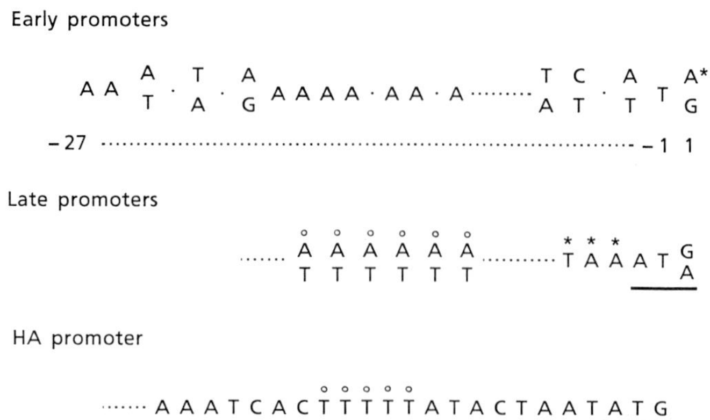


**FIGURE 3.** Consensus sequences proposed for vaccinia virus promoters and the sequence of the HA promoter region. The figure was reconstituted from the results published by Mars and Beaud (1987), Hanggi *et al.* (1986), Rosel *et al.* (1986), and Shida (1986a). Asterisk (*), start sites of transcription; °s, cluster of adenine or thymine residues located in the late promoter region; underlines, the codon for the initiation methionine.

### 3.3. Oligosaccharide Processing

Since the HA protein is a typical membrane glycoprotein, its oligosaccharides should follow general processing ways (Kornfeld *et al.*, 1978). The glycoproteins extruded into the RER lumen are exported to the Golgi apparatus and then to the plasma membrane (Blobel *et al.*, 1979; Kreil, 1981). During import into the RER, a high mannose oligosaccharide is transferred *en bloc* to certain asparagine residues of nascent glycoproteins via the *N*-glycosidic bond (Katz *et al.*, 1977; Rothman and Lodish, 1977). When the glycoproteins reach the medial cisternae of the Golgi apparatus through Cis cisternae, the some mannose residues are trimmed and the GlcNAc residues associate (Dunphy *et al.*, 1985), after which sialic acids (SA) and galactoses (Gal) attach to GlcNAc residues of the glycoprotein in Trans cisternae to complete the *N*-glycan moiety (Roth and Berger, 1982; Roth *et al.*, 1985) (Fig. 4).

Viral glycoproteins, such as the G protein of VSV, were used extensively to establish this scheme, on the assumption that the glycosylation of the viral glycoprotein is performed by cellular machinery. The system using the viral glycoprotein has a couple of advantages. First, the virus infection leads the cellular machinery to synthesize only the viral protein so as to simplify the experimental result. Second, the virus synchronously induces large amounts of viral glycoproteins.

Consistent with the scheme of general N-linked oligosaccharide biosynthesis, are the following observations on the HA glycoprotein. The translation products made *in vitro* acquire a higher molecular weight in the presence of microsomes than in their absence (H. Shida, unpublished observation). Even the newly synthesized form of the HA protein labeled for 5 min with [$^3$H]leucine possessed the high-mannose oligosaccharides sensitive to Endo H and a subsequent chase converted the HA-associated oligosaccharides to Endo H resistant, implying the generation of complex-type oligosaccharide. All the mutant HAs that accumulate in the RER harbored the high-mannose oligosaccharides (Shida and Matsumoto, 1983).

In spite of the established biosynthetic pathway of N-type oligosaccharide, that of the O-linked oligosaccharide was less clear, although initiation had been suggested by the addition of GalNAc residues to Ser or Thr residues, followed by the sequential addition of sugars (e.g., Gal, GlcNAc, Fuc, and SA) (reviewed by Kornfeld and Kornfeld, 1980). Whereas the glycosyltransferases involved in the biosynthesis of O-linked oligosaccharide were purified from the smooth microsome fractions including Golgi apparatus (Hanover *et al.*, 1980), a couple of reports suggested that the transfer of initial sugars, GalNAc, occurred in the RER co-translationally (Jokinen *et al.*, 1981; Strous, 1981).

The VV HA protein that possesses both O- and N-linked carbohydrates provided a system suitable for examining the biosynthesis of O-linked oligosaccharides, as VV inhibits host protein synthesis severely, as well as inducing

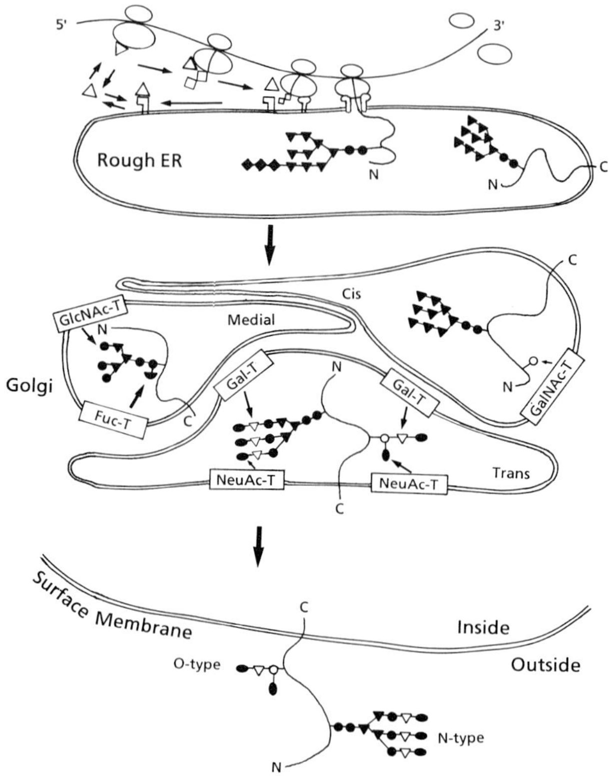


**FIGURE 4.** Biosynthetic pathway of membrane glycoproteins. (Based on Blobel, 1980; Blobel *et al.*, 1979; Kornfeld *et al.*, 1978; Kornfeld and Kornfeld, 1980; Sabatini *et al.*, 1982; and Walter *et al.*, 1984). △, signal-recognition particle; ◇◇, signal sequence; ⊓, docking protein (SRP receptor); ⚲, ribosome receptor; ⊓⊓, pore structure for translocation; ●, *N*-acetylglucosamine; ▼, mannose; ◆, glucose; ▽, galactose; ⬬, sialic (*N*-acetylneuraminic) acid; ○, *N*-acetylgalactosamine; ◗, fucose; GalNAc-T, *N*-acetylgalactosamine transferase; GlcNAc-T, *N*-acetylglucosamine transferase; Fuc-T, fucose transferase; Gal-T, galactose transferase; NeuAc-T, *N*-acetylneuraminic acid transferase.

relatively large amounts of HA protein and, in addition, mutants defective in the transport of the HA out of the ER were available (Shida and Matsumoto, 1983).

Two lines of evidence supported that O-linked oligosaccharide chains of the HA attached to the protein moiety in the Golgi apparatus exclusively. First, the kinetic analysis indicated that the high mannose type conversion to complex-type N-type oligosaccharide was concomitant with the association of the carbohydrate with the HA protein moiety in the presence of tunicamycin. Second, the mutant HAs accumulating in the RER did not incorporate any radioactivity from [$^3$H]glucosamine in the presence of tunicamycin (Shida and Matsumoto, 1983).

It was reported that glucosamine added exogenously could be efficiently converted to the galactosamine in cells so as to be incorporated into glycoproteins as a GalNAc residue (Rossignol *et al.*, 1969; Cummings *et al.*, 1983).

Similar results were obtained for the kinetics of the O-glycosylation of human chorionic gonadotropin (hCG) (Hanover *et al.*, 1982) and herpesvirus glycoprotein (Johnson and Spear, 1983). Subsequently, cytochemical studies at the electron microscope level indicated that terminal GalNAc were located in *cis*- and *trans*-Golgi cisternaes but not in RER (Roth, 1984). An analysis of carbohydrates associated with the precursor form of the low density lipoprotein receptor suggested that the addition of GalNAc residues to Ser/Thr residues preceded the conversion of N-linked high-mannose-type oligosaccharides to complex-type structures (Cummings, 1983). In addition, it was reported that monensin, a carboxylic ionophore, which inhibit newly synthesized proteins passing through the Golgi cisternaes, prevents terminal glycosylation of the N- and O-linked oligosaccharides of HLA-DR-associated invariant chain (Machamer and Cresswell, 1984). On the whole, the results strongly suggested that O-glycosylation generally consisted of a following stepwise process: the initial GalNAc residue attached to the Ser or Thr residues in the *cis*-Golgi followed by the sequential addition of terminal sugars such as Gal and SA residues when the glycoproteins reached *trans*-Golgi, where the appropriate glycosyltransferases resided (Fig. 4).

## 3.4. Structural Requirement for Intracellular Transport

As all glycoproteins are synthesized at RER-bound polyribosomes and then transported to various compartments where they function (e.g., lysosome, Golgi apparatus) (Harris *et al.*, 1978; Neufeld *et al.*, 1975, Palade, 1975), the information directing their final destination (sorting signal) should be included in their structures (Blobel, 1980). Recently, Rothman and co-workers indicated that the tripeptide comprising the acceptor of N-linked carbohydrates can be secreted from the ER very rapidly. This result could be interpreted that since the tripeptide is too small to carry the sorting signal, intracellular transport to the cell surface may not require any positive signals but the signal-mediated diversions from the bulk flow to the cell surface determine the final location of glycoproteins (Wieland *et al.*, 1987; Pfeffer and Rothman *et al.*, 1987).

However, many membrane glycoproteins located at the cell surface have common structures such as the transmembrane domain consisting of a cluster of hydrophobic amino acids followed by a cytoplasmic domain including basic amino acid-rich sequences. Thus, membrane glycoproteins should require some structural fitness for efficient transport. Structural requirements for the transport were examined by using glycosylation inhibitors and various mutants which have defects in intracellular transport (Rose and Bergman, 1982, 1983; Doyle *et al.*, 1985; Adams and Rose, 1985; Hardwick *et al.*, 1986; Davis and Hunter, 1987).

To examine the involvement of *N*-glycosydicadally linked oligosaccharides, tunicamycin was used to synthesize this type of carbohydrate-free HA protein. The HA protein synthesized in the presence of the drug was normally inserted into the RER with concomitant cleavage of the signal sequence and then transported to the cell surface, although it was somewhat inefficient (Shida and Matsumoto, 1983). Thus, *N*-glycosylation is not a prerequisite for intracellular transport of the HA protein. Involvement of the O-type carbohydrates has not been examined.

Tunicamycin impaired, however, exit from the RER of some glycoproteins examined thus far. In these cases, unglycosylated proteins aggregated and decreased their solubilities, probably because of structural alterations (Leavitt *et al.*, 1977; Gibson *et al.*, 1979). Thus, it is accepted generally that each glycoprotein has a different oligosaccharide requirement for proper polypeptide folding and solubility, which are essential for intracellular transport.

Examination of a number of HA$^-$ mutants isolated may shed some light into the structural requirement of the protein moiety for the intracellular transport. These mutants were classified into four groups with regard to their locations (see Table I). The first group contained the HA protein located at the plasma membrane. B1 HA, a member of this group, was transported as efficiently as the wild-type HA, while another member, A2 HA, was done inefficiently; approximately one half of the HA molecules were located at the cell surface, and the

**Table I**
**HA$^-$ Mutants and HA$^+$ Revertants Isolated from a Vaccinia Virus[a]**

| Group | Virus | HA activity | Cell fusion | HA protein: Location | HA protein: $M_r$ (kDa) |
|---|---|---|---|---|---|
| Wild type | IHD-J | + | − | CS | 85 |
| — | IHD-W[b] | − | ++ | SER | 41 |
| I | B1 | − | − | CS | 85 |
| | A2 | − | ++ | CS + NE | 85 + 67 |
| II | B2 | − | ++ | Med + RER | 75 + 59 |
| | G1 | − | ++ | Med + RER | 45 + 31 |
| III | A1, A3, C3, F2, F3, F4, G2 | − | + | NE + RER | 68 |
| IV | H1 | ND | ND | RER + Golgi | 75 + 67 |
| RIII-1[c] | R11, R51 | + | − | CS | 85 |
| RIII-2[c] | R1, R2, R21 | + | − | CS + RER | 85 + 66 |

[a]$M_r$, molecular weight; kDa, kilodalton; CS, cell-surface membrane; SER, smooth endoplasmic reticulum; NE, nuclear envelope; Med, medium; RER, rough endoplasmic reticulum; ND, not determined.
[b]A spontaneous mutant derived from IHD-J (Ichihashi and Dales, 1971; Shida and Dales, 1982).
[c]Revertants isolated from A1 mutant (Shida, 1986; Shida and Matsumoto, 1983).

remainder stayed in the RER. The A2 mutant HA has a Tyr residue in place of the Cys residue, which is located at the central region of the ectoplasmic domain of the wild-type HA, leading to a comparably gross conformational change, as suggested by the fact that this one point mutation affected the two biological functions simultaneously. Thus, proper structure and conformation of the ecto-domain are important for the HA to be transported efficiently.

The second group contained in the HA proteins are secreted into medium but not retained at the cell surface (unpublished data). Nevertheless, a fraction remained in the RER. When the kinetics of intracellular transport of the B2 mutant HA were explored by pulse-chase experiments, the conversion of its high-mannose-type oligosaccharides to the complex type was slower than that of the wild type. The molecular mass of the B2 HA was ~7000 kDa smaller than the wild type. Although sequence analysis had not been performed, it seemed to retain the amino-terminal half but lose the transmembrane and cytoplasmic domain, because it was glycosylated similarly to the wild type. Another mutant G1 HA has a smaller molecular mass but essentially the same properties as B2. These results suggest that the transmembrane and/or cytoplasmic domain may be required for efficient transport.

The third group consisted of seven mutants that showed the same phenotype; their HA proteins accumulated in the RER and nuclear envelope, judged by immunofluorescence microscopy and the co-distribution pattern with fluorescene-conjugated Con A, a marker of the RER. All these mutants have been shown to have exactly the same mutations at the boundary between the transmembrane and cytoplasmic domain, resulting in a completely different cytoplasmic domain from the wild type (Shida and Matsumoto, 1983; Shida, 1986b). Two kinds of revertants were isolated from one of these mutants. One was a true revertant showing exactly the same phenotype and nucleotide sequence as the wild type. Another type of revertant showed inefficient transport from the RER to the Golgi, but that from the Golgi to the plasma membrane seemed efficient, similar to the wild-type HA. A sequencing analysis showed that these revertants lack their cytoplasmic domain completely (Shida, 1986b). These results suggest that the cytoplasmic portion of the HA protein may function in transport out of the RER to Golgi.

The fourth group consisted of one mutant designated H1. This mutant HA was located in the RER and early Golgi, but did not reach *trans*-Golgi, as judged by the extent of N-linked carbohydrate processing. Sequencing analysis demonstrated that the Ile residue located at the center of the transmembrane domain was replaced by an Arg residue (M. Seki and H. Shida, unpublished observation), suggesting the importance of the transmembrane domain in transport. Similar results have been reported recently (Adams and Rose, 1985; Davis and Hunter, 1987).

Figure 2 illustrates the structure of the HA protein and mutated regions.

Many mutants described above that have complete or partial defects in exiting from the ER contained mutated regions close to or within the transmembrane domain, although these mutants were isolated by the HA$^-$ phenotype and not by a transport defect. Thus, the correct structure of the region closely connected with the membrane seemed to be required for the efficient transport of the HA out of the RER to the Golgi.

In analogy with influenza HA, and immunoglobulin (Ig), the VV HA protein may associate with heavy-chain-binding protein (BiP) in the nascent state just after insertion into the RER luminal side (Gething *et al.*, 1986; Bole *et al.*, 1986). BiP has been reported to associate with many proteins that are improperly folded, such as unassembled Ig, causing a retention of proteins in the ER (Dorner *et al.*, 1987). When glycosylation and protein folding occur properly, the association with BiP is transient and the secretory pathway to the Golgi apparatus is traversed. The trimerization of influenza HA and VSV G protein, as a prerequisite for export from the ER, is a matter of controversey (Gething *et al.*, 1986; Yewdell *et al.*, 1988). It is possible that the mutations in the VV HA protein affected their conformations to form a stable complex with BiP or lead to delayed dissociation from the BiP, as the case of the A2 HA. However, it should also be considered that the cytoplasmic domain may function in efficient transport through the interaction with the components in the cytoplasm. These speculation, of course, should be ascertained by empirical methods. Clearly, correct structure and conformation are necessary for membrane glycoproteins to be transported efficiently.

## 4. ACTIVITIES AND ROLES OF HA PROTEIN

### 4.1. Hemadsorption Activity

Hemagglutination is a phenomenon based on the aggregation of RBC bridged by polyvalent substances having an affinity with the cell. The requirement of glycosylation for hemadsorption activity of VV HA was examined by using tunicamycin and glycosidases. Whereas tunicamycin did not affect the hemadsorption activity of VV J virus-infected cells, treatment of these cells with a glycosidase mixture capable of digesting both N- and O-linked carbohydrates abolished the biological activity. These results imply that O-linked chains rather than N-linked varieties participate in the expression of the HA activity (Shida and Dales, 1981).

The role of the protein moiety in the HA activity was examined using HA$^-$ mutants. Two mutants belonging to group I seemed to be glycosylated normally but contained a one-point mutation in the centers of their ectoplasmic domains (see Fig. 2). Since B1 HA protein can be transported normally and still retain cell fusion inhibition activity, another biological activity of the HA protein described in the Section 4.2, i.e., conformation of the B1 HA, may not be disordered extensively, suggesting that the Glu residue at 121 position may participate to constitute an active center for hemadsorption activity.

## 4.2. Inhibition of Cell Fusion

Spontaneous $HA^-$ variants of VV have been observed to be correlated with the capacity to cause cell–cell fusion or polykaryocytosis identifying these mutants as fusion positive ($F^+$). In dual infections with $HA^+$ and $HA^-$ VV, the $HA^+$ phenotype is dominant over the $F^+$ phenotype, in line with the view that the presence of HA and polykaryocytosis are mutually exclusive (Ichihashi and Dales, 1971).

Analyses using a number of $HA^-$ mutants (Table I) proved that the HA protein itself is involved in the inhibition of cell fusion. All the $HA^-$ mutants that lacked the HA molecules on the cell-surface membrane induced polykaryocytosis. By contrast, none of the $HA^+$ revertants isolated from one of the HA mutants induced cell fusion.

B1 HA lost its hemagglutination activity but still retained fusion-inhibition activity, suggesting that these active sites are different in the HA molecule. By contrast, the A2 HA lost both the hemagglutination and fusion-inhibition activity and was transported inefficiently, suggesting that gross conformational change is caused by the single-point mutation of Cys residue at the 103 position. Since VV HA does not form either homo- or hetero-oligomers through disulfide bonds, this Cys residue may play an important role in these functions by maintaining the proper tertiary structure of the HA molecule by forming intramolecular disulfide bonds.

Two groups have reported the identification of components involved in the syncytiogenesis by using the antibodies that inhibit cell fusion (Stern and Dales, 1976; Rodriguez *et al.*, 1987). One is the surface tubular element (STE) of virions and another component is a 14-kDa glycoprotein present on surfaces of both virions and infected cells. Although the contradiction between these reports has not been resolved, both antibodies used inhibited polykaryocytosis and neutralized virus infectivity, suggesting that these components play important roles in viral infection. It has been reported that the antibody reacting with the 14-kDa glycoprotein inhibited the viral penetration but not attachment to the cell surface (Rodriguez *et al.*, 1987).

Since the HA protein inhibited the polykaryocyte formation, it profoundly influenced the mechanism for spreading of the progeny viruses. This implies that cell–cell dissemination with the $HA^-$ strain takes place in cell culture despite the presence of neutralizing antibody in the overlay medium (Dales *et al.*, 1976). Since the 14-kDa glycoprotein, a fusogenic factor, seems to function in viral penetration mediated by the fusion of the viral envelope with the cell membrane (Rodriguez *et al.*, 1987), it is possible to speculate that the HA protein located on the cellular membrane wrapping the VV virion (EEV) may influence the route of EEV entry into the cell; EEV with HA protein may be infected by endocytosis followed by the fusion of membranes in endosome as described in the case of SFV (Helenius *et al.*, 1980; Lenard and Miller, 1982).

### 4.3. Effect on Virus Propagation

Although the HA protein may influence the spread and penetration modes of VV, $HA^-$ mutants can propagate in cell culture as well as the $HA^+$ wild type (Ichihashi and Dales, 1971). Plaque sizes and temperature sensitivities of $HA^-$ mutants were also similar to the wild type. This may be attributed to the multiple pathways of VV entry into host cells, as demonstrated by the fact that epithelial growth factor receptor, one of the receptors for VV, is not the only route for VV infection (Epstein *et al.*, 1985; Stroobant *et al.*, 1985). Although the HA trait evidently does not affect the propagation of VV in cell culture, it influences the replication of VV in bodies in some cases. An early report showed that $HA^-$ mutants could be enriched in Ehrlich ascites tumor cells cultured in mouse ascite fluid, probably because of the preferential propagation of $HA^-$ mutants, whereas $HA^+$ wild-type virus replicated dominantly over $HA^-$ mutants in the chorioallantoic fluid of eggs (Cassel, 1957).

The virulence and propagation of $HA^-$ variants of various VV strains in mice and rabbits were compared with each $HA^+$ wild type. Although inactivation of the HA gene led to poorer propagation of all strains tested in skin, attenuation effects on the propagation in other organs including brain, liver, and spleen were different among VV strains. Lister strain and its derivatives used as vaccines against smallpox were affected less than the WR strain, a more virulent strain, which propagated better in brain (Shida *et al.*, 1988; Flexner *et al.*, 1987). These results suggest that involvement of the HA in VV propagation may depend on the organs infected, implying further that each VV may use different modes of viral entry and dissemination in each organ.

## 5. HA GENE AS A SITE TO RECEIVE FOREIGN GENES

Pertinent to this discussion is the character of the HA gene as a site to insert foreign genes. As background information, the general characteristics of the VV expression vectors are described. Finally, the use of the VV vector (VVV) to examine the molecular mechanism of gene expression in human T-cell leukemia virus type I (HTLV-I) and construct a potential new vaccine against HTLV-I will be presented as examples of VVV applications.

### 5.1. Vaccinia Virus Vector

The VV vector, as a method of introducing genes into eukaryotic cells, has several advantages over small DNA virus vectors, although the genetic engineering of VV is more difficult. VVV has the potential of retaining complete infectivity in a broad range of cell types, has a greater capacity for foreign DNA, and expresses foreign genes very efficiently (reviewed by Mackett *et al.*, 1985; Mackett and Smith, 1986). The inability of mRNA splicing in VV-infected cells (Venkatesan *et al.*, 1981) may be advantageous in some cases to examine the mechanism of gene expression. In addition, the fact that VVV can express foreign genes in significant amounts in various animals because it can propagate to some extent in their bodies provides a means of investigating immunological responses to one component of pathogenic agents and indicates the potential of the VVV as new live vaccines. Many of the features of VV that were used successfully as vaccines for the eventual global eradication of smallpox seem to be retained by recombinant VV. The vaccine is economical to manufacture, easy to administer, is stable without refrigeration, is potent as a single inoculation, and stimulates both antibody and cell-mediated immune responses. A disadvantage was the rare occurrence of side effects following vaccination (reviewed by Arita and Fenner, 1985). The VV genome, 187 base pairs (kb) encodes more than 150 polypeptides, including the enzymes responsible for RNA synthesis, post-transcriptional modification, and DNA replication that enable the virus to propagate entirely in the cytoplasm of infected cells (Moss, 1985). A variable amount of DNA can be packaged into virus particles (Smith and Moss, 1983). The biological properties of VV must be taken into account when developing it as expression vectors.

The large size of the VV genome makes it difficult to construct recombinant genomes *in vitro*. Furthermore, since the VV RNA polymerase recognizes its own unique regulatory sequences (Puckett and Moss, 1983; Rohrmann and Moss, 1985; Foglesong, 1985; Golini and Kates, 1985), it is necessary to use VV promoters for the efficient expression of foreign genes in VV. Thus, a plasmid is, at first, constructed that contains the foreign gene placed downstream of the VV promoter element and flanked by segments of the VV gene. This chimeric gene is then inserted into the VV genome by homologous recombination between the virus and plasmid DNA in VV-infected cells (Fig. 5).

The chimeric gene should be constructed to produce the authentic foreign protein by the recombinant VV. To avoid potential problems arising from improper codon phasing or the generation of fusion proteins, it is better to use the

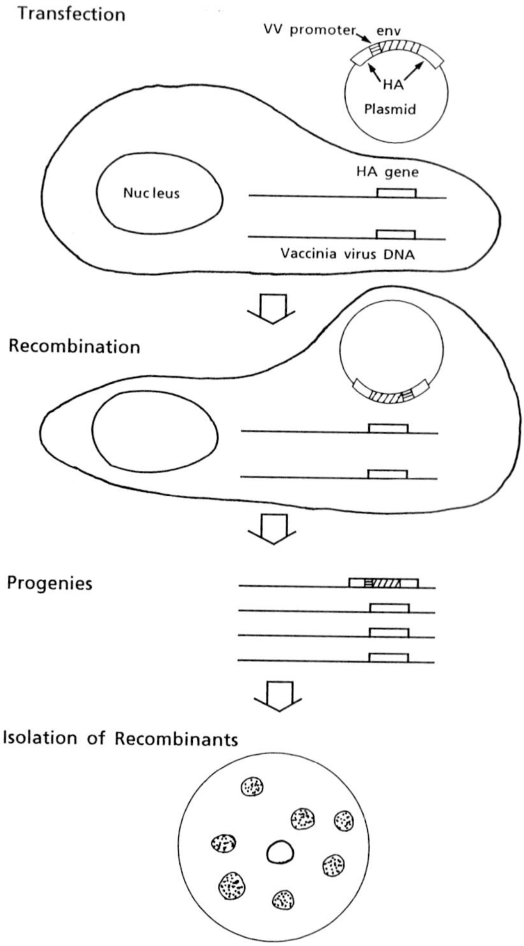


**FIGURE 5.** Schematic presentation for generation and isolation of vaccinia virus recombinants containing a foreign gene in the middle of the HA gene. The plasmid that harbors a foreign gene placed downstream of the VV promoter and flanked by the segment of the VV HA gene is transfected into the cells infected with VV. Homologous recombination takes place between the HA-gene sequences of the plasmid and the virus genome, resulting in the insertion of the foreign gene into virus genome. The recombinant VV produced is isolated from the total progeny viruses based on the color difference upon staining by overlay of the solution containing red blood cells.

VV-promoter element that retains the transcriptional start site but then removes the translational start site. One well-characterized promoter, p7.5, was designated for this purpose (Cochran *et al.*, 1985). To achieve higher levels of expression of the foreign genes, the promoters of late genes encoding the most abundant virus proteins may be used (Wittek *et al.*, 1984; Funahashi *et al.*, 1988). Although their translation initiation codons are usually located very close to the RNA start sites (Rosel *et al.*, 1986; Hanggi *et al.*, 1986), the mutation of ATG triplet to ATA, which constitutes the initiation codon, does not seem to affect its promoter activity significantly, as demonstrated in the case of the promoter of the gene encoding the A-type inclusion body (S. Funahashi and H. Shida, unpublished observation), and allows the synthesis of authentic foreign proteins. One exceptional case may be the type I glycoprotein. Co-translational cleavage of a signal sequence can occur even if it has an extra sequence at its amino-terminus, thus resulting finally in authentic foreign proteins (Rottier *et al.*, 1987; unpublished observations). The efficiency of cleavage and the properties of the amino-terminal extra sequence have to be checked in each case.

Since the virus DNA that flanks the chimeric gene determines the site at which the foreign gene is inserted, the foreign gene must be inserted into a nonessential region of the VV genome. Although several nonessential regions have been reported (Perkus *et al.*, 1985; Perkus *et al.*, 1986), the one accompanying the selection marker distinguishing between the recombinant and wild-type VV is much more useful, because with the use of standard procedures with wild-type VV, at best only 0.1% of total progeny viruses are recombinants. Many recombinant VV expressing foreign genes that are inserted into the middle of thymidine kinase (TK) gene are constructed because the recombinant can be selected on the basis of the $TK^-$ phenotype in the presence of 5-bromodeoxyuridine (BUdR) (Mackett *et al.*, 1982). Another nonessential region accompanying the selection marker is the HA gene.

## 5.2. HA Site to Insert the Foreign Genes

The method for isolating $HA^-$ mutants among the $HA^+$ wild-type viruses by different plaque colors (Oda, 1965; Shida and Matsumoto, 1983) can be applied in selecting the recombinant virus harboring foreign genes within the body of the HA gene, as insertion of the foreign gene inactivates the HA gene. As the difference in colors of $HA^-$ plaques from the $HA^+$ varieties upon staining by chicken erythrocytes is striking, it is possible to find one $HA^-$ plaque existing among more than 1000 $HA^+$ plaques in one petri dish. Our experiences demonstrated that most of the $HA^-$ plaques are actually recombinant viruses, while the others are spontaneous $HA^-$ mutants. The advantage of this screening marker is that recombinants can be found in all the cell lines that support the growth of VV, in contrast to the necessity of using the $TK^-$ cell line to select $TK^-$ recombinant VV (Mackett *et al.*, 1982). More important points to be considered when using the HA site to insert the foreign genes are the expression efficiency of the foreign genes inserted and the alteration of the properties of VV led by the destruction of the HA gene.

Although the expression modes of VV genes have been demonstrated to be determined by the 5′ region upstream of the coding sequences (reviewed by Moss, 1985), it has been reported that the position where the promoter element was translocated affected the expression efficacy of the gene (Hanggi *et al.*,

1986). Regulatory elements, such as enhancer and silencer, may be conceivable, albeit not yet reported. Comparison of the recombinant VV that harbor the envelope gene of bovine leukemia virus (BLV) placed downstream of the p7.5 promoter in the middle of the TK and HA genes was performed. The immunoprecipitation analysis indicated that the BLV envelope protein was synthesized 3- to 10-fold more efficiently from the env gene flanked by the segments of the HA gene than those of the TK gene, whereas HA and TK recombinant VV propagated similarly (M. Morita and K. Sekikawa, personal communication). These results suggested that the HA gene supported more efficient gene expression when directed by the p7.5 promoter than by the TK gene.

Both the inactivation of the TK and HA gene led to the attenuation of the VV WR strain (Buller *et al.*, 1985; Flexner *et al.*, 1987), which has been used extensively in molecular biological studies but has never been licensed for use as a vaccine against smallpox. However, the inactivation of the HA gene attenuated vaccine strains, Lister and LC16mO, occurs to a much lesser degree. These strains have been injected into a number of people. In particular, the LC16mO has been shown to have a very low neurovirulence but to be able to propagate relatively well in skin (Hashizume *et al.*, 1985). Although the effect of the TK gene destruction on these strains has not been studied in detail, comparative studies of the recombinant VV expressing the BLV envelope gene, which were derived from the vaccine strains, revealed that the $HA^-$ recombinants elicited neutralizing antibodies against BLV, while the $TK^-$ varieties failed to do so and that the former induced higher titers of anti-VV antibodies than the latter (M. Morita and K. Sekikawa, personal communication). These results suggested that the lower attenuation caused by the inactivation of the HA gene than that of the TK gene and the more efficient expression of the envelope gene from the middle of the HA gene worked cooperatively, resulting in a marked difference in the induction of the neutralizing antibodies.

The facts mentioned above may be important in choosing the proper site in the VV gene for inserting foreign genes. Although several recombinant VV based on the WR strain have successfully protected vaccinated animals from challenge with pathogenic viruses, recombinant VV derived from the vaccine strain should be safer in the application of the vaccine to human beings. However, a recombinant virus based on the attenuated VV strain may not be sufficiently antigenic to elicit protective immunity, and in some cases, it would not be desirable to involve gene disruption leading to further attenuation. Other cases may require further attenuation to reduce the rare adverse complications. The differences between the HA and the TK sites concerning the effects on the gene expression and attenuation of VV may provide one standard for choosing the site in each case. Discovery and characterization of the more nonessential regions for VV growth are necessary to construct a more effective and safer vaccine.

### 5.3. Examples of Use of Vaccinia Virus Vector

Recently, the VV vector has been widely used to investigate the molecular mechanism of gene expression and protein localization in cell culture as well as to make new types of vaccines (reviewed by Mackett and Smith, 1986). Its broad host range and ability for efficient expression of foreign genes has been demonstrated to be useful in examining the polarized transport of various viral glycoproteins to the apical or basolateral surface membrane, since the viruses that contain these glycoproteins can grow in limited host cells which are not suitable to be studied in detail (McQueen *et al.*, 1986; Stephens and Compans, 1986; Stephens *et al.*, 1986). One remarkable result obtained using VVV in the field of immunology is identification of the components that elicit cytotoxic T cells playing the most major role in recovering from viral diseases. The inner viral proteins have been shown to be the major components in many viruses, in contrast to the surface glycoproteins that play a major role in inducing humoral antibodies (Yewdell *et al.*, 1985; 1986; Bennink *et al.*, 1987).

It is very important to examine the properties of the human retroviruses, which cause fatal diseases in human beings, as great difficulties are being encountered in developing vaccines and drugs using conventional methods (Newmark, 1986; Ezzell, 1987). A more detailed understanding of these viruses at molecular, and cellular levels should promote the development of treatments to control these diseases. The VVV has been expected to offer an attractive means for developing the vaccine against viruses and to examine the molecular mechanism of their gene expression.

The following sections describe our use of VVV to investigate the gene expression and development of a vaccine to human T-cell leukemia virus type I (HTLV-I) that is an etiological agent of adult T-cell leukemia (ATL) (Hinuma *et al.*, 1981; Kalyanaraman *et al.*, 1982; Yoshida *et al.*, 1984).

#### 5.3.1. Location of pX Proteins

First, I will describe the properties of HTLV-I briefly. Detailed descriptions of HTLV-I are given elsewhere (Yamamoto and Hinuma, 1985; Yoshida, 1987). As HTLV-I belongs to the retrovirus group, its RNA genome is converted to the DNA form by its own reverse transcriptase, and then integrated in the chromosome of the host lymphocytes. The viral components are synthesized from the integrated genome and assemble to constitute viral particles. The HTLV-I genome contains *gag, pol,* and *env* genes in an order like other retroviruses and an extra region called "pX" (Fig. 6). The *gag* gene codes for proteins constituting the nucleocapsid, the *pol* gene encodes reverse transcriptase, and the *env* gene encodes the envelope glycoproteins covering the viral particles. Three viral

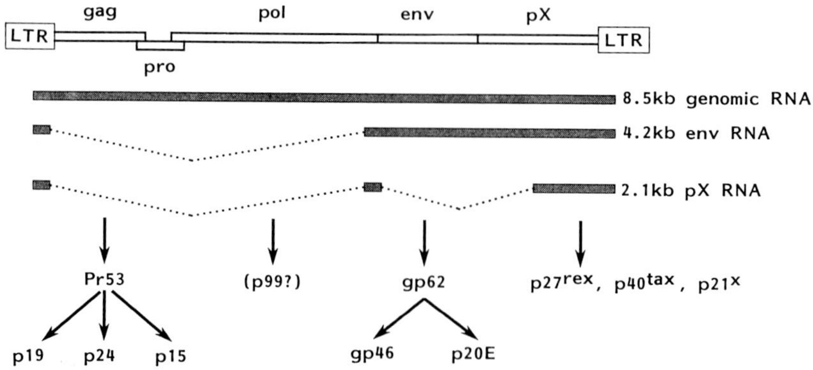


**FIGURE 6.** Genome, RNAs, and proteins of HTLV-I.

mRNA—8.5, 4.2, and 2.1 kb—are produced during HTLV-I replication by a splicing mechanism from a single species of transcript (Seiki *et al.*, 1985). The 8.5-kb RNA is the primary transcript enclosed in the progeny virion, and this genomic RNA is also used for the expression of *gag* and *pol* proteins; the 4.2-kb RNA formed by single splicing directs the synthesis of the *env* protein, and the 2.1-kb RNA, generated by double splicing, is used for the expression of three pX proteins—$p40^{tax}$, $p27^{rex}$, and $p21^{x}$. $p40^{tax}$ is encoded by a frame different from $p27^{rex}$ and $p21^{x}$. $p27^{rex}$ is translated from the first AUG codon and the translation of $p21^{x}$ is initiated from the fourth AUG (Nagashima *et al.*, 1986). $p40^{tax}$ is a *trans* -activator of the viral and some cellular genome at transcription level (Seiki *et al.*, 1986). $p27^{rex}$ is a post-transcriptional regulator that increases the unspliced 8.5-kb RNA and the 4.2-kb RNA level in contrast decreases the amount of 2.1-kb RNA (Inoue *et al.*, 1987; Hidaka *et al.*, 1988). Thus, $p27^{rex}$ seems to regulate the splicing event of HTLV-I RNA, although detailed molecular mechanism is still obscure. The function of $p21^{x}$ is unknown so far.

The use of VVV expressing authentic pX proteins individually in the African green monkey kidney cell line (CV-1 cell) showed that $p40^{tax}$ and $p21^{x}$ are located in the nuclear matrix and that $p27^{rex}$ is resident in the nucleolus (Siomi *et al.*, 1988) (Fig. 7). While the location of $p40^{tax}$ is plausible in light of its function, that of $p27^{rex}$ is unexpected because the sole function of the nucleolus has been reported to be involved in the generation of ribosome, mainly the transcription of ribosomal RNA and assembly of ribosome particles (Sommerville, 1986). Keeping the function of $p27^{rex}$ in mind, these results raise the possibility that the nucleolus may be also involved in the generation of mRNA in some way. After the first discovery in recombinant VV-infected cells, the nucleolar location of $p27^{rex}$ was confirmed in a HTLV-I-producing lymphoma cell

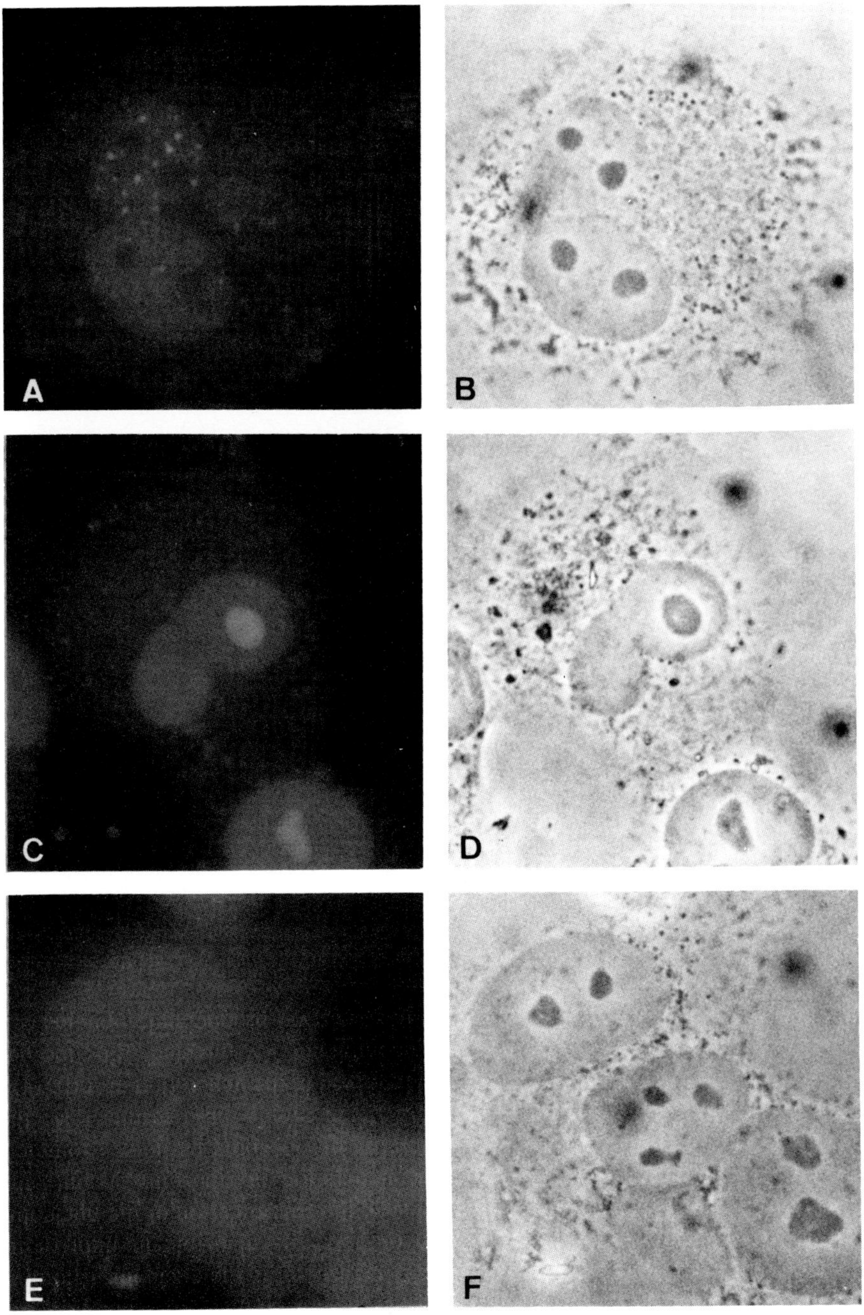


**FIGURE 7.** Localization of pX proteins synthesized in the recombinant vaccinia virus-infected cells. (A,C,E) Immunofluorescence micrographs showing the locations of p40$^{tax}$, p27$^{rex}$, and p21$^{x}$, respectively. (B,D,F) Phase-contrast micrographs of the corresponding cells.

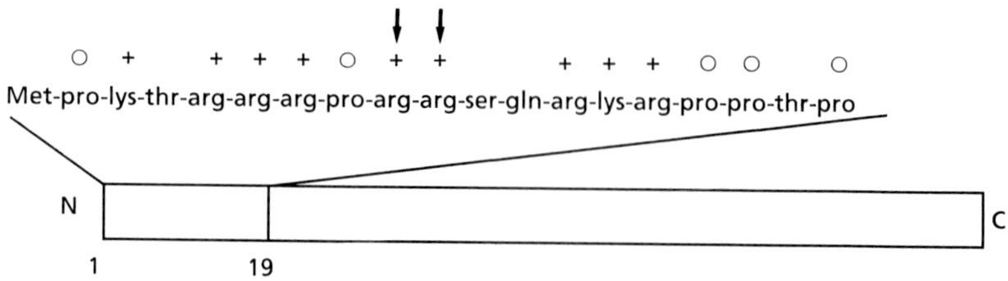


**FIGURE 8.** Amino acid sequence of putative signal of p27ˣ to locate in nucleolus. °s, Pro residues; +s, Lys and Arg residues. Arrows, Arg residues at 9 and 10 positions from the amino-terminus of p27rex.

line (MT-2 cell) and human osteosarcoma cell line (HOS cell) transfected with plasmid that expresses functional pX proteins. Apparently, a wide host range of VV initiated this finding, as it has been difficult to identify the precise location of pX proteins within the nucleus in lymphomas at the light-microscopic level (Kiyokawa *et al.*, 1985).

Efficient expression of $p27^{rex}$ in the VV vector system allowed the identification of sequence requirements for the nucleolar localization signal (Siomi *et al.*, 1988). Site directed mutagenesis methods revealed that the amino terminal region containing the Arg residue-rich sequence was responsible. Two adjacent Arg residues (positions 9 and 10) are especially important (Fig. 8). Although the transport signals of other proteins residing in the nucleolus have not yet been identified, this finding may be the initial step in discovering the general features of such signals.

### 5.3.2. Processing of *gag* Protein

When viral proteins assemble into mature virions, 53-kDa *gag* precursor protein is processed to three 19-kDa, 24-kDa, and 15-kDa proteins (Hattori *et al.*, 1984). Nucleotide sequence analysis suggested that the protease gene responsible for this processing was present just behind the coding region of the *gag* proteins. This open reading frame, however, does not have an initiation ATG codon leading to two hypotheses (Nam and Hatanaka, 1986). One explanation is that a splicing event is necessary to generate the mature *gag* mRNA and another is that a ribosomal slip during translation suppressed the terminal codon of the *gag* sequence to generate a fused protein of *gag* and protease and that subsequent cleavage produced free protease, which processed the *gag* precursor protein.

The VVV expressing the *gag* gene offered a convenient means to test these hypotheses because RNAs expressed from the VV genome are not spliced (Moss, 1985). If the *gag* precursor protein synthesized from the recombinant VV is processed into mature proteins, the latter hypothesis is supported. The result obtained indicated this. Further analyses using *in vitro* mutagenesis clearly demonstrated that the open reading frame really encoded the protease, the existence

**Table II**
**Detection of HTLV-I Antigen in Lymphocytes of Vaccinated Rabbits After Challenge with HTLV-I[a]**

| Vaccine | Rabbit number | HTLV-I antigen in cultured lymphocytes (week) | | | | | | | | | | | | | | | | | | | | | | | | | | | | | |
|---|---|---|---|---|---|---|---|---|---|---|---|---|---|---|---|---|---|---|---|---|---|---|---|---|---|---|---|---|---|---|---|
| | | 1 | 2 | 3 | 4 | 5 | 6 | 7 | 8 | 9 | 10 | 11 | 12 | 13 | 14 | 15 | 16 | 17 | 18 | 19 | 21 | 22 | 23 | 26 | 27 | 31 | 32 | 35 | 39 | 40 | 44 |
| WR-proenvl | 1 | – | – | – | – | – | – | – | – | – | – | – | – | – | – | – | – | – | – | – | – | | – | | – | – | | – | | – | –[b] |
| | 3 | – | – | – | – | – | – | – | – | – | – | – | – | – | – | – | – | –[c] | | | | | | | | | | | | | |
| WR | 9 | – | – | – | – | – | – | – | – | + | + | + | – | + | + | +[b] | | | | | | | | | | | | | | | |
| mO-proenv | 21 | – | – | – | – | – | – | – | – | – | – | – | – | | – | | – | | – | | | – | | – | | | – | | –[b] | | |
| | 23 | – | – | – | – | – | – | – | – | – | – | – | – | | – | | – | | – | | | – | | – | | | – | | –[b] | | |
| LC16mO | 24 | – | – | – | – | – | – | + | + | + | + | + | + | | + | | + | | –[b] | | | | | | | | | | | | |
| | 25 | – | – | – | – | – | + | + | + | + | + | + | + | | – | | + | | –[b] | | | | | | | | | | | | |
| LO-proenv | 26 | – | – | – | – | – | – | – | – | – | – | – | – | | – | | – | | – | | | – | | – | | | – | | –[b] | | |
| | 27 | – | – | – | – | – | – | – | – | – | – | + | + | | – | | – | | –[b] | | | | | | | | | | | | |
| | 28 | – | – | – | – | – | – | – | – | – | + | + | + | | + | | + | | –[b] | | | | | | | | | | | | |
| LO-1 | 29 | – | – | – | – | – | – | + | + | + | + | + | + | | + | | – | | –[b] | | | | | | | | | | | | |
| | 30 | – | – | – | – | – | + | + | + | + | + | + | + | | + | | – | | –[b] | | | | | | | | | | | | |

[a]Rabbits vaccinated with the recombinant VV were challenged by HTLV-I-producing cells. Infection of HTLV-I was detected by the appearance of the rabbit lymphocytes having HTLV-I antigen.
[b]Killed.
[c]Died by accident.

of the fusion protein, and the sequence involved in the ribosomal slip (Nam *et al.*, 1988).

### 5.3.3. Vaccine against HTLV-I

To develop an effective vaccine against HTLV-I, the *env* glycoprotein of HTLV-I should be the first candidate considered, for several reasons. The *env* glycoprotein is the major antigen recognized by the serum antibodies of persons infected with HTLV-I (Schupbach *et al.*, 1984). The antibody to the HTLV-I *env* protein was demonstrated to inhibit the fusion activity of HTLV-I by which HTLV-I is thought to be transmitted from cell to cell (Hoshino *et al.*, 1983; Nagy *et al.*, 1983; Kiyokawa *et al.*, 1984). Thus, the VV expressing the *env* proteins from the middle of the HA gene were constructed and tested to determine whether it could protect rabbits from HTLV-I infection. The recombinant VV induced a high titer of anti-*env* antibody and prevented infection. This was the first demonstration that the effective vaccine against a human retrovirus could be made (Shida *et al.*, 1987). Next, the various VV strains were examined to construct an effective and safe vaccine applicable to human beings. Among the WR, Lister LO1, and LC16mO strains examined, LC16mO-derived recombinants were effective and had the lowest neurovirulence. By contrast, LO1-derived recombinants had insufficient antigenicity but a higher neurovirulence than the LC16mO-derived one (Shida *et al.*, 1988) (Table II). The fact that the propagation and its RVV in brain were poorer than that of LO1 and its RVV, although LC16mO replicated in other organs better than did LO1, suggested that a different gene may affect the replication ability of VV in the skin and brain. In spite of our success, failure to prevent infection by a few pathogens with the recombinant VV based on other vaccine strains, such as Wyeth, and using other sites to insert the foreign gene have been reported (M. A. Epstein and B. Roizman, personal communication). These facts indicate that the selection of the VV strain and the site supporting the efficient expression of the foreign gene are very important in constructing an effective vaccine.

ACKNOWLEDGMENTS. The author thanks all those who generously provided preprints of their work and greatly facilitated the writing of this review. The authors' work was supported by grants from the Ministry of Education, Science and Culture, and Public Welfare.

## 6. REFERENCES

Adams, D. A., and Rose, J. K., 1985, Incorporation of a charged amino acid into the membrane-spanning domain blocks cell surface transport but not membrane anchoring of a viral glycoprotein, *Mol. Cell. Biol.* **5**:1442–1448.

Arita, I., and Fenner, F., 1985, Complications of smallpox vaccination, in: *Vaccinia Virus as Vectors for Vaccine Antigens* (J. Quinnan, ed.), pp. 49–60, Elsevier, New York.

Bennink, J. R., Yewdell, J. W., Smith, G. L., and Moss, B., 1987, Anti-influenza virus cytotoxic T lymphocytes recognize the three viral polymerases and a nonstructural protein: Responsiveness to individual viral antigens is major histocompatibility complex controlled, *J. Virol.* **61:**1098–1102.

Bertholet, C., Drillien, R., and Wittek, R., 1985, One hundred base pairs of 5′ flanking sequence of a vaccinia virus late gene are sufficient to temporally regulate late transcription, *Proc. Natl. Acad. Sci. USA* **82:**2096–2100.

Bhavanandan, V. P., and Katlic, A. W., 1979, The interaction of wheat germ agglutinin with sialoglycoproteins: The role of sialic acid, *J. Biol. Chem.* **254:**4000–4008.

Blackman, K. E., and Bubel, H. C., 1972, Origin of the vaccinia virus hemagglutinin. *J. Virol.* **9:** 290–296.

Blobel, G., Walter, P., Chang, C. N., Goldman, B. M., Erikson, A. H., and Lingappa, V. R., 1979, Translocation of proteins across membranes: The signal hypothesis and beyond, *Symp. Soc. Exp. Biol.* **33:**9–36.

Blobel, G., and Dobberstein, B., 1975, Transfer of proteins across membranes. I. Presence of proteolytically processed and unprocessed nascent immunoglobin light chains on membrane bound ribosomes of murine myeloma, *J. Cell Biol.* **67:**835–851.

Blobel, G., 1980, Intracellular protein topogenesis, *Proc. Natl. Acad. Sci. USA* **77:**1496–1500.

Bole, D. G., Hendershot, L. M., and Kearney, J. F., 1986, Posttranslational association of immunoglobulin heavy chain binding protein with nascent heavy chains in nonsecreting and secreting hybridomas, *J. Cell. Biol.* **102:**1558–1566.

Briggs, M. S., and Gierasch, L. M., 1986, Molecular mechanisms of protein secretion: The role of the signal sequence, *Adv. Protein Chem.* **38:**109–180.

Buller, R. M. L., Smith, G. L., Cremer, K., Notkins, A. L., and Moss, B., 1985, Decreased virulence of recombinant vaccinia virus expression vectors is associated with a thymidine kinase-negative phenotype, *Nature (Lond.)* **317:**813–815.

Cassel, W. A., 1957, Multiplication of vaccinia virus in the Ehrlich ascites carcinoma, *Virology* **3:** 514–526.

Chu, C. M., 1946, Studies on vaccinia hemagglutinin. I. Some physico-chemical properties, *J. Hyg.* **46:**42–48.

Cochran, M. A., Puckett, C., and Moss, B., 1985, *In vitro* mutagenesis of the promoter region for a vaccinia virus gene: Evidence for tandem early and late regulatory signals, *J. Virol.* **54:**30–37.

Cummings, R. D., Kornfeld, S., Schneider, W. S., Hobgod, K. K., Tolleschang, H., Brown, M. S., and Goldstein, J. L., 1983, Biosynthesis of N- and O-linked oligosaccharides of the low density lipoprotein receptor, *J. Biol. Chem.* **258:**15261–15273.

Dales, S., Stern, W., Weintraub, B. S., and Huima, T., 1976, Genetically controlled surface modifications by poxviruses influencing cell–cell and cell–virus interactions, in: *Cell Membrane Receptors for Viruses, Antigens and Antibodies, Polypeptide Hormones, and Small Molecules* (R. F. Beers, Jr., and E. G. Bassett, eds.), pp. 253–270, Raven, New York.

Davis, G. L., and Hunter, E., 1987, A charged amino acid substitution within the transmembrane anchor of the Rous sarcoma virus envelope glycoprotein affects surface expression but not intracellular transport, *J. Cell Biol.* **105:**1191–1203.

Dorner, A. J., Bole, D. G., and Kaufman, R. J., 1987, The relationship of N-linked glycosylation and heavy chain-binding protein association with the secretion of glycoproteins, *J. Cell Biol.* **105:**2665–2674.

Doyle, C., Roth, M. G., Sambrook, J., and Gething, M.-J., 1985, Mutations in the cytoplasmic domain in the influenza virus hemagglutinin affect different stages of intracellular transport, *J. Cell Biol.* **100:**704–714.

Dunphy, W. G., Brands, R., and Rothman, J. E., 1985, Attachment of terminal N-acetylglucosamine to asparagine linked oligosaccha rides occurs in central cisternae of the Golgi stack, *Cell* **40:**463–472.

Epstein, D., Marsh, Y. V., Schreiber, A. B., Newman, S. R., Todaro, G. J., and Nestor, J. J., Jr., 1985, Epidermal growth factor receptor occupancy inhibits vaccinia virus infection, *Nature (Lond.)* **318:**663–665.

Ezzell, C., 1987, Trials of vaccine against AIDS to begin in humans, *Nature (Lond.)* **328:**747.

Fields, S., Winter, G., and Brownlee, G. G. 1981, Structure of the neuraminidase gene in human influenza virus A/PR/8/34, *Nature (Lond.)* **290:**213–217.

Flexner, C., Hügin, A., and Moss, B., 1987, Prevention of vaccinia virus infection in immunodeficient mice by vector-directed IL-2 expression, *Nature (Lond.)* **330:**259–262.

Foglesong, P. D., 1985, *In vitro* transcription of a cloned vaccinia virus gene by a soluble extract prepared from vaccinia virus-infected HeLa cells, *J. Virol.* **53:**822–826.

Funahashi, S., Sato, T., and Shida, H., 1988, Cloning and characterization of the gene encoding the major protein of the A-type inclusion body of cowpox virus, *J. Gen. Virol.* **69:**35–47.

Garoff, H., 1985, Using recombinant DNA techniques to study protein targeting in the eucaryotic cell, *Ann. Cell Biol.* **1:**403–445.

Gething, M.-J., McCammon, K., and Sambrook, J., 1986, Expression of wild-type and mutant forms of influenza hemagglutinin: The role of folding in intracellular transport, *Cell* **46:**939–950.

Gibson, R., Schlesinger, S., and Kornfeld, S., 1979, The nonglycosylated glycoprotein of vesicular stomatitis virus is temperature-sensitive and undergoes intracellular aggregation at elevated temperatures, *J. Biol. Chem.* **254:**3600–3607.

Golini, F., and Kates, J. R., 1985, A soluble transcription system derived from purified vaccinia virions, *J. Virol.* **53:**205–213.

Hanggi, M., Bannwarth, W., and Stunnenberg, H. G., 1986, Conserved TAAAT motif in vaccinia virus late promoters: Overlapping TATA box and site of transcription initiation, *EMBO J.* **5:** 1071–1076.

Hanover, J. A., Lennarz, W. J., and Young, J. D., 1980, Synthesis of N- and O-linked glycopeptides in oviduct membrane preparations, *J. Biol. Chem.* **255:**6713–6716.

Hanover, J. A., Elting, J., Mintz, G. R., and Lennarz, W. J., 1982, Temporal aspects of the N- and O-glycosylation of human chorionic gonadotropin, *J. Biol. Chem.* **257:**10172–10177.

Hardwick, J. M., Shaw, K. E. S., Wills, J. W., and Hunter, E., 1986, Amino-terminal deletion mutants of the Rous sarcoma virus glycoprotein do not block signal peptide cleavage but can block intracellular transport, *J. Cell Biol.* **103:**829–838.

Harris, J. R., 1978, The biochemistry and ultrastructure of the nuclear envelope, *Biochim. Biophys. Acta* **515:**55–104.

Hashizume, S., Yoshizawa, H., Morita, M., and Suzuki, K., 1985, Properties of attenuated mutant of vaccinia virus, LC16m8 derived from Lister strain, in: *Vaccinia Viruses as Vectors for Vaccinia Antigens* (J. Quinnan, ed.), pp. 87–99. Elsevier, New York.

Hattori, S., Kiyokawa, T., Inakawa, K., Shimizu, F., Hashimura, E., Seiki, M., and Yoshida, M., 1984, Identification of *gag* and *env* gene products of human T-cell leukemia virus (HTLV), Virology **136:**338–347.

Heijne, G. V., 1984, How signal sequences maintain cleavage specificity, *J. Mol. Biol.* **173:**243–251.

Helenius, A., Kartenbeck, J., Simons, K., and Fries, E., 1980, On the entry of Semliki Forest virus into BHK-21 cells, *J. Cell Biol.* **84:**404–420.

Hidaka, M., Inoue, J., Yoshida, M., and Seiki, M., 1988, Post-transcriptional regulator (rex) of HTLV-I initiates expression of viral structural proteins but suppresses expression of regulatory proteins, *EMBO J.* **1:**519–523.

Hinuma, Y., Nagata, K., Hanaoka, M., Nakai, M., Matsumoto, T., Kinoshita, K., Shirakawa, S., and Miyoshi, I., 1981, Adult T cell leukemia: Antigen in an ATL cell line and detection of antibodies to the antigen in human sera, *Proc. Natl. Acad. Sci. USA* **78:**6476–6480.

Hoshino, H., Shimoyama, M., Miwa, M., and Sugimura, T., 1983, Detection of lymphocytes producing a human retrovirus associated with adult T-cell leukemia by syncytia induction assay, *Proc. Natl. Acad. Sci. USA* **80:**7337–7341.

Ichihashi, Y., 1977, Vaccinia-specific hemagglutinin, *Virology* **76:**527–538.

Ichihashi, Y., and Dales, S., 1971, Biogenesis of poxviruses: interrelationship between hemagglutinin production and polykariocytosis, *Virology* **46:**533–543.

Ichihashi, Y., Matsumoto, S., and Dales, S., 1971, Biogenesis of poxviruses: role of A-type inclusions and last cell membranes in virus dissemination, *Virology B46:*507–532.

Ikuta, K., Miyamoto, H., and Kato, S., 1979, Two components specifically responsible for hemagglutination in vaccinia and cowpox viruses, *Virology* **96:**327–331.

Inoue, J., Yoshida, M., and Seiki, M., 1987, Transcriptional ($p40^{x}$) and post-transcriptional ($p27^{x\text{-III}}$) regulators are required for the expression and replication of human T-cell leukemia virus type I genes, *Proc. Natl. Acad. Sci. USA* **84:**3653–3657.

Johnson, D. C., and Spear, P. G., 1983, O-linked oligosaccharides are acquired by Herpes simplex virus glycoproteins in the Golgi Apparatus, *Cell* **32:**987–997.

Jokinen, M., Ulmanen, I., Anderson, L. C., Kaariainen, L., and Gahmberg, C. G., 1981, Cell-free synthesis and glycosylation of the major human-red-cell sialoglycoprotein, glycophorin A, *Eur. J. Biochem.* **114:**393–397.

Kalyanaraman, V. S., Sarngadharan, M. G., Nakao, Y., Ito, Y., Aoki, T., and Gallo, R. C., 1982, Natural antibodies to the structural core protein (p24) of the human T cell leukemia (lymphoma) retrovirus found in sera of leukemia patients in Japan, *Proc. Natl. Acad. Sci. USA* **79:**1653–1657.

Katz, F. N., Rothman, J. E., Lingappa, V. R., Blobel, G., and Lodish, H. F., 1977, Membrane assembly in vitro: synthesis, glylcosylation, and asymmetric insertion of a transmembrane protein, *Proc. Natl. Acad. Sci. USA* **74:**3278–3282.

Kiyokawa, T., Seiki, M., Iwashita, S., Imagawa, K., Shimizu, F., and Yoshida, M., 1985, $p27^{x\text{-III}}$ and $p21^{x\text{-III}}$, proteins encoded by the pX sequence of human T-cell leukemia virus type I, *Proc. Natl. Acad. Sci. USA* **82:**8359–8363.

Kopito, R. R., and Lodish, H. F., 1985, Primary structure and transmembrane orientation of the murine anion exchange protein, *Nature (Lond.)* **316:**234–238.

Kornfeld, S., and Kornfeld, R., 1980, Structure of glycoproteins and their oligosaccharide units, in: *Biochemistry of Glycoproteins and Proteoglycans,* (W. J. Lennarz, ed.), pp. 1–34, Plenum, New York.

Kornfeld, S., Li, E., and Tabas, I., 1978, The synthesis of complex-type oligosaccharides. II. Characterization of the processing intermediates in the synthesis of the complex oligosaccharide units of the vesicular stomatitis virus G protein, *J. Biol. Chem.* **253:**7771–7778.

Kreil, G., 1981, Transfer of proteins across membranes, *Annu. Rev. Biochem.* **50:**317–348.

Kuo, S. C., and Lampen, J. O., 1976, Tunicamycin inhibition of [$^3$H] glucosamine incorporation into yeast glycoproteins. Binding of tunicamycin and interactions with phospholipids, *Arch. Biochem. Biophys.* **172:**574–581.

Leavitt, R., Schlesinger, S., and Kornfeld, S., 1977, Impaired intracellular migration and altered solubility of nonglycosylated glycoproteins of vesicular stomatitis virus and sindbis virus, *J. Biol. Chem.* **252:**9018–9023.

Lenard, J., and Miller, D. K., 1982, Uncoating of enveloped viruses, *Cell* **28:**5–6.

Lingappa, V. R., Katz, F. N., Lodish, H. F., and Blobel, G., 1978, A signal sequence for the insertion of a transmembrane glycoprotein, *J. Biol. Chem.* **253:**8667–8670.

Lodish, H. F., and Froshauer, S., 1977, Binding of viral glycoprotein mRNA to endoplasmic reticulum membranes is disrupted by puromycin, *J. Cell Biol.* **74:**358–364.

Machamer, C. E., and Cresswell, P., 1984, Monensin prevents terminal glycosylation of the N- and O-linked oligosaccharides of the HLA-DR-associated invariant chain and inhibits its dissociation from the α–β chain complex, *Proc. Natl. Acad. Sci. USA* **81:**1287–1291.

Mackett, M., and Smith, G. L., 1986, Vaccinia virus expression vectors, *J. Gen. Virol.* **67:**2067–2082.

Mackett, M., Smith, G. L., and Moss, B., 1982, Vaccinia virus: A selectable eukaryotic cloning and expression vector, *Proc. Natl. Acad. Sci. USA* **79:**7415–7419.

Mackett, M., Smith, G. L., and Moss, B., 1985, The construction and characterization of vaccinia virus recombinants expressing foreign genes, in: *DNA Cloning: A Practical Approach* pp. 191–211, (D. M. Glove, ed.), IRL Press, Oxford.

Mars, M., and Beaud, G., 1987, Characterization of vaccinia virus early promoters and evaluation of their informational content, *J. Mol. Biol.* **198:**619–631.

Marshall, R., 1972, Glycoproteins, *Annu. Rev. Biochem.* **41:**673–702.

McQueen, N., Nayak, D. P., Stephens, E. B., and Compans, R. W., 1986, Polarized expression of a chimeric protein in which the transmembrane and cytoplasmic domains of the influenza virus hemagglutinin have been replaced by those of the vesicular stomatitis virus G protein, *Proc. Natl. Acad. Sci. USA* **83:**9318–9322.

Min-Jou, W., Verhoeyen, M., Devos, R., Saman, E., Fang, R., Huylebroeck, D., Fiers, W., Threlfall, G., Barber, C., Carey, N., and Emtage, S., 1980, Complete structure of the hemagglutinin gene from the human influenza A/Victoria/3/75 (H3 N2) strain as determined from cloned DNA, *Cell* **19:**683–696.

Morrison, T. G., and Lodish, H. F., 1975, Site of synthesis of membrane and nonmembrane proteins of vesicular stomatitis virus, *J. Biol. Chem.* **250:**6955–6962.

Moss, B., 1985, Replication of poxviruses, in: *Virology* (B. N. Fields, R. M. Chanock, and B. Roizman, eds.), pp. 685–703, Raven, New York.

Nagashima, K., Yoshida, M., and Seiki, M., 1986, A single species of pX mRNA of human T-cell leukemia virus type I encodes trans-activator p40$^{x}$ and two other phosphoproteins, *J. Virol.* **60;** 394–399.

Nagler, F. P. O., 1942, Application of Hirst's phenomenon to titration of vaccinia virus and vaccinia immune serum, *Med. J. Aust.* **1:**281–283.

Nagy, K., Clapham, P., Cheingsong-Popov, R., and Weiss, R. A., 1983, Human T-cell leukemia virus type I: Induction of syncytia and inhibition by patients' sera, *Int. J. Cancer* **32:**321–328.

Nam, S. H., and Hatanaka, M., 1986, Identification of a protease gene of human T-cell leukemia virus type I (HTLV-I) and its structural comparison, *Biochem. Biophy. Res. Commun.* **139:**129–135.

Nam, S.-H., Kidokoro, M., Shida, H., and Hatanaka, M., 1988, Processing of gag precursor polyprotein of Human T-cell leukemia virus type I (HTLV-1) by virus-encoded protease *J. Virol.* **62:**3718–3728.

Neufeld, E. F., Tim, T. W., and Shapiro, L. J., 1975, Inherited disorders of lysosomal metabolism, *Annu. Rev. Biochem.* **44:**357–376.

Newmark, P., 1986, Problems with AIDS vaccines, *Nature (Lond.)* **324:**304–305.

Noda, M., Ikeda, T., Kayano, T., Suzuki, H., Takeshima, H., Kurasaki, M., Takahashi, H., and Numa, S., 1986, Existence of distinct sodium channel messenger RNA in rat brain, *Nature (Lond.)* **320:**188–192.

Oda, M., 1965, Rescue of dermovaccinia abortive infection by neurovaccinia virus in L cells, *Virology* **25:**664–666.

Oesch, B., Westaway, D., Walchi, M., McKinley, M. P., Kent, S. B. H., Aebersold, R., Barry, R. A., Tempst, P., Teplow, D. B., Hood, L. E., Prusiner, S. B., and Weissmann, C., 1985, A cellular gene encodes scrapie PrP 27-30 protein, *Cell* **40:**735–746.

Palade, G. L., 1975, Intracellular aspects of the process of protein synthesis, *Science* **189:**347–358.

Panicali, D., and Paoletti, E., 1982, Construction of poxviruses as cloning vectors: insertion of the thymidine kinase gene of herpes simplex virus into the DNA of infectious vaccinia virus, *Proc. Natl. Acad. Sci. USA* **79**:4927–4931.

Payne, G. L., 1979, Identification of the vaccinia hemagglutinin polypeptide from a cell system yielding large amounts of extracellular enveloped virus, *J. Virol.* **31**:147–155.

Pennington, T. H., 1974, Vaccinia virus polypeptide synthesis: Sequential appearance and stability of pre- and postreplicative polypeptides, *J. Gen. Virol.* **25**:433–444.

Perkus, M. E., Piccini, B. R., Lipinskas, B. R., and Paoletti, E., 1985, Recombinant vaccinia virus: Immunization against multiple pathogens, *Science* **229**:981–984.

Perkus, M. E., Panicali, D., Mercer, S., and Paoletti, E., 1986, Insertion and deletion mutants of vaccinia virus, *Virology* **152**:285–297.

Perlman, D., and Halvorson, H. O., 1983, A putative signal peptidase recognition site and sequence in eukaryotic and prokaryotic signal peptide, *J. Mol. Biol.* **167**:391–409.

Pfeffer, S. R., and Rothman, J. E., 1987, Biosynthetic protein transport and sorting by the endoplasmic reticulum and Golgi, *Annu. Rev. Biochem.* **56**:829–852.

Puckett, C., and Moss, B., 1983, Selective transcription of vaccinia virus genes in template dependent soluble extracts of infected cells, *Cell* **35**:441–448.

Rodriguez, J. F., Paez, E., and Esteban, M., 1987, A 14,000-$M_r$ envelope protein of vaccinia virus is involved in cell fusion and forms convalently linked trimers, *J. Virol.* **61**:395–404.

Rohrmann, G., and Moss, B., 1985, Transcription of vaccinia virus early genes by a template-dependent soluble extract of purified virions, *J. Virol.* **56**:349–355.

Rose, J. K., and Bergman, J. E., 1982, Expression from cloned cDNA of cell surface and secreted forms of the glycoprotein of vesicular stomatitis virus in eukaryotic cells, *Cell* **30**:753–762.

Rose, J. K., and Bergman, J. E., 1983, Altered cytoplasmic domains affect intracellular transport of the vesicular stomatitis virus glycoprotein, *Cell* **34**:513–524.

Rose, J. K., Welch, W. J., Sefton, B. M., Esch, F. S., and Ling, N. C., 1980, Visicular stomatitis virus glycoprotein is anchored in the viral membrane by a hydrophobic domain near the COOH terminus, *Proc. Natl. Acad. Sci. USA* **77**:3884–3888.

Rosel, J. L., and Moss, B., 1985, Transcriptional and translational mapping and nucleotide sequence analysis of a vaccinia virus gene encoding the precursor of the major core polypeptide, *J. Virol.* **56**:830–838.

Rosel, J. L., Earl, P. L., Weir, J. P., and Moss, B., 1986, Conserved TAAATG sequence at the transcriptional and translational initiation sites of vaccinia virus late genes deduced by structural and functional analysis of the Hind III H genomic fragment, *J. Virol.* **60**:436–449.

Rossignol, B., Herman, G., and Clauser, H., 1969, Tentative identification of N-acetylgalactosamine incorporating membranes from ovine submaxillary glands (OSG), *Biochem. Biophys. Res. Commun.* **34**:111–119.

Roth, J., 1984, Cytochemical localization of terminal N-acetyl-D-galactosamine residues in cellular compartments of intestinal goblet cells: Implications for the topology of O-glycosylation, *J. Cell Biol.* **98**:399–406.

Roth, J., and Berger, E. G., 1982, Immunocytochemical localization of galactosyltransferase in HeLa cells: codistribution with thiamine pyrophosphate in trans-Golgi cisternae, *J. Cell Biol.* **92**:223–229.

Roth, J., Taatjes, D. J., Lucocq, J. M., Weinstein, J., and Paulson, J. C., 1985, Demonstration of an extensive trans-tubular network continuous with the Golgi apparatus stack that may function in glycosylation, *Cell* **43**:287–295.

Rothmann, J. E., and Lodish, H. F., 1977, Synchronized transmembrane insertion and glycosylation of a nascent membrane protein, *Nature (Lond.)* **269**:775–780.

Rottier, P. J. M., Florkiewicz, R. Z., Shaw, A. S., and Rose, J. K., 1987, An internalized amino-terminal signal sequence retains full activity in vivo but not in vitro, *J. Biol. Chem.* **262:**8889–8895.

Sabatini, D. D., Kreibich, G., Morimoto, T., and Adesnik, M., 1982, Mechanism for the incorporation of proteins in membranes and organelles, *J. Cell Biol.* **92:**1–22.

Satake, M., Coligan, J. E., Elango, N., Norrby, E., and Venkatesan, S., 1985, Respiratory syncytial virus envelope glycoprotein (G) has a novel structure, *Nucl. Acids Res.* **13:**7795–7812.

Schupbach, J., Sangadharan, M. G., and Gallo, R. C., 1984, Antigens on HTLV-infected cells recognized by leukemia and AIDS sera are related to HTLV viral glycoprotein, *Science* **224:** 607–610.

Seiki, M., Hikikoshi, A., Taniguchi, T., and Yoshida, M., 1985, Expression of the pX gene of HTLV-I: general splicing mechanism in the HTLV family, *Science* **228:**1532–1534.

Seiki, M., Inoue, J., Takeda, T., and Yoshida, M., 1986, Direct evidence that $p40^x$ of human T-cell leukemia virus type I is a trans-acting transcriptional activator, *EMBO J.* **5:**561–565.

Shida, H., 1986a, Nucleotide sequence of the vaccinia virus hemagglutinin gene, *Virology* **150:**451–462.

Shida, H., 1986b, Variants of vaccinia virus hemagglutinin altered in intracellular transport, *Mol. Cell. Biol.* **6:**3734–3745.

Shida, H., and Dales, S., 1981, Biogenesis of vaccinia: Carbohydrates of the hemagglutinin molecule, *Virology* **111:**56–72.

Shida, H., and Dales, S., 1982, Biogenesis of vaccinia: Molecular basis for the hemagglutination-negative phenotype of the IHD-W strain, *Virology* **117:**219–237.

Shida, H., and Matsumoto, S., 1983, Analysis of the hemagglutinin glycoprotein from mutants of vaccinia virus that accumulates on the nuclear envelope, *Cell* **33:**423–434.

Shida, H., Tochikura, T., Sato, T., Konno, T., Hirayoshi, K., Ito, Y., Hatanaka, M., Hinuma, Y., Sugimoto, M., T-Nishimaki, F., Maruyama, T., Miki, K., Suzuki, K., Morita, M., Sashiyama, H., Yoshimura, N., and Hayami, M., 1987, Effect of the recombinant vaccinia viruses that express HTLV-I envelope gene on HTLV-I infection, *EMBO J.* **6:**3379–3384.

Shida, H., Hinuma, Y., Hatanaka, M., Morita, M., Kidokora, M., Suzuki, K., Maruyama, T., T-Nishimaki, F., Sugimoto, M., Kitamura, R., Miyazawa, T., and Hayami, M., 1988, *J. Virol.* **62:**4474–4489.

Siomi, H., Shida, H., Nam, S. H., Nosaka, T., Maki, M., and Hatanaka, M., 1988, Sequence requirements for nucleolar localization of human T cell leukemia virus type I pX protein, which regulates viral RNA processing, *Cell* **55:**197–209.

Smith, G. L., and Moss, B., 1983, Infectious poxvirus vectors have capacity for at least 25,000 base pairs of foreign DNA, *Gene* **25:**21–28.

Smith, C. E., Pratt, C. B., and Baxby, D., 1973, In vitro dissociation and reconstitution of poxvirus hemagglutinin, *J. Gen. Virol.* **18:**111–118.

Sommerville, J., 1986, Nucleolar structure and ribosome biogenesis, *Trends Biochem. Sci.* **11:**438–442.

Spiro, G. R., 1966, Characterization of carbohydrate units of glycoproteins, *Methods Enzymol.* **8:** 26–52.

Stephens, E. B., and Compans, R. W., 1986, Nonpolarized expression of a secreted murine leukemia virus glycoprotein in polarized epithelial cells, *Cell* **47:**1053–1059.

Stephens, E. B., Compans, R. W., Earl, P., and Moss, B., 1986, Surface expression of viral glycoproteins is polarized in epithelial cells infected with recombinant vaccinia viral vectors, *EMBO J.* **5:**237–245.

Stern, W., and Dales, S., 1976, Biogenesis of vaccinia: Isolation and characterization of a surface component that elicits antibody suppressing infectivity and cell–cell fusion, *Virology* **75:**232–241.

Stone, J. D., 1946a, Inactivation of vaccinia and ectromelia virus hemagglutinins by lecithinase, *Aust. J. Exp. Biol. Med. Sci.* **24:**191–196.

Stone, J. D., 1946b, Lipid hemagglutinins, *Aust. J. Exp. Biol. Med. Sci.* **24:**9–13.

Stone, J. D., and Burnet, F. M., 1946, The production of vaccinia hemagglutinin in rabbit skin, *Aust. J. Exp. Biol. Med. Sci.* **24:**9–13.

Stroobant, P., Rice, A., Gullick, W. J., Chang, D. J., Kerr, I. M., and Waterfield, M. D., 1985, Purification and characterization of vaccinia virus growth factor, *Cell* **42:**383–393.

Strous, G. J. A. M., 1981, Initial glycosylation of proteins with acetylgalactosaminylserine linkages, *Proc. Natl. Acad. Sci. USA* **76:**2694–2698.

Strubin, M., Mach, B., and Long, E. O., 1984, The complete sequence of the mRNA for the HLA-DR-associated invariant chain reveals a polypeptide with an unusual transmembrane polarity, *EMBO J.* **3:**869–872.

Takatsuki, Y., Kohono, K., and Tamura, G., 1975, Inhibition of biosynthesis of polyisoprenol sugars in chick embryo microsomes by tunicamycin, *Agric. Biol. Chem.* **39:**2089–2091.

Tarentino, A. L., and Maley, F., 1974, Purification and properties of an endo-β-*N*-acetylglucosaminidase from *Streptomyces* griseus, *J. Biol. Chem.* **249:**811–817.

Tarentino, A. L., Plummer, T. H., Jr., and Maley, F., 1974, The release of intact oligosaccharides from specific glycoproteins by endo-β-*N*-acetylglucosaminidase H, *J. Biol. Chem.* **249:**818–824.

Tkacz, J. S., and Lampen, J. O., 1975, Tunicamycin inhibition of polyisoprenyl *N*-acetylglucosaminyl pyrophosphate formation of calf liver microsomes, *Biochem. Biophys. Res. Commun.* **65:**248–257.

Vassef, A., 1987, Conserved sequences near the early transcription start sites of vaccinia virus, *Nucl. Acids Res.* **15:**1427–1443.

Venkatesan, S., Baroudy, B. M., and Moss, B., 1981, Distinctive nucleotide sequences adjacent to multiple initiation and termination sites of an early vaccinia virus gene, *Cell* **125:**805–813.

Walter, P., Gilmore, R., and Blobel, G., 1984, Protein translocation across the endoplasmic reticulum, *Cell* **38:**5–8.

Weintraub, S., and Dales, S., 1974, Biogenesis of poxviruses: genetically controlled modifications of structural and functional components of plasma membrane, *Virology* **60:**96–127.

Weir, J. P., and Moss, B., 1983, Nucleotide sequence of the vaccinia virus thymidime kinase gene and the nature of spontaneous frame shift mutations, *J. Virol.* **46:**530–537.

Weir, J. P., and Moss, B., 1987, Determination of the transcriptional regulatory region of a vaccinia virus late gene, *J. Virol.* **61:**75–80.

Wickner, W. T., and Lodish, H. F., 1985, Multiple mechanisms of protein insertion into and across membranes, *Science* **230:**400–407.

Wieland, F. T., Gleason, M. L., Serafini, T. A., and Rothman, J. E., 1987, The rate of bulk flow from the endoplasmic reticulum to the cell surface, *Cell* **50:**289–300.

Wittek, R., Richner, B., and Hiller, G., 1984, Mapping of the genes coding for the two major vaccinia virus core polypeptides, *Nucl. Acids Res.* **12:**4835–4848.

Yamamoto, N., and Hinuma, Y., 1985, Viral aetiology of adult T-cell leukemia, *J. Gen. Virol.* **66:** 1641–1660.

Yewdell, J. W., Bennink, J. R., Smith, G. L., and Moss, B., 1985, Influenza A virus nucleoprotein is a major target antigen for cross-reactive anti-influenza A virus cytotoxic T lymphocytes, *Proc. Natl. Acad. Sci. USA* **82:**1785–1789.

Yewdell, J. W., Bennink, J. R., Mackett, M., Lefrancois, L., Lyles, D. S., and Moss, B., 1986, Recognition of cloned vesicular stomatitis virus internal and external gene products by cytotoxic T lymphocytes, *J. Exp. Med.* **163:**1529–1538.

Yewdell, J. W., Yellen, A., and Bachi, T., 1988, Monoclonal antibodies localize events in the folding, assembly, and intracellular transport of the influenza virus hemagglutinin glycoprotein, *Cell* **52:**843–852.

Yoshida, M., 1987, Expression of the HTLV-I genome and its association with a unique T-cell malignancy, *Biochim. Biophys. Acta* **907:**145–161.

Yoshida, M., Seiki, M., Yamaguchi, K., and Takatsuki, K., 1984, Monoclonal integration of human T-cell leukemia provirus in all primary tumors of adult T-cell leukemia suggests causative role of human T-cell leukemia virus in disease, *Proc. Natl. Acad. Sci. USA* **81:**2534–2537.

Zerial, M., Huylebroeck, D., and Garoff, H., 1987, Foreign transmembrane peptides replacing the internal signal sequence of transferrin receptor allow its translocation and membrane binding, *Cell* **48:**147–155.

# Index